FORAGES

Volume II: **The Science of Grassland Agriculture**

GRASS ■ MEADOW ■ PASTURE
FROM GENESIS TO REVELATION

Genesis 1:12 ■ The land produced vegetation [GRASS]: plants bearing seed according to their kinds. . . . And God saw that it was good.

Genesis 47:4 ■ They also said to him, "We have come to live here awhile, because the famine is severe in Canaan and your servants' flocks have no PASTURE."

Deuteronomy 11:15 ■ I will provide GRASS in the fields for your cattle, and you will eat and be satisfied.

2 Samuel 22:3-4 ■ The God of Israel spoke . . . "When one rules over men in righteousness . . . he is like the light of morning at sunrise on a cloudless morning, like the brightness after rain that brings the GRASS from the earth."

1 Kings 4:22-23 ■ Solomon's daily provisions were . . . ten head of stall-fed cattle, twenty of PASTURE-fed cattle and a hundred sheep and goats, as well as deer, gazelles, roebucks and choice fowl.

Job 5:25 ■ You will know that your children will be many, and your descendants like the GRASS of the earth.

Psalm 23:2 ■ He makes me lie down in green PASTURES, he leads me beside quiet waters, he restores my soul.

Psalm 37:2 ■ ...for like the GRASS they will soon wither, like green plants they will soon die away.

Psalm 65:12-13 ■ The GRASSLANDS of the desert overflow; the hills are clothed with gladness. The MEADOWS are covered with flocks and the valleys are mantled with grain; they shout for joy and sing.

Psalm 72:16 ■ Let its fruit flourish like Lebanon; let it thrive like the GRASS of the field.

Psalm 103:15 ■ As for man, his days are like GRASS; he flourishes like a flower of the field.

Psalm 104:14 ■ He makes GRASS grow for the cattle, and plants for man to cultivate—bringing forth food from the earth.

Psalm 147:8 ■ He covers the sky with clouds; he supplies the earth with rain and makes GRASS grow on the hills.

Proverbs 19:12 ■ A king's rage is like the roar of a lion, but his favor is like dew on the GRASS.

Proverbs 27:25 ■ When the HAY is removed and new growth appears and the GRASS from hills is gathered in . . .

Isaiah 40:8 ■ "The GRASS withers and the flowers fall, but the word of our God stands forever."

Isaiah 44:4 ■ They will spring up like GRASS in a MEADOW, like poplar trees by flowing streams.

Isaiah 51:12 ■ I, even I, am he who comforts you. Who are you that you fear mortal men, the sons of men, who are but GRASS?

Joel 1:18-20 ■ How the cattle moan! The herds mill about because they have no PASTURE; even the flocks of sheep are suffering. To you, O Lord, I call, for fire has devoured the open PASTURES and flames have burned up all the trees of the field. Even the wild animals pant for you; the streams of water have dried up and fire has devoured the open PASTURES.

Matthew 6:30 ■ "If that is how God clothes the GRASS of the field, which is here today and tomorrow is thrown into the fire, will he not much more clothe you, O you of little faith?" (See also Luke 12:28.)

Matthew 14:19 ■ And he directed the people to sit down on the GRASS.

1 Peter 1:24 ■ For, "All men are like GRASS, and all their glory is like the flowers of the field; the GRASS withers and the flowers fall."

Revelation 9:4 ■ They were told not to harm the GRASS of the earth or any plant or tree, but only those people who did not have the seal of God on their foreheads.

—From the *New International Version Bible*

FIFTH EDITION

FORAGES

Volume II: The Science of Grassland Agriculture

UNDER THE EDITORIAL AUTHORSHIP OF

Robert F Barnes, ASA, CSSA, SSSA
Darrell A. Miller, University of Illinois
C. Jerry Nelson, University of Missouri

WITH 42 CONTRIBUTING AUTHORS

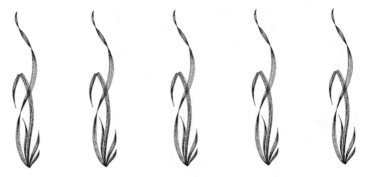

IOWA STATE UNIVERSITY PRESS, Ames, Iowa, USA

Copyright is not claimed for Chapters 5, 6, 9, 13, 16, and 17, which are in the public domain.

Photograph used throughout front matter courtesy of Robert F Barnes.

Authorization to photocopy items for internal or personal use, or the internal or personal use of specific clients, is granted by Iowa State University Press, provided that the base fee of $.10 per copy is paid directly to the Copyright Clearance Center, 27 Congress Street, Salem, MA 01970. For those organizations that have been granted a photocopy license by CCC, a separate system of payments has been arranged. The fee code for users of the Transactional Reporting Service is 0-8138-0683-6/95 $.10.

∞ Printed on acid-free paper in the United States of America

First and second editions © 1951 and 1962 by H.D. Hughes, Maurice E. Heath, Darrel S. Metcalfe, and Iowa State University Press; third edition © 1973 by Maurice E. Heath, Darrel S. Metcalfe, Robert F Barnes, and Iowa State University Press; fourth edition © 1985 by Maurice E. Heath, Robert F Barnes, and Darrel S. Metcalfe, and Iowa State University Press

Fifth edition, volume 2, 1995

Library of Congress Cataloging-in-Publication Data

Forages.
    "In memoriam: Professor Maurice E. Heath, 1910–1989"—Pref.
    Includes bibliographical references and indexes.
    Contents: v. 1. An introduction to grassland agriculture—v. 2. The science of grassland agriculture.
    1. Forage plants. 2. Forage plants—United States. I. Barnes, Robert F. II. Miller, Darrell A. III. Nelson, C. J. (Curtis J.) IV. Heath, Maurice E.
ISBN 0-8138-0681-X (v. 1: acid-free paper)
ISBN 0-8138-0683-6 (v. 2)
SB193.F64   1995   633.2   94-37719

The authors and editors dedicate this volume

To the many committed scientists, educators, and professionals who have spent their lives and careers researching and applying the scientific principles of grassland agriculture. Their endeavors have enhanced the quality of life by providing food, fiber, tranquility, and recreation through the economic development and environmental stewardship of our renewable forage, grassland, and range resources.

To the foresight of forage researchers who have given us new technologies and have integrated forages into many agricultural, ecological, and social systems. Such research and practice have demonstrated the value of managing our renewable resources to support recreation, wildlife habitats, and, via animals, the use of forages to sustain a productive, high-quality, and safe food supply.

To the young researchers who are learning from their mentors and realizing that the pursuit of knowledge, while challenging, is a major contribution to mankind. This volume is based on the many developed and developing topics regarding forage systems, many of which will be surpassed or become outdated by the discovery and development of new technologies. In that sense, the dynamics of the youth in science will be an inspiration and stimulus for continued change in the years to come.

CONTENTS

PREFACE

FRONTIERS OF FORAGE RESEARCH

New frontiers are emerging in forages. Many major developments and new research findings have occurred in the last 10 years in forage crop production and utilization. The number of farmers and ranchers has decreased to the levels prevalent more than 100 years ago, yet the land managed by each has expanded several fold. Government policies are reflecting shifts in land management from production of a food supply to include a quality environment and a sustainable system. In this matrix of social, economic, and political agendas, forages are used to enhance the stability of agricultural production systems by improving soil productivity, reducing soil and water losses, and ensuring livestock production. Forages are the basic component in feeding programs for dairy cattle, beef cattle, and sheep, as well as an important feed for horses, goats, other classes of livestock, and wildlife. Forages contribute to recreation and to economic development of farming systems, and help to provide a cleaner environment. Therefore, it is important that the latest research and technologies are accessible to help integrate forages into profitable systems.

EMERGING TECHNOLOGIES

Recent advancements have made it possible for forage researchers to incorporate results and technologies from basic research into goals of applied research. These emerging technologies are often built on past research findings but are combined with very recent research discoveries, resulting in new, exciting viewpoints of forage development, production, and utilization. History suggests that research efforts to maintain and enhance the contributions of forages to more intensive agricultural production systems must be strengthened in response to new problems and opportunities that are continuing to arise. With these challenges, the authors and co-authors have put forth and evaluated the most recent emerging technologies that will set the standards for future forage production and farming systems.

INTEGRATED SYSTEMS

Forages have a dual purpose in our society. Forages, as a vegetative resource, are an integral part of grassland and range resources, but they are also considered commercial crops in their own right and play a central role in livestock production and conservation systems. In this book basic management phases of a forage crop are presented in a stepwise sequence that can easily be followed and put into practice in integrated systems. Integrated pest management or integrated forage management begins with the establishment of the crop and extends to its management as a means of enhancing the use of available forage by cutting or grazing by desirable herbivores, and of reducing its devastation by diseases and pests and use by other herbivory. Biosystem management with economic gain occurs.

Biotechnologists are working closely with geneticists and plant breeders to develop pest-resistant cultivars and, at the same time, to produce high-quality, highly digestible cultivars for domestic ruminants. This integrated system will also involve the quality analyst and the animal nutritionist. Development of near infrared spectoscopy analysis is an example of how the quality of the final product can be determined in a relatively short time. Another part of the overall system involves agricultural engineering in determining the most efficient methods of harvesting, packaging, storing, and utilizing the forage. Economics plays a dominant role when integrating all aspects of forage production. Based on the latest research, the chapters in this text provide the basic knowledge for such integration.

ECOSYSTEMS

Perennial forage crop production involves a complex ecosystem that is much more unique and intricate than annual grain crop production. Literally thousands of forage ecosystems exist, many in the field or pasture. A grazing ecosystem involves many factors, such as grazing animals, other herbivores, morphological development of the forage species, the soil resource and its conservation, pest management, and environ-

mental concerns. Rangeland and agroforestry ecosystems involve a much broader aspect of the environment, plus the integration of various tree species into the system. Another ecosystem concern is the growth and management of forages specifically for soil conservation and stabilization, or for water quality management. Forages play a very important role in maintaining wildlife and providing recreational areas. All of these ecosystems are presented in this reference text.

LIVESTOCK SYSTEMS

The existence of forages, especially perennial forages, enable ruminant livestock enterprises to exist and be economically profitable. Currently, over half the feed units for milk production come from forages, over 70% for beef, and over 80% for sheep. Many different forage species exist. Which species or combination of species is best for each animal species? How should each forage be managed? Quality and digestibility also are extremely important. All of these aspects are presented in the text and are based on the latest research findings.

REFERENCES

The authors have provided references documenting the latest research. Full titles are given to help readers evaluate which references to pursue further. The scientific names are up-to-date as of the time of publication. Readers are referred to the Crop Science Society of America or to the USDA Genetic Resources Information Network (GRIN) for an updated record. The in-depth index furnishes ready accessibility to the specific technical information.

AUTHORSHIP

The authors of this book are recognized as authorities in each particular subject matter area. We are grateful for their contribution and leadership in developing such thorough and up-to-date reference chapters. The various subject matter areas are interwoven into a framework that allows the reader to synthesize and develop a clear systems approach to forage production and utilization.

FORAGES—A BOOK IN TRANSITION

This fifth edition of *Forages: The Science of Grassland Agriculture* continues the strong science-based tradition of the first four editions. Recognizing the need for an authoritative text on forages, H. D. Hughes, Maurice E. Heath, and Darrell S. Metcalfe conceived and brought to reality in 1951 the first edition of *Forages,* as the book is commonly known. At that time the science base for forage management was rapidly emerging, and there was a need for a well-referenced scientific treatise on forages in the US. The expanded and revised second edition by the same editors was published in 1962. Robert F Barnes replaced H. D. Hughes for the third edition in 1973 and assumed more leadership in the fourth edition in 1985. With the deaths of H. D. Hughes and Maurice E. Heath and the retirement of Darrell S. Metcalfe, Robert F Barnes took the lead. The addition of Darrell A. Miller and C. Jerry Nelson completed the editorial transition for the fifth edition.

Each of the three original editors had a far-reaching vision for the role of forages and the capability to put together the most recognized book on grassland agriculture in the US. One can scarcely realize the magnitude of their contributions. The editors, authors, and readers of the fifth edition owe these three forage pioneers resounding gratitude. They led the way in documenting, interpreting, and utilizing research findings for grassland improvement based on science. The goal of the current editors and authors is to honor them by continuing that quest.

FORAGES ARE FOREVER

Using perennial and annual forages together is one way of ensuring that *"forages are forever."* One can take this phase literally and advisably, in that forages are necessary in our ecosystems. They are needed to maintain our environment, to sustain our farming operations, and to provide recreation. They are needed largely for economic reasons and provide the very basis for an integrated and profitable agriculture. Our desire is that this text will contribute to a healthy environment for all humankind and to a more sustainable world.

Robert F Barnes

Darrell A. Miller

C. Jerry Nelson

PART **1**

Principles

of

Forage

Management

1

Forage Crop Management: Application of Emerging Technologies

JEFFREY J. VOLENEC
Purdue University

C. JERRY NELSON
University of Missouri

TO predict forage plant responses to changes in management strategies we often construct conceptual models to help understand the complex interactions among forages species or cultivars, the environment, and the available management options. Though accuracy in conceptual models is the long-term goal, often we must begin the process at lower levels of understanding, then generate research data to add more details and precision. In many situations understanding characteristics through simple correlation with forage performance is sufficient to effectively manage forage crops. In other situations, however, more detailed mechanisms are needed to understand why plants respond as they do to changes in environment and/or management. Accurate knowledge of the physiology, morphology, and anatomy of forage plants is especially important as we attempt to create better-adapted forages using modern genetic techniques, to expand production of forage species to more stressful environments, and to explore alternative management systems to optimize plant performance within an environment.

Our intent is to examine several emerging concepts related to forage management, espe-

cially those where useful models and a skeleton data base currently exist. We perceive continued improvement in the data base, development of more completeness in the conceptual model, and a likelihood of application to forage management. We do not review the literature exhaustively or cover all topics where significant scientific discoveries are likely to occur. Our goal is to heighten awareness regarding some current developments, and we expect readers to be better prepared to understand these concepts as they are tested more thoroughly and moved into practice.

GENOTYPE BY ENVIRONMENTAL INTERACTIONS

Forage crop performance depends on the forage species and cultivar, i.e., the genotype, and the environmental conditions in which the species is grown (Fig. 1.1). Thus, performance can be improved by species and cultivar selection or by altering the environment through management. Genetic makeup of the crop, i.e., the genotype, influences how it will respond to the environment. Timing of practices such as planting, cutting or grazing, and addition of fertilizer nutrients are examples of how management alters the environment.

Prevailing environmental factors over which we may have little control, such as change in temperature or water stress, can affect the expression of genetic information in the crop. In addition, previous exposure to environmental stress can condition plants, making them more productive (or less productive) in the prevailing environment. For example, short days and cool temperatures of autumn stimulate hardening of perennial plants such as alfalfa (*Medicago sativa* L.), enabling them

JEFFREY J. VOLENEC is Professor of Agronomy at Purdue University. He received the MS and PhD degrees from the University of Missouri. His research focuses on carbohydrate and nitrogen metabolism associated with regrowth and persistence of forage plants.

C. JERRY NELSON is Curators' Professor in the Department of Agronomy, University of Missouri. He received the MS degree from the University of Minnesota and the PhD degree from the University of Wisconsin. His research is on regulation of growth processes in perennial forage grasses and persistence of forage legumes.

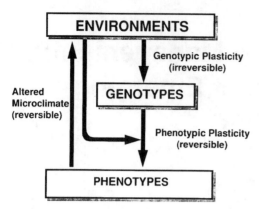

Fig. 1.1. *Genetic plasticity* refers to the change in the genotypic composition of the population in response to the environment. The environment also affects the way each genotype in the population expresses its information to alter the phenotype, i.e., expression of phenotypic plasticity. The phenotype can influence the environment around the plant to make the system dynamic.

to survive winter. However, some alfalfa cultivars respond less to hardening conditions because they lack the genetic information necessary to perceive and/or respond to autumnlike conditions. This may restrict their adaptation to warm areas. Thus, plant genotype and environment interact to give the plant the appearance and performance we routinely see, i.e., the phenotype.

Genotypic Plasticity. *Genotypic plasticity* refers to the ability to change the frequency and character of genes within a population (Fig. 1.1). In plants, genotypic plasticity depends on the genetic makeup of individuals, and a genetic shift may or may not lead to noticeable changes in the appearance of plants in the population. Many forage species are naturally outcrossed; such cultivars are heterogeneous populations, and each plant within the cultivar is unique in terms of its genetic makeup. This feature aids adaptation because the population or cultivar has the capacity to undergo natural selection in the prevailing environment, leading to a shift in its genetic base, as the most aggressive or best-adapted plants survive.

The management conditions can also differentially influence survival of plants within a population. For example, frequent and close defoliation of perennial grasses favors those plants in the population or cultivar that produce short leaves and tiller profusely. Plants with long leaves and few tillers will be suppressed or may die with frequent defoliation.

Conversely, if the same base population is fertilized with nitrogen (N) and cut infrequently, plants with long leaves and sparse tillering will be favored.

Reseeding legumes such as birdsfoot trefoil (*Lotus corniculatus* L.) also express genotypic plasticity through differential seed production, seed dormancy characteristics, and seedling vigor among the plants comprising the population that extends production from 1 yr to the next. For these same reasons, seed of alfalfa is generally produced in the same latitude as the forage production area. This restricts genetic drift in the population toward plants with high seed production and a change in winterhardiness (Smith et al. 1986).

Forage breeders generally select new cultivars for uniformity (homogeneity) within the population but not to the extent that genetic plasticity is lost. Genetic plasticity in a cultivar is especially important in areas where climate and soils are variable or when management flexibility is desired. For example, genetic plasticity for insect resistance or winterhardiness within the cultivar may be advantageous, even if genes for these characters are linked with undesirable traits such as low yield or poor forage quality. In cases where stresses are minimal, the high-performing plants survive and constitute much of the yield.

In general, genotypic plasticity of a stand of a nonreseeding species is not reversible because the genotypes that are not adapted to the prevailing environment are lost from the population (Fig. 1.1). In the worst case, prevailing stresses are so extreme that genetic plasticity is not sufficient to permit survival of any plants in the population. Some reversibility can occur in cross-pollinated species if they reseed and new plants are established. The best recourse, however, is to choose a cultivar that has sufficient genetic plasticity to allow some genotypes in the population to survive the most severe stress anticipated, while retaining sufficient genetic diversity to remain plastic during periods of less severe stress.

Phenotypic Plasticity. *Phenotypic plasticity* differs from genotypic plasticity in that it refers to the capacity of a given plant, i.e., a genotype, to adjust in growth form (Fig. 1.1). Exposed to different management scenarios the same plant may produce different phenotypes because the microclimate, i.e., the climate close to the plant, is altered. For example, Gibson et al. (1992) compared annual rye-grass (*Lolium multiflorum* Lam.) and dallis-grass (*Paspalum dilatatum* Poir.) growing in swards of different plant density. They found leaf blades in annual ryegrass were displayed nearly vertical when plant density was high but were largely horizontal when density was low. This response caused leaf area to be placed at different positions in the canopy depending on plant density, and it represents phenotypic plasticity. In contrast with annual ryegrass, leaf blades of dallisgrass remained nearly horizontal at all plant densities, reflecting little phenotypic plasticity. These characters may alter the competitive ability of the stand, especially for light (Rhodes and Stern 1978).

In contrast with genotypic plasticity, phenotypic plasticity is reversible. Thus, when management practices are changed, plants can again alter their growth habit in response to the new environment. This is advantageous if plant survival is enhanced or competitiveness is maintained. For example, bermudagrass (*Cynodon dactylon* [L.] Pers.) is usually propagated vegetatively by transplanting rhizomes or stolons such that plants in the new stand are genetically identical. This means the plasticity observed in response to changes in management is all phenotypic. Most other forage species are propagated by seed, often have both genotypic and phenotypic plasticity, and may allow more flexibility in management.

Over the past few decades grassland managers in the eastern and western areas of the US have developed different philosophies regarding use of plasticity. In general, those in the East favor a minimal number of species in the sward, each represented by an improved cultivar with a relatively narrow range of genetic plasticity. This emphasizes the need for relevant management practices to minimize stresses on the genetic plasticity in the population, and it challenges the phenotypic plasticity to allow adaptation to protect the genotypes when environments are severe or alternate management strategies are used. If phenotypic plasticity of individual plants is not sufficient to adjust, some plants may be weakened and die, further narrowing the genetic plasticity.

In contrast, managers in the West place emphasis on broad genetic diversity in their grasslands and rangelands, which consist of a large number of species grown together that offer enhanced genetic plasticity. Each species is usually indigenous to the area, and main-taining the broad range of natural genetic diversity is a goal. Reseeding extensive areas is rarely a viable alternative, so diversity is retained by judicious management, which means the species and genotypes with greatest phenotypic plasticity allow individual plants to survive stressful environments and management practices. Emphasis on sustainable agriculture and diversity in agroecosystems will likely modify strategies in the eastern US toward those in the western US (Nelson and Moser 1994).

ADOPTION OF TECHNOLOGY

Most farmers or ranchers have a general understanding of the forage species and, especially, the production potential of the soil resource they are using. Further, they are perceptive observers, have a financial goal and a conservation ethic, and are generally willing to assume some risk. They are also interconnected with a social and service network that is a source of ideas and new products. Thus, the forage or range producer is an individual who is looking for better or easier ways to meet objectives. But, at the same time, interest and knowledge are likely to be stronger regarding row crops or animals than they are about forages. Forages and forage-livestock systems are complicated to manage, are often located on the lower-productivity sites, and are not as high in priority for use of time and capital as are other enterprises.

Experience-based Decision Making. Farmers and ranchers are innovators and entrepreneurs, often discussing, trying, and working out the best management practices for a particular forage crop based on keen observation, trial, and a balance of successes and failures (Fig. 1.2). Livestock operators quickly learned that leaves are higher in quality than stems, and legumes are higher in quality than grasses. They noted plants respond differently to sheep grazing than to cattle grazing, and legumes enrich the soil with N whereas grasses increase the soil organic matter and improve soil structure. The scientific reasons for these differences are often unknown to the manager, however, and this lack of information restricts efforts to optimize forage productivity genetically or through management.

Thousands of years ago, largely through trial and error, the basic systems of forage management were first described. Mainly during this century, scientists have conducted research to determine why these systems work, and the reasons give insight as to how the

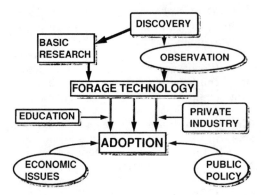

Fig. 1.2. New technology is based on discovery, but its adoption often depends on whether economic conditions or public policies encourage its use. Some technologies are adopted directly by education, whereas others require developing a specific product.

processes and systems may be changed genetically or through management. Some have argued this is retrospective science, but the knowledge gained has directed researchers to more or better-focused assessments that have led to new management and breeding strategies.

Even today, however, adoption of new technology can get ahead of the science base very quickly. A specific case is the rapid acceptance and use of the large round baler. This labor-saving method of hay packaging and storage became available in the early 1970s and was quickly adopted by producers with virtually no scientific data base. Only later, after several million bales had been made and stored in a range of conditions, were basic principles developed for forming, storing, and feeding the bales. In many cases the first information put together was not research data but an assemblage of methods that tended to work in the field for innovative farmers and ranchers.

Other cases have not worked out as well. The world is blessed with thousands of potential forage species, many of which have been tried, and many more than once. Sometimes, a species works well in a specific environment or niche, but it is not widely adapted. For example, comfrey (*Symphytum officinale* L.), prairie brome (*Bromus catharticus* M. Vahl), and cup-plant (*Silphium perfoliatum* L.) are species of recent interest that have a low science base in the US. Each has merits but is not a panacea—just as it was several years ago when kudzu (*Pueraria lobata* [Willd.] Ohwi), johnsongrass (*Sorghum halepense* [L.] Pers.), and timothy (*Phleum pratense* L.) were introduced. In each case there was a history

in a previous environment, an enthusiasm in the new environment, but a lack of scientific information on adaptation, management, and forage quality in the new environment. Some are successful, some escape as weeds, some are not adopted even though they are shown to have potential, while others wait to have their interest recycled by another generation.

Forages tend to be more subject than other crops to these interests and influxes; plant introductions have contributed greatly to the list of useful species in the US. Today, however, the Natural Resources Conservation Service of the USDA has several Plant Material Centers that collect and evaluate new plants for adaptation to a range of sites and environments (Chap. 23, Vol. 1; Chap. 17, Vol. 2). This data base helps decision making about potential value of the species. New species are not easily introduced successfully. For example, the first attempts with alfalfa failed. Only after experience and a small science base was developed on soil requirements and cutting management did alfalfa gain in popularity and have its potential realized.

Research-based Decision Making. Most often a new practice or product is discovered, is researched for its effectiveness, reliability, and transferability from one situation to another, and is generally used in several farmer demonstrations before it is adopted. This allows a systematic data base to be developed and used to understand the basic processes, pitfalls to be minimized or avoided, and industry or some other marketer to develop the educational programs. A new herbicide, cultivar, or piece of farm machinery would fit this pattern.

Discovery of the knowledge is only part of the solution, however. While the technology can be made available, there are powerful forces that alter or influence its rate and extent of adoption (Fig. 1.2). These include perceived economics for the producer to adopt the technology directly, economics for industry to develop the machinery or products, i.e., to package and market the technology, and governmental policies that restrict or enhance adaptation. Ethical and social considerations are also involved in the decision-making process.

Patents and Intellectual Property Rights. Whether technology flows through public channels or through industry often depends on its patent capability, i.e., it provides the owner with exclusive rights to the technology for a period of time, generally 17 yr. This is clear for technologies such as a new pesticide or piece of hay-handling equipment. Once a patent is granted, an industry can risk the financial inputs necessary to manufacture and market the product. This is often done in consultation and cooperation with public agencies after protection is granted.

A newer concept is recognition of intellectual property rights associated with the genetic makeup of a new plant cultivar or germplasm. Today, many forage cultivars are protected through the federal Plant Variety Protection Act, which was passed in 1970 and amended in 1994. Since 1985, it has also been possible to protect new, improved seed-producing forage cultivars through patenting similar to what an engineer might do with a new invention. Protecting new cultivars through the Plant Variety Protection Act is far more common than patenting of a new cultivar. When the private or public plant breeder protects a cultivar, exclusive ownership is given for a specified period of time to the breeder or his or her institution/company, and the breeder has control over its use and multiplication.

The protected cultivar is described in specific terms, often including genetic fingerprinting using DNA markers, so it can be kept pure and can be tracked during distribution, use, and marketing. Protected cultivars have strict regulations on production and marketing of seed. These regulations allow the owner to use fair market value to generate a return on the investment made in developing the new cultivar.

Selling Seed of Protected Cultivars. A major problem with protected forage cultivars is what is known as *brown-bagging,* a means of illegally marketing seed of a protected cultivar. Legal charges can be pressed against farmers or seed sellers who do not market the protected cultivar within the guidelines established, including paying royalties to the owner. Nearly all alfalfa cultivars are protected through the Plant Variety Protection Act, but brown-bagging is generally not a major problem because seed production occurs in geographic locations that are far removed from forage-producing areas. This spacial separation allows a degree of ownership control. Also, alfalfa seed production requires very specialized management to be economically competitive. Thus, nearly all new alfalfa cultivars are protected, and seed is produced and marketed by industry.

Brown-bagging is a serious deterrent to cultivar development and marketing in most other forage species, however, and has thwarted many industry attempts to develop and market superior cultivars. For example, much of the tall fescue (*Festuca arundinacea* Schreb.), orchardgrass (*Dactylis glomerata* L.), timothy, and red clover (*Trifolium pratense* L.) seed is produced in the Midwest as a coproduct in the same fields normally used for forage production. Based on the farmer's need for forage, and the perceived price of seed, a decision is made as to which product to harvest.

In the past it was legal to sell seed of a protected cultivar as long as it was not advertised or offered through a third party. But, as of April 1995, selling seed of a cultivar protected through the Plant Variety Protection Act to another farmer is illegal. By the revision in the Plant Variety Protection Act in 1994, farmers may continue to save protected seed for their own use, but any sale is prohibited without the written permission of the protection owner. Seed companies with protected cultivars are very serious about enforcing the law, and laws are likely to become even stricter.

IMPROVING FORAGE QUALITY

Currently, considerable research attention is being given to improved forage quality. This has occurred partly due to the advent of new near infrared reflectance spectroscopy methodologies (Chap. 8) that have greatly expanded the possibilities for routine forage quality analysis. In addition, the importance of producing and feeding a quality forage has been clearly established, and the potential for genetic improvement, even genetic engineering, of plant constituents associated with quality is particularly appealing. The potential use of forages for biomass and energy production has also emphasized quality of the product and has contributed to the strong interest.

Forage Protein Quality. Knowledge concerning deficiencies in forage protein quality is an example where our improved understanding obtained from years of research will likely lead to new opportunities for forage improvement. Research in the early 1960s demonstrates that wool growth of sheep is frequently limited by the supply of the sulfur (S) containing amino acids, methionine and cysteine. Feeding these amino acids enhanced wool production up to 35%. Surprisingly, follow-up research reveals that proteins in forage from most legumes are S-rich, but most of the S-containing amino acids are converted in the rumen to low-S-containing microbial protein. What was clearly needed was a source of S-rich proteins that resist degradation by rumen microflora.

Findings of Spencer et al. (1988) reveal that pea (*Pisum sativum* L.) seeds contain water-soluble, S-rich proteins that resist degradation in the rumen. Work is now underway to move the corresponding gene(s) into forage legume species using molecular genetic techniques. Such approaches seem reasonable alternatives, especially when all other conventional means for addressing a problem have been exhausted. When finally developed, the process, processes, or cultivar will likely be patented or protected in order to charge royalties to recoup the investments. Whether such protection is controlled by a public agency or an industry and what the effect may be on seed cost remain to be seen. Ultimately, public acceptance will determine the role of biotechnology in modern forage agriculture.

Multifoliolate Leaf Alfalfa. An example of a discovery whose true utility remains a topic of discussion is the multifoliolate leaf trait of alfalfa. The "multileaf" characteristic, as it has come to be known, is a mutation identified in the 1930s that results in production of more than three leaflets per leaf. Keen interest in this trait exists because it could be one method of increasing leafiness of alfalfa and improving forage quality and enhancing animal performance. Research suggests that the increased leaf:stem ratio of highly multileaf plants can lead to small increases in forage digestibility, but timely harvesting remains as the critical factor currently influencing alfalfa quality (Volenec and Cherney 1990).

Clearly, factors in addition to leafiness influence alfalfa quality, and there is considerable interest in developing alfalfa cultivars with improved levels of protein and digestibility, especially improved stem digestibility (Buxton and Casler 1993). Nevertheless, the multileaf trait has made it to the marketplace and is considered a value-added feature of many modern alfalfa cultivars. Adoption of new technology, in this case the multileaf character, despite the lack of supporting scientific evidence suggests that experience-based decision making (that high leafiness of alfalfa leads to improved forage quality) remains a powerful force for educators to contend with.

Grass Tetany. Another problem being addressed with some new solutions involves grass tetany, or hypomagnesemia, that is caused by low magnesium (Mg) concentrations in forages, mainly grasses (Chap. 9). Several approaches to solving the problem are being actively pursued. For example, genetic studies have shown that Mg concentration of tall fescue can be improved by breeding, permitting development of germplasms with low tetany potential (Sleper 1979). These germplasms accumulate high Mg over a range of temperatures (Sleper et al. 1980) and fertilizer regimes (Brown and Sleper 1980), perhaps because genetically improved plants are better able to remove Mg from the soil and accumulate it in the leaves. This technology is likely available in other grasses and could be bred into specific cultivars, perhaps by industry.

At the same time, Reinbott and Blevins (1991) have discovered that phosphorus (P) fertilization stimulates Mg uptake by grasses from low-Mg soils. The exact mechanism is unknown, but in extensive field studies with wheat (*Triticum aestivum* L. emend. Thell.) and tall fescue they found Mg concentrations were increased and tetany potential was reduced by modest applications of P to low-P soils (Reinbott and Blevins 1994). The genetic and management solutions to the problem will both require additional testing and evaluation. Other methodologies for control of tetany are emerging. Cultivar protection, economic issues, and public policies such as water quality regulations on fertilizer applications may influence which solution to the problem is preferred.

REDEFINING STORED ORGANIC RESERVES

Few topics in forage physiology and management have been studied more extensively than the role of total nonstructural carbohydrates (TNC) in shoot regrowth and plant persistence. From a classic study, Graber et al. (1927) found nonstructural carbohydrate concentrations in roots of legumes and stem bases of grasses decline in spring as plants resume growth. Nonstructural carbohydrates in storage organs also decline after defoliation and are restored before the next harvest. The concept of storage and reuse of reserve carbohydrates has found extensive use as *the* basis for harvest management decisions.

Carbohydrates appear to be used as a source of carbon (C) for spring growth and regrowth after cutting, but does the TNC content of storage organs determine the rate of shoot regrowth or level of plant persistence?

Alternatively, does something else regulate shoot regrowth, with TNC merely being used to support the process? Largely overlooked in the classical report, however, is another important finding about alfalfa: significant declines in taproot N concentrations also occurred in early spring as growth resumed and following harvest in summer (Fig. 1.3). Graber et al. (1927) concluded that both taproot TNC and N reserves are important for regrowth and persistence of forages. While the concept that TNC reserves mediate forage crop regrowth and persistence has been widely accepted (Smith 1964; Brown et al. 1972; Heichel et al. 1988), little further attention has been given to the role of N reserves in growth and persistence of forages.

Carbohydrate Reserves in Legumes. May (1960) questioned the cause-effect relationship between nonstructural carbohydrate depletion in storage organs and shoot regrowth because the direct association between taproot TNC concentrations and growth rate or stress tolerance had not been unequivocally established. For example, Jung and Smith

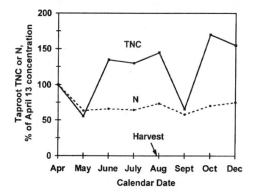

Fig. 1.3. Trends of total nonstructural carbohydrates (TNC) and nitrogen (N) in taproots of alfalfa growing in Wisconsin. Plants were harvested at the seedpod stage of growth on August 9. Data are expressed as percentage of concentrations present on April 13. (Adapted from Graber et al. 1927.)

(1961) found little association between taproot TNC concentrations and freezing tolerance of alfalfa or red clover (Table 1.1), but taproots with high concentrations of total sugars and several nitrogenous fractions were more freezing tolerant than were taproots having low concentrations of these compounds. Recent studies on fall management of alfalfa also show little relationship between taproot TNC levels in fall, plant survival over winter, and spring growth (Edmisten et al. 1988; Brink and Marten 1989; Brown et al. 1990).

Genetic differences in rates of alfalfa shoot regrowth after defoliation also are not closely associated with taproot TNC levels (Volenec 1985). Even though genotypes differed up to fourfold in taproot TNC concentrations, shoot regrowth rates after defoliation were similar (Fankhauser et al. 1989; Habben and Volenec 1990). Repeated defoliations essentially exhausted taproot TNC of low-TNC lines but did not reduce herbage regrowth rates (Boyce and Volenec 1992). These results are not consistent with the perceived direct role of TNC reserves in agronomic performance of forages.

Nitrogen Reserves in Legumes. It is becoming clear that other physiological factors, in addition to stored TNC, are involved in regrowth and persistence of forages. Few studies have critically evaluated the role of stored N in regrowth and stress tolerance of forages. In the report of Graber et al. (1927), changes in taproot N concentrations in alfalfa were not as great as those for taproot TNC (Fig. 1.3). This may have been interpreted as an indication that N reserves are not as important as TNC reserves in shoot regrowth of alfalfa. Now, however, it is well-known that young regrowth is very high in N (protein) and N_2 fixation declines dramatically after cutting (Heichel et al. 1988), to the point where rates of N_2 fixation could not meet the N needs of regrowing shoots.

Turnover studies using the stable isotope, ^{15}N, have been used to verify the transport of

Table 1.1. Correlation between electrolyte leakage (an indicator of freezing damage) and taproot organic constituents for red clover and alfalfa.

Taproot Constituent	Red Clover	Alfalfa
Soluble nonprotein nitrogen	−0.72[b]	−0.83[b]
Soluble protein nitrogen	−0.47[a]	−0.83[b]
Reducing sugars	0.28	0.25
Nonreducing sugars	−0.93[b]	−0.90[b]
Starch	0.44	0.29
Total nonstructural carbohydrates	−0.06	−0.13

Source: Jung and Smith (1961).
[a,b]Significant at 5% and 1% probability levels, respectively.

N from storage organs to regrowing shoot tissues after harvest. In perennial ryegrass (*L. perenne* L.), essentially all N in regrowing leaves 6 d after defoliation was organic N remobilized from roots and stubble (Ourry et al. 1988). In subterranean clover (*T. subterraneum* L.), from 100% to 143% of N present in crowns and roots turned over, i.e., was replaced, between the first and fourth harvests (Phillips et al. 1983). These authors suggest that root protein degradation provided N required for leaf growth after defoliation because of reduced fixation of N_2 and/or low uptake of soil N. Culvenor and Simpson (1991) report most N found in new leaves during early regrowth of subterranean clover was mobilized from roots and vegetative tissues, largely from soluble proteins. These results generally agree with those of Kim et al. (1993) who report that 80% of the N found in alfalfa shoots after 10 d of regrowth came from N reserves and that most of this N had been stored in the roots.

Several studies have reported an association between accumulation of taproot N and stress tolerance of forage legumes. For example, alfalfa, red clover, and yellow sweetclover (*Melilotus officinalis* Lam.) all accumulate high concentrations of taproot N in late fall (Bula and Smith 1954). Crude protein levels in taproots were positively correlated with winterhardiness of these species. Further, taproot N concentrations for alfalfa and red clover in Wisconsin declined nearly 50% during the last week of April and the first week of May because stored N was utilized to support new shoot growth. Similar results were noted with sweetclover cultivars differing in cold tolerance (Hodgson and Bula 1956) and with winter-hardy and nonhardy alfalfa cultivars (Bula et al. 1956).

Nitrogen and Grasses. Several studies have evaluated grasses. Again, it is well-known that stored carbohydrate, often as fructan in cool-season grasses, decreases during early stages of regrowth and then is restored as plants develop (Volenec 1986). But physiological studies suggest rate of leaf elongation and early dry matter accumulation during growth of tall fescue are generally not carbohydrate limited (Wilhelm and Nelson 1978; Schnyder and Nelson 1989). Further, leaves of grasses fertilized with high N often grow faster and have lower concentrations of carbohydrate in the elongating tissue than those receiving low N (Volenec and Nelson 1984b).

Grass leaves respond to N fertilization by increasing rates of cell division, but rate of cell elongation and final cell size are largely unaffected by N (MacAdam et al. 1989). Thus, N regulates the number of cells, but each cell subsequently requires large amounts of carbohydrate for wall synthesis during its elongation and for adding secondary cell wall material after elongation ceases. On a dry weight basis, fully expanded grass leaves usually contain 12% to 18% protein (about 2% to 3% N), 75% to 80% carbohydrate (cell wall), and 5% to 7% mineral. Thus, while the N requirement for growth is small, its concentration appears to mediate regrowth or growth rate of grass leaves (Gastal and Nelson 1994). In contrast, unless severely starved (Moser et al. 1982), carbohydrate is adequate and is largely utilized as a substrate to support growth of the cells that are produced largely in response to N.

Nitrogen Storage Forms. What forms of N might be stored in stem bases of forage grasses and roots of forage legumes? Nitrate and proteins are the N compounds most often accumulated in storage organs of herbaceous plants (Millard 1988). Millard further suggests that, quantitatively, ribulose bisphosphate carboxylase, i.e., Rubisco, the key enzyme associated with photosynthesis (Chap. 3, Vol. 1), is the most important protein accumulated for N storage in leaves. Rubisco is known to be degraded as leaves age, releasing N in the form of amino acids for transport to meristems for synthesis of new tissue and thus serving in a storage role.

Less is known about proteins in roots of legumes and, especially, in stem bases of grasses. In alfalfa taproots, amino acids undergo a cyclic pattern of depletion and reaccumulation after defoliation (Fig. 1.4). The most abundant amino acids, aspartate and asparagine, are rapidly depleted from taproots after harvest (Hendershot and Volenec 1993b).

Proteins also accumulate in taproots of forage legumes as plants harden for winter, with the greatest accumulation occurring in alfalfa and the least in red clover (Fig. 1.5). During hardening, several new proteins appear in taproots, and birdsfoot trefoil and alfalfa accumulate large amounts of specific proteins. The patterns of accumulation and disappearance of these specific proteins are consistent with a role as vegetative storage proteins. Preliminary evaluations indicate taproot vegetative storage proteins from alfalfa accumulate during fall as plants harden for winter and are extensively depleted from taproots when meristems on crowns are reactivated in

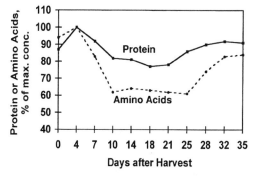

Fig. 1.4. Relative concentrations of taproot amino acids and proteins after alfalfa harvest. Data are expressed as percentage of the highest concentrations observed (Day 4). (Adapted from Hendershot and Volenec 1993b.)

spring (Fig. 1.5) and after defoliation in summer (Hendershot and Volenec 1993a, 1993b). These depletion-reaccumulation patterns for taproot proteins are consistent with those of taproot TNC reserves.

Our understanding of what these proteins are and how they contribute to regrowth of forage legumes is far from complete. Vegetative storage proteins are the most abundant soluble proteins in alfalfa taproots. Regulation of their accumulation in fall and their degradation in spring and after harvest in summer are important considerations. Additional investigations of N accumulation in storage organs of other legume species, and especially forage grasses, are needed to help clarify the complementary roles of stored N and TNC in mediating regrowth and persistence of forages.

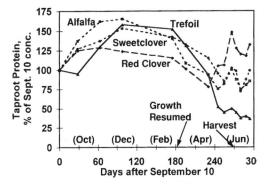

Fig. 1.5. Relative protein concentrations in taproots of four legumes growing in Indiana. Shoots emerged above the soil surface on March 3. Plants were harvested on June 2. Data are expressed as percentage of concentrations on September 10.

YIELD COMPONENTS

Yield of forages can be calculated using two yield components: tiller (or shoot) density and tiller (or shoot) weight (Nelson and Sleper 1977). Improving forage production depends on understanding the relative contribution of each yield component to herbage yield and determining the physiological constraints on tiller growth and tiller production. These yield components are, unfortunately, negatively correlated with each other in both forage grasses and legumes. Nevertheless, results of recent studies have shed light on the interactions between tiller (shoot) production and tiller (shoot) growth rates, especially in grasses, but critically important issues remain unresolved.

Control of Tiller Production. Tillers of grasses and shoots of legumes arise from axillary buds present in the axils of lower leaves (Fig. 1.6). Each leaf has associated with it one axillary bud that has the potential to develop into a tiller. As a result, tiller production per plant is controlled, in part, by leaf production per plant (Simons et al. 1973). In tall fescue, the number of tillers produced per plant is closely related to the number of leaves produced per plant during vegetative growth (Zarrough et al. 1984), but not all potential tillers develop as anticipated (Fig. 1.7). Genotypes with rapid leaf growth rates initiate new leaves at a slow rate (i.e., there is a longer time interval between appearances of successive leaves), but each leaf is large, which also causes axillary buds to be produced at a slow rate. In addition, the genotype with a slow rate of leaf appearance also has a low rate of site usage; i.e., a low proportion of the axillary buds produced actually develop into a tiller compared with the high-tillering genotype. This suggests that both rate of leaf appearance (axillary bud production) and site usage (proportion of potential buds that develop into tillers) affect tiller production of grasses.

In both tall fescue genotypes studied (Fig. 1.7), site usage for tiller development decreased as the plants developed. Site usage may be controlled indirectly by leaf length and canopy structure (Skinner and Nelson 1992), which, in turn, is influenced by forage management decisions. Tall dense canopies, created when grasses are well fertilized and harvested infrequently, often have lower tiller densities when compared with frequently clipped or intensively grazed canopies. For a long time this was considered to be due to apical dominance or a competition for minerals

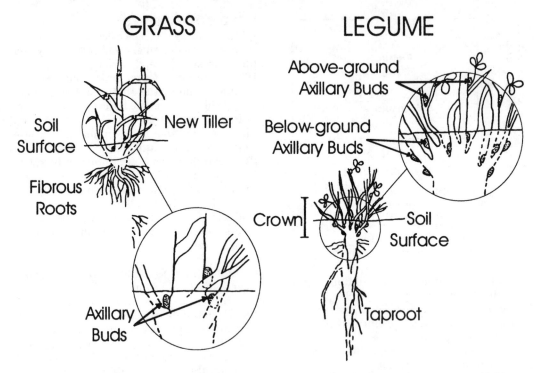

GRASS

LEGUME

Fig. 1.6. Buds present in axils of leaves give rise to new tillers (grasses) and shoots (legumes). These can be located on stubble at or above the soil surface, or in the case of forage legumes, on crown tissues several centimeters below the soil surface. *Purdue Univ. drawing.*

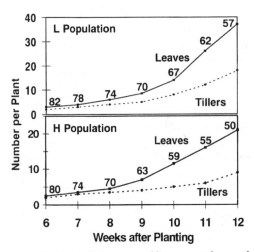

Fig. 1.7. Time course of leaves per plant and tillers per plant during seedling development of tall fescue populations selected for high (*H*) and low (*L*) rates of leaf growth. Numbers indicate percentages of tiller sites at each sampling date that had produced a tiller. (From Zarrough et al. 1984.)

and photosynthate, which reduced tillering (Murphy and Briske 1992). Now, however, control of site usage and tiller production in several grasses is believed to involve changes in light quality (Casal et al. 1987). Sunlight, with its balanced spectral quality at the top of the canopy, undergoes a shift as it passes through the canopy, being depleted in red wavelengths by photosynthetic pigments to a much greater degree than other wavelengths. Thus, light passing through a full canopy to reach the soil surface, the location where tillering occurs, is deficient in red light, and tiller development is reduced. Tillering is active if the canopy is kept short or open so red light can penetrate to near the soil level.

Addition of red light near the crown of a grass plant in a dense canopy can stimulate tillering. For example, tillers per plant of dallisgrass doubled after Deregibus et al. (1985) used red light-emitting diodes to increase the proportion of red light around the crowns. A similar trend occurred for smutgrass (*Sporobolus indicus* [L.] R. Br.), but the influence of red light was more complex. Red light increased tiller production and delayed tiller death, but only during certain times of the year. The involvement of red light in tillering strongly suggests a role for phytochrome, a chromophore that is involved in numerous signaling responses in higher plants, but the exact mechanism is not clear. Work is now underway to determine the tissues involved in sensing changes in spectral composition (Robin et al. 1994) and to elucidate the interactions among light quality, leaf elongation rates, and tillering (Skinner and Simmons 1993). However, increasing the abundance of red light reaching the base of the canopy while maintaining light interception and its influence on productivity will be challenging.

Tiller production of grasses is also affected by N fertilization, especially when the stand is thin. Applying 90 kg ha^{-1} of N annually increased productivity of several tall fescue genotypes, largely due to increased tiller density with only a small change in weight per tiller (Fig. 1.8). At higher rates of N tiller density approached a maximum, and yield increase was due largely to increased weight per tiller. Weight per tiller is closely related to leaf growth per tiller. Thus, phenotypic plasticity is expressed in response to N as the genotype is held constant for all N rates, but the nature of the growth response depends on tiller density.

More information is needed relative to how light quality reaching the tiller sites and N supply interact to alter the growth response of grasses. This may have special significance in management of legume-grass mixtures in which the legume component contributes to both shading and N supply of the grass. For example, if N supply from the legume is high,

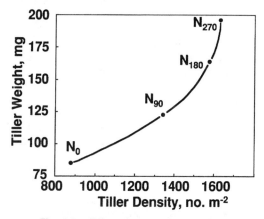

Fig. 1.8. Effect of nitrogen (N) fertilization at 0, 90, 180, and 270 kg ha^{-1} annually on the relationship between tiller density and tiller weight for six genotypes of tall fescue grown in Missouri. (From Nelson and Zarrough 1981.)

the tillering responses of the grass may shift to increase its competitiveness to again reduce the shading and N contribution of the legume, after which the cycle can be repeated. All of this contributes to the dynamic nature of the ecosystem that is managed for forage, pasture, or range production.

It may be possible to alter the relationship between leaf elongation rate and leaf appearance rate of grasses by selecting for shorter leaf blades, i.e., to increase number of leaves produced, while retaining a rapid elongation rate to maintain leaf size. This should result in forage grasses that possess rapid leaf growth rates, high tillering rates, and enhanced yield potential, especially if the canopy is kept short to allow light penetration. This plant type would have more growing points and more leaf meristems per unit land area to develop leaf area and yield. The feasibility of this approach remains untested. Especially critical is the ability of these altered canopies to perform well during grazing and in mixtures with legumes.

LEAF GROWTH AND TILLER WEIGHT

As with tillering, factors influencing tiller weight have been best characterized in forage grasses. Most research on tiller growth and yield has been conducted during vegetative growth stages. In this context, virtually all the harvestable mass of a grass tiller is leaf tissue. Mechanisms regulating growth of grass leaves are poorly understood despite their importance. Traditionally, reserve carbohydrates have been identified as a prime factor limiting grass leaf growth after harvest (Ward and Blaser 1961; Booysen and Nelson 1975). Upon close analysis, however, it has become clear that in most environments leaf meristems of cool-season grasses contain an abundance of sugars and nonstructural polysaccharides (Volenec and Nelson 1984a), and it is unlikely that these constituents limit grass leaf growth after harvest. Most likely this also explains why selection programs for enhanced leaf photosynthesis, i.e., a process that increases the carbohydrate supply, have not resulted in higher yield of forage grasses (Nelson 1988).

In tall fescue, as in other grasses, leaf growth occurs in an intercalary meristem, a growth zone, located in the basal 25 to 30 cm of elongating leaf blades (Fig. 1.9). These leaf growth zones contain regions of active cell division and cell elongation that, acting together, result in leaf elongation (MacAdam et al. 1989). Nonstructural carbohydrate may exceed 40% of dry weight of the growth zone, and much of the carbohydrate accumulates as fructan polymers (Volenec and Nelson 1984a). Protein concentration in the dividing cells of the meristem is two to three times that found in fully expanded leaf blades. The rapid synthesis and turnover of proteins in dividing cells and developing tissues result in very high rates of dark respiration. Later, during cell elongation, protein concentration is diluted by water influx and rapid cell wall synthesis. In contrast with levels of carbohydrate, N content in the leaf growth zone is positively related to leaf elongation rate (Gastal and Nelson 1994).

Several studies with tall fescue suggest that cell division rate within leaf meristems may ultimately limit leaf growth rates of this species. For example, addition of N to tall fescue plants grown in N-depleted soil doubles leaf elongation rates but has little effect on leaf cell size (Table 1.2). The rapid growth of leaves of N-fertilized plants is due to increased cell division rates, which nearly dou-

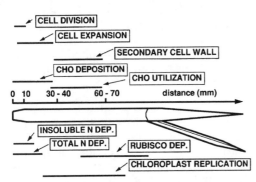

Fig. 1.9. Growth zones and areas of carbohydrate (CHO) and nitrogen (N) deposition in an elongating leaf blade of tall fescue. The blade, surrounded by sheaths of previous leaves, elongates by cell division and elongation at the base. Nitrogen and large amounts of water-soluble carbohydrate are deposited at the base. Much of the carbohydrate is stored, then used later for secondary cell wall synthesis, whereas proteins (N) used for cell production are reused for developing chloroplasts and photosynthetic enzymes. (Adapted from MacAdam et al. 1989; Allard and Nelson 1991; Nelson et al. 1992; and Gastal and Nelson 1994.)

ble with high N, while length of epidermal cells remains similar (MacAdam et al. 1989). Interestingly, it was found that the time required per cell division was similar with added N, but it appeared each cell divided more times before the cells were pushed from the cell division region of the meristem into the cell elongation zone. It is unclear what regulates the number of divisions a cell will make before it stops.

Addition of N could be expected to deplete TNC in leaf growth zones because of their rapid consumption to support faster leaf growth processes. However, sugar and fructan concentrations in the growing tissue remain largely unchanged (Table 1.2). The abundance of meristem carbohydrate, despite a twofold difference in leaf growth rates, suggests that C availability in meristems is adequate and that other physiological processes may limit leaf growth of this species (Volenec and Nelson 1984b; Schnyder and Nelson 1989). While cell division rates appear to be an important feature influencing leaf elongation rates of tall fescue, the physiological and biochemical limitations to cell division in leaf meristems remain unknown.

The mechanisms controlling alfalfa shoot growth rates, which influence yield per shoot, also remain to be elucidated. The concept that taproot sugar and starch concentrations mediate shoot regrowth rates after harvest is an oversimplification. For example, alfalfa leaf expansion declines markedly if the dark period is extended by only a few hours, even though taproots contain high concentrations of TNC (Hendershot and Volenec 1989). A close association was observed, however, between leaf expansion rates and TNC concentrations of expanding leaves. Leaf growth diminished rapidly as leaf TNC concentrations declined below 25 g kg^{-1} of tissue dry weight. Therefore, abundant TNC supply in taproots does not equate with rapid growth of shoot tissues, because leaf expansion may be carbohydrate limited.

Recent findings indicate much of the TNC that disappears from taproots of alfalfa after harvest is consumed in respiratory processes, and only a small proportion of the total is ac-

Table 1.2. Influence of nitrogen fertilization on leaf elongation rate (LER), cell characteristics, concentration of total nonstructural carbohydrates (TNC, dry wt. basis), and protein (fresh wt. basis).

Nitrogen	LER	Cell length	Cell production	TNC	Protein
(kg/ha)	(mm/d)	(μm/cell)	(no./d)	(% dry wt.)	(mg/g fr.wt.)
22	8.1	113	72	43.8	4.2
336	15.3	111	137	39.5	9.4

Source: Volenec and Nelson (1983, 1984b).

tually incorporated into herbage (Ta et al. 1990). This supports an earlier suggestion by May (1960). Results of grafting studies also suggest that shoot characteristics, and not characteristics found in roots, have the greatest impact on alfalfa shoot growth rates (Fankhauser and Volenec 1989). Other grafting studies comparing fall dormant and fall nondormant cultivars have shown that herbage morphology was influenced to the greatest extent by scion source (Heichel and Henjum 1990). In legumes, therefore, it appears leaf expansion is mostly related to carbohydrate supply in the leaf. This could be supplied from the taproot or by photosynthesis of the leaf itself. The sources of C and N for leaf growth of legumes need to be evaluated.

CONTROL OF STAND PERSISTENCE

Persistence of forages is critical because many are perennials and all are defoliated periodically during the growing season. Yet our understanding of plant characteristics that confer stand persistence is far from complete (Beuselinck et al. 1994). The concept of persistence varies with the context in which it is applied. To a producer, persistence is evaluated in terms of perenniality of a stand or community of plants. Whether or not specific members of a stand persist is generally less important to a producer than whether or not the stand as a whole persists. To many scientists focusing on forage improvement, however, persistence of perennials applies principally to the survival of individual plants, with less emphasis on community-level concepts of stand persistence.

Stand persistence of crown-forming legumes such as alfalfa or red clover depends on survival of the individual plants, i.e., plant persistence, because the plants are not managed to allow reseeding. Further, autotoxicity in alfalfa restricts the ability of young seedlings to become established in thinning stands (Tesar 1993). With annuals such as common lespedeza (*Kummerowia striata* [Thunb.] Schindler) or biennials such as sweetclover, some reseeding and establishment of new plants must occur regularly. The basic principles regarding seed production, seed dormancy, and seedling vigor of annuals and short-lived perennials are being evaluated. Some short-lived perennials such as birdsfoot trefoil have the ability to develop rhizomes for plant propagation, which reduces dependence on reseeding (Beuselinck et al. 1994).

Interest in using alfalfa for continuous stocking has provided a renewed opportunity to understand the physiological and morphological adaptations that exist in plants that persist under this severe stress. Selection for alfalfa cultivars tolerant of near-continuous stocking has been successful. Prostrate plant growth (decumbency) and prolific production of axillary buds on crowns are two traits that have been consistently associated with persistence under grazing (Brummer and Bouton 1991, 1992). These attributes taken together allow parts of the plant to escape being grazed and increase the amount of leaf area retained on the crowns, a trait also associated with grazing tolerance (Briske 1991). When clipped frequently, taproot TNC concentrations were depleted in some alfalfa cultivars with grazing tolerance, but not in all. For example, taproot TNC concentrations and root masses of 'Alfagraze', a grazing tolerant cultivar, remained high when intensively stocked (Smith and Bouton 1993). This suggests large quantities of TNC in taproots may aid persistence of intensively grazed alfalfa, but TNC is not the sole determinant. The role of taproot N reserves in grazing tolerance has not been evaluated.

Disease resistance also is thought to aid alfalfa persistence, especially under intensive grazing systems where plants are weakened and animal traffic can damage crowns, thereby increasing disease incidence (Lodge 1991). However, alfalfa persistence under intensive grazing pressure is not closely associated with conventional disease resistance ratings (Smith et al. 1992). Alfagraze possesses only limited resistance to five major pathogens of alfalfa yet survives under intensive stocking better than other cultivars with higher overall levels of disease resistance. Additional selection for grazing tolerance within five alfalfa cultivars did not alter disease resistance in a consistent manner when the grazing tolerant germplasms were compared with original cultivars. Thus, disease-resistance levels present in the original cultivars may be sufficient for survival when these plants undergo stresses associated with intensive, continuous stocking.

Other questions remain unanswered regarding forage plant persistence. (1) Why do alfalfa cultivars that exhibit nondormant growth in fall often winterkill? Because of rapid regrowth after cutting and more substantial herbage growth during fall, improving winterhardiness and persistence of nondormant alfalfas in the north central US appears desirable. (2) What is the threshold level of stored reserves necessary for plant persistence? Could amounts beyond this threshold level be partitioned into harvestable yield? (3) Can plants be selected and cultivars developed that break winter dormancy in spring in response to photoperiod instead of temperature? Temperature fluctuations during late winter and spring often stimulate premature bud break and shoot growth of legumes. These shoots are extended above the protective soil and are often killed during a late spring freeze. Regrowth may occur, but plant persistence can be compromised.

A NEW LOOK AT ROOTS

Underground organs have never received research attention comparable with their role or relative to that given top growth, but new frontiers are being established concerning how roots influence growth rate aboveground and morphological development of shoots. In many crop plants, and in nearly all forages, a part or all of the top growth is harvested as the economic product. Conceptually, the root merely has to be large enough and active enough to provide anchorage and to absorb the needed water and minerals. An overinvestment by the plant in root growth could be considered a liability, in that resources that could be partitioned into economic product (shoots) are inefficiently used for growth and maintenance of roots. Nevertheless, roots must survive and function for forage plants to thrive.

In general, cool-season grasses invest a lower proportion of their C resources in root growth than do native warm-season grasses such as big bluestem (*Andropogon gerardii* Vitman) or indiangrass (*Sorghastrum nutans* [L.] Nash). Genetic studies reveal that root characters are heritable but that root phenotype can be influenced markedly by soil environmental conditions (Brick and Barnes 1982; Pederson et al. 1984). New questions have arisen, however, with regard to the roles of roots and the necessity of having a well-managed root system in order to achieve biological potential.

Environmental factors alter the relative growth rates of roots and shoots. For example, adding N to grasses generally stimulates top growth more than root growth (Belanger et al. 1992), top growth is favored in plants experiencing low light (Kephart et al. 1992), and root growth is favored in plants experiencing drought stress (Davies and Zhang 1991). Defoliation has been shown to stop root growth

of alfalfa (Rapoport and Travis 1984). Rapoport and Travis found root growth resumed 3 wk after defoliation and was correlated with deposition of TNC into root tissues. This suggests shoots have priority over roots regarding growth after defoliation. Similar responses occur with grasses.

To date, these root growth responses are not fully understood, but progress is being made that will have strong implications for forage management. Cytokinins produced in the roots have an effect on tillering (Briske 1991), and abscisic acid produced in the roots affects shoot growth and water relations (Davies and Zhang 1991). These are only a few highlights of the changes occurring regarding the role of roots. Undoubtedly, in the next few years, more attention will be given to root management to improve our understanding and ability to manage the top growth.

GLOBAL CLIMATE IMPLICATIONS

Few topics have been as controversial or uncertain during the past few years as the concern and alarm about gradual changes in global climate (Gates 1993). This change is due to two major factors: (1) a release of "greenhouse gases" to the atmosphere that trap radiation lost from the earth to cause global warming, i.e., the greenhouse effect, and (2) a gradual breakdown of the ozone layer in the upper atmosphere, which filters out ultraviolet radiation. Ultraviolet radiation causes skin cancer in humans and can have detrimental effects on domestic animals and wildlife. Ultraviolet radiation generally causes stunted or reduced plant growth and can cause mutations in plants.

The increase in ultraviolet radiation as a result of damage to the ozone layer will likely be alleviated by environmental policies and regulations that limit the release of fluorocarbons and nitrous oxides into the atmosphere. These components react with ozone, causing it to break down. The recovery process will be slow, however, and complete restoration of the ozone layer may not be possible. In the meantime, the public needs to be concerned about possible effects of ultraviolet radiation on plants and animals, especially in higher latitudes above which the ozone layer is thinnest.

Greenhouse gases include carbon dioxide (CO_2), which has gradually increased from a concentration near 290 ppm in 1890 to near 355 ppm in 1990 (Fig. 1.10) but the rate of change has become faster. The increase in the proportion of total world C being associated with the atmosphere is attributed largely to burning of coal and oil reserves, which releases CO_2 to the atmosphere, and to deforestation, which releases C sequestered in trees and soil organic matter. More recently, other greenhouse gases, namely methane, nitrous oxide, ozone, and chlorofluorocarbons (freons), have also increased in the atmosphere. Although CO_2 is generally referred to most in this context, on a molar basis the other gases contribute much more to the greenhouse effect than does CO_2, and their concentrations are increasing at even faster rates. Currently, however, CO_2 is responsible for about 60% of the total effectiveness of greenhouse gases (Gates 1993). Second in importance is methane, which arises largely from rice (*Oryza sativa* L.) paddies, wetlands, and bacterial fermentation during digestion of feedstuffs by wild and domestic ruminants.

Although agreement is not unanimous (Schneider 1990; Kellogg 1991), a widely held belief based on global circulation models is that the earth's surface and air temperature will increase gradually, slightly near the equator, but by 5° to 6°C at higher latitudes due to the way air circulates around the globe. Estimates are that air temperature will likely increase an average of 3°C in the central US over the next 75 yr or so. This change will be accompanied by more precipitation in winter and less in summer and will displace our current climates about 250 km northward, to correct for temperature, about 35 km per decade, and slightly eastward to correct for rainfall. Current climates will progress about 500 m upward in elevation in mountainous regions.

Similar situations will likely occur in Europe. Jones and Carter (1992) predict only a 2°C increase in temperature due to the closeness of the ocean, which will moderate the

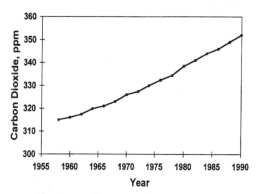

Fig. 1.10. The atmospheric carbon dioxide concentration measured at Mauna Loa, Hawaii, shows a gradual increase. Concentration was 290 ppm in 1890.

rate of change, a 10% increase in winter rainfall, and a 10% decrease in summer rainfall. Their model for production of perennial ryegrass, an important C_3 pasture grass, shows annual yield will increase throughout Europe but mostly in the southern regions, especially near the Atlantic, due to enhanced winter growth. Yields in summer in the mountainous areas are expected to increase due to more favorable temperatures. Similar to expectations in the US, at times when temperatures are near optimum, yield of perennial ryegrass in northern and eastern areas of Europe is expected to increase due to enhanced photosynthesis of the C_3 species in the high CO_2 environment.

On a positive side, the CO_2 increase will reduce photorespiration of C_3 crops (Chap. 3. Vol. 1), which will increase their net photosynthesis and productivity slightly (Ryle et al. 1992). Evidence suggests many C_3 forage plants will respond to higher CO_2 concentrations, up to about 600 ppm, by an increase in photosynthesis and biomass productivity, but the increase will depend on the species. The plants must have tillering sites and active meristems to have sufficient sinks, i.e., areas of growth, to use the extra photosynthate for growth. Otherwise, most of the extra growth will probably accumulate in roots, but leaves will be thicker, and protein content of leaves will be lower because of dilution with more structural carbohydrate (Bazzaz 1990). Thus, the expected benefits in terms of producing higher yields of quality forage in the high CO_2 environment may not be fully realized, even with C_3 species.

There will likely be little effect on C_4 species because of the nature of the CO_2 fixation process (Chap. 3, Vol. 1). Conversely, N_2 fixation by forage legumes may be enhanced due to higher supplies of photosynthate to the roots and nodules, even though dry matter production is only marginally increased. This may make more N available for the legume and for associated grasses.

With the climate change most growing areas will experience a longer, drier summer period, which will be more stressful than at present for cool-season grasses and legumes. High CO_2 environments cause stomata to open less than in low CO_2 environments (stomata opening less conserves water by lowering transpiration). This, along with the resulting larger root system should increase efficiency of water use, but many researchers suggest that the accompanying temperature increase will offset much of the water savings.

This will shift interest in some localities from cool-season C_3 species to warm-season C_4 grasses that are more tolerant of the longer, summerlike environments.

Flowering in many temperate forages is affected by photoperiod, which will remain unchanged, but the higher temperatures will likely hasten flowering somewhat for most species. Thus, grasses and legumes will mature at earlier dates in the spring, and forage quality may be lower due to the higher temperatures. Summer regrowths may be lower in forage quality than at present. Fall growth of cool-season grasses will likely be delayed until later months, lengthening the summer dormant period. Winters will likely be less cold, but with higher winter rainfall, so overwintering strategies may change to emphasize retention of dormancy in perennials, resistance to frost heaving, and resistance to diseases, in contrast with the current need for absolute ability to survive low temperatures. If species in mid- or southern latitudes of the US are normally nondormant, their winter production may be enhanced.

The reduced rainfall and higher temperatures through a longer summer period will require higher levels of forage and pasture management to ensure persistence of cool-season C_3 species. This will make it more feasible to use grazing systems based on C_4 grasses for summer pasture and forage production, while the C_3 grasses are rested during summer to enhance their persistence. Some cool-season forage grasses such as tall fescue may have enhanced fall growth after a long summer dormancy period (Horst and Nelson 1979).

Ruminants will also be exposed to higher temperatures, adding stress to their metabolic and reproductive systems. It is well-known that forage quality is lower at high temperatures (Buxton and Casler 1993) and that quality declines more rapidly during exposure to high temperatures. Thus, season-long animal gains on pasture may remain similar to present-day gains even though more feed may be available during the longer grazing season, but animal performance per day will likely be lower due to lower forage quality.

In some areas there will likely be a strong interest, and perhaps government support, in agroforestry programs, whereby trees will be planted in pastures to provide shade for animals and to provide a means to sequester CO_2 in long-term storage products, i.e., wood, to reduce the amount in the atmosphere. It is probable special government programs will be initiated to encourage management practices for forages that will increase the soil organic matter content, again as a means of sequestering CO_2 and increasing the long-term productivity of the soil resource.

Overall, forage and grassland managers will likely have major roles in abating changes in global climate.

QUESTIONS

1. What is the difference between genotypic and phenotypic plasticity? How do they contribute to adaptation range of a cultivar?
2. What is the role of plant materials centers?
3. Why is "brown-bagging" illegal?
4. How do government policies influence adoption of technology? Give examples.
5. Why would reserve levels for nitrogen be lower than for carbohydrate?
6. What is a vegetative storage protein?
7. What factors regulate tiller production in a grass plant?
8. What is the relationship between plant persistence and stand persistence for alfalfa? For birdsfoot trefoil? For orchardgrass? For common lespedeza?
9. Why will forage quality likely be altered by global climate change?
10. Why have roots been studied less than shoots?

REFERENCES

Allard, G, and CJ Nelson. 1991. Photosynthate partitioning in basal zones of tall fescue leaf blades. Plant Physiol. 95:663-68.

Bazzaz, FA. 1990. The response of natural ecosystems to the rising global CO_2 levels. Annu. Rev. Ecol. Syst. 21:167-96.

Belanger, G, F Gastal, and FR Warembourg. 1992. The effects of nitrogen fertilization and the growing season on carbon partitioning in a sward of tall fescue (*Festuca arundinacea* Schreb.). Ann. Bot. 70:239-44.

Beuselinck, PR, et al. 1994. Improving legume persistence in forage ecosystems. J. Prod. Agric. 7:311-22.

Booysen, P deV, and CJ Nelson. 1975. Leaf area and carbohydrate reserves in regrowth of tall fescue. Crop Sci. 15:262-66.

Boyce, PJ, and JJ Volenec. 1992. Taproot carbohydrate concentrations and stress tolerance of contrasting alfalfa genotypes. Crop Sci. 32:757-61.

Brick, MA, and KD Barnes. 1982. Inheritance and anatomy of root bark area in alfalfa. Crop Sci. 22:747-52.

Brink, GE, and GC Marten. 1989. Harvest management of alfalfa-nutrient yield vs. forage quality, and relationship to persistence. J. Prod. Agric. 2:32-36.

Briske, DD. 1991. Developmental morphology and physiology of grasses. In RK Heitschmidt and JW Stuth (eds.), Grazing Management. An Ecological Perspective. Portland, Oreg.: Timber Press, 85-108.

Brown, JR, and DA Sleper. 1980. Mineral concentration in two tall fescue genotypes grown under variable soil nutrient levels. Agron. J. 72:742-45.

Brown, LG, CS Hoveland, and KJ Karnok. 1990.

Harvest management effects on alfalfa yield and root carbohydrates in three Georgia environments. Agron. J. 82:267-73.

Brown, RH, RB Pearce, DD Wolf and RE Blaser. 1972. Energy accumulation and utilization. In AA Hanson, DK Barnes, and RR Hill, Jr. (eds.), Alfalfa Science and Technology, Am. Soc. Agron. Monogr. 15. Madison, Wis., 143-84.

Brummer, EC, and JH Bouton. 1991. Plant traits associated with grazing-tolerant alfalfa. Agron. J. 83:996-1000.

———. 1992. Physiological traits associated with grazing-tolerant alfalfa. Agron. J. 84:138-43.

Bula, RJ, and D Smith. 1954. Cold resistance and chemical composition in overwintering alfalfa, red clover and sweetclover. Agron. J. 46:397-401.

Bula, RJ, D Smith, and HJ Hodgson. 1956. Cold resistance in alfalfa at two diverse latitudes. Agron. J. 48:153-56.

Buxton, DR, and MD Casler. 1993. Environmental and genetic effects on cell wall composition and digestibility. In HG Jung et al. (eds.), Forage Cell Wall Structure and Digestibility. Madison, Wis.: American Society of Agronomy, 685-714.

Casal, JJ, RA Sanchez, and VA Deregibus. 1987. Tillering response of L. multiflorum plants to changes in red/far-red ratios typical of sparse canopies. J. Exp. Bot. 38:1432-39.

Culvenor, RA, and RJ Simpson. 1991. Mobilization of nitrogen in swards of Trifolium subterraneum L. during regrowth after defoliation. New Phytol. 117:81-90.

Davies, WJ, and J Zhang. 1991. Regulation of growth and development of plants in drying soil. Annu. Rev. Plant Physiol. Plant Mol. Biol. 42:55-76.

Deregibus, VA, RA Sanchez, JJ Casal, and MJ Trlica. 1985. Tillering responses to enrichment of red light beneath the canopy in a humid natural grassland. J. Appl. Ecol. 22:199-206.

Edmisten, KL, DD Wolf, and M Lentner. 1988. Fall harvest management of alfalfa. I. Date of fall harvest and length of the growth period prior to fall harvest. Agron. J. 80:688-93.

Fankhauser, JJ, Jr., and JJ Volenec. 1989. Root vs shoot effects on herbage regrowth and carbohydrate metabolism of alfalfa. Crop Sci. 29:735-40.

Fankhauser, JJ, Jr., JJ Volenec, and GA Brown. 1989. Composition and structure of starch from taproots of contrasting genotypes of Medicago sativa L. Plant Physiol. 90:1189-94.

Gastal, F, and CJ Nelson. 1994. Nitrogen use within the growing leaf blade of tall fescue. Plant Physiol. 105:191-97.

Gates, DM. 1993. Climate Change and Its Biological Consequences. Sunderland, Mass.: Sinauer Assocs.

Gibson, D, JJ Casal, and VA Deregibus. 1992. The effect of plant density on shoot and leaf lamina angles in Lolium multiflorum and Paspalum dilitatum. Ann. Bot. 70:69-73.

Graber, LF, NT Nelson, WA Luekel, and WB Albert. 1927. Organic Food Reserves in Relation to Growth of Alfalfa and Other Perennial Herbaceous Plants. Wis. Agric. Exp. Stn. Res. Bull. 80.

Habben, JE, and JJ Volenec. 1990. Starch grain distribution in taproots of defoliated Medicago sativa L. Plant Physiol. 94:1056-61.

Heichel, GH, and KI Henjum. 1990. Fall dormancy response of alfalfa investigated using reciprocal cleft grafts. Crop Sci. 30:1123-27.

Heichel, GH, RH Delaney, and HT Cralle. 1988. Carbon assimilation, partitioning, and utilization. In AA Hanson, DK Barnes, and RR Hill, Jr. (eds.), Alfalfa and Alfalfa Improvement, Am. Soc. Agron. Monogr. 29 Madison, Wis., 195-228.

Hendershot, KL, and JJ Volenec. 1989. Shoot growth, dark respiration, and nonstructural carbohydrates of contrasting alfalfa genotypes. Crop Sci. 29:1271-75.

———. 1993a. Taproot nitrogen accumulation and use in overwintering alfalfa (Medicago sativa L.). J. Plant Physiol. 141:68-74.

———. 1993b. Nitrogen pools in taproots of alfalfa (Medicago sativa L.) after defoliation. J. Plant Physiol. 141:129-35.

Hodgson, HJ, and RJ Bula. 1956. Hardening behavior of sweetclover (Melilotus spp.) varieties in a subarctic environment. Agron. J. 48:157-60.

Horst, GL, and CJ Nelson. 1979. Compensatory growth of tall fescue following drought. Agron. J. 71:559-63.

Jones, MB, and TR Carter. 1992. European grassland in a changing climate. In Proc. 14th Gen. Meet. Eur. Grassl. Fed., Lahti, Finland, 97-110.

Jung, GA, and D Smith. 1961. Trends in cold resistance and chemical changes over winter in the roots and crowns of alfalfa and medium red clover. I. Changes in certain nitrogen and carbohydrate fractions. Agron. J. 53:359-64.

Kellogg, WW. 1991. Response to skeptics of global warming. Bull. Am. Meteorol. Soc. 74:499-511.

Kephart, KD, DR Buxton, and SE Taylor. 1992. Growth of C_3 and C_4 perennial grasses under reduced irradiance. Crop Sci. 32:1033-38.

Kim, TH, A Ourry, J Boucaud, and G Lemaire. 1993. Partitioning of nitrogen derived from N_2 fixation and reserves in nodulated Medicago sativa L. during regrowth. J. Exp. Bot. 44:555-62.

Lodge, GM. 1991. Management practices and other factors contributing to the decline in persistence of grazed lucerne in temperate Australia: A review. Aust. J. Exp. Agric. 31:713-24.

MacAdam, JW, JJ Volenec, and CJ Nelson. 1989. Effects of nitrogen on mesophyll cell division and epidermal cell elongation in tall fescue leaf blades. Plant Physiol. 89:549-56.

May, LH. 1960. The utilization of carbohydrate reserves in pasture plants after defoliation. Herb. Abstr. 30:239-45.

Millard, P. 1988. The accumulation and storage of nitrogen by herbaceous plants. Plant Cell Physiol. 11:1-8.

Moser, LE, JJ Volenec, and CJ Nelson. 1982. Respiration, carbohydrate content, and leaf growth of tall fescue. Crop Sci. 22:781-86.

Murphy, JS, and DD Briske. 1992. Regulation of tillering by apical dominance: Chronology, interpretive value, and current perspective. J. Range Manage. 45:419-29.

Nelson, CJ. 1988. Genetic associations between photosynthetic characteristics and yield: Review of the evidence. Plant Physiol. Biochem. 26:543-54.

Nelson, CJ, and LE Moser. 1994. Plant factors affecting forage quality. In GC Fahey et al. (eds.), Forage Quality, Evaluation, and Utilization.

Madison, Wis.: American Society of Agronomy, 115-54.

Nelson, CJ, and DA Sleper. 1977. Morphological characters associated with productivity of tall fescue. In E Wojahn and H Thons (eds.), Proc. 13th Int. Grassl. Congr., Leipzig, Germany, 177-79.

Nelson, CJ, and KM Zarrough. 1981. Tiller density and tiller weight as yield determinants in vegetative swards. In CE Wright (ed.), Plant Physiology and Herbage Production, Br. Grassl. Soc. Occup. Symp. 13. Hurley, UK, 25-29.

Nelson, CJ, SY Choi, F Gastal, and JH Coutts. 1992. Nitrogen effects on relationships between leaf growth and leaf photosynthesis. In N Murata (ed.), Research in Photosynthesis, vol. 4. Dordrecht, Netherlands: Kluwer, 789-92.

Ourry, A, J Boucaud, and J Salette. 1988. Nitrogen mobilization from stubble and roots during regrowth of defoliated perennial ryegrass. J. Exp. Bot. 39:803-9.

Pederson, GA, RR Hill, Jr., and WA Kendall. 1984. Genetic variability for root characters in alfalfa populations differing in winterhardiness. Crop Sci. 24:465-68.

Phillips, DA, DM Center, and MB Jones. 1983. Nitrogen turnover and assimilation during regrowth in *Trifolium subterraneum* L. and *Bromus mollis* L. Plant Physiol. 71:472-76.

Reinbott, TM, and DG Blevins. 1991. Phosphate interaction with uptake and leaf concentration of magnesium, calcium, and potassium in winter wheat seedlings. Agron. J. 83:1043-46.

———. 1994. Phosphorus and temperature effects on magnesium, calcium, and potassium in wheat and tall fescue leaves. Agron. J. 86:523-29.

Robin, CH, MJM Hay, PCD Newton, and DH Greer. 1994. Effect of light quality (red:far-red ratio) at the apical bud of the main stolon on morphogenesis of white clover (*Trifolium repens* L.). Ann. Bot. 74:119-29.

Rapoport, HF, and RL Travis. 1984. Alfalfa root growth, cambial activity, and carbohydrate dynamics during the regrowth cycle. Crop Sci. 24:899-903.

Rhodes, I, and WR Stern. 1978. Competition for light. In JR Wilson (ed.), Plant Relations in Pastures. East Melbourne, Australia: CSIRO, 175-89.

Ryle, GJA, CE Powell, and W Tewson. 1992. Effect of elevated CO_2 on the photosynthesis, respiration and growth of perennial ryegrass. J. Exp. Bot. 43:811-13.

Schneider, SH. 1990. The global warming debate heats up: An analysis and perspective. Bull. Am. Meteorol. Soc. 71:1291-1304.

Schnyder, H, and CJ Nelson. 1989. Growth rate and assimilate partitioning in the elongation zone of tall fescue leaf blades at high and low irradiance. Plant Physiol. 90:1201-6.

Simons, RG, A Davies, and A Troughton. 1973. Effect of spacing on the growth of genotypes of perennial ryegrass. J. Agric. Sci. Camb. 80:495-502.

Skinner, RH, and CJ Nelson. 1992. Estimation of potential tiller production and site usage during tall fescue canopy development. Ann. Bot. 70:493-99.

Skinner, RH, and SR Simmons. 1993. Modulation of leaf elongation, tiller appearance and tiller senescence in spring barley by far-red light. Plant Cell Environ. 16:555-62.

Sleper, DA. 1979. Plant breeding, selection, and species in relation to grass tetany. In VV Randig and DL Grunes (eds.), Grass Tetany. Madison, Wis.: American Society of Agronomy, 63-77.

Sleper, DA, GB Garner, CJ Nelson, and JL Sebaugh. 1980. Mineral concentration of tall fescue genotypes grown under controlled conditions. Agron. J. 72:720-22.

Smith, D. 1964. Winter injury and the survival of forage plants. Herb. Abstr. 34:203-9.

Smith, D, RJ Bula, and RP Walgenbach. 1986. Forage Management in the North. 5th ed. Dubuque, Iowa.: Kendall/Hunt.

Smith, SR, and JH Bouton. 1993. Selection within alfalfa cultivars for persistence under continuous stocking. Crop Sci. 33:1321-28.

Smith, SR, JH Bouton, and CS Hoveland. 1992. Persistence of alfalfa under continuous grazing in pure stands and in mixtures with tall fescue. Crop Sci. 32:1259-64.

Spencer, D, TJV Higgins, M Freer, H Dove, and JB Combe. 1988. Monitoring the fate of dietary proteins in rumen fluid using gel electrophoresis. Br. J. Nutr. 60:241-47.

Ta, TC, FDH MacDowall, and MA Faris. 1990. Utilization of carbon and nitrogen reserves of alfalfa roots in supporting N_2-fixation and shoot regrowth. Plant Soil 127:231-36.

Tesar, MB. 1993. Delayed seeding of alfalfa avoids autotoxicity after plowing or glyphosate treatment of established stands. Agron. J. 85:256-63.

Volenec, JJ. 1985. Leaf area expansion and shoot elongation of diverse alfalfa germplasms. Crop Sci. 25:822-27.

———. 1986. Nonstructural carbohydrates in stem base components of tall fescue during regrowth. Crop Sci. 26:122-27.

Volenec, JJ, and JH Cherney. 1990. Yield components, morphology and forage quality of multifoliolate alfalfa phenotypes. Crop Sci. 30:1234-38.

Volenec, JJ, and CJ Nelson. 1983. Responses of tall fescue leaf meristems to nitrogen fertilization and harvest frequency. Crop Sci. 23:720-24.

———. 1984a. Carbohydrate metabolism in leaf meristems of tall fescue. I. Relationship to genetically altered leaf elongation rates. Plant Physiol. 74:590-94.

———. 1984b. Carbohydrate metabolism in leaf meristems of tall fescue. II. Relationship to leaf elongation rates modified by nitrogen fertilization. Plant Physiol. 74:595-600.

Ward, CY, and RE Blaser. 1961. Carbohydrate reserves and leaf area in regrowth of orchardgrass. Crop Sci. 1:366-70.

Wilhelm, WW, and CJ Nelson. 1978. Leaf growth, leaf aging and photosynthetic rate of tall fescue genotypes. Crop Sci. 18:769-72.

Zarrough, KM, CJ Nelson, and DA Sleper. 1984. Interrelationships between rates of leaf appearance and tillering in selected tall fescue populations. Crop Sci. 24:565-69.

2

Forage Crop Development: Application of Emerging Technologies

EDWIN T. BINGHAM
University of Wisconsin

BOB V. CONGER
University of Tennessee

EMERGING technologies in plant genetics and breeding are based on deoxyribonucleic acid (DNA), the major component of chromosomes and regulator of genetic information. DNA is made up of four different nucleotides. Each nucleotide consists of a nitrogenous base, a five-carbon (C) sugar in a ring, and a phosphate group. The nitrogenous bases, adenine (A), guanine (G), thymine (T), and cytosine (C), account for the differences among the four nucleotides (Fig. 2.1). The genetic code is read in groups of three sequential nucleotides. Each group of three is a codon for one amino acid. For example, CGA codes for alanine and TGC codes for threonine. Other triplet combinations are codons for other amino acids.

The following features make DNA a powerful research tool in forage crops.

1. DNA is the hereditary material of all plants and animals.
2. Inherited traits are controlled by genes, and genes are made of DNA.
3. DNA of specific genes can often be isolated or made (synthesized) by scientists.

4. DNA of specific genes can be transferred from one organism to another (genetic engineering).
5. DNA of specific genes can be genetically

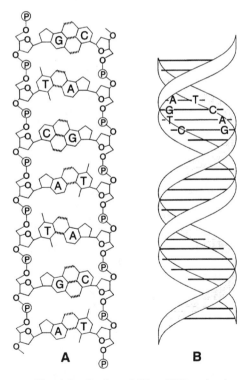

Fig. 2.1. Portion of (*A*) a DNA molecule uncoiled and (*B*) the coiled double helix model. The molecule has a ladderlike structure with the uprights composed of alternating sugar and phosphate groups and the cross rings composed of paired nucleotides.

EDWIN T. BINGHAM is Professor of Agronomy at the University of Wisconsin, Madison. He holds BS and MS degrees from Utah State University and a PhD in plant breeding from Cornell University. His research emphasizes alfalfa genetics and biotechnology.

BOB V. CONGER is the Austin Distinguished Professor of Agriculture, the University of Tennessee, Knoxville. He holds a BS in agronomy from Colorado State University and a PhD in genetics from Washington State University. His research emphasizes cell and tissue culture and breeding of forage grasses.

engineered into bacteria or yeast and multiplied into millions of copies (gene cloning).

6. DNA of cloned genes has many uses in emerging technologies.

This chapter describes emerging technologies that are being used for highly focused research in and development of forage crops. Forage crops are benefiting along with all other crops from the application of these powerful new methods. Emerging technologies based on DNA are not replacing conventional breeding and development methods; rather, they are adding new dimensions to conventional development and producing unprecedented knowledge about the genetic makeup of plants. This opens new avenues and opportunities for plant breeding and crop development. The many ways genetic material of a plant can be studied and used by a combination of traditional and emerging technologies are illustrated in Figure 2.2.

PLANT GENOMES

The total DNA inherited from an ancestor of a forage crop is called a *genome* and is contained in a set of chromosomes. Some forage crops are based on one ancestor and one genome, e.g., alfalfa (*Medicago sativa* L.), red clover (*Trifolium pratense* L.), corn (*Zea mays* L.), sorghum (*Sorghum bicolor* [L.] Moench), orchardgrass (*Dactylis glomerata* L.), and perennial ryegrass (*Lolium perenne* L). When a plant has two copies of the same genome, it is called a *diploid* (red clover, corn, sorghum, and perennial ryegrass). When it has four copies of the same genome it is called an *autotetraploid* (alfalfa and orchardgrass). Several other forage crops are hybrids of two or more ancestors and have two or more different genomes, e.g., smooth bromegrass (*Bromus inermis* Leyss.), tall fescue (*Festuca arundinacea* Schreb.), timothy (*Phleum pratense* L.), and oat (*Avena sativa* L.). When a plant has two copies of each of two *different* genomes, it is called an *allotetraploid* (smooth bromegrass and tall fescue). When it has two copies of each of three *different* genomes, it is called an *allohexaploid* (timothy and oat).

Think of the mule as a familiar example of a hybrid with two ancestors. The mule is sterile because it has half of its diploid genome from the horse and half from the donkey. The reason the mule is sterile is because the horse and donkey chromosomes do not pair to form gametes. Similarly, when cultivated perennial alfalfa is crossed with its wild annual relative *M. marina,* the hybrid also is sterile (McCoy and Bingham 1988).

Many forage plants are hybrids but are fertile because plant genomes tolerate doubling (usually as a reproductive accident) whereas animals do not. Doubling results in diploid genomes and fertility. Triticale, which is used for both grain and forage, is a man-made hybrid between wheat (*Triticum aestivum* L. emend. Thell.) and rye (*Secale cereale* L.). It possesses diploid genomes from each of the parental species.

There is much new plant genome research that uses DNA methods. The ability to analyze the actual DNA in each plant genome permits tracking genes and genomes in the breeding program during development of new crop cultivars. DNA maps indicating the exact order and arrangement of genes in genomes are being developed for most crops, including forages.

DNA MARKERS

Fragments that are cut from DNA molecules at precise sites are potential DNA markers. The ability to cut DNA at precise sites is based on the discovery of enzymes isolated from bacteria that cut DNA molecules. Scientists already have isolated hundreds of these enzymes known as *restriction-site enzymes,* and there are more to be discovered. An important class of DNA markers is called *restriction fragment length polymorphisms,* or RFLPs (Tanksley et al. 1989). The restriction fragment lengths of the DNA can be compared by placing the DNA pieces on a gel and exposing them to an electric field (electrophoresis), causing different lengths of pieces to migrate at different speeds. After a period of time the DNA pieces separate to produce bands at different positions (polymorphisms). Restriction fragment length polymorphisms were the first DNA markers and are used for examples in this chapter. Several other types of DNA markers have been developed recently (Williams et al. 1990). DNA markers are essentially unlimited in number and can always be seen in DNA preparations.

When each individual in a group of plants with a given trait, such as resistance to a particular disease, always has a specific DNA marker or RFLP, then the RFLP can be used to identify the trait in the laboratory. That RFLP can be followed from parent to progeny and used to track the trait in breeding programs. This is called *marker-assisted selection* (MAS). MAS is used along with conventional cross-pollination and plant breeding to shorten development time.

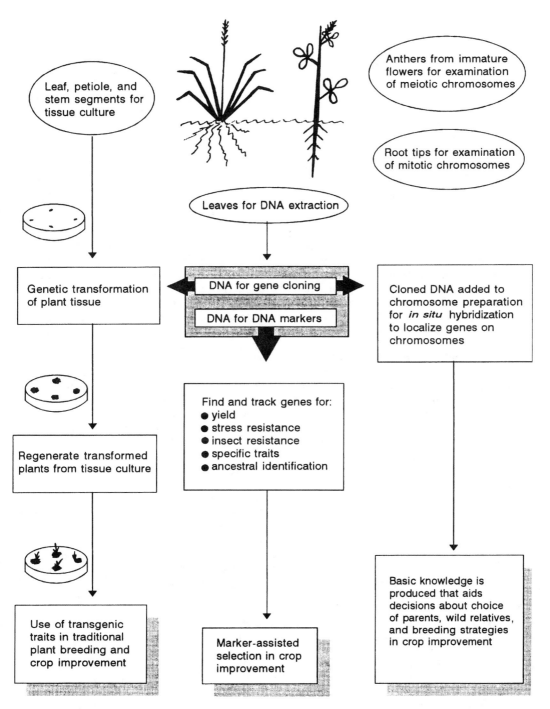

Fig. 2.2. Forage plant research and improvement using traditional and merging technologies.

Some applications of DNA markers include

1. Locating and following genes for simple (single gene) traits such as resistance to a specific disease or insect.

2. Locating and understanding genes for complex (multiple genes) traits such as yield and stress resistance.

3. Understanding ancestral genomes to help make decisions about which ancestors to

use as sources of genetic traits in crop improvement.

4. Understanding plant structure and composition and how to change them to make forage plants more productive or more nutritious.

5. "Fingerprinting" cultivars based on their DNA to monitor purity and accurately identify cultivars that otherwise look alike.

GENETIC TRANSFORMATION

Genetic transformation, genetic engineering, gene transfer, recombinant DNA technology, and *biotechnology* are all names of the process that involves isolating the DNA of a specific gene in one organism, transferring that DNA to another organism, and ensuring that the transferred DNA is incorporated into the recipient genome (Peacock 1993). Agronomically useful traits that have been genetically engineered into plants include herbicide resistance, insect resistance, viral disease resistance, fungal disease resistance, and male sterility.

Genetic transformation is a special process because it is done without sexual crosses. Theoretically, DNA can be transferred from any organism to any other organism. For example, genes from a virus, bacteria, and animal have been transferred into alfalfa (see the following section). Plant, animal, and bacterial genes are routinely transferred first to the common gut bacterium *Escherichia coli* (*E. coli*) or to a common soil bacterium *Agrobacterium tumefaciens* for cloning. In gene cloning, a specific gene is transferred to bacteria or to yeast, and the microbes are multiplied into millions or billions of microbes each containing a copy of the gene. Hence, the gene is cloned into an unlimited number of copies. This huge surplus of gene copies is needed for transformation.

Methods of plant transformation all involve transferring the specific cloned gene into recipient plant cells. Currently, the three most common current transfer methods are (1) *Agrobacterium,* where the microbe infects the plant and delivers DNA to plant cells; (2) particle bombardment, where the DNA is literally shot into plant cells using microscopic particles of gold, tungsten, or fibers coated with cloned DNA; and (3) direct uptake of cloned DNA, where DNA is absorbed into plant cells.

Agrobacterium works well in the dicot forages such as alfalfa, clover, and birdsfoot trefoil (*Lotus corniculatus* L.). *Agrobacterium* works well because it naturally infects many dicots and naturally transfers its DNA into plant cells as a plant pathogen. Therefore, if the *Agrobacterium* DNA has the specific desired gene cloned into it, it proceeds to infect the plant cell with DNA containing the desired gene. Most importantly, the *Agrobacterium* DNA is incorporated into the plant genome by a process that already occurs in nature.

Genetic Transformation of Forage Legumes. The basic strategy for alfalfa transformation with *Agrobacterium* involves special strains of *Agrobacterium* that are resistant to the antibiotic kanamycin and that do not produce plant tumors (Shahin et al. 1986; Blake et al. 1991). Not all alfalfa is a good recipient; alfalfa lines developed to regenerate plants from cells in tissue cultures are best for transformation (Bingham 1991).

Alfalfa stem segments or pieces of leaves or petioles are placed in special media cultures to form callus colonies. Alfalfa callus then is inoculated with the proper *Agrobacterium* strain carrying a gene for the desired trait. Alfalfa cells that become infected and transformed carry *Agrobacterium* DNA with kanamycin resistance as well as the desired gene for transformation. The alfalfa callus is then transferred to culture medium-containing kanamycin, which kills the nontransformed cells. Transformed cells with kanamycin resistance selectively develop into kanamycin-resistant colonies that also contain the desired transformed gene. These callus colonies are then transferred to a new culture medium for regeneration of whole alfalfa plants that are genetically transformed. These plants, if fertile, should transfer the traits onto progeny through the normal sexual means. This also allows the breeder to use the new gene in conventional ways in the breeding program.

Transformation of alfalfa is now routine, but two challenges remain. One is identifying and cloning genes that may be of interest in alfalfa improvement or in the production of value-added products. The other challenge is obtaining expression of transformed traits that is high enough to achieve desired results. Even though the DNA has been transferred, it may be only partially effective in changing the plant's metabolism. For example, alfalfa has been transformed to produce a toxin with insecticidal properties (Bingham 1992). However, the insecticidal activity in transformed alfalfa currently is not high enough to give resistance to damaging insects. Hence, continuing research is necessary to enhance gene expression in transformed plants in general, including alfalfa.

Several traits of agronomic interest that have been transferred to alfalfa by genetic transformation are still in the research stage. Virus resistance was incorporated in alfalfa by transferring the coat protein gene of the alfalfa mosaic virus (AMV) into the alfalfa genome by *Agrobacterium* transformation (Hill et al. 1991). The transgenic alfalfa plants containing the virus coat protein are resistant to AMV. This form of virus resistance is called *cross protection* and is being pursued in several crops.

A gene from beans for production of beta-phaseolin seed protein was transferred into alfalfa by *Agrobacterium* transformation (Bagga et al. 1992). There is interest in transforming forage crops to cause them to synthesize and to store "normal" seed proteins in their biomass to make the forage even more nutritious. In the case of the beta-phaseolin protein, it was expressed in leaves, but it did not accumulate to a high level.

In another attempt to add a nutritious new protein to alfalfa leaves, a chicken ovalbumin gene, i.e., one associated with protein synthesis in the egg, was transferred to three alfalfa cultivars (Schroeder et al. 1991). Expression of the ovalbumin gene was low, but research eventually should find ways to enhance expression of protein synthesis genes to improve nutritional quality of forage crops.

Resistance to the herbicide glufosinate was transferred to alfalfa with *Agrobacterium* (D'Halluin et al. 1990). Transgenic alfalfa tested in the field was resistant to postemergence treatment with glufosinate whereas the nontransformed control was not resistant. This transformation event appeared to occur close to application, since expression of the transgene gave the desired results. A goal is to develop plants to tolerate a safer group of herbicides. This would reduce restrictions for use of the pesticide.

The concept of transforming forage crops to produce value-added traits is very appealing (Bingham 1992). Ideally, a high-value product is produced from the forage, and the residue remaining is used for feed as usual. Value-added traits under development in alfalfa include an enzyme used to make biodegradable detergent, an enzyme that breaks down wood fibers in the manufacture of paper, and pharmaceutical products (Bingham 1992). Such new products would add value to the existing forage. Moreover, after the value-added product is removed, the plant residue could be used for animal feed.

Perennial forage crops are ideal for certain value-added products because they permit multiple harvests without reestablishment, the legumes provide their own nitrogen (N) through symbiosis, they are good for soil conservation, and the residue can be used as feed. The value-added concept has great potential but is still in developmental stages. Full use of the concept depends on regulatory approval to feed plant residue containing recombinant DNA to livestock.

Genetic Transformation of Forage Grasses. Genetic transformation in the grass family (including cereals) has been slower because *Agrobacterium* does not normally infect these species. Also, the technology has been impeded by the difficulty in regenerating gramineous plants from cell and tissue cultures. This is a fundamental requirement for genetic transformation.

The two most common methods used to transform grasses and cereals are direct uptake of DNA by protoplasts (cells in which the walls have been removed by special enzymes) and particle bombardment. The first demonstration of gene transfer in a forage grass utilized an orchardgrass genotype with high regeneration capacity (Conger and Hanning 1991) and direct uptake of a hygromycin-resistant gene by protoplasts (Horn et al. 1988). Whole plants resistant to this antibiotic were then regenerated from these cells. The hygromycin-resistant gene was subsequently used to also demonstrate genetic transformation in tall fescue (Wang et al. 1992). Neither of these demonstrations has practical application.

Particle bombardment is a mechanical process for inserting DNA into plant cells. Currently there exist one commercially available device and at least four others designed and built by individual laboratories (Gray and Finer 1993). For a detailed description of these "guns," the reader is referred to Gray and Finer 1993. Particle bombardment has been used to produce genetically engineered herbicide-resistant corn plants (Fromm et al. 1990; Gordon-Kamm et al. 1990).

WIDE HYBRIDIZATION

Being able to hybridize (cross) plants that are distantly related and ordinarily do not cross has particular appeal in forage crops. This is because such wide hybrids are often robust forage producers. Although they sometimes are sterile or have poor fertility, in some cases they can be vegetatively propagated, e.g., 'Coastal' bermudagrass (*Cynodon dactylon* [L.] Pers.). Many forage grasses are wide hybrids that arose naturally, and scientists would like to accelerate the process. Wide hybridization combines whole genomes whereas genetic transformation or engineering adds a single gene to a genome.

Sometimes the wide hybrid can be produced by making a large number of crosses and isolating an occasional viable seed. When the hybrid cannot be obtained by making crosses, there are two laboratory methods of producing wide hybrids in certain forage crops. They are (1) somatic cell hybridization and (2) embryo rescue. Somatic cell hybridization fuses protoplasts of vegetative cells of two different species and then cultures the hybrid product into a whole plant by plant tissue culture (Pupilli et al. 1992). Briefly, protoplasts are cells without a cell wall that are produced by digesting away the cell wall with special enzymes. Protoplasts of two different species are mixed, and some fuse. Then the DNA from two protoplasts is combined into a single protoplast. The new hybrid protoplast needs to be cultured and regenerated into a plant to be used in breeding.

Embryo rescue requires making a wide cross with conventional pollination, dissecting out the early embryo a few days after pollination, and culturing it into a plant (McCoy and Bingham 1988). If the embryo is not removed, it dies in the shriveled seed of the wide cross.

Even though these methods have been in use for several years, they have not been used to produce new forage crops directly. This is because some of the wide hybrids have been sterile and because development time often requires several years. Nonetheless, the methods have contributed basic knowledge about species relationships (McCoy and Bingham 1988).

MOLECULAR CYTOGENETICS

Cytogenetics is the study of the chromosomes on which DNA is organized. The organization of DNA and the chromosome is very specific. Specific genes for specific traits, and specific DNA sequences that code for the traits, are located at specific positions on the chromosomes. When DNA markers are known to be linked to a particular trait, the marker segment is also involved in the regulation of the trait. DNA markers can be used to locate the position of the gene for that trait on the chromosome. In effect, many disciplines are coming together in molecular cytogenetics. There is much new knowledge coming from this unprecedented union of research approaches (Heslop-Harrison 1991).

It is now possible to hybridize specific genes, cloned in the presence of radioactive DNA, with whole chromosome preparations in the laboratory. The precise position of the specific gene on the chromosome can be determined by using photographic film to detect the radioactive site. The exact site on the chromosome can be seen on the film by examining it under the microscope. The cloned radioactive DNA hybridizes with the identical DNA in the chromosome. Researchers are matching DNA maps with the physical maps of individual chromosomes. Through this matching process it is possible to identify pieces of DNA that have descended from wild relatives of our crops. Sometimes the crop genome(s) has several segments of DNA from wild relatives, and sometimes there is none. Therefore, it is possible to know the nature and source of the DNA now contained in forage crops and to predict the wild relatives we should use as genetic resources in future forage crop development.

The same strategy of hybridizing cloned DNA to specific sites on chromosomes reveals where transformed DNA is incorporated in the genome. From this we can learn how DNA inserted at certain positions on the chromosome affects expression of introduced genes. The manner in which a specific gene functions or "expresses" in different insertion sites helps researchers understand much about chromosome structure.

The emerging technologies are interrelated and serve an important function in uniting all aspects of forage crop development. This integrated approach, which is based on great knowledge of the genome, ensures a bright and productive future in forage crop development.

QUESTIONS

1. Cite at least four reasons why DNA is a powerful tool in research on forage crop development.
2. Define *plant genome*.
3. Define *DNA markers*, and list at least four applications of DNA markers.
4. Describe genetic transformation, and list an ex-

ample of its use in grasses and an example of its use in legumes.

5. Describe wide hybridization. Why are wide hybrids sometimes appealing in forage crops?

REFERENCES

Bagga, S, D Sutton, JD Kemp, and C Sengupta-Gopalan. 1992. Constitutive expression of the ß-phaseolin gene in different tissues of transgenic alfalfa. Plant Mol. Biol. 19:951-58.

Bingham, ET. 1991. Registration of alfalfa hybrid Regen-SY germplasm for tissue culture and transformation research. Crop Sci. 31:1098.

———. 1992. Breeding value-added traits in alfalfa. Am. Soc. Agron. Abstr. Madison, Wis., 90.

Blake, NK, RL Ditterline, and RG Stout. 1991. Polymerase chain reaction used for monitoring multiple gene integration in *Agrobacterium*-mediated transformation. Crop Sci. 31:1686-88.

Conger, BV, and GE Hanning. 1991. Registration of Embryogen-P orchardgrass germplasm with a high capacity for somatic embryogenesis from in vitro cultures. Crop Sci. 31:855.

D'Halluin, K, J Botterman, and W De Greef. 1990. Engineering of herbicide-resistant alfalfa and evaluation under field conditions. Crop Sci. 30:866-71.

Fromm, ME, F Morrish, C Armstrong, R Williams, J Thomas, and TM Klein. 1990. Inheritance and expression of chimeric genes in the progeny of transgenic maize plants. Bio/Technol. 8:833-39.

Gordon-Kamm, WJ, TM Spencer, ML Mangano, TR Adams, RJ Daines, WG Start, JV O'Brien, SA Chambers, WR Adams, Jr., NG Willetts, TB Rice, CJ Mackey, RW Krueger, AP Kausch, and PG Lemaux. 1990. Transformation of maize cells and regeneration of fertile transgenic plants. Plant Cell 2:603-18.

Gray, DJ, and JJ Finer. 1993. Development and operation of five particle guns for introduction of DNA into plant cells. Plant Cell, Tissue Organ Cult. 33:219.

Heslop-Harrison, J. 1991. The molecular cytogenetics of plants. J. Cell Sci. 100:15-21.

Hill, KK, N Jarvis-Eagan, EL Halk, KJ Krahn, LW Liao, RS Methewson, DJ Merlo, SE Nelson, KE Rashka, and LS Loesch-Fries. 1991. The development of virus-resistant alfalfa, *Medicago sativa* L. Bio/Technol. 9:373-77.

Horn, ME, RD Shillito, BV Conger, and CT Harms. 1988. Transgenic plants of orchardgrass (*Dactylis glomerata* L.) from protoplasts. Plant Cell Rep. 7:469-72.

McCoy, TJ, and ET Bingham. 1988. Cytology and cytogenetics of alfalfa. In AA Hanson, DK Barnes, and RR Hill, Jr. (eds.), Alfalfa and Alfalfa Improvement, Am. Soc. Agron Monogr. 29. Madison, Wis., 737-36.

Peacock, WJ. 1993. Genetic engineering for pastures. In Proc. 17th Int. Grassl. Congr., Palmerston North, New Zealand, and Australia, 29-32.

Pupilli, F, GM Scarpa, F Damiani, and S Arcioni. 1992. Production of interspecific somatic hybrid plants in the genus *Medicago* through protoplast fusion. Theor. Appl. Genet. 84:792-97.

Schroeder, HE, M Rafigul, I Khan, WR Knibb, D Spencer, and TJ Higgins. 1991. Expression of a chicken ovalbumin gene in three lucerne cultivars. Aust. J. Plant Physiol. 18:495-505.

Shahin, EA, A Spielmann, K Sukhapinda, RB Simpson, and M Yashar. 1986. Transformation of cultivated alfalfa using disarmed *Agrobacterium tumefaciens*. Crop Sci. 26:1235-40.

Tanksley, SD, ND Young, AH Paterson, and MW Bonierbale. 1989. RFLP mapping in plant breeding: New tools for an old science. Bio/Technol. 7:257-64.

Wang, ZY, T Takamizo, VA Iglesias, M Osusky, J Nagel, I Potrykus, and G Spangenberg. 1992. Transgenic plants of tall fescue (*Festuca arundinacea* Schreb.) obtained by direct gene transfer to protoplasts. Bio/Technol. 10:667-74.

Williams, J, A Kubelik, K Livak, J Rafalski, and S Tingey. 1990. DNA polymorphisms amplified by arbitrary primers are useful as genetic markers. Nucl. Acid Res. 18:6531-35.

3

Forage Establishment and Renovation

LESTER R. VOUGH
University of Maryland

A. MORRIS DECKER
University of Maryland

TIMOTHY H. TAYLOR
University of Kentucky

HIGH-YIELDING, high-quality forage crops provide the foundation for profitable feeding programs for dairy and other livestock. Thick, vigorous stands are essential for high yields. Obtaining such stands is dependent initially upon proper seeding practices and favorable seedbed and environmental conditions. Proper soil pH, fertility, and seedbed preparation, crop sequences to avoid herbicide residues from previous crops, selection of high-quality seed, seeding at the right time, good seeding techniques with equipment precisely adjusted for seeding rate and depth, and adequate control of weeds and insects are among the key factors for obtaining thick, vigorous stands.

PLANNING FOR NEW SEEDINGS

Preparations for forage establishment need to begin as much as 2 yr prior to the actual

LESTER R. VOUGH is Forage Crops Extension Specialist at the University of Maryland. He received the MS degree from the University of Minnesota and the PhD from Purdue University. He was Extension and Research Agronomist at Oregon State University from 1972 to 1978. His extension and research programs are primarily concerned with production, management, and utilization of forages.

A. MORRIS DECKER is Professor Emeritus of Agronomy, University of Maryland. He received the MS degree from Utah State University and the PhD from the University of Maryland. His research has been in forage crop management, plant environment, municipal sludge utilization, and cover crop systems.

TIMOTHY H. TAYLOR is Professor Emeritus of Agronomy, University of Kentucky. He received the MS degree from the University of Kentucky and the PhD from Pennsylvania State University. His research interests were in the area of forage crop establishment and management of simple and complex grassland plant communities.

seeding, especially for no-till seedings. The more productive the soil and the better the seedbed conditions, the greater the potential for successful establishment of highly productive stands. Broadleaf weeds need to be eliminated before seeding, either through tillage or herbicide application. Residual herbicides from previous crops must also be considered in this planning. Problems of triazine carryover may be encountered with forage seedings following corn (*Zea mays* L.), particularly in fields following 2 yr or more of no-till corn (Fink and Fletchall 1963). It may be necessary to reduce triazine herbicide rates in the final year of corn or to switch to shorter residual herbicides such as cyanazine (2-{[4-chloro-6-(ethylamino)-1,3,5-triazin-2-yl]amino}-2-methylpropionitrile), alachlor [2-chloro-2'-6'-diethyl-N-(methoxymethyl)-acetanilide], and metolachlor [2-chloro-N-(2-ethyl-6-methylphenyl)-N-(2-methoxy-1-methylethyl) acetamide]. In a rotation of 2 yr or more of no-till corn followed by small grain and then no-till forage, lime should be applied to correct any pH problems prior to the last year of corn. This releases triazine herbicides bound in the soil while the field is still in corn (Lowder and Weber 1982). Delaying the lime application until immediately prior to seeding the forage crop can result in triazine toxicity to forage grasses and legumes and stand failure. Some type of tillage prior to seeding the small grain helps to move lime and fertilizer into the plow layer as well as reduce potential herbicide injury.

SPECIES AND CULTIVAR SELECTION

The selection of a species, or a mixture of species, and cultivar(s) to plant is extremely important because it will affect efficiency of production for the entire life of the stand. Decisions must be made in the context not only of individual farms but even individual fields because each farm and field is usually unique. Species may respond differently due to differences in soil type, depth, and drainage. For example, fields with poor soil drainage, low pH, and fertility should not be seeded to alfalfa (*Medicago sativa* L.). Fields with shallow, droughty soils should not be seeded to timothy (*Phleum pratense* L.). Species characteristics and adaptations should be matched to specific soil and climatic conditions and intended forage use (e.g., hay/silage, spring/ summer grazing, fall/winter grazing). The kind and age of livestock being fed (beef versus dairy [*Bos taurus*] cows, mature dry cows versus young growing or fattening animals), the time of year in which forage is to be used, and the seasonal distribution of forage growth all have an influence in determining which forage species or combination of species to plant.

Pure grass or legume stands are generally easier to manage for hay and silage production than mixtures. A higher level of management is required to maintain a proper balance of legume and grass in a mixture. For example, there are several herbicides available for weed control in pure stands, but few are available for use in mixed legume-grass stands. Cutting management may be more difficult with mixtures, especially if the recommended stages of maturity for cutting each species do not coincide. And if not properly managed, the grass portion of the mixture can become dominant and lower the feeding value of the hay or silage.

Legume-grass mixtures are more difficult to manage, but they offer many advantages over pure stands because (1) Legumes fix atmospheric nitrogen (N) and thus reduce the need for N fertilizer. (2) Legumes increase the protein concentration and other quality factors over pure grass. (3) Legumes extend the grazing of pastures further into the warm, dry midsummer period than cool-season grasses alone. (4) Once established, perennial legume-grass mixtures are more competitive with weeds than pure stands. (5) Mixtures provide better cover for winter protection against plant heaving and freezing and against soil erosion than pure stands of legumes. (6) Mixtures are usually higher yielding than either legumes or grasses alone. (7) Mixtures are easier to cure as hay or to en-

sile than pure legume stands. (8) Mixtures tolerate wider differences in soil conditions. (9) Mixtures generally ensure a better stand; if the legume fails, the grass alone, with added N, will ensure a crop. (10) Mixtures tend to reduce bloat in grazing cattle when the stand is 40% or more grass. (11) Mixtures tend to reduce the possibility of nitrate poisoning and grass tetany if the stand is 40% or more legume. (12) Grasses reduce lodging of legumes.

Principles for composing mixtures include (1) Keep the mixture simple—one grass and one legume are often sufficient. Seldom are more than four species justified. (2) Plants should have similar maturity dates, have compatible growth characteristics, and be adapted to the intended mixture use. Species that germinate quickly and have vigorous seedling development ensure early production, but too much of that species can crowd out others. (3) Pasture mixtures should include not only species that mature together but those similar in palatability, since unpalatable species will soon dominate.

High-quality seed is essential for good forage stands. Certified seed should be used to ensure cultivar purity and high seed quality. Newer cultivars provide superior agronomic characteristics that economically justify their selection over older, lower-cost cultivars that have less disease and insect resistances and lower yield capabilities.

LIME AND FERTILITY REQUIREMENTS

Forages are no different than other crops in that satisfactory stands and yields are obtained only when the crop is adequately limed and fertilized. Persistent, high-yielding stands are always associated with a favorable pH and high fertility. The key to establishing and maintaining productive forage species is adequate fertilization for that particular forage mixture based on soil test recommendations (see Chap. 6, Vol. 1, and Chap. 5, this volume).

Analysis of a representative soil sample is the best method by which existing soil nutrient levels and fertilizer recommendations can be determined. Soil pH and existing levels of nutrients can be used with the field history to develop a sound fertility program. For conventionally tilled seedbed seedings, soil samples should be taken to the depth of the plow layer. For no-till seedings, two sets of soil samples should be taken—one from the 0- to 5-cm depth to determine surface pH and fertility and one to the normal plow depth. The 0- to 5-cm sample is a must for fields that have been in no-till corn and will not be plowed following the last year of corn. Surface applications of N fertilizers to the corn frequently result in this layer being quite acidic.

Lack of adequate lime on acid soils is the greatest deterrent to high forage production since it affects the efficient utilization of other materials used. Lime corrects soil acidity and supplies calcium (Ca) or Ca and magnesium (Mg), depending upon the liming material used. Higher pH not only increases the availability of practically all the essential plant nutrients but promotes the growth of desirable microorganisms and reduces the toxic effects of aluminum and manganese (Mn). Lime should ideally be applied 6-12 mo prior to seeding and thoroughly incorporated into the plow layer to neutralize soil acidity. With no-till seedings, surface applications should be made 1-2 yr ahead of seeding to allow for movement into the soil profile and to avoid triazine carryover problems. In comparison with incorporated lime and conventional establishment, surface liming and no-till seeding was found to result in less vigorous forage seedlings and slower establishment (Koch and Estes 1986). Yield of legumes and total forage the seeding year were greater with incorporated than with surface-applied lime. In succeeding years, however, annual yields and 6-yr total yields from 4.5 Mg/ha lime rate were not significantly different due to liming method.

Phosphorus (P) level is especially critical during establishment. Banding 15-30 kg P/ha 2.5-5.0 cm directly below the seed at seeding helps promote rapid root development and seedling establishment, especially on soils low to medium in P (Sheard et al. 1971; Sheard 1980). Direct contact between fertilizer and seed should be avoided as it can inhibit germination. Broadcast application of P is not as effective as banded. It takes at least four times as much broadcast P to give the same response as banded.

Application of fertilizer N at establishment has been demonstrated to increase yields when soils are low in N (less than 15 ppm soil nitrate) or organic matter is less than 1.5% (Hannaway and Shuler 1993). When the soil nitrate levels were greater than 15 ppm and conditions were favorable for effective nodulation (soil pH 6.2-7.5 and high populations of appropriate *Rhizobium* bacteria present), using preplant N fertilizer in legume establishment did not result in economical yield increases. However, rates in excess of 34 kg generally have detrimental effects on rhizobial infection and N fixation. (Ward and Blaser 1961; Vough et al. 1982). Nitrogen should never be applied in renovation seedings as it quickly stimulates growth of the existing grass and weeds, leading to excessive shading of young seedlings and undue competition for water, light, and soil nutrients. The demand for K by young seedlings is relatively low. It becomes much more important for yield and persistence once stands are established.

SEEDING GUIDELINES

Seeding at Proper Time. Seeds require moisture, oxygen (O), and minimal soil temperature to germinate and grow. The two primary seeding periods for cool-season species are late winter-spring (late February to early May) and late summer-fall (August to mid-October). Spring seedings are most common, particularly in the northern half of the US. Spring moisture is generally good, evaporation is less, and soil moisture is retained longer during the establishment period than with late summer seedings. However, seeding too early in cold, wet soils can result in poor germination, seedling loss due to fungal diseases, and weak stands. On the other hand late spring or summer seedings often fail due to stress from high temperature and lack of moisture. The surface 2.5-5.0 cm of soil can dry quickly, and young seedlings can desiccate.

Advantages of late summer-fall seedings include (1) Such seedings have less competition from weeds. Herbicides are not normally needed. (2) Seedings can be made after early-harvested crops such as barley (*Hordeam vulgare* L.), wheat (*Triticum aestivum* L. emend. Thell.), or vegetables and have full production the following year. (3) Late summer-fall seedings lessen the spring workload. (4) Liming, fertilization, and tillage are done during drier weather, thus reducing risk of soil compaction. (5) Damping-off diseases (*Pythium* spp. and *Rhizoctonia* spp.) are not usually a problem.

In addition to adequate soil moisture, late summer-fall seedings depend upon sufficient time and heat unit accumulation before a killing frost. These seedings should be made early enough to allow at least 6 wk for growth after germination; seedlings should be 7.5-10.0 cm tall before the killing frost. This means seeding July 20 to August 1 in areas such as the northern portions of Minnesota, Wisconsin, Michigan, and New York to as late as October 15 in parts of the southern and southwestern US. While damping-off diseases are not usually a problem, late summer-fall legume seedings are more susceptible to Sclerotinia crown and stem rot (*Sclerotinia trifoliorum* Eriks). New seedings may be completely destroyed when conditions are favorable for disease development. Seedings should be made at the earliest possible date so that seedlings are well established by the time infection occurs.

Seeding Depth. Seeding depth varies with soil type (sandy, clay, or loam), soil moisture availability, time of seeding, and firmness of seedbed. Since small-seeded forage grasses and legumes have a very small supply of stored energy to support the developing seedling, proper seeding depth is very important in getting good stands. Seeds placed too deep are not likely to emerge. Seeds placed on the surface or at a very shallow depth or in a loose or cloddy seedbed often do not have adequate seed-soil contact. In those cases, dry soil conditions following seeding usually result in desiccation and death of the seedlings.

A firm seedbed is essential for proper seed placement, good seed-soil contact, and successful establishment. Seed should be covered with enough soil to provide moist conditions for germination but not so deep that the shoot cannot reach the surface. In humid areas best results are obtained when seed placement is between 0.5 and 1.5 cm (Table 3.1). In the more arid areas a slightly greater seeding depth may be required. Seed placed deeper than 2.5 cm may not emerge or may be so weakened that survival is reduced. Generally, the optimum seeding depths are 0.6-1.2 cm on clay and loam soils and 1.2-2.5 cm on sandy soils (Table 3.1). Shallower depths are better for early spring seedings—moisture is usually abundant, and soil temperature is warmer near the surface. Deeper depths are recommended for late spring and summer seedings when moisture conditions are less favorable.

Seeding Rates. Recommended seeding rates for any given species or mixture of species vary from state to state, and in some cases even within states, due to differences in soils, climate, and establishment methods. Rates are usually given in ranges. For example, the recommended rate for pure seedings of alfalfa in Maryland is 16.8-20.2 kg/ha, Iowa 13.4-22.4, central California 22.4-33.6, and Southern California 28.0-39.2. Selecting specific seeding rates within the recommended ranges depends upon factors such as

1. *Soil type and fertility.* Lower rates can be used when seeding on light, sandy soils because seedling emergence is easier than in heavy soils. Higher rates should be used on low-fertility soils as more seedling losses are likely to occur and plants will be smaller with less spread of the crowns.

2. *Amount and distribution of rainfall.* In areas of limited rainfall and on light, sandy soils, lower seeding rates are used since soil moisture may not be sufficient to support dense populations due to competition for moisture. Also, seedling losses due to damping-off diseases are generally less under these conditions.

3. *Condition of the seedbed and seeding method.* Lower rates can be used when seeding into well-prepared seedbeds where uniform seed coverage and good seed-soil contact can be achieved. Use the higher part of the recommended ranges when seedbed preparation and seeding techniques are less than optimum.

4. *Seed quality.* Higher rates are necessary if the seed lot has low germinability, a high percentage of hard seed, or a low percentage of pure seed. The term *pure live seed* (PLS) is used to express the quality of seeds, even though it is usually not listed on a seed analysis label. Pure live seed is expressed as a percentage of a seed lot that is pure seed that will germinate. The percentage is determined by multiplying the percentage of pure seed by the percentage of germination and dividing by 100.

TABLE 3.1. Average stands produced from 100 seeds at various depths on three soils in precision-depth hand-seeded plots.

Species	Depth (cm)											
	Sand				Clay				Loam			
	1.2	2.5	3.8	5.0	1.2	2.5	3.8	5.0	1.2	2.5	3.8	5.0
	Number of plants											
Alfalfa	71.4	72.6	54.8	40.1	51.9	48.4	28.1	13.1	59.3	54.9	31.5	15.6
Red clover	67.3	65.9	53.1	27.1	40.1	35.1	14.2	7.2	46.7	45.3	24.2	12.6
Bromegrass	70.5	64.2	48.0	29.1	56.1	37.4	17.2	5.5	68.2	50.2	30.9	19.2
Orchardgrass	61.2	56.4	30.1	12.6	60.1	25.9	6.3	1.2	55.5	38.7	28.1	15.6

Source: Sund et al. (1966).

Percentage of PLS provides a way to compare the quality of different seed lots, but it has weaknesses, and the seed analysis labels should be examined closely before purchasing seed. For example, if one lot of seed is 98% pure and has 90% germination, the PLS is 88.2%. Another lot with 90% pure seed and 98% germination would also have 88.2% PLS. However, the first lot has 2% of something other than pure seed; the second lot has 10%. The label will show whether the other component consists of other crop seeds, weed seeds, or inert matter.

5. *The desired composition of the stand if seeding a legume-grass mixture.*

6. *Whether seeding with or without a companion crop.* Higher rates are recommended when seeding without a companion crop since more plants are needed to obtain high yields the seeding year and to reduce weed invasion (Wakefield and Skoland 1965; Brown and Stafford 1970).

It is common for no more than a third of the sown seed to produce seedlings and only half of those to survive the first year (Sprague et al. 1963; Vough et al. 1982). Seedling survival frequently is as low as 20% (Brown et al. 1960).

Inoculation of Legume Seeds. All legume seed should be inoculated with the proper strains of N-fixing bacteria (*Rhizobium* spp.) before seeding. Much of the alfalfa and clover (*Trifolium* spp.) seed being marketed today is preinoculated. However, if the seed is not preinoculated or the seeding date is beyond the expiration date for the inoculant, the seed should be inoculated with a fresh culture prior to seeding. It is critical that inoculant with the proper strain of rhizobia be used. A strong specific relationship exists between bacteria and the host plant.

New innovations in legume seed inoculation have been developed in recent years. These innovations have advantages over commonly used methods such as applying water, milk, and colas to the surface of the seed for adherence of inoculant to the seed. With the common methods, much of the inoculant initially adhering to the seed will slough off when the seed dries. Also, the use of soft drinks as a wetting agent can be detrimental to the rhizobia since bacteria are sensitive to acidic conditions. Successful legume inoculation under adverse conditions generally depends upon coating large amounts of inoculant on the seed, keeping it there until the seed is in the ground, and ensuring survival of the rhizobia in the soil until the young

seedlings are infected.

Some commercial inoculation materials provide a sticker supplement that is used to moisten the seed and bind the humus inoculant to the seed. It has been estimated that some of these new techniques coat each seed with up to 12 times more rhizobia than are coated by conventional methods. The sticker supplement also provides nutrients that help the rhizobia survive in the soil.

A sugar water solution (ratio of 450 g sugar per liter water) can be used as an effective sticker supplement at an economical cost. The seed is thoroughly moistened with this solution before adding 50 g inoculum/kg seed. This is much more inoculum than most growers are accustomed to applying, but the purpose is to increase the number of rhizobia per seed so that enough will survive for effective nodulation even under adverse conditions. With good soil moisture and a properly prepared seedbed, the amount of inoculum could be reduced. Cornstarch can be added after thoroughly mixing the inoculant with the seed. The cornstarch dries the seed so that it will flow through the seeder and seeding can begin immediately. It also provides an energy (carbohydrate) source for the rhizobia. Dry or drill box treatments usually result in poor inoculant adhesion and are not recommended.

Heat, direct sunlight, and drying are detrimental to survival of the rhizobia inoculant. Inoculant should be stored in a refrigerator from time of purchase to time of use. The seed dealer should also have stored the inoculant in a refrigerator or cooler. Also, always check the expiration date before purchasing any inoculant materials or any preinoculated seed.

Other Seed Treatments. Other seed treatments include clay and lime coatings and application of fungicides. Lime-coated seed is used in some areas of acid soils to improve the environment immediately adjacent to the seed. It has been found to be beneficial in the establishment of aerially seeded alfalfa on acid soils in New Zealand (White 1970). On soils that are more suitable, it has not been helpful in improving stands or yields (Barnes 1978; Tesar and Huset 1979).

Applying a systemic fungicide such as metalaxyl [N-(2,6-dimethylphenyl)-N-(methoxyacetyl)-alanine methyl ester] provides seedling protection against damping-off and seed and root rot organisms (*Phytophthora* spp.).

SEEDING METHODS

Common terminology used to describe forage seeding methods include *conventional, no-till, broadcast, drilled, band,* and *culti-packer*. These terms are not mutually exclusive. For example, conventional seedings might be broadcast seeded or drill seeded. Seeding methods are categorized for this discussion by type of seedbed tillage and by no-tillage. The erosion potential of a field needs to be considered before choosing a tillage method. Primary tillage tools like the moldboard plow, chisel plow, and heavy disks bury much of the surface residue, leaving bare soil subject to runoff and erosion, especially on sloping fields.

Tilled Seedbed Seedings. Tilled seedbed seedings are sometimes referred to as *conventional seedings* since conventional tillage practices (plowing, disking, harrowing, etc.) are used to prepare the seedbed. The purposes of tillage are to loosen the soil, eliminate existing vegetation, turn under surface weed seeds, incorporate lime and fertilizer into the soil, and provide a smooth surface for harvesting operations. Any tillage sequence that controls weeds and provides a firm seedbed with just enough loose surface soil for shallow seed placement with good seed-soil contact is satisfactory.

Tillage that leaves some residue on the surface will provide a better environment for developing seedlings than an overworked seedbed with no residue or mulch. Too much surface residue or trash may result in too shallow seed placement due to seeding units riding on top of the residue. Cloddy or trashy seedbeds are usually too rough or uneven for uniform depth control and seed placement and are too coarse for good seed-soil contact. Overworking the soil results in fluffy, powdery seedbeds that dry out quickly and may be too fine, increasing the potential for surface crusting following rainfall and poor seedling emergence. Crusting is particularly a problem with small-seeded legumes. Small clods or soil granules can be beneficial to prevent soil crusting. Major problems with conventionally tilled seedbed seedings are soil moisture loss during tillage and soil erosion potential until the crop is established.

BROADCAST FROST SEEDING. Seed may be broadcast in late winter on the soil surface of fall-sown cereals. Freezing and thawing action (honeycombing of the soil surface with ice crystals) plus rain will cover seed to about the right depth. Frost-seeded red clover (*T. pratense* L.) establishes better than most species. Frost seeding is successful only during short periods when soil and climatic conditions are right. If sowing is delayed until after freezing and surfaces are dry, drilling into a prepared seedbed is preferred.

CULTI-PACKER SEEDING. Culti-packer seeders consist of two sets of corrugated rollers with seed-metering boxes mounted between them. The first set of rollers firms the soil into shallow corrugations, seed is then dropped, and the second set of rollers splits the ridges of the corrugations, covering the seed and firming the soil around it (Fig. 3.1). These seeders have been a common method for seeding forages on prepared seedbeds for many years because of optimum seed placement and good seed-soil contact if seedbeds are properly prepared.

On medium- and heavier-textured soils, some of the seed remains on the top and sides of the ridges as well as at the bottom of the corrugations. Since these corrugations are split by the second set of rollers, the seed is distributed across a range of depths from 0.5-2.5 cm. On sandy soils, most of the seed falls to the bottom of the corrugations, and deeper coverage is obtained as is desirable. Culti-packer seeders should not be used on heavy soils having a moist surface because crusting is likely. And adequate soil coverage may not be obtained on seedbeds having heavy crop residues or with lightweight grass seeds such as smooth bromegrass (*Bromus inermis* Leyss.) and wheatgrasses (*Agropyron* spp.)

The principles of a culti-packer seeding can be accomplished without the specific use of a culti-packer seeder. A single corrugated roller can be used to firm the soil into shallow corrugations. Seed can be distributed with spinner seeders, grain drills having small-seeded legume and grass seed attachments, or sprayers, or it can be broadcast from the air. Grain drills with boxes for small seeds often drop the seed from a height of about 60 cm. If using a grain drill, seeding should be done with the disks, shovels, or hoes raised out of the soil. Distributing the seed through sprayers is a relatively new technique often referred to as *fluid or suspension seeding*. It is a very effective way of broadcasting seed uniformly over large areas in a short time, usually by custom applicators. The seed is then covered with a second, sometimes even a third, trip across the field with a culti-packer.

DRILL SEEDING. Grain drills with boxes for small seeds and seed tubes extending to ground level can accurately meter the seed, but controlling the depth of seeding can be

Fig. 3.1. Culti-packer seeder: The first roller firms soil into shallow corrugations, seed is dropped, and the second roller splits ridges and firms soil around seed.

difficult. A considerable amount of seed may be covered too deeply, especially if the furrow openers (disks, shovels, or hoes) are set too deeply and the seed is dropped before or near the openers. Seed that falls beneath soil thrown up by the openers usually is covered with too much soil for the seedlings to emerge. If the seed furrow is too deep, seed falling in the furrow may be too deep for emergence, especially if additional soil is washed into the furrow by rain. Heavier seeding rates should be used to compensate for these losses.

Drills with presswheels generally provide excellent results if a uniform shallow depth can be maintained. They also work better than culti-packer seeders on fields with crop residue. Drills without presswheels should be followed with a culti-packer. In some areas producers lightly harrow rather than culti-pack, but harrowing doesn't provide the desired seed-soil contact as well as culti-packing.

BAND SEEDING. Some grain drills are equipped for band seeding. Band seeding is the placement of seed in rows directly above, but not in contact with, a band of fertilizer. It is among the more reliable methods of getting good stands, especially on soils low in available P. A band of fertilizer is placed 3-6 cm deep. Seed tubes extend to within 5-10 cm of the soil surface about 30-40 cm behind the openers to place the seed in a band on top of the soil directly over the band of fertilizer. Seed is dropped well behind the openers so that soil can flow back into the furrow to cover the fertilizer, thus separating it from the seed. Best results are obtained with drills having presswheels to firm the soil over the seed and provide good seed-soil contact (Fig. 3.2); this is especially critical on light soils as firming establishes capillary flow to help keep the surface soil and the seed moist. Culti-packing after band seeding is usually not desirable, especially if the seeder has press-wheels.

Band seeding is superior to broadcast seeding only on soils with low fertility levels and during seasons when environmental conditions were unfavorable (Brown et al. 1960).

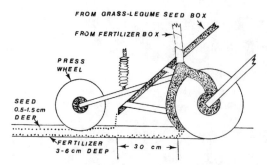

Fig. 3.2. Schematic for converting grain drill to band seeder. Flexible tubing from seed box is attached to place seed about 30 cm behind openers placing fertilizer. The presswheel, which firms soil around seed, is especially helpful on light soils with limited moisture.

SEEDING WITH A COMPANION CROP. A companion crop of small grain, primarily oat (*Avena sativa* L.), is often used with spring seedings of legumes and grasses in the northern US. A companion crop provides quick ground cover, helping to reduce wind and water erosion and resist invasion of weeds during forage establishment. It also provides a usable crop for grain, bedding, silage, or pasture.

Companion crops compete with the forage seeding for light, moisture, and nutrients (Bula et al. 1954; Klebesadel and Smith 1960; Tesar 1984). Spring-seeded small grains should not be sown heavier than 75% of the usually recommended rate. This reduces competition, especially in dry seasons, and decreases the likelihood of lodging in wet seasons. Top-dressing small grains with high rates of N should be avoided. Early removal of the companion crop as silage or pasture will usually favor forage establishment. Any spring-seeded small grain can be used, but oat is the most satisfactory. If weeds are present, especially broadleaf types, competition may be more severe and may last longer than competition from companion crops (Laskey and Wakefield 1978).

Fall-seeded small grains are usually more of a threat to spring forage establishment than spring grains. Reduced seeding rates are not practical because of the long period for development, where any reasonable rate eventually produces about the same number of tillers per unit area. When excessive spring growth occurs, pasturing will reduce competition and improve forage stands. Small grains are excellent quality feed, but grazing must be discontinued before jointing if they are to be saved for grain.

Straw should be removed as soon as the grain is harvested to avoid smothering of forage plants. In seasons when forage reaches the early bloom stage, a stubble-hay harvest can be made.

South of approximately 39°N latitude, seeding perennial forages during late August and September after a small grain harvest is often more practical than seeding with a spring companion crop. Such a seeding does not have to compete with companion crops or weeds, and the small grain does not have to be managed to favor the forage. Success of such a seeding depends on adequate rain between grain harvest and forage seeding. Ideally, the field is tilled immediately after the grain harvest. Surface weeds germinate and are destroyed with one or more tillage operations. Winter annual weeds such as common chickweed (*Stellaria media* [L.] Vill.) and henbit (*Lamium amplexicaule* L.) can be a problem with late summer seedings, especially during mild winters. When this occurs, herbicides may be needed (see Chap. 7, Vol. 1), and if so, they need to be applied at an early stage.

CLEAR SEEDINGS WITH INCORPORATED PRE-EMERGENCE HERBICIDES. Preemergence herbicides are replacing the traditional companion crop for spring seedings. These herbicides, followed by postemergence treatment when necessary, will give weed control and higher forage yields in the seeding year than normally obtained with companion crops. When herbicides eliminate weed competition, forage seeding rates can often be reduced (Hoveland 1970; Moline and Robinson 1971). However, presently available preemergence herbicides cannot be used with grass-legume mixtures.

MANAGEMENT OF NEW SEEDINGS. Weeds often invade new plantings, and the stand may be reduced if weeds are not controlled. Clipping such stands may be necessary, but it should not be done too early, or only the tops of the weeds will be removed, leaving active buds on the stubble to produce new branches and even more competition. Sufficient weed growth should be allowed so that most active buds are removed when clipped. Most forages regrow from crown buds (sweetclover [*Melilotus* spp.] and lespedeza [*Kummerowia* spp.] are exceptions) and are usually not seriously damaged by low cutting. Clipping too frequently can reduce seedling development as well as forage yields the following year (Sprague et al. 1963). More productive stands result in the absence of weeds without the need for clipping, especially during periods of moisture stress (Klebesadel and Smith 1958).

Pasturing or clipping new seedings can be helpful, but both should end 4-6 wk before a killing frost to allow a buildup of reserves for winter. Less harm will occur with moderate pasturing just before or after frost than by continuous grazing to freeze-up. Avoid grazing during wet periods. Top growth can serve as a protective mulch for young plants during the winter. Harvesting or grazing management of seedling stands should be no more severe than for mature stands of the same species.

No-Till Seedings. Technology and equipment are available to consistently establish excellent forage stands without tillage. No-till seeding not only reduces soil erosion, thus leaving the field in better physical condition, but also conserves soil moisture for germination and new seedling growth. Additional benefits are reduced fuel, labor, and time requirements, avoiding the problem of soil crusting frequently encountered with conventional seedings, and seeding on an already firm seedbed. On soils where stones and rocks are a problem, no-till allows them to remain on or below the soil surface, thus reducing the need to pick them before and/or after seeding. Plantings can be made into a variety of ground covers. Late summer seeding into winter barley or wheat stubble is one of the more reliable methods and one that has great flexibility in terms of matching seeding dates, soils, and weather conditions.

Following harvest of the winter barley or wheat, allow time for as many weed seeds to germinate as practical so that maximum weed kill will be accomplished with application of a nonselective herbicide such as paraquat (1,1'-dimethyl-4,4'-bipyridinium ion) or glyphosate [isopropylamine salt of N-(phosphono-methyl) glycine]. This will generally be 2-3 wk following grain harvest. A second herbicide application should be made at seeding after a second flush of weeds has germinated. Seeding can be made anytime within a 3- to 6-wk period when soil moisture and other conditions are right. Since soil tillage is not required, seedings can be made much sooner following a rain, and more soil moisture will remain for the new seeding than with a conventionally tilled seedbed. The stubble reduces soil erosion and provides protection for the seedlings from intense sunlight and damage from blowing sand or soil particles.

Seedings can be made into the killed stubble of summer annual grasses, particularly sudangrass (*Sorghum bicolor* [L.] Moench) or sorghum-sudangrass hybrids and foxtail millet (*Setaria italica* [L.] Beauv.). These summer annual grasses grow vigorously, and weeds are generally not able to compete. Allelopathic toxins may also inhibit weed germination and development. Thus they are good smother crops in preparation for forage seedings. However, since they do grow vigorously during the summer, they can deplete soil moisture if not killed 3-6 wk prior to the forage seeding.

When perennial broadleaf weeds are present, selective translocated herbicides such as 2,4-D (2,4-dichlorophenoxyacetic acid) and dicamba (2-methoxy-3,6-dichlorobenzoic acid) will provide control, but treatment must be at least 30 d prior to seedings; 6 mo is better (Campbell 1976). Where both broadleaf and grass perennials are present, a translocated herbicide such as glyphosate should be used; with these herbicides a seeding delay of 1-3 wk will give increased sward control and improved forage stands (Mueller-Warrant and Koch 1980; Welty et al. 1981).

Allelopathic toxins have been implicated in reduction of new forage stands (Toai and Linscott 1979; Peters and Zam 1981; Young and Bartholomew 1981; Cope 1982). Such toxins may explain some of the difficulty in establishing birdsfoot trefoil (*Lotus corniculatus* L.) in tall fescue (*Festuca arundinacea* Schreb.) sod (Grant and Sallans 1964; Peters 1968; Luu et al. 1982). This suggests that it may be desirable to grow one or more crops between a sod crop and no-till legumes. Luu (1980) found that toxins associated with tall fescue swards were largely eliminated by burning the sod prior to seeding birdsfoot trefoil.

When an old sod is destroyed with a herbicide in preparation for no-till seeding, the feed supply and habitat of the insect population are drastically reduced. Young no-till seedlings are especially vulnerable since they may now be the major insect food supply. Stands are often improved when insecticides are applied at seeding (Braithwaite et al. 1958; Decker et al. 1964; Kalmbacker et al. 1979; Vough et al. 1982). Insect populations vary considerably, and the potential problem is difficult to predict; no hard and fast rules are available. Insecticides are considered by some to be justified insurance; the decision must be made on individual cases based on recent field history, climatic conditions, and economic implications.

NO-TILL PASTURE RENOVATION. Much of the permanent pasture in the humid US is not very productive. This is due mainly to the lack of adequate fertilization, poor grazing management, and the nature of the forage species present. Renovation of these pastures could improve production more than twofold, depending upon soil characteristics and the condition of existing sod (Decker et al. 1969, 1976; Taylor et al. 1969; Decker and Dudley 1976). Renovation is the improvement of a pasture by the partial or complete destruction of the sod, plus liming, fertilizing, weed control, and seeding as may be required to establish or re-establish desired forage plants without an intervening crop.

Recent no-till technology enables farmers to renovate old pastures without plowing or disking. Herbicides such as paraquat and glyphosate permit suppression of existing vegetation without tillage. No-till drills can place forage seeds at the proper depth and ensure good seed-soil contact.

Suppressing Competition from Existing Vegetation. Pastures to be sod-seeded should receive intensive, close grazing or close clipping 4-6 wk prior to a late summer seeding or during the fall prior to a late winter seeding. This will reduce top growth and surface residue, assisting in suppressing competition from existing vegetation. This is one time in pasture management when overgrazing is beneficial. The sward should then be allowed to green up in order to obtain effective suppression with the herbicide. There are two methods for applying the herbicide—band and broadcast. The method to be used depends upon the composition and condition of the existing sod, whether the renovation is a completely new seeding or only the introduction of a legume, and the time of year of the seeding.

If an adequate stand of productive grasses such as orchardgrass (*Dactylis glomerata* L.), smooth bromegrass, or tall fescue is present and the renovation is being made in late winter/early spring primarily for introducing legumes, then a band application of paraquat at seeding is recommended. If productive species are not present in the existing sod and both grass and legume species are being seeded in late summer, then broadcast applications of paraquat or glyphosate are recommended.

Band Application for Early Spring Seeding. Band application of herbicide entails applying a band of spray over the seed rows so that about half of the surface area is sprayed; for example, with 10-cm bands over 20-cm rows, unsprayed bands are left between each row (Fig. 3.3). The spray boom is mounted on the drill, generally ahead of the seeding units.

When productive species are already present, band application of the herbicide has several advantages. First, band application generally results in fewer weeds in the new seeding. Whenever crop growth is suppressed or destroyed, weed seeds in the soil tend to germinate. This is especially true for broadleaf weeds and summer annual grasses that are common problems in spring seedings. Band spraying allows about half of the sod to remain actively growing, thus helping to retard weed growth, much like companion crops in conventional seedings.

Second, the remaining sod provides pasture for grazing much earlier than when all of the sod is suppressed or killed with a broadcast application of herbicide. Since about half of the original sward remains intact, pastures can produce significant grazing 30-60 d after seeding; this will normally be within the grazing restrictions of the herbicide. Even though the spring flush of growth is reduced by more than half with the banded herbicide, total production for the seeding year is not reduced since the seeded legumes provide increased yields of higher-quality forage later in the season compared with the untreated sward (Taylor and Jones 1982; Badger 1983). Animal gains are almost double those of unimproved swards in the seeding year (Decker et al. 1976).

Third, if productive species are present, there is no need to kill them and reseed the same species again. Fourth, there tends to be less insect feeding on new seedlings since there is an existing band of vegetation for them to feed on.

Banded paraquat should be applied within recommended rates to that half of the area actually sprayed. This means that for each hectare of pasture being renovated, only half the normal volume of material will be used.

An alternative method for introducing red clover into pasture is frost seeding in late winter before grass initiates spring growth. Seed is broadcast when alternate freezing and thawing of the soil surface is occurring. This works best when grass stands are thin and where the grass sward was heavily grazed the previous fall and early winter. On well-drained soils, where puddling is not a problem, the trampling effect of animals on the pasture can improve seed-soil contact, but an-

Fig. 3.3. Banded contact herbicide-protective mulch in killed strips (10-12 cm) provides ideal conditions for developing seedlings. Live grass strips behave like companion crops to control weeds and to allow for earlier, more productive grazing than when the entire sward is killed.

imals should be removed when legume germination begins. While frost seeding generally works quite well with red clover, it is not usually recommended for grasses and other legumes.

Broadcast Application for Late Summer Seeding. It is important that all existing vegetation be eliminated 4-6 wk prior to seeding so that it will not be depleting soil moisture that should be conserved for the new seeding nor be competitive with the young seedlings during establishment. The use of either paraquat or glyphosate and the application rate are determined by the type of vegetation present. If the pasture has been closely grazed or mowed, allow 2.5-5.0 cm of regrowth before spraying to obtain more effective control. This will be followed by an application of paraquat at the time of seeding for control of weeds that germinated after the herbicide application 4-6 wk earlier.

Selecting Legumes for Pasture Renovation. Red clover is more easily established than most legumes and should usually be a part of most mixtures (Decker and Dudley 1976). It produces high yields the year of seeding but seldom lasts more than 3 yr under grazing. Either sod-seeding or frost seeding red clover every third year in late winter can be used to maintain stands. Species like birdsfoot trefoil are slower to establish, but stands are longer lived. A combination of such species should provide earlier production with longer-lasting results. Since red clover is so competitive, no more than 2.5 kg seed/ha should be used in mixtures with other legumes. One technique that has worked to reduce species competition is to seed species in alternate rows (Decker et al. 1976). This will require a seeder with two legume hoppers or dividers in a single box. Seeding rates for sod-seeding will normally be the same as for conventional seedings. Increased rates may be justified when competition is high and available moisture is low (Groya and Sheaffer 1981; Sheaffer and Swanson 1982).

NO-TILL SEEDING EQUIPMENT. The availability of no-till seeding equipment increased dramatically between 1970 and the mid-1980s. While these machines differ widely in price and design, all produce satisfactory stands when properly adjusted and operated.

No-till drills should be heavy enough to ensure proper thatch and soil penetration. Features that make seedings more reliable include (1) rolling or power-driven coulters to cut through mulch covers, (2) double-disc or other types of seed placement units that line up precisely with the slit left by the cutting coulter, (3) depth bands, wheels, or another method for controlling seed placement on each seeding unit, (4) independently operating units to follow the soil terrain, and (5) presswheels to ensure good seed-soil contact (Fig. 3.4). The importance of the presswheel

increases as soil texture and soil moisture decrease; with heavy-textured soils high in moisture, it can even be detrimental. In these cases it may be desirable to remove the presswheel and leave the furrow open for the seed to germinate and to develop at the bottom of the open slit. If the slit is closed with a presswheel, the seed may be covered too deeply and be unable to emerge.

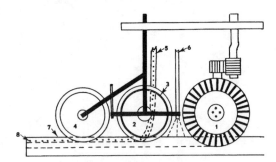

Fig. 3.4. No-till sod-seeder: (1) coulter to cut through surface thatch, (2) seed placement unit (single or double disc or rotating cutter wheel), (3) depth band or other depth control device, (4) presswheel to ensure good seed-soil contact, (5) seed tube, and (6) tube for banding contact herbicide. Seed should be placed 1.25-2.5 cm (8) below the soil surface (7).

NO-TILLING ANNUALS INTO PASTURES. Both winter and summer annual species can be used to increase yields and extend the grazing season (Decker et al. 1969; Decker 1970; Hart et al. 1971; Hoveland and Carden 1971; Templeton and Taylor 1975; Hoveland et al. 1978; Belesky et al. 1981). The practice works best with cool-season annuals seeded into warm-season perennial grasses. Seeding should be delayed until cool night temperatures slow the growth of the warm-season grass; this will be 2-5 wk before the first killing frost in most cases. In more northern areas the big advantage from annuals is for early spring grazing, while fall and winter grazing is more feasible farther south. A lack of soil moisture at seeding may limit this practice in some cases (Robinson 1963). Seeding annuals into heavy grass swards makes it possible to continue grazing during extended periods of precipitation without the soil damage that often occurs when annuals are seeded into tilled fields (Decker 1970).

Winter annuals sod-seeded into bermudagrass (*Cynodon dactylon* [L.] Pers.) swards work especially well since there is minimum competition between the two species (Fribourg and Overton 1973; Decker et al. 1974). Thus management of such combinations is relatively simple, and yield potentials are often higher. A disadvantage is that seedings must be made each fall when soil moisture may be limiting. It is possible to sod-seed cool-season perennial grasses and legumes into dormant warm-season grass swards (Wilkinson et al. 1968; Fribourg and Overton 1979; Fribourg et al. 1979). By seeding two perennials with contrasting growth patterns, extension of the grazing season can be realized without annual sod-seeding, but the management of such mixtures, including times and rates of N application, is more complicated than with annuals.

Successful seeding of summer annuals into perennial cool-season swards can provide improved forage distribution, but more restrictions often exist. Such seedings must be made at a time of year when the existing sward often has depleted soil moisture; close grazing and/or herbicide application to the existing sward may be required (Hart et al. 1971; Belesky et al. 1981). Access to irrigation facilities would certainly be desirable.

While most pasture improvements using sod-seeding have taken place on humid pastures of the eastern US, the technique has also been effectively used on some western rangelands (Gomm 1962; Dudley et al. 1966; Willard and Schuster 1971).

WHY DO FORAGE SEEDINGS FAIL?

A Partial Guide in Analyzing Forage Establishment Problems. If asked why a seeding failed, the average person may say "poor seed," "dry season," or "winterkill," but such general reasons seldom tell the whole story. Many more seeds are sown than numbers of plants needed. Why do some plants live but so many die? A partial answer is that losses occur over a long period from initial germination, early seedling establishment, and growth and development of seedlings into a mature stand. Lack of germination of live seed occurs because of

1. *Impermeable seed coat.* Scarification may be needed.
2. *Insufficient air.* Seed sown too deeply (especially in wet, heavy soils) may not have enough O to germinate.
3. *Insufficient moisture.* Alternating temperature and moisture levels (too low for complete germination) can lower seed viability and result in death.

Establishment after germination may fail because of

1. *Drying.* Seed placed in loose surface soil may germinate after a light rain but dry out and die before developing sufficient roots for establishment.
2. *Freezing.* Seeds are especially sensitive to freezing as the young root breaks the seed coat; temperatures below −3°C are lethal. Soil coverage reduces the likelihood of injury, and once rooted, seedlings can withstand much lower temperatures.
3. *Coverage too shallow.* Soil cover or mulch protects against both drying and freezing; without it, seed establish only when the soil surface remains moist for extended periods.
4. *Coverage too deep.* More seed probably are wasted in this way than any other.
5. *Crusted soil surface.* This can prevent emergence, especially when seed are sown too deeply on fine-textured soils.
6. *Toxicity.* Seed in direct contact with banded fertilizer, improper use of herbicides, or herbicide carryover from the previous crop can damage seed and/or young seedlings.

Growth of seedlings after establishment may stop because of

1. *Undesirable pH.* Lime should be applied according to soil test to provide desirable pH; Ca and Mg should be applied as nutrients.
2. *Low fertility.* A soil test should be used to ensure adequate P, potassium (K), or other nutrients.
3. *Inadequate legume inoculation.*
4. *Poor drainage.* Water accumulation on the surface or in the soil profile can limit growth.
5. *Drought.* This is the most commonly given reason for stand failures.
6. *Competition from companion crops.* Cereals compete with forage seedlings for water, light, and nutrients.
7. *Competition from weeds.* Weeds are similar to companion crops, but competition may be more severe and last longer.
8. *Insects.* Pests such as alfalfa weevil (*Hypera postica* [Gyllenhal]), potato leafhopper (*Empoasca fabae* [Harris]), and grasshoppers (*Melanoplus* and *Camnula* spp.) can eliminate new stands.
9. *Diseases.* Pathogens such as damping-off and Sclerotinia crown and stem rot can be fatal.
10. *Winter-killing.* Seeding too late in the

fall or seeding poorly adapted cultivars can increase winter-kill.

QUESTIONS

1. Why are thick, vigorous forage stands essential to profitable forage production?
2. What are the advantages of seeding grass-legume mixtures? How would species for mixtures be selected?
3. What are the requirements of a good seedbed?
4. How critical is seeding depth with the various seeding methods? How can it best be controlled?
5. What are the reasons for using a companion crop? Under what conditions would you not use a companion crop?
6. How critical is early removal of a companion crop? Nitrogen fertilization? Seeding rate of a companion crop?
7. How would you manage a new forage seeding where weeds have invaded the stand?
8. Discuss the relative importance of species selection and seeding method in terms of available soil resource, climate of the area, and intended forage use.
9. How would you describe the advantages and disadvantages of conventional and no-till forage establishment?
10. What are the major similarities and differences between no-till hay crop seeding techniques and those for no-till pasture renovation?

REFERENCES

Badger, TH. 1983. Evaluation of sod-seeded pasture renovation. MS thesis, Univ. of Maryland, College Park.

Barnes, DK. 1978. Effects of treated seed on inoculation, establishment and yield of alfalfa. In Proc. 8th Annu. Alfalfa Symp. Cert. Alfalfa Seed Counc., Davis, Calif., 52-55.

Belesky, DP, SR Wilkinson, RN Dawson, and JE Elsner. 1981. Forage production of a tall fescue sod intercropped with sorghum × sudangrass and rye. Agron. J. 73:657-60.

Braithwaite, BM, A Jane, and FG Swain. 1958. Effect of insecticides on sod sown sub clovers. J. Aust. Inst. Agric. Sci. 24:155-57.

Brown, BA, AM Decker, MA Sprague, HA MacDonald, MR Teel, and JB Washko. 1960. Band and Broadcast Seeding of Alfalfa-Bromegrass in the Northeast. Md. Agric. Exp. Stn. Bull. A-108 and Northeast Reg. Publ. 41.

Brown, CS, and RF Stafford. 1970. Get top yields from alfalfa seedings. Better Crops with Plant Food 54(2):16-18.

Bula, RJ, D Smith, and EE Miller. 1954. Measurements of light beneath a small grain companion crop as related to legume establishment. Bot. Gaz. 115:271-78.

Campbell, MH. 1976. Effect of timing of glyphosate and 2,2-DPA application on establishment of surface-sown pasture species. Aust. J. Exp. Agric. Anim. Husb. 16:491-99.

Cope, WA. 1982. Inhibition of germination and seedling growth of eight forage species by leachates from seeds. Crop Sci. 22:1109-11.

Decker, AM. 1970. Crop combinations produce feed more efficiently. Better Crops with Plant Food 54(1):6-9.

Decker, AM, and RF Dudley. 1976. Minimum tillage establishment of five forage species using five sod-seeding units and two herbicides. In J Luchok, JD Cawthon, and MJ Breslin (eds.), Hill Lands, Proc. Int. Symp., 3-9 Oct., Morgantown, W.Va. Morgantown: W.Va. Univ. Books, 140-46.

Decker, AM, HJ Retzer, and FG Swain. 1964. Improved soil openers for the establishment of small-seeded legumes in sod. Agron. J. 56:211-14.

Decker, AM, HJ Retzer, FG Swain, and RF Dudley. 1969. Midland Bermudagrass Forage Production Supplemented by Sod-seeded Cool-Season Annual Forages. Md. Agric. Exp. Stn. Bull. 484.

Decker, AM, HJ Retzer, and RF Dudley. 1974. Cool-season perennials vs. cool-season annuals sod seeded into a bermudagrass sward. Agron. J. 66:381-83.

Decker, AM, JH Vandersall, and NA Clark. 1976. Pasture renovation with alternate row sod-seeding of different legume species. In J Luchok, JD Cawthon, and MJ Breslin (eds.), Hill Lands, Proc. Int. Symp., 3-9 Oct., Morgantown, W.Va. Morgantown: W.Va. Univ. Books, 146-49.

Dudley, RF, EB Hudspeth, Jr., and CW Gant. 1966. Bushland range interseeder. J. Range Manage. 19:227-29.

Fink, RJ, and OH Fletchall. 1963. Forage crop establishment in soil containing atrazine or simazine residues. Weeds 11:81-83.

Fribourg, HA, and JR Overton. 1973. Forage production on bermudagrass sods overseeded with tall fescue and winter annual grasses. Agron. J. 65:295-98.

———. 1979. Persistence and productivity of tall fescue in bermudagrass sods subjected to different clipping managements. Agron. J. 71:620-24.

Fribourg, HA, JB McLaren, KM Barth, JM Bryan, and JT Connell. 1979. Productivity and quality of bermudagrass and orchardgrass-ladino clover pastures for beef steers. Agron. J. 71:315-20.

Gomm, FB. 1962. Reseeding Studies at a Small High-Altitude Park in Southwestern Montana. Mont. Agric. Exp. Stn. Bull. 568.

Grant, EA, and WG Sallans. 1964. Influence of plant extracts on germination and growth of eight forage species. J. Br. Grassl. Soc. 19:191-97.

Groya, FL, and CC Sheaffer. 1981. Establishment of sod-seeded alfalfa at various levels of soil moisture and grass competition. Agron. J. 73:560-65.

Hannaway, DB, and PE Shuler. 1993. Nitrogen fertilization in alfalfa production. J. Prod. Agric. 6:80-85.

Hart, RH, HJ Retzer, RF Dudley, and GE Carlson. 1971. Seeding sorghum and sudangrass hybrids into tall fescue sod. Agron. J. 3:478-80.

Hoveland, CS. 1970. Establishing Sericea Lespedeza at Low Seeding Rate with a Herbicide. Ala. Agric. Exp. Stn. Circ. 174.

Hoveland, CS, and EL Carden. 1971. Overseeding winter annual grasses in sericea lespedeza. Agron. J. 63:333-34.

Hoveland, CS, WB Anthony, JA McGuire, and JG Starling. 1978. Beef cow-calf performance on

Coastal bermudagrass overseeded with winter annual clovers and grasses. Agron. J. 70:418-20.

Kalmbacker, RS, DR Minnick, and FG Martin. 1979. Destruction of sod-seeded legume seedlings by the snail (*Polygyra cereola*). Agron. J. 71:365-68.

Klebesadel, LJ, and D Smith. 1958. The influence of oat stubble management on the establishment of alfalfa and red clover. Agron. J. 50:680-83.

———. 1960. Effects of harvesting an oat companion crop at four stages of maturity on the yield of oats, on light near the soil surface, on soil moisture, and on the establishment of alfalfa. Agron. J. 52:627-30.

Koch, DW, and GO Estes. 1986. Liming rate and method in relation to forage establishment—crop and soil chemical responses. Agron. J. 78:567-71.

Laskey, BC, and RC Wakefield. 1978. Competitive effects of several grass species and weeds on the establishment of birdsfoot trefoil. Agron. J. 70:146-48.

Lowder, SW, and JB Weber. 1982. Atrazine efficacy and longevity as affected by tillage, liming, and fertilizer type. Weed Sci. 30:273-80.

Luu, KT. 1980. Characterization of allelopathic effects of tall fescue on birdsfoot trefoil. PhD diss., Univ. of Missouri, Columbia (Diss. Abstr. 8108824).

Luu, KT, AG Matches, and EJ Peters. 1982. Allelopathic effects of tall fescue on birdsfoot trefoil as influenced by N fertilization and seasonal changes. Agron. J. 74:805-8.

Moline, WJ, and LR Robinson. 1971. Effects of herbicides and seeding rates on the production of alfalfa. Agron. J. 63:614-17.

Mueller-Warrant, GW, and DW Koch. 1980. Establishment of alfalfa by conventional and minimum-tillage seeding techniques in a quackgrass-dominant sward. Agron. J. 72:884-89.

Peters, EJ. 1968. Toxicity of tall fescue to rape and birdsfoot trefoil seeds and seedlings. Crop Sci. 8:650-53.

Peters, EJ, and AHBM Zam. 1981. Allelopathic effects of tall fescue genotypes. Agron. J. 73:56-58.

Robinson, RR. 1963. Rainfall distribution in relation to sod seeding for winter grazing. Agron. J. 55:307-8.

Sheaffer, CC, and DR Swanson. 1982. Seeding rates and grass suppression for sod-seeded red clover and alfalfa. Agron. J. 74:355-58.

Sheard, RW. 1980. Nitrogen in the P band for forage establishment. Agron. J. 72:89-97.

Sheard, RW, GJ Bradshaw, and DL Massey. 1971. Phosphorus placement for the establishment of alfalfa and bromegrass. Agron. J. 63:922-27.

Sprague, MA, MM Hoover, Jr., MJ Wright, HA MacDonald, BA Brown, AM Decker, JB Washko, VG Sprague, and KE Varney. 1963. Seedling Management of Grass-Legume Associations in the Northeast. N.J. Agric. Exp. Stn. Bull. 804 and Northeast Reg. Publ. 42.

Sund, JM, GP Barrington, and JM Scholl. 1966.

Methods and depths of sowing forage grasses and legumes. In Proc. 10th Int. Grassl. Congr., Helsinki, Finland, 319-23.

Taylor, TH, and LT Jones, Jr. 1982. Persistence and productivity of sod-seeded legumes compared with nitrogen fertilized grass sod. In Am. Soc. Agron. Abstr. Madison, Wis., 129.

Taylor, TH, EM Smith, and WC Templeton. 1969. Use of minimum tillage and herbicide for establishing legumes in Kentucky bluegrass (*Poa pratensis* L.) swards. Agron. J. 61:761-66.

Templeton, WC, Jr., and TH Taylor. 1975. Performance of big flower vetch seeded into bermudagrass and tall fescue swards. Agron. J. 67:709-12.

Tesar, MB. 1984. Good Stands for Top Alfalfa Production. Mich. Coop. Ext. Serv. Bull. 1017.

Tesar, MB, and D Huset. 1979. Lime coating and inoculation of alfalfa. In 16th Cent. Alfalfa Improv. Conf., 1.

Toai, TV, and DL Linscott. 1979. Phytotoxic effect of decaying quackgrass (*Agropyron repens)* residues. Weed Sci. 27:595-98.

Vough, LR, AM Decker, and RF Dudley. 1982. Influence of pesticide, fertilizers, row spacings, and seeding rates on no-tillage establishment of alfalfa. In JA Smith and VW Hays (eds.), Proc. 14th Int. Grassl. Congr., 15-24 June 1981, Lexington, Ky. Boulder, Colo.: Westview, 547-50.

Wakefield, RC, and N Skoland. 1965. Effects of seeding rate and chemical weed control on establishment and subsequent growth of alfalfa (*Medicago sativa* L.) and birdsfoot trefoil (*Lotus corniculatus* L.). Agron. J. 57:547-50.

Ward, CY, and RE Blaser. 1961. Effect of nitrogen fertilizer on emergence and seedling growth of forage plants and subsequent production. Agron. J. 53:115-20.

Welty, LE, RL Anderson, RH Delaney, and PF Hensleigh. 1981. Glyphosate timing effects on establishment of sod-seeded legumes and grasses. Agron. J. 73:813-17.

White, JGH. 1970. Establishment of lucerne (*Medicago sativa* L.) in uncultivated country by sod seeding and eversowing. In MJT Norman (ed.), Proc. 11th Int. Grassl. Congr., 13-23 Apr., Surfers Paradise, Queensland, Australia. St. Lucia: Univ. Queensland Press, 134-38.

Willard, EW, and JL Schuster. 1971. An evaluation of an interseeded sideoats grama stand four years after establishment. J. Range Manage. 24:223-26.

Wilkinson, SR, LF Welch, GA Hillsman, and WA Jackson. 1968. Compatibility of tall fescue and Coastal bermudagrass as affected by nitrogen fertilization and height of clip. Agron. J. 60:359-62.

Young, CC, and DP Bartholomew. 1981. Allelopathy in a grass-legume association: I. Effects of *Hemarthria altissima* (Poir.) Stapf. and Hubb. root residues on the growth of *Desmodium intortum* (Mill.) Urb. and *Hemarthria altissima* in a tropical soil. Crop Sci. 21:770-74.

4

Integrated Pest Management in Forages

GARY W. FICK
Cornell University

WILLIAM O. LAMP
University of Maryland

MANAGEMENT of agricultural systems is complex and challenging. That is especially true of grassland agriculture, where the interactions between forages and livestock must be carefully controlled to achieve the twin goals of profitability and sustainability. Forage scientists and producers alike have been seeking ways to improve management in grassland agriculture, and one means for doing this is integrated pest management.

Integrated pest management (IPM) developed out of the need for alternatives to pesticides. Following World War II, new pesticide materials offered such simple and initially effective pest control that they were quickly adopted and widely used. As pests adapted through natural selection and became resistant to the new materials, and as scientists, farmers, and the public in general became more aware of the environmental risks and increasing economic costs associated with their use, it became clear that more effective

GARY W. FICK is Professor of Agronomy in the Department of Soil, Crop, and Atmospheric Sciences at Cornell University, Ithaca, New York. His degrees are from the University of Nebraska, Massey University in New Zealand, and the University of California at Davis. He teaches several courses, including "Forage Crops," and leads a research program focusing on quality, development, and integrated pest management of forages.

WILLIAM O. LAMP is Associate Professor of Entomology at the University of Maryland, College Park, where he teaches and conducts research on integrated pest management. He received graduate training from Ohio State University and the University of Nebraska and has postdoctoral experience from the University of Illinois. His research emphasizes crop protection from insects through the integration of crop management practices and insect-plant interactions.

and less hazardous options were needed (Stern et al. 1959). By the 1970s, major research and extension programs developed a new synthesis of old and new pest control methods that became known as *IPM* (Huffaker 1980). The emphasis was to minimize, but not necessarily eliminate, the use of pesticides. Alfalfa (*Medicago sativa* L.) was the first perennial forage crop considered, and it now provides the model for forage IPM programs.

DEFINITIONS

The primary goals of forage crop management are to achieve high yields, high nutritive quality, and long stand life. *Pests* are biological agents that interfere with these goals and consequently cause economic losses. As shown in Table 4.1, pests have a broad range of natural diversity. In nature, each organism has essential biological functions, and organisms are not sorted as harmful or beneficial (Harlan 1992). Classification as a pest is based on human values. A challenge for pest management is to control what is harmful without harming what is beneficial, and nature has not made that easy.

To integrate means to bring together all parts so that something is made whole or complete. Integrated pest management brings together several control methods while seeking to minimize environmental hazards and maximize economic benefits for producers and consumers (Bottrell 1979). Thus, the IPM approach is more comprehensive than the pesticide-oriented programs it is intended to replace. It is also an evolving concept (Frisbie and Smith 1989) that needs to be placed in a larger context of numerous management

TABLE 4.1. A biological classification of the main kinds of agricultural pests

Biological class
    Subclass: example

Pathogens or diseases
    Viruses and mycoplasmalike agents: alfalfa mosaic virus
        Bacteria: bacterial wilt of alfalfa, *Clavibacter michiganense*
                                    subsp. *insidiosum* (McCull.) Davis et al.
        Fungi: Phytophthora root rot of alfalfa, *Phytophthora megasperma* Drechs. f. sp. *medicaginis*
        Nematodes: root lesion nematodes, *Pratylenchus* spp.

Arthropods
    Insects: alfalfa weevil, *Hypera postica* (Gyllenhal)
                                potato leafhopper, *Empoasca fabae* (Harris)
    Mites: two-spotted spider mite, *Tetranychus urticae* (Koch)

Mollusks
    Slugs and snails: gray garden slug, *Deroceras reticulatum* Müller

Vertebrates
    Birds: blackbirds, *Agelaius phoeniceus* L.
    Mammals: rodents, rabbits, wild ruminants

Weeds
    Monocots: yellow foxtail, *Setaria glauca* (L.) Beauv.
                                yellow nutsedge, *Cyperus esculentus* L.
    Dicots: pigweed, *Amaranthus retroflexus* L.
                                dandelion, *Taraxacum officinale* Weber in Wigg.

practices that must be integrated with the time, labor, and capital resources available to individual managers (Fick and Power 1992).

PRINCIPLES AND PRACTICES

Detailed discussion of the principles of IPM is beyond the scope of this chapter (Lamp and Armbrust 1994), but key concepts related to forage pest management can be summarized:

1. *Many species of organisms inhabit forage ecosystems, but only a few are pests.* Only specific combinations of host crops, environments, and potential pests can cause economic loss. Since most species are actually beneficial in some way, pest control methods should target real pests and conserve other organisms in the forage environment.

2. *Populations of all pests are naturally suppressed, and control measures should take advantage of natural suppression.* Biotic factors limit population growth of all species.

Crop management practices that encourage natural enemies and competitors will aid in the suppression of pest populations.

3. *Crop ecosystems can be managed to enhance pest suppression.* The condition of a crop and its physical environment influence pest populations, and crop management practices often affect losses induced by pests. Knowledge of pest ecology is therefore critical in predicting how changes in crop management will alter pest problems.

4. *Each pest population can be managed by various methods, and generally the optimal approach is to integrate two or more methods.* Because of the complexity and unpredictability of forage systems, any one approach is rarely sufficient for permanent suppression of key pests below economic levels. Instead, IPM strategies combine pest control methods, including both preventive and responsive practices (Table 4.2).

5. *When a pest population density warrants*

TABLE 4.2. Methods and examples of pest control

Method	Example
Natural (none)	No human-directed changes; reliance is on natural processes that maintain the "balance of nature"
Biological control	Change the pest's biological environment so that natural enemies of the pest are increased (sometimes by raising and releasing the natural enemies)
Resistance breeding	Change the pest's host crop through plant breeding so that it is less affected by the pest (usually with improved forage cultivars)
Cultural control	Change the pest's physical environment so that the pest is less successful (crop rotations, fertilization, early harvesting, soil drainage, etc.)
Direct control	Change the pest so that it dies quickly or dies without completing its life cycle, either by chemicals (pesticides, hormones, etc.) or by physical means (trapping, sterilization, cultivation, etc.)

additional suppression, responsive control measures may be applied. The density of a pest that justifies the cost of responsive control is called the *economic injury level* (EIL) (Pedigo et al. 1986). When densities are expected to exceed the EIL, action may be taken to prevent any additional damage. The pest density at which control should be implemented is called the *economic threshold* or *action threshold.* Development of EILs and thresholds has become a standard part of IPM, yet it is important to recognize that pesticide use should be limited to emergency situations. The emphasis of IPM is on the prevention of significant losses to pests.

6. *Control measures may produce unexpected and undesirable results.* Examples are noneffective control practices, development of pest resistance to pesticides or to defense mechanisms of crop cultivars, pest resurgence and outbreaks of minor pests following pesticide applications, and harm to nontarget organisms, including humans.

7. *Pesticide usage results in pest evolution so that pesticides become ineffective.* Every major category of pests has now developed genetic resistance to pesticides that were once effective (Georghiou 1986). Effective pesticides are important in pest control emergencies, but their use increases selection pressure and speeds the development of resistance against them. Therefore, usage should be limited in order to retain these options for future exigencies.

These principles of IPM show that specific pest management practices may or may not alleviate the negative impact of pests on crop yield, quality, and persistence. Knowledge of the effect of key pests on forage crops, critical periods leading to losses, and factors leading to pest outbreaks is needed to design optimal IPM programs for protecting forage crops. Crop management practices, such as species and cultivar selection, fertilization, harvesting schedule, and crop rotation, may also be manipulated to significantly reduce pest losses.

STEPS IN AN IPM PROGRAM

Based on the principles listed in the preceding, the following steps are necessary components of IPM programs:

1. *Recognition of the diversity of organisms within the forage agroecosystem.* The attitude that all insects, weeds, and molds are pests to be eliminated must be corrected with understanding and education. Most noncrop organisms in an agroecosystem function in maintaining a favorable balance and are beneficial in terms of human goals.

2. *Identification of pest species.* Pest management is necessary because a few organisms can become very serious pests. We must be able to identify these organisms and to assess their potential for economic injury in individual crops and locations.

3. *Understanding pest biology.* Effective pest control with minimum impact on nontarget organisms requires a knowledge of pest biology and the related effects of environment. Certain life stages are more subject to control than others. Certain environments foster or retard development of a pest population. The more we know about a pest, the more we know about how to manage it.

4. *Selection and use of available preventive control measures.* Many preventive pest control methods are available. Examples include sowing pest-resistant forages, rotating crops to break pest life cycles, fertilizing with needed nutrients to promote crop vigor, and limiting pesticides to allow buildup of natural biological controls. Using IPM concepts involves anticipating problems and properly implementing preventive control methods so that pest emergencies are usually avoided.

5. *Monitoring pest populations and applying responsive controls.* Scouting or sampling for pests allows growers to anticipate problems while there is still time to prevent economic losses. Many IPM programs involve sample collection and analysis procedures, some based on weather data and computer models. When action is indicated, options for forages usually include early harvesting or pesticide applications. Professional crop management consultants provide systematic monitoring of pests and interpretation of scouting results for an increasing number of growers. Such professional services often provide additional information relevant to soil, crop, and livestock management.

6. *Evaluation and refinement.* The results of a pest management program need to be evaluated with regard to the producer's objectives. Successes and failures need to be reviewed and adjustments made that correct past deficiencies and incorporate new developments. "Check plots" can help demonstrate what might have happened with some other management, and good field records can pinpoint problems in space and time.

CASE STUDIES

Insects. Some insect pests can be so damaging that stands are killed or persistence reduced. However, damage caused by most insects is limited to reduced yield and quality.

ALFALFA WEEVIL (*Hypera postica* [Gyllenhal]). The alfalfa weevil is a snout beetle with a complete metamorphosis and normally only one generation per year. Larvae and adults feed mainly on alfalfa, and most injury is caused during spring by chewing damage to leaves near the end of the larval stage of development (Fig. 4.1). In northern states (e.g., Minnesota and New York), the weevil survives the winter only as an adult. In more-southern states (e.g., Maryland and Oklahoma), eggs deposited in alfalfa stubble in autumn and winter may also survive. Adults do a limited amount of feeding before laying eggs. The first obvious larval feeding causes "pinholes" in the unfolding leaves at the growing point, but as larvae grow, they consume increasingly larger and more mature leaves. An alfalfa stand can be almost completely defoliated in 2 to 5 d with four or more small larvae per stem (Liu and Fick 1975).

In the southern US, larvae hatched from overwintering eggs emerge when alfalfa is less than 20 cm tall and may cause serious defoliation while the crop is still vegetative. This compounds injury to the first growth.

Fig. 4.1. The insect larvae of alfalfa weevil remove leaf tissue and thus reduce the yield and quality of alfalfa herbage. Natural enemies, early harvesting, and limited use of insecticides are effective in control. *G. W. Fick photo.*

Where eggs do not overwinter, damage usually peaks sometime during bud stage, and both yield and quality are reduced (Liu and Fick 1975; Fick and Liu 1976). If harvesting occurs while significant numbers of more mature larvae are present, enough larvae can survive to feed on new shoots so that regrowth is delayed by up to 10 d (Fick 1976). A short pupal stage in early summer is followed by aestivation of adults. The related Egyptian alfalfa weevil (*H. brunneipennis* [Boheman]) has a similar life cycle and causes similar damage to alfalfa in irrigated areas of the western US (Summers et al. 1985).

In the 1960s, alfalfa weevil became so serious in the eastern US that insecticide applications were essential to produce alfalfa. However, a successful IPM program gradually brought the insect under control without primary reliance on insecticides (Grau et al. 1985). All categories of pest control were applied to the solution. 'Weevilchek' and 'Team' were released as moderately resistant alfalfa cultivars in the late 1960s (Hanson and Davis 1972). There was also a major effort launched by USDA to introduce natural enemies of the alfalfa weevil into infested regions. A tiny parasitic wasp without a common name, *Bathyplectes curculionis* (Thoms.), was widely publicized and proved effective, but a whole arsenal of natural enemies now numbering over 15 species is more effective than any single parasite (Armbrust et al. 1985).

Cultural controls are also effective against the alfalfa weevil. In southern states, it is sometimes possible to decrease or delay larval damage in the early spring by decreasing the amount of stubble where eggs are laid the previous fall. Selection of fall dormant alfalfa cultivars (Reid et al. 1989) or late fall harvesting or grazing (Dowdy et al. 1992) can accomplish this. In areas like Wisconsin where the weevil overwinters only as adults, peak damage occurs in the bud stages of development. Harvesting at bud stage just before severe defoliation occurs provides effective control. Field scouting, which is typically based on weather records, usually starts in mid-May, and intervention is recommended at 40% tip-feeding damage (Undersander et al. 1991). If the crop has reached bud stage, it can be harvested in the next few days, usually without any insecticide application being necessary to avoid economic loss. If harvesting must be delayed because alfalfa is vegetative or other work has higher priority, an insecticide can be used to prevent economic losses. If an insecticide is used, it is necessary to delay the harvest for a strictly defined period, which allows the chemical residues to dissipate to safe levels in the harvested feed. As soon as possible after harvest, the stubble should be checked for surviving larvae. If more than 50% of the growing shoots show damage, or if larvae are found but no regrowth has occurred in 3 or 4 d, the stubble should be sprayed as soon as possible.

Computer simulation modeling was used to develop and verify the IPM program for alfalfa weevil (Onstad and Shoemaker 1984). Models showed that early harvesting and natural enemies should be effective most of the time in controlling this pest, but occasional economic losses might occur because of differential effects of weather on the pest and its natural enemies. Thus, scouting is a key aspect of the program to locate both the "hot spots" and to time the early harvests.

POTATO LEAFHOPPER (*Empoasca fabae* [Harris]). The potato leafhopper is native to North America and gains pest status from its ecological traits as well as from the feeding injury it causes. The leafhopper develops by incomplete metamorphosis, has several generations per year, and feeds on many species of plants (Manglitz and Ratcliffe 1988). It is a primary pest of alfalfa from June through August in the north central and northeastern US and adjacent Canada. Although this insect does not survive winter in the north, annual migrations from the southern US ensure the pest is present every year. However, insect populations perform differently each year for a variety of reasons, and thus outbreaks are difficult to predict (Lamp et al. 1991). For alfalfa and other forage legumes, the injury, often called *hopperburn*, is caused by the sucking habit of feeding with a combination of probing and ingestion behaviors that disrupt translocation through vascular tissue (Nielsen et al. 1990). Subsequently, injured plants have reduced rates of photosynthesis and transpiration (Flinn et al. 1990). Damage results in reduced rates of growth and lost forage quality, especially crude protein (Hutchins and Pedigo 1989; Oloumi-Sadeghi et al. 1989).

Pest management of the leafhopper is difficult because of its high mobility, rapid reproduction rate, and the relatively small populations needed to cause serious damage. Arrival dates make new spring seedings especially vulnerable to injury, and whole shoots as well as leaves can be killed on seedling plants. By the time injury is visible, pesticide applica-

tions are no longer economical. As a consequence, control options are applied either to prevent long-term development of damaging populations or to respond to the short-term expectation of economic densities.

Plant damage is often prevented by maintenance of a vigorous forage stand, with proper fertility management and cutting schedules being particularly important (Lamp and Nielsen 1988; Kitchen et al. 1990). Grass-alfalfa mixtures generally have less injury than pure alfalfa (Lamp 1991). Breeding for host-plant resistance to potato leafhopper has not been effective in alfalfa, although it is for some other crops. Because of the unpredictability of outbreaks, scouting programs are widely used. When the locally accepted threshold is exceeded, the options are to apply an insecticide or to cut the field as soon as possible. Thus, the general strategy for managing potato leafhopper is (1) to plant locally adapted cultivars with methods that give vigorous early growth, (2) to apply crop management practices that maintain crop vigor and health, (3) to maintain a summer scouting program that will determine when emergency insecticides are needed, and (4) to harvest the alfalfa crop in the bud stages, thereby reducing buildup of the leafhopper population by removing its food supply. If a crop harvest is approaching, harvesting instead of spraying insecticide is the desirable alternative.

Diseases. Most plant diseases are infectious, i.e., caused by spreading, internal parasitism. Yield and quality reductions are often followed by the death of infected plants, and shortened stand life is a typical consequence of forage diseases.

PHYTOPHTHORA ROOT ROT (*Phytophthora megasperma* Drechs. F. sp. *medicaginis*). Phytophthora root rot of alfalfa is an important disease on soils that are periodically waterlogged (Alva et al. 1986). The disease is found in most of the US and Canada where alfalfa is grown. Infected seedlings appear water soaked and then collapse and die. More mature plants develop gray or brownish lesions on tap and branch roots in saturated soil. An active infection causes a "wet rot," but the disease is often not discovered until after the roots have been severed (Fig. 4.2) and plants, which had appeared healthy, start to wilt as the soil dries out (Stuteville and Erwin 1990).

The primary cultural control for this disease is to select sites with good soil drainage. However, disease problems may develop in

Fig. 4.2. Normal taproot development of alfalfa grown in moderately well-drained soil (*left*) compared with taproots damaged by phytophthora root rot in an adjacent somewhat poorly drained soil. Soil drainage or resistant cultivars provide effective control. *G. W. Fick photo.*

rainy periods even on better-drained soils (Mueller and Fick 1987). Selection of resistant cultivars is thus the other element of the IPM program to control this disease. Disease resistance is now commonly available and reported as an important characteristic of alfalfa cultivars. In fact, multiple resistance is the hallmark of modern alfalfas, with 10 disease and insect resistances now routinely being ascertained for new cultivars as they are released (CASC 1992).

Weeds. Weeds in forage crops are primarily competitors for light, soil nutrients, and water. At the time of forage establishment, weed competition can be so severe that forage seedlings cannot survive. In established stands, weeds may be relatively "passive opportunists," filling in gaps caused by the death of forage plants. However, some species, especially those with vigorous rhizomes like quackgrass (*Elytrigia repens* [L.] Nevski) or stolons like common chickweed (*Stellaria media* [L.] Cyrillo) can become very aggressive and reduce yield, forage quality, and stand life all at once.

YELLOW FOXTAIL (*Setaria glauca* [L.] Beauv.). Yellow foxtail is an annual grass that grows and produces seed in the warmest part of the growing season. It is native to Europe and is now widespread in agricultural areas of the world. It causes serious competition with forage seedings sown in late spring or early summer. In addition, a biotype has become the most serious weed in established stands of alfalfa in the central valley of California (Schoner et al. 1978; Jacobsen and Wallace 1986). To maximize forage quality in that zone, cutting intervals are typically short (less than 30 d), and irrigation is applied right after harvest. Collectively, this provides a moist and open environment for weed establishment. The yellow foxtail biotype adapted to California alfalfa fields is relatively prostrate, thus minimizing defoliation by mowing. Its bristly seed heads are noxious contaminants of the forage because they cause irritation and ulcerations of the mouths of livestock. Thus, hays contaminated with foxtail may be unmarketable.

Because of the seriousness of the foxtail problem, control by herbicides has received considerable attention (Orr 1988). However, cultural control also is possible (Norris and Ayres 1991). Delaying irrigation until 1 or 2 wk following cutting often reduces encroachment by yellow foxtail, especially in the first full year of production. This practice becomes less effective as stands age, probably because alfalfa plants are less vigorous in older stands and foxtail can compete more successfully regardless of when irrigation occurs.

The most effective practice in reducing yellow foxtail invasion of alfalfa in California is to lengthen the cutting interval (Norris and Ayres 1991). An interval of 37 d had less foxtail than an interval of 31 d, and 31 d had less foxtail than an interval of 25 d. However, the producer faces an economic tradeoff based on the goals of maximizing yield (greatest at 37 d), maximizing forage quality (greatest at 25 d), and minimizing yellow foxtail in the forage (realized at 37 d). Economic analysis showed that optimal management was to harvest at a 31-d interval (Norris and Ayres 1991). The IPM program for yellow foxtail control can thus be based on cultural practices related to harvest and irrigation management.

DANDELION (*Taraxacum officinale* Weber in Wigg.). Dandelion is a common, spring-blooming member of the aster family introduced to North America from Europe. It is now distributed worldwide in temperate zones (Lorenzi and Jeffery 1987). It forms a rosette of leaves that can become over 30-cm tall in alfalfa fields in the spring. Its bright yellow flowers followed by white globes of "parachuted" seed heads disfigure vegetative alfalfa fields in early spring, causing conscientious growers to worry that their fields are being taken over by weeds (Fig. 4.3). Ecological studies of dandelion invasion into alfalfa fields show that this weed is mostly opportunistic (Janke 1987). Seeds of dandelion are usually dispersed when alfalfa is at late vegetative and early bud stages of development. If the first harvest of alfalfa is taken at that time, dandelion seedlings can become established. Some seedling dandelions survive in the dense shade of alfalfa, waiting for disease or mismanagement to weaken the forage. When

that happens, dandelions grow quickly to take over the open space and typically appear in abundance.

Except for lower crude protein, dandelions are comparable to alfalfa in forage quality (Marten et al. 1987), but they have a high moisture content and may slow the drying rate of alfalfa hay and increase the probability of rain damage (Doll 1984). By the time dandelions are an obvious problem in alfalfa fields, the stand may already be too thin to respond to herbicides that remove broadleaf weeds and thus reduce competition.

Cultural practices can certainly slow the encroachment and development of dandelion problems. Maintenance of alfalfa vigor by proper fertility and harvest management is effective (Janke 1987). Including a perennial grass in the forage mixture, especially a high-quality, spreading type such as low-alkaloid reed canarygrass (*Phalaris arundinacea* L.), can fill open space and inhibit dandelion invasion (Karsten 1992). High stocking rates with sheep are effective in keeping dandelions out of New Zealand pastures, partly because the weed is intolerant of treading. Biological

Fig. 4.3. Dandelion encroachment was increased in these 3-yr-old alfalfa plots by harvesting the previous year at the early bud stage in the spring (*foreground*). Plots in the background were harvested at late bud to early bloom. *G. W. Fick photo.*

control of dandelion with pathogens is showing promise in lawns of Kentucky bluegrass (*Poa pratensis* L.) and may become an option for forage stands in the future (Riddle et al. 1991).

Many forages in addition to alfalfa provide examples of pest control without the use of pesticides. Proper fertilization of Kentucky bluegrass provides cultural control of white grubs (*Phyllophaga* spp.) (Tashiro 1987). An endophytic fungus (*Acremonium lolii* Latch, Christ. & Sam.) provides biological control of the Argentine stem weevil (*Listronotus bonariensis* [Kuschel]) in perennial ryegrass (*Lolium perenne* L.) (Prestidge and Gallagher 1988). Resistance breeding for several viruses is effective with white clover (*Trifolium repens* L.) in the southern US (Pederson and McLaughlin 1988). Such information can be applied in the development of IPM programs for many forage species.

SUMMARY

Integrated pest management is an approach to pest control that utilizes multiple control methods and encompasses all categories of pests. It is based on ecological principles and understanding of the biology of key organisms in ecosystems that are being managed. It has goals of minimizing immediate and long-term risks to the environment while maximizing economic returns through pest management. For IPM practices to be adopted, they must be compatible with the total management system used on farms and ranches.

Cultural practices that accomplish several management goals simultaneously are especially appropriate, for example, early harvesting to maximize forage quality and to minimize pest damage. Selecting resistant cultivars is effective and can be easily utilized with forages that are regularly resown. However, many forage species are not resown frequently or managed so intensively. Natural biological controls apply in many nearly natural forage systems, and enhanced biological control through deliberate release of natural enemies has been very effective for control of some introduced pests. Pesticides in the context of IPM are viewed as emergency controls. It is now clear that insects, plant pathogens, and weeds develop resistance to pesticides so that routine use not only causes environmental risks but also reduces long-term options in pest management emergencies.

QUESTIONS

1. Why are pesticides initially so appealing to crop producers, and why do they lose their appeal?
2. Why are pesticides sometimes used in integrated pest management?
3. Is integrated pest management possible without pesticides? Why?
4. Explain why there are no pests in natural ecosystems that are not utilized by humans.
5. Give an example of a beneficial organism in each of the biological classes mentioned in Table 4.1.
6. Define and give examples of biological pest control, cultural pest control, and resistance breeding.
7. Why is early harvesting at bud stage preferable to insecticide treatment in the control of several insect pests of alfalfa?
8. Dandelion is sometimes called a "cosmetic pest" in alfalfa. Explain why this is an appropriate or inappropriate term.
9. Some resistance breeding alters the food quality of the host for the pest. Why must this be done with caution with the forage crops?
10. Many diseases of perennial forages shorten stand life. Why does stand life have such a strong influence on the economics of forage production?

REFERENCES

Alva, AK, LE Lanyon, and KT Leath. 1986. Production of alfalfa in Pennsylvania soils of differing wetness. Agron. J. 78:469-73.

Armbrust, EJ, JV Maddox, and MR McGuire. 1985. Controlling alfalfa pests with biological agents. In Integrated Pest Management on Major Agricultural Systems, Tex. Agric. Exp. Stn. MP-1616. College Station, 424-43.

Bottrell, DG. 1979. Integrated Pest Management. Washington, D.C.: Council on Environmental Quality.

Certified Alfalfa Seed Council (CASC). 1992. Alfalfa Varieties. Davis, Calif.: CASC.

Doll, JD. 1984. Are dandelions in forages important and can they be managed? In Proc. Wis. Forage Counc., 80-84.

Dowdy, AK, RC Berberet, JF Stritzke, JL Caddel, and RW McNew. 1992. Late fall harvest, winter grazing, and weed control for reduction of alfalfa weevil (Coleoptera: Curculionidae) populations. J. Econ. Entomol. 85:1946-53.

Fick, GW. 1976. Alfalfa weevil effects on regrowth of alfalfa. Agron. J. 68:809-12.

Fick, GW, and BWY Liu. 1976. Alfalfa weevil effects on root reserves, development rate, and canopy structure of alfalfa. Agron. J. 68:595-99.

Fick, GW, and AG Power. 1992. Pests and integrated control. In Ecosystems of the World—Field Crop Ecosystems. Amsterdam: Elsevier, 59-83.

Flinn, PW, AA Hower, and DP Knievel. 1990. Physiological response of alfalfa to injury by *Empoasca fabae* (Homoptera: Cicadellidae). Environ. Entomol. 19:176-81.

Frisbie, RE, and JW Smith, Jr. 1989. Biologically

intensive integrated pest management: The future. In Progress and Perspectives for the 21st Century. Lanham, Md.: Entomology Society of America, 151-64.

Georghiou, GP. 1986. The magnitude of the resistance problem. In Pesticide Resistance. Washington, D.C.: National Academy Press, 14-43.

Grau, CR, GC Brown, and BC Pass. 1985. Implementing IPM in alfalfa. In Integrated Pest Management on Major Agricultural Systems, Tex. Agric. Exp. Stn. MP-1616. College Station, 604-18.

Hanson, CH, and RL Davis. 1972. Highlights in the United States. In AA Hanson, DK Barnes, and RR Hill Jr. (eds.) Alfalfa Science and Technology, Am. Soc. Agron. Monogr. 15. Madison, Wis., 35-51.

Harlan, JR. 1992. What is a weed? In Crops and Man, 2d ed. Madison, Wis.: American Society of Agronomy, 83-99.

Huffaker, CB (ed.). 1980. New Technology of Pest Control. New York: Wiley.

Hutchins, SH, and LP Pedigo. 1989. Potato leafhopper-induced injury on growth and development of alfalfa. Crop Sci. 29:1005-11.

Jacobsen, SE, and RW Wallace. 1986. Life cycle and population dynamics of yellow foxtail in established alfalfa. In Proc. West. Soc. Weed Sci. 39:179-85.

Janke, RR. 1987. Weed management in established alfalfa based on cultural control of crop-weed interactions. PhD thesis, Cornell Univ., Ithaca, N.Y.

Karsten, HD. 1992. Weed invasion and forage quality of alfalfa (*Medicago sativa* L.) sown with reed canarygrass (*Phalaris arundinacea* L.) and white clover (*Trifolium repens* L.). MS thesis, Cornell Univ., Ithaca, N.Y.

Kitchen, NR, DD Buchholz, and CJ Nelson. 1990. Potassium fertilizer and potato leafhopper effects on alfalfa growth. Agron. J. 82:1069-74.

Lamp, WO. 1991. Reduced *Empoasca fabae* (Homoptera: Cicadellidae) density in oat-alfalfa intercrop systems. Environ. Entomol. 20:118-26.

Lamp, WO, and EJ Armbrust (eds.). 1994. Protection of Perennial Forage Crops. New York: Wiley.

Lamp, WO, and GR Nielsen. 1988. Pest impact on alfalfa potential. In Proc. 18th Natl. Alfalfa Symp., 2-3 Mar., St. Joseph, Mo. Davis, Calif.: Certified Alfalfa Seed Council, 48-57.

Lamp, WO, GR Nielsen, and GP Dively. 1991. Insect pest-induced losses in alfalfa: Patterns in Maryland and implications for management. J. Econ. Entomol. 84:610-18.

Liu, BWY, and GW Fick. 1975. Yield and quality losses due to alfalfa weevil. Agron. J. 67:828-32.

Lorenzi, HJ, and LS Jeffery. 1987. Weeds of the United States and Their Control. New York: Van Nostrand Reinhold.

Manglitz, GR, and RH Ratcliffe. 1988. Insects and mites. In AA Hanson, DK Barnes, and RR Hill, Jr. (eds.), Alfalfa and Alfalfa Improvement, Am. Soc. Agron. Monogr. 29. Madison, Wis., 671-704.

Marten, GC, CC Sheaffer, and DL Wyse. 1987. Forage nutritive value and palatability of perennial weeds. Agron. J. 79:980-86.

Mueller, SC, and GW Fick. 1987. Response of susceptible and resistant alfalfa cultivars to phytophthora root rot in the absence of measurable flooding damage. Agron. J. 79:201-4.

Nielsen, GR, WO Lamp, and GW Stutte. 1990. Potato leafhopper (Homoptera: Cicadellidae) feeding disruption of phloem translocation in alfalfa. J. Econ. Entomol. 83:807-13.

Norris, RF, and D Ayres. 1991. Cutting interval and irrigation timing in alfalfa: Yellow foxtail invasion and economic analysis. Agron. J. 83:552-58.

Oloumi-Sadeghi, H, LR Zavaleta, G Kapusta, and WO Lamp. 1989. Effects of potato leafhopper (Homoptera: Cicadellidae) and weed control on alfalfa yield and quality. J. Econ. Entomol. 82:923-31.

Onstad, DW, and CA Shoemaker. 1984. Management of alfalfa and the alfalfa weevil (*Hypera postica*): An example of systems analysis in forage production. Agric. Syst. 14:1-30.

Orr, JP. 1988. Post-emergence yellow foxtail control in established alfalfa. In Res. Prog. Rep. West. Soc. Weed Sci., 182.

Pederson, GA, and MR McLaughlin. 1988. Performance of southern regional virus resistant (SRVR) white clover germplasm. In Proc. 10th Trifolium Conf. College Station: Texas Agricultural Experiment Station, 57-58.

Pedigo, LP, SH Hutchins, and LG Higley. 1986. Economic injury levels in theory and practice. Annu. Rev. Entomol. 31:341-68.

Prestidge, RA, and RT Gallagher. 1988. Endophyte fungus confers resistance to ryegrass: Argentine stem weevil larval studies. Ecol. Entomol. 13:429-36.

Reid, JL, RC Berberet, and JL Caddel. 1989. Effects of alfalfa dormancy on egg and larval population levels of the alfalfa weevil (Coleoptera: Curculionidae). J. Econ. Entomol. 82:264-69.

Riddle, GE, LL Burpee, and GJ Boland. 1991. Virulence of *Sclerotina sclerotiorum* and *S. minor* on dandelion (*Taraxacum officinale*). Weed Sci. 39:109-18.

Schoner, CA, Jr., RF Norris, and W Chilcote. 1978. Yellow foxtail (*Setaria lutescens*) biotype studies: Growth and morphological characteristics. Weed Sci. 26:632-36.

Stern, VM, RF Smith, R Bosch, and KS Hagen. 1959. The integrated control concept. Hilgardia 29:81-101.

Stuteville, DL, and DC Erwin (eds.). 1990. Compendium of Alfalfa Diseases. 2d ed. St. Paul, Minn.: American Phytopathology Society Press.

Summers, CG, DG Gilchrist, and RF Norris. 1985. Integrated Pest Management for Alfalfa Hay, Univ. Calif. Div. Agric. Nat. Res. Publ. 3312. Oakland, Calif.

Tashiro, H. 1987. Turfgrass Insects of the United States and Canada. Ithaca, N.Y.: Cornell Univ. Press.

Undersander, D, NP Martin, D Cosgrove, KA Kelling, M Schmitt, J Wedberg, R Becker, CR Grau, and JD Doll. 1991. Alfalfa Management Guide. Madison, Wis.: American Society of Agronomy.

Nutrient Management of Forages

5

RONALD F. FOLLETT
Agricultural Research Service, USDA

STANLEY R. WILKINSON
Agricultural Research Service, USDA

FERTILIZATION and improved soil fertility management practices usually improve forage production and increase feed for animal agriculture. Top yields of high-quality forage and a nutrient-sufficient status for forage crops requires that soil fertility and soil pH levels are adequate. Forage crops often are grown on poorer soils, and growers seldom manage them as well as they do their more readily marketable cash crops. "Soil fertility" is the capacity of soil to supply nutrients for plant growth, whereas "soil productivity" is the soil's ability to produce a crop. Productivity is a function of a soil's natural fertility (plus nutrients from fertilizer, manure, or other sources), soil physical properties, climate, management, and other noninherent factors used to produce crops. With good management, maximum use is made of soil fertility.

The productivity potential of a soil will not be reached if even one essential nutrient is "limiting." Nitrogen (N), phosphorus (P), and potassium (K) limit fertility in many soils. Im-

proved mineral nutrition practices have been, are, and will continue to be of great importance for optimum forage and animal production. The objective of this chapter is to describe principles involved in wise fertilization of forages rather than fertilizer practices for specific forage and soil combinations.

PLANT NUTRIENT MANAGEMENT PLANNING

Predicting forage response to fertilization is fundamental to determining fertilizer profitability. Fertilizer requirement is the quantity of nutrients that need to be applied in either inorganic or organic form (in addition to that supplied by the soil) to increase plant growth to the desired or optimum, but not necessarily maximum, level. Fertilizing to achieve maximum production can be unprofitable and may result in environmental degradation. Nutrient management planning is essential for effective pasture and forage production and has three parts. These are (1) Assessment of the nutrient status and needs of the forage system. Components of this assessment include assessment of quantity and form of plant nutrients that are available for the producer to apply, identification of soil type and its suitability for forage production, and cultural management requirements of the forage crop. (2) The establishment of economic input thresholds and possible financial returns. The availability and cost of nutrients are important to establishing the economic input threshold. (3) Strategies for ensuring adequate nutrition of the plant and animal components of the pasture system while protecting soil and water resources. All as-

RONALD F. FOLLETT is Soil Scientist and Research Leader, Soil-Plant-Nutrient Research Unit, ARS, USDA, Fort Collins, Colorado. He earned his BS and MS degrees from Colorado State University and his PhD from Purdue University. Dr. Follett's research career interests include plant mineral nutrition, nutrient management, nitrogen cycling, irrigation and drainage, groundwater quality protection, and soil and crop management systems.

STANLEY R. WILKINSON is Soil Scientist, Southern Piedmont Conservation Research Center, ARS, USDA, Watkinsville, Georgia. He received the MS and PhD degrees from Purdue University. Dr. Wilkinson's research interests have been in soil fertility, mineral cycling, and plant nutrition.

pects of nutrient management need to be defined to achieve an economic and environmentally sustainable crop production system.

Nutrients can be added or returned to soil for crop production through fertilizer, manure, crop or other organic residue, rainfall, and (for N) biological fixation. Inadequate nutrient availability results in decreased crop yields and poor economic returns. Excessive plant nutrient additions and/or poor management may result in nutrient losses into the environment. Agricultural practices affect the fate of plant nutrients already present in the soil and those added or returned to soil by influencing the amounts of plant nutrients harvested, lost to the environment, and/or remaining in the soil at the end of a given time period (Fig. 5.1). Soil supplies 14 of the 17 elements essential for crop growth (nickel [Ni] is recently established as an essential plant element [Brown et al. 1987a, 1987b, 1990]). Nitrogen, P, and K are most commonly deficient. Secondary and micronutrient deficiencies are well documented in some soils, with sulfur (S), zinc (Zn), and boron (B) the most common. Plant essential nutrients must be in adequate supply to maintain maximum economic yields.

Soil and Plant Analyses. Soil and plant nutrient analyses are used to avoid plant nutrient stress and optimize yield or quality. Sampling, sample analysis, interpretation, and formulation of recommendations are the steps involved in determining optimum soil or plant nutrient concentrations. Soil sampling usually involves only the surface 15 cm, but subsoil sampling may be helpful for deep-rooted perennial forages. Soil analyses assess quantity and availability of soil nutrients and potential chemical limitations to plant root growth. Successful soil testing is based upon good correlations between soil analyses and crop response to applied nutrients. Usefulness of plant or soil analyses and interpretation is greatly improved when both are conducted. A plant analysis indicates the current state of mineral nutrient sufficiency. Causes for deficiencies may be revealed by a soil analysis; e.g., deficiency may result from a low soil nutrient level or reduced uptake caused by nutrient interactions, antagonism, or imbalance.

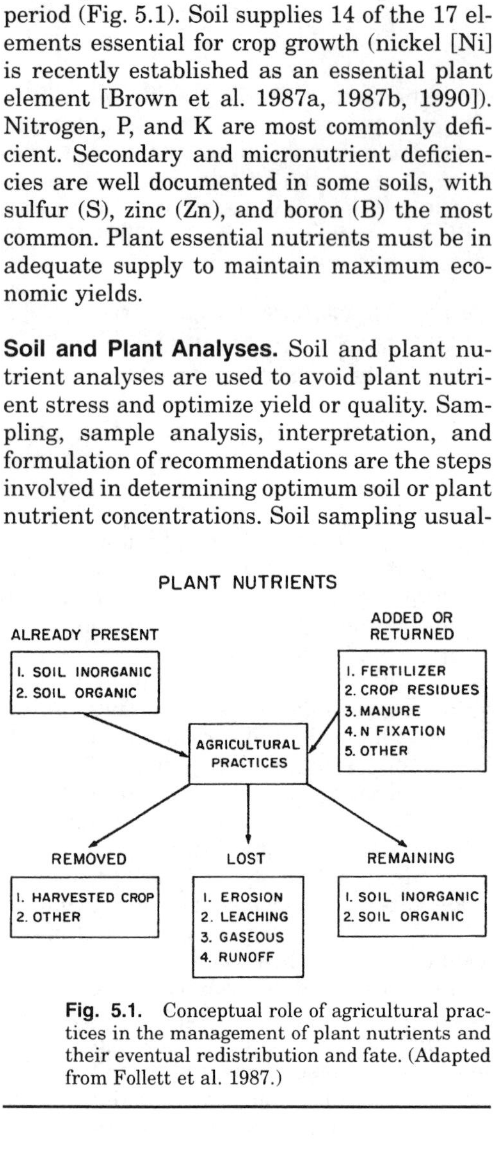

Fertilizer requirement is the amount of a nutrient needed (in addition to that supplied by the soil) to increase plant growth to a desired or optimum level. External nutrient requirements are those concentrations of available nutrients needed in the soil from indigenous sources or applied fertilizer for a given yield level. Fertilizer is defined as "any organic or inorganic material of natural or synthetic origin which is added to a soil to supply certain elements essential for the growth of plants" (SCSA 1982). Internal nutrient requirements are those concentrations needed in plant tissue for a given yield level. Yield levels often are expressed as a percentage of maximum (for that environment). Crop yield and quality responses, as a function of nutrient input or plant concentration, may be separated into four zones: *deficiency; critical nutrient range* (CNR), where near-maximum yields are obtained with minimum amounts of nutrients; *adequacy,* where no further yield increases or decreases occur; and *yield depression,* where yield decreases occur with increasing nutrient concentrations (Fig. 5.2). Yield depression may be caused by nutrient toxicity, imbalance, or antagonism leading to deficiency of another nutrient. Plant and soil analyses help prevent yield and stand losses while identifying optimum fertilization practices to permit the farmer to achieve full potential of the soil, crop, and environment. Such analyses help with the diagnosis of nutrient excesses; they are also increasingly important because of the need to conserve nonrenewable resources and the realization that overfertilization can have negative impacts on the environment.

Another assessment concept is the diagnostic recommendation integration system

PLANT NUTRIENTS

Fig. 5.1. Conceptual role of agricultural practices in the management of plant nutrients and their eventual redistribution and fate. (Adapted from Follett et al. 1987.)

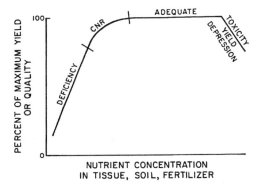

Fig. 5.2. Crop response as a function of nutrient input.

(DRIS), which develops reference nutrient concentrations and their ratios and uses these "norms" to assess adequacy of mineral nutrition in other populations of the same species (Walworth and Sumner 1987). This technique is gaining in usefulness with increased availability of rapid, sophisticated, analytical instrumentation and computers.

Protection of Water and Soil Resources. Wise fertilization requires that water and soil resources are not degraded in achieving maximum forage and livestock production. Excessive use of nutrients can result in ecological imbalances in estuaries, lakes, reservoirs, bays, and slower streams. This process, known as *eutrophication,* occurs when overenrichment by nutrients (primarily N and P) alters the biological composition of ecological communities and stimulates growth of algae. Nitrogen is generally the nutrient limiting algal growth in the saline waters of estuaries and bays. Phosphorus tends to be the limiting factor in lakes, reservoirs, and slower streams. Nutrients may degrade water quality when they are carried by eroding sediments and water runoff into surface waters and are leached into soil and through the plant root zone by water infiltration. If infiltrating water moves laterally and returns to the surface (interflow), its dissolved nutrient load is added to overland flow. Otherwise, continued downward infiltration of water, with its dissolved nutrient load, will eventually reach groundwater.

Nitrogen is the nutrient that is of concern in the contamination of groundwater, primarily as a result of nitrate (NO_3^-) leaching. Nitrate is completely water soluble, is readily leached into the soil profile, can potentially move below the bottom of the root zone, and

may eventually leach into groundwater supplies. The ammonium (NH_4^+) form of N usually does not leach and is strongly adsorbed by soil except for sands and soils having low-retention (cation exchange) capacities. Thus, NH_4^+ can readily be transported by eroding sediments into surface water.

Soil erosion is important to plant nutrient movement into surface waters. Nutrients are sorbed to surfaces of clays and finer sediments or to soil organic matter. Soluble nutrients on or near the soil surface usually move into the soil with infiltrating water during the first part of a storm. Productive soils retain large amounts of nutrients in the upper root zone, especially with an actively growing forage crop. When water-soluble fertilizers are added to soil, P is converted to forms having very low solubility that react with other chemical compounds in the soil or are sorbed to clay, unless the soil is sand or a soil having low-P retention capacity. For example, to reduce runoff of soluble P fertilizer from bahiagrass (*Paspalum notatum* Flugge) pastures in southern Florida required a much lower rate of P application than currently recommended; forage yield or quality were not significantly altered (Rechcigl et al. 1992). For management practices to decrease the effect of soil erosion processes on the production and transport of sediment-associated nutrients, they must directly influence the processes involved. Production of forages and practices that leave crop residues on the soil surface are generally accepted as a means to protect against soil-borne nutrient losses.

Strategies. Optimizing nutrient uptake while minimizing nutrient losses to the environment requires (1) optimum use of the crop's ability to compete with processes whereby nutrients are lost from the soil-plant system and (2) direct lowering of the rate and duration of the loss processes themselves (Follett and Walker 1989). Two key parts of the first approach are to ensure vigorous crop growth and nutrient assimilation capacity and to apply nutrients (especially mobile nutrients such as N) in phase with crop demand. An example of the second approach, primarily for N and P, might include refraining from manure application on top of snow-covered or frozen soil, thereby directly lowering potential runoff losses. In addition, farmers must be advised to select and fertilize for realistic yield goals. Olson (1985) emphasizes that a realistic yield goal would be no more that 10% above recent average yield for a given field or farm. Such a yield goal will still likely be difficult to achieve because of limitations imposed by environmental factors and the farmer's operational skills.

ACIDITY, ALKALINITY, AND SALINITY. Procedures to test for soil acidity, alkalinity, and salinity (salty soil) to optimize crop growth are well established. Soil samples can be analyzed by private or public laboratories to determine possible problems and corrective treatments. In addition, the potential exists to develop forage grasses and legumes that are tolerant to extreme soil acidity, alkalinity, and salinity (Foy and Fleming 1978; McKimmie and Dobrenz 1991; Spain et al. 1975). In areas with heavy rainfall and natural leaching, salinity is usually not a problem; in low-rainfall or irrigated areas, salts applied in irrigation water and from high rates of fertilizers or manure must be limited, especially for forage species of low salt tolerance. Salinity of irrigation water is reported to be more highly correlated to soil-plant-water relations than is soil salinity (Bauder et al. 1992; Devitt 1989); therefore, water management is important. The US Salinity Laboratory recommends high-frequency, low-volume irrigation when irrigating with saline water to maintain proper leaching requirements (Hoffman et al. 1980; USDA 1954). Such a practice, where it can be applied, minimizes oscillations in soil water content, which in turn minimizes wide oscillations in soil salinity; irrigating beyond evapotranspiration (ET) ensures the downward flux of salts from the active root zone (Maas and Hoffman 1977; Martin et al. 1991; Gilbertson et al. 1979).

TOXIC ELEMENTS. The importance of essential nutrient elements for plant and animal nutrition has long been recognized. However, response of either plants or animals to the insult of toxic concentrations of these and other elements must also be understood. To be toxic, an element must be "available" to plants or animals—that is, it must exist in a form that can enter the plant or animal's tissues. In addition to some of the more well-known toxic, but nonessential, elements (i.e., aluminum [Al] to plants and lead [Pb], cadmium [Cd], and mercury [Hg] to animals), some essential elements are also toxic when taken up or consumed in excess amounts. Examples of essential elements that are toxic to plants, when taken up in excess amounts, include B, copper (Cu), manganese (Mn), and Zn. Examples of

elements taken up by plants and essential to animals, but that may be toxic to animals under some conditions, include Cu, Mn, molybdenum (Mo), and selenium (Se) (Allaway 1975; Gough et al. 1979; Underwood 1977).

SOIL PHYSICAL CONDITIONS. Soil physical characteristics are its texture, structure, consistence, aeration, tilth, and moisture (Donahue et al. 1983). The effects of these characteristics upon plant-available water, soil aeration, and physical impediment determine the suitability of a soil for plant growth. A good environment for plant roots is largely determined by the size and distribution of pore spaces among soil particles. A compacted layer within the soil profile, such as a traffic pan, may restrict root growth and access to water and nutrients.

Forage plants can contribute to improved soil physical conditions as well as to soil fertility. Reports of increased cultivated crop yields in rotation with forages are legion. Grasses and sod crops improve soil structure, as do legumes, which can also fix atmospheric N. However, legumes deplete other nutrients, particularly P and K, and this must be recognized to maintain an adequate legume production level and for production of succeeding crops (Griffin 1974).

THE PLANT

Soils anchor plants and provide water and essential mineral nutrients for plant growth. Plants absorb nutrients from the soil solution, which is replenished from exchangeable and nonexchangeable nutrient pools. Soil solution nutrient concentration may be decreased by plant uptake, fixation by clay minerals, precipitation in insoluble forms, and immobilization during microbial processes. Soluble fertilizers are a major source of replenishment of nutrient ions in the soil solution under intensive forage production systems, but a proportion of soluble fertilizers will end up in the absorbed or exchangeable phases. Rate of nutrient replenishment in soil solution is extremely important in determining the best size of the available nutrient pool. Plant nutrients arrive at the absorbing surface of the root by mass flow, diffusion, and extension of roots into new soil volumes. Mass flow of nutrients refers to nutrients contained in the water that moves to the root as a result of transpiration. As a mechanism it is nearly always inadequate to meet P and K needs of plants but may be adequate for N (as NO_3^-), calcium (Ca), magnesium (Mg), S, B, Cu, Zn, iron (Fe), and chlorine (Cl). Quantity is determined from the volume of flow and concentration of nutrients in soil solution. Nutrient diffusion results from the concentration gradient established when a root absorbs nutrients from the soil solution. The quantity of a nutrient diffusing per unit of time depends on the diffusion coefficient of the particular nutrient, concentration gradient between the soil mass and root surface, soil moisture content, porosity and tortuosity of the soil, and total root surface available for absorption. Root extension into unexplored soil areas increases both amount and rate of nutrient supply.

After the nutrient is brought from the environment into the plant (absorption by the roots), it must be moved (translocation) to the site of assimilation in the plant (metabolism). Movement in the plant occurs in the vascular system. Movement can be via the xylem (apoplastic) or the phloem (symplastic). Mobility within the plant varies with elements; i.e., N, P, and K are relatively mobile while Ca, Fe, Zn, and B are not mobile. Magnesium and many of the other elements are intermediate in their mobility within the plant. Biochemical functions of mineral elements fall into the following general categories: (1) structural constituents, enzymatic and oxidation-reduction reactions (carbon [C], hydrogen [H], oxygen [O], N, S); (2) functions concerned primarily with energy transfer and translocation (P and B); (3) functions concerned with enzyme reactions, establishing osmotic potentials, balancing anions, having membrane permeability and integrity (K, sodium [Na], Mg, Ca, Mn, Cl, Ni); and (4) functions concerning enzyme activity and enabling electron transport (Fe, Cu, Zn, Mo). The last group may also be absorbed in the form of chelates from the soil solution (Mengel and Kirkby 1978).

Amount of plant growth for a given environment depends on quantity and balance of growth-determining factors, with the least optimum factor limiting growth the most. This concept is termed the "Principle of Limiting Factors" or "Liebig's Law of the Minimum" and is extremely valuable. However, two additional aspects must be considered. First, Liebig's Law of the Minimum applies to conditions where inflows and outflows of energy, minerals, and other factors are balanced (steady state condition); e.g., where forage growth is limited by N, a sudden increase in available N may remove N as the limiting factor. During the transitional period to a new production level, the next limiting factor or factors may be difficult to identify until a new steady state condition is established. Second, factors interact to modify effects of individual factors. Thus when solar radiation, temperature, and soil water are nonlimiting, fertilizer requirements will be higher than when such factors are limiting. Pasture-, range-, or forage land productivity is controlled primarily by temperature, water (rainfall), soil fertility, and defoliation (grazing) management.

Essential elements for plant growth and their relative abundances in plant tissue are given in Table 5.1. Differences arise in the abilities of various forage plants to extract nutrients from the soil in amounts required because of differences in plant responses to the wide range of soil conditions encountered for forage production. Knowledge of amounts of various nutrient elements removed by a particular forage is helpful in planning an effective fertilizer program. In Table 5.2, information is provided on macronutrients removed by some typical forages. Yield times concentration equals uptake. Yield is usually the most important factor in nutrient removal by forage crops.

Plant Stress Adaptation. The potential exists for improved efficiencies in essential element use and for plant adaptation to a number of plant mineral stress conditions such as Al and Mn toxicities, salinity, Fe deficiency, and B and certain heavy metal toxicities. Thus, a largely untapped potential exists to develop improved forage grasses and legumes tolerant of extreme soil acidity and low-fertility conditions that occur in vast expanses of tropical soils (Spain et al. 1975). Plant adaptation also needs to be developed for other stress conditions such as droughtiness, shallow root zones, fragipans, and tillage pans. Such adaptations may occur by stress avoidance or tolerance mechanisms. For further reading, see Gerloff 1987.

NUTRIENT CYCLING

Essential plant nutrients for pasture production may cycle from soil to plant to animal to the atmosphere and back to the soil. The extent and rate of return of nutrients back to the pool of available soil nutrients greatly affect fertilizer requirements of grazed pastures. Inputs to the cycle occur from fertilizers and manures, the atmosphere (biological N fixation and deposition), soil minerals, and organic matter. Losses may occur through harvest of animal or plant products, transfer of nutrients within the pasture with animal

TABLE 5.1. "Normal" concentrations of chemical elements arranged in order of the relative abundance in plant tissue.

Chemical element	Symbol	Probable form for absorption	Concentration[a] (mg/kg)	Relative abundance[a] (atom basis)
Micronutrients				
Nickel	Ni	Ni^{++}	0.10	1
Molybdenum	Mo	MoO_4^{--}	1.00	1
Selenium[b]	Se	SeO_3^{--}	0.15	1
Cobalt[b]	Co	Co^{++}	0.20	2
Chromium[b]	Cr	Cr^{6+}, Cr^{3+}	0.20	2
Iodine[b]	I	I^-	3.00	14
Chlorine	Cl	Cl^-	50.00	29
Zinc	Zn	Zn^{++}	15.00	135
Copper	Cu	Cu^{++}, Cu^+	15.00	139
Manganese	Mn	Mn^{++}	50.00	535
Iron	Fe	Fe^{++}, Fe^{+++}	100.00	1,053
Boron	B	H_3BO_3	20.00	1,088
Silicon[b]	Si	SiO_4^{4-}	400.00	8,398
Macronutrients				
Sulfur	S	SO_4^{--}	2000.00	36,720
Phosphorus	P	$H_2PO_4^-, HPO_4^{--}$	2500.00	47,461
Magnesium	Mg	Mg^{++}	2500.00	60,547
Calcium	Ca	Ca^{++}	4000.00	60,742
Potassium	K	K^+	20,000.00	300,781
Nitrogen	N	NH_4^+, NO_3^-	25,000.00	1,048,828
Oxygen[c]	O	O_2, H_2O	420,000.00	15,442,383
Carbon[c]	C	CO_2, HCO_3^-	450,000.00	22,041,016
Hydrogen	H	H_2O	55,000.00	32,089,844

[a]General concentrations in plant tissue where nutrient is nonlimiting to growth.
[b]Elements that have not been shown to be essential for plants but are essential for animals whose principal diet is composed of these plants.
[c]Primarily uptake in gaseous form from the air.

TABLE 5.2. Nutrients removed by several forage crops

Crop	Dry matter yield	Nitrogen	Phosphorus	Potassium	Magnesium	Sulfur
			(kg/ha)			
Legumes						
Alfalfa	17,920	505	39	447	45	45
Red clover	8,064	196	26	247	18	. . .
Korean lespedeza	6,720	185	26	154	31	25
Cool-season grasses						
Orchardgrass	13,440	335	49	349	28	39
Smooth bromegrass	11,200	186	32	236	11	22
Timothy	8,960	168	27	233	11	18
Tall fescue	8,960	174	37	195	17	22
Western wheatgrass	1,500	22	2	29	2	3
Crested wheatgrass	1,500	18	2	. . .	. . .	. . .
Warm-season grasses						
Coastal bermudagrass	22,400	560	68	391	50	50
Carpetgrass	3,394	54	9	271	6	. . .
Bluegramma	500	10	1	9	1	. . .
Pensacola bahiagrass	11,200	174	26	112	34	34
Tropical grasses						
Paragrass	26,880	344	48	429	88	50
Napiergrass	28,224	338	72	564	71	84
Guineagrass	25,760	323	49	407	111	52
Pangolagrass	26,880	342	54	409	76	52
Annual crops						
Corn silage	26,880	269	49	279	56	34
Sorghum-sudan	17,920	364	61	442	54	34

Source: From Follett and Wilkinson (1985).

excreta, fixation and precipitation of nutrients in soil, volatilization, leaching, soil erosion, and surface runoff.

Mineral nutrient cycling in a pasture ecosystem is portrayed in Figure 5.3. Essential features of nutrient cycles are soil, plant, animal and atmospheric nutrient pools; rate and quantity of nutrients moving between these pools; and inputs and outputs. Mineral nutrients cycle on global, regional, and pasture ecosystem scales; they also cycle within soil, plant, animal, and atmospheric pools (Haynes and Williams 1993; Russelle 1992; Wilkinson and Lowrey 1973).

The soil compartment for pasture ecosys-

tems includes a labile pool of nutrients available for plant root uptake in dynamic equilibrium with nutrients in residues and in unavailable forms (inorganic and organic). Plant roots absorb nutrients from the available soil pool and translocate them to herbage. Nutrients in herbage consumed by grazing animals are either utilized by the animal or excreted as feces or urine and returned to the soil. When nutrients are released from excreta and herbage residues to the available nutrient pool in the soil, the nutrients have been recycled. Specific information relating to N, P, and K cycling is presented later.

In the real world, effects of energy flow

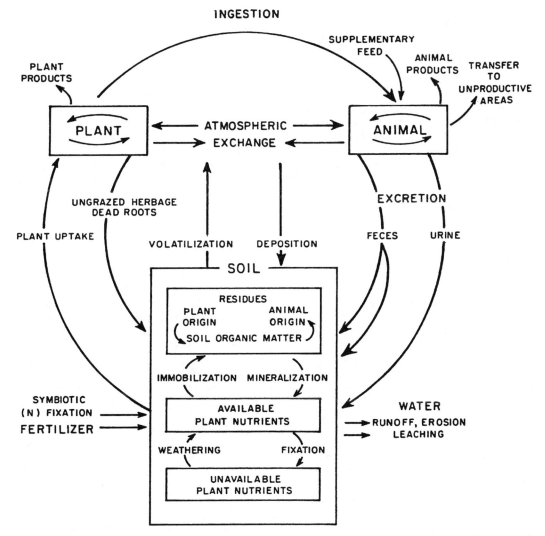

Fig. 5.3. Simplified mineral nutrient cycle for pasture ecosystems. (Modified from Wilkinson and Lowry 1973.)

(temperature, light energy, and potential for biomass accumulation), hydrologic cycles (flows and storage of water), and nutrient cycles are interconnected and interdependent. Climate and weather patterns affect energy flows, water movement and use, and nutrient use and movement. This interconnectedness and interdependence underscore the complexity of pasture ecosystems and confirm that changes in pasture ecosystem management may have unforeseen effects. Holistic (whole) approaches are necessary for appropriate management of pasture ecosystems. Such approaches are required to accurately determine the plant nutrient requirements of the vast number and types of pasture ecosystems that exist or that may be developed.

Role of Soil Organisms. Soil microfauna and microflora have a major role in nutrient cycling. Release of nutrients from plant and animal residue is dependent on microbial activity. Soil bacteria utilize the more readily available soluble or degradable organic substrates. Fungi and actinomycetes decompose resistant materials such as cellulose, hemicellulose, and lignin. Dung beetles (Fincher et al. 1981), earthworms, and other soil fauna increase the decomposition rates of feces and plant litter by mixing them with soil. Presence of vesicular arbuscular mycorrhizae (VAM) in association with roots can effectively increase root surface area, thereby increasing nutrient- and water-scavenging ability. VAM infection of roots is considered more helpful for taprooted forage species such as legumes than for fibrous-rooted species such as grasses. VAM may also be more important for warm-season than for cool-season grasses (Hetrick et al. 1988). Soil microorganism activity is a function of soil temperature and moisture. Microbial and soil fauna activity, with sufficient substrate (food or energy), is much higher in moist subtropical regions than in semiarid temperate regions. Activity of microflora is more likely to be limited by N in grassland and by C in cultivated cropland soils (Schimel 1986). The role and importance of nonsymbiotic and symbiotic N fixation is covered in Chapter 4 of Volume 1. At any one time, soil-microbial biomass contains much of the actively cycling N of the soil and represents a relatively available pool of nutrients capable of rapid turnover (Bristow and Jarvis 1991).

Role of the Grazing Animal. Grazing animals in pasture ecosystems affect primary productivity (plant growth) by defoliation, traffic patterns, herbage fouling, partitioning of ingested nutrients to body weight, feces, and urine, dispersion and redistribution of herbage nutrients in excreta, and nutrient turnover rates. Defoliation by grazing animals prevents senescence of plant tissue, removes nutrients in animal products, changes the nutrient pathways from internal plant recycling or leaf fall to return as feces and urine, increases light penetration into the canopy by partial defoliation, and through selective grazing may promote one species over another, thereby altering the botanical composition. Animal traffic compacts soil, sometimes making its physical characteristics for plant growth less desirable. Fouling of herbage with feces reduces its acceptability for grazing, thereby resulting in increased forage maturity and reduced quality and/or degree of consumption by grazers. Urine does not cause herbage to be unacceptable for grazing. Nutrient turnover rates and microbial activity may be reduced or enhanced by the redistribution of nutrients associated with the mobility of the grazing animal.

PARTITIONING OF INGESTED FORAGE N-P-K TO ANIMAL PRODUCTS. Nutrient balances within animals are determined by measuring nutrient intake in the forage eaten minus that retained in products (milk, liveweight gain, wool) and that excreted in dung and urine. Nutrient retention is greatest in actively growing livestock and least in mature livestock. A general rule of thumb for removal of N in livestock products from a grazed pasture is 27.2 g N/kg liveweight gain (LWG) and 0.6 g N/kg milk produced. Phosphorus and K contents of LWG gain are 6.8 g P/kg and 1.5 g K/kg, and 1.0 g P/kg milk and 1.2 g K/kg milk. Nitrogen retention estimates as a percentage of dietary intake range from approximately 8% for LWG to 20% for high milk-producing cows. Such estimates are only approximate because factors such as level of dietary N and age and type of animal also impact N retention.

A 250-kg steer ingesting 6.0 kg forage containing 3% N/d and gaining 0.8 kg per day may ingest 180 g N/d, retain about 20 g in LWG (12% retention), and excrete the remainder, about 160 g N/d. Nitrogen excretion is in both urine and feces. The proportion of total N excreted in urine increases linearly with increasing N consumption. The relationship appears similar for legume or grass N. Jarvis et al. (1989) found that excretion of N between urine and feces was similar for grass/clover (N conc. = 2.47%) and grass fertilized with 210 kg N/ha (N conc. = 2.40%) at 55%-60% of total N excreted in urine, but when fed grass fertilized with 420 kg N/ha (N conc. = 3.59%), about 74% of the total N was excreted in the urine. Nitrogen excreted in urine is water soluble and immediately available for plant uptake and growth, as is K either from feces or urine.

Phosphorus is excreted mainly in feces. The proportion of total P excreted as organic P is relatively constant over the range of 0.1% to 0.4% P in the diet, while the proportion of inorganic P increases. Therefore, the higher the P in the diet, the greater the concentration of inorganic P in the feces, and the higher its availability for plant growth. Potassium is mainly excreted in urine (50%-90%).

Sulfur excretion patterns in relation to S concentrations of forage eaten are similar to those of N. About 0.11 g S/100 g of feed eaten appears in feces, while for N about 0.8 g N/100 g of feed eaten appears in feces. Thus increasing amounts of S, as sulfate (SO_4^{--}), in the urine, increase the availability of S for plant uptake or leaching. Grazing animals increase recycling rates of S and also accelerate losses of S from the ecosystem by leaching, particularly of urinary forms of S (Nguyen and Goh 1992).

ANIMAL TYPE, BEHAVIOR, AND DISTRIBUTION OF EXCRETA. Animal species, age, size, and sex affect herbage and nutrient retention and the ability to graze closely and selectively (Haynes and Williams 1993). Animal mobility and behavior complicate the return of nutrients in animal residue. Sheep tend to be more gregarious than cattle and enhance localization of excretal nutrient returns. Also, sheep may utilize more of the forage grown.

Animals on range utilize forage to a higher degree near watering points (Arnold and Dudzinski 1978). Greater density of dung in areas near watering and shade points is frequently observed (Haynes and Williams 1993; Wilkinson et al. 1989). Wilkinson et al. (1989) found annual transport of K to areas near watering points equivalent to 59% of the fertilizer K applied when steers were grazing endophyte-infected tall fescue (*Festuca arundinacea* Schreb.). Similar trends in soil profile NO_3^- were also observed. West et al. (1989) documents large accumulations of extractable P and exchangeable K near watering points. Rowarth et al. (1992) found that sheep behavior patterns resulted in P accumulations in relatively flat areas or depletion in steeper parts of hill pastures.

Even without transfer to unproductive areas such as woods, shade, watering points, fence lines, and cow paths, consumption and excretion of nutrients by ruminants results in gathering of nutrients from large areas of the pasture and returning them to smaller areas. This concentrating effect frequently means that nutrients cycled through livestock cannot be used efficiently by forage plants in these smaller areas.

On an annual basis, less than 35% of the pasture area receives excretal N; some areas receive one or more applications (overlapping of excreta). This uneven distribution means some of the pasture will be underfertilized (depletion) and some overfertilized (accumulation). This uneven spatial distribution of excreta on pasture productivity is analogous to an uneven fertilizer distribution on yields. Factors affecting the use efficiency of uneven fertilizer distribution are described by Welch et al. (1964). Uneven distribution of excreta also occurs with rotational stocking, but the magnitude of the unevenness is decreased compared with nonrotational stocking systems. Set-stocked animals may transfer more fertility to stock camps than rotationally grazed animals (Quin 1982). Supplemental feeds fed grazing livestock may add a significant amount of additional nutrients via the excreta and thereby increase nutrient cycling and availability.

Effectiveness of Nutrient Cycling. Potential control points for improving effectiveness of nutrient cycling in meeting pasture nutrient requirements involve ways of (1) increasing available nutrient pool size (gains in the cycling pool), (2) increasing transport rate between component pools (turnover rates), and (3) decreasing the losses of nutrients from the cycling pool of nutrients. Potential management means for controlling nutrient cycling in pasture ecosystems involve the following: soil selection, soil and pasture fertilization, soil management, pasture crop selection and management systems, and animal management systems. Nutrient recoveries are much higher for machine cut and harvested forage than for grazed pastures on a field basis. However, overall nutrient recoveries may be considerably lower when forage is transported and fed to cattle and manure nutrients are improperly recycled. We further discuss the concepts of effectiveness and efficiency later.

Nutrient Mass Balance. Mineral nutrients are neither created nor destroyed but may be added, lost, transformed, or recycled. Nutrient balances are used in connection with nutrient cycles to account for amounts and distribution of nutrients. Nutrient balance is obtained when nutrient input minus nutrient output plus or minus changes in ecosystem nutrient pool sizes are equal to zero. Nutrient balance approaches help us understand that everything is connected to everything else, that everything must go somewhere, and that "there is no such thing as a free lunch"—not even with the help of sophisticated chemical and biological technology. Determining nutrient mass balances is an essential step in evaluating resource quality, resource use efficiency, conservation of nonrenewable resources, and agroecosystem sustainability.

NITROGEN

Forage crops that receive too little N have decreased yield and potentially poor economic returns, while too much N applied in either inorganic or organic forms may result in the accumulation within the soil of mineral forms of N. Plant roots absorb N from soil solution in inorganic forms. In legume and legume-grass pastures, symbiotic biological N fixation may account for large amounts of N in legume growth; with time, large reserves of organic N in residues and soil organic matter can accumulate. Nonsymbiotic (free-living) biological N fixation can contribute toward reduction of N fertilizer requirements but will seldom, if ever, eliminate the need for N fertilizers in improved pastures. Currently, soil tests have not been calibrated to predict amount of N fertilizer required for a given level of forage yield. Consequently, N fertilizer (inorganic or organic) requirements are based on expected yield level, N removal, and experience with specific pasture ecosystems.

The Nitrogen Cycle. Nitrogen accounts for 78% of the atmosphere as elemental N (N_2) and is not directly available for plant uptake and metabolism (Fig. 5.4). The N taken up by plants can originate from indigenous organic or inorganic forms already in the soil. Organic N occurs naturally as part of the soil's organic matter fraction; it can also be added to the soil from manure, symbiotic and nonsymbiotic biological N fixation, plant residues, and other sources and is tranformed to NH_4^+ by the process of ammonification. Inorganic (mineral) forms of N include NH_4^+ and NO_3^-, both readily taken up by crops, and nitrite (NO_2^-), which occurs as an intermediate form during mineralization of NH_4^+ to NO_3^-. Ammonium, a cation, can be sorbed to negatively charged sites on soil colloids, the cation exchange capacity (CEC), incorporated (fixed)

into clay and other complexes, or immobilized back into organic form by soil microbial processes and plant uptake; it can be lost during sediment erosion or, under certain conditions, can volatilize into the atmosphere as ammonia (NH_3). Gaseous NH_3 can return to the soil-plant system by direct uptake into plant leaves or by dissolving in precipitation.

Usually NO_2^- does not accumulate in soils because it is rapidly transformed to NO_3^- or denitrified to gaseous N (N_2), nitrous oxide (N_2O), nitric oxide (NO), or one of the other gaseous nitrogen oxide (NO_x) compounds. Immobilization of NO_2^-, and NO_3^- back to organic forms can also occur through enzymatic activities associated with plant or microbial N uptake and N utilization processes. Nitrate, a water-soluble anion, is not sorbed to the CEC of the soil, is very mobile, and moves readily with percolating water (leaching). Nitrate can also be lost to the atmosphere through denitrification processes. Products of incomplete denitrification (nitrous oxides) may contribute to the greenhouse effect and thinning of the ozone layer.

Nitrogen Models. The type of model that is most appropriate for a specific application depends upon the needs of the model user. An example of user needs might be N management decisions made at the farm, state, or national level. Because of the wide range possible, it is not the intent of this chapter to discuss individual models. However, to be effective, models must provide useful answers for the purpose to which they are applied. For example, soil fertility management and fertilizer recommendations will increasingly involve use of computer models as part of the total technical support system. Such models must, in general, account for the biological, physical, and chemical processes that influence N availability to crops. Difficulties arise because the overall fate of N in soil is determined by complex relationships. Also, soils may be variable, or limited site-specific data for use in computer models may be all that is available. There is the need to account for the balance between the concurrent mineralization and immobilization (MI) processes of the N cycle. The N and lignin content and biomass allocation between roots and tops of plants may affect N mineralization potential (Wedin and Tilman 1990). Inputs of C (from plant and animal residues) impact MI processes for the supply and timing of nutrient release (N, P, S, and others) to crops. Knowledge of these processes and their controls provide the basis for improved manage-

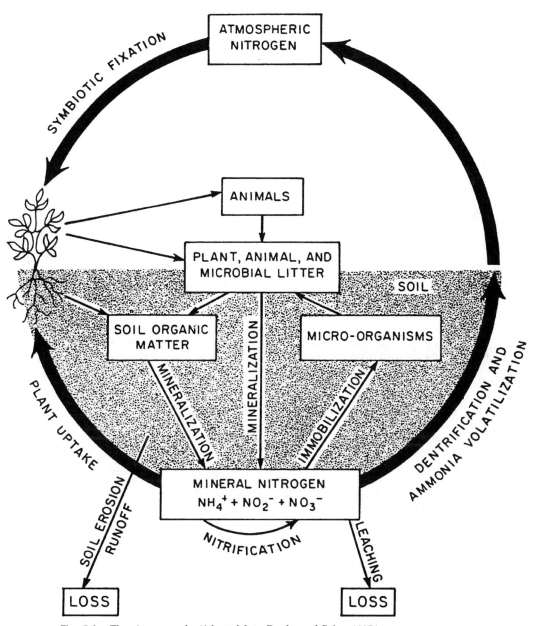

Fig. 5.4.   The nitrogen cycle. (Adapted from Reeder and Sabey 1987.)

ment models and practices to maintain or increase soil productivity for forage production.

Forage Establishment. To ensure establishment and early growth of grass or legume seedlings, knowing the N status of the soil is important. Improved procedures for assessing N status are needed. Small applications of N may be desirable on soils expected to be low in available N. Caution is needed when fertilizing legumes with N because symbiotic N fixation may be depressed. Assessment of N sta-

tus of the soil is difficult, but chemical tests for mineral N, mineralization rate tests, and cropping history are being used.

Maintenance of Forage Production. Quantities of N available from rainfall, nonsymbiotic N fixation, and mineralization of soil organic matter are seldom adequate for optimum production. Well-nodulated legumes usually fix sufficient N to meet their own requirements. Nitrogen fertilizer needed for grasses depends on production goal, yield potential of grass species, absence of other limiting factors, past fertilization, cropping history, effective amounts of precipitation, and length of season. Grasses with C_4 metabolism, such as 'Coastal' bermudagrass (*Cynodon dactylon* [L.] Pers.), appear to be more efficient in N utilization than are C_3 grasses such as orchardgrass (*Dactylis glomerata* L.) (Brown 1978). Differences in yield response reflect this C_4 and C_3 relationship, but longer growing seasons in subtropical and tropical regions also contribute to greater N response in C_4 grasses. Yield response to N fertilization of napiergrass (*Pennisetum purpureum* Schumach.) in Puerto Rico, irrigated Coastal bermudagrass in Texas, orchardgrass in Indiana, and tall fescue in Texas and Wisconsin are shown in Figure 5.5.

Effects of N applied on percentage of maximum yield, N use efficiency, fertilizer N recovery, plant tissue NO_3^- N accumulation, and NO_3^- N accumulation in the root zone are illustrated by data from an ongoing experiment with 'Tifton 44' bermudagrass at Watkinsville, Georgia (unpublished information, S. R. Wilkinson). Use of percentage maximum yield as a function of N application rate permits evaluation of N requirements for different yield goals (Fig. 5.6A). Efficiency of N use in producing forage, expressed as kg of forage produced/kg of N applied, is shown in Figure 5.6B; efficiency of N use increased and then declined with N rate for Tifton 44 bermudagrass. Efficiency of N in producing yield allows the determination of the price:benefit ratios. If 1 kg N costs \$0.50, then one must produce at least 10 kg forage at \$0.05/kg to return the cost of fertilizer.

The percentage of N recovery in yield from fertilizer is determined by dividing the differ-

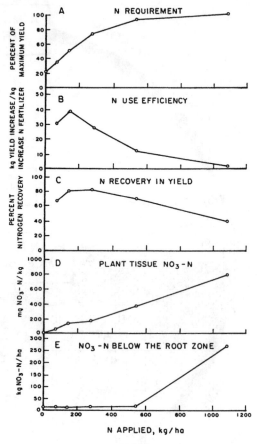

Fig. 5.6. Nitrogen requirement, nitrogen use efficiency, nitrogen recovery, NO_3 nitrogen in plant tissue, and NO_3 nitrogen accumulation below the root zone of Tifton 44 bermudagrass. (Unpublished data, S. R. Wilkinson)

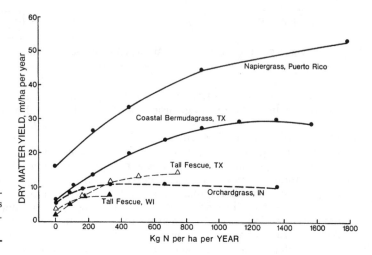

Fig. 5.5. Effect of nitrogen fertilization on annual dry matter yields of cool- and warm-season grasses. (Dougherty and Rhykerd 1985)

ence in N uptake by fertilized and unfertilized grasses by the amount of fertilizer applied (×100). As N applied increases, recovery percentages may initially increase because plant uptake of N increases more rapidly than biological immobilization and loss. With further increases in N fertilization rates, percentage recoveries decline; e.g., N recovery percentages declined above about 300 kg fertilizer N (Fig. 5.6C).

Plant tissue NO_3^- N increases slowly at first and then more rapidly with increasing N fertilizer application rate (Fig. 5.6D). Nitrate accumulation is important in two aspects: first, diagnosis of deficiency, adequacy, or excessive levels in the environment and, second, the avoidance of potentially toxic NO_3 accumulation in plant tissue. Adequacy of N nutrition, judged from NO_3^- content, is situation dependent and beyond the scope of this chapter. However, it appears that for bermudagrass harvested at 4-wk intervals, NO_3^- content of 700 mg N/kg was associated with near-maximum yields.

Fertilizer N not recovered by crop uptake is either retained in the system with potential residual value or lost by leaching below the root zone, surface runoff, volatilization, or denitrification. Figure 5.6E illustrates for Tifton 44 bermudagrass that leaching losses are minor until the higher N application rate is used. Long-term application of N fertilizer at +400 kg N/ha resulted in unacceptable NO_3^- concentrations in shallow groundwater (greater than the EPA drinking water standard of 10 mg NO_3^- N/L) (Dudzinski et al. 1983).

Plant tissue NO_3^- may also be a useful indicator not only of N deficiency but of the presence of excessive NO_3^- in the soil root zone. Excessive NO_3^- in the soil root zone is a prerequisite to leaching or loss by denitrification. Field and modeling research suggests that 50% maximum yield may be the point of greatest N use efficiency and for the long-term may approximate the upper limit of N fertilization, the level where groundwater is not contaminated and where fertilizer N is efficiently used (Overman and Wilkinson 1992). As energy resources become scarcer and more expensive, fertilizer losses become an increasingly important economic factor. The N lost by leaching and surface runoff can pollute water resources, and this should be avoided (Follett and Walker 1989). Nitrogen fertilization should achieve high yields, N use efficiency, and N recoveries in harvested products—and low losses of N to the environment.

Timing and Method of Application. Timing of N fertilization is mainly determined by growth characteristics of the forage, temperature, water (rainfall and irrigation), management (defoliation), and need for forage; need is modified by the capability of the forage species to yield at the time of livestock consumption. Split application will result in better forage yield distribution and will help avoid excessive NO_3^- accumulation in herbage if drought occurs (Overman et al. 1989). However, high rates may not increase yields because the application of N during the highest growth rate period may not be adequate.

Method of application depends on form of fertilizer. Anhydrous NH_3 must be injected into the soil to prevent gaseous N losses. Liquid manures may be either applied to the surface or injected to prevent gaseous N losses. When urea or animal wastes are applied to the surface, and subsequent weather conditions are warm and dry, losses of N by NH_3 volatilization can be very high (Russelle 1992). Such losses are avoided if these N forms are injected or incorporated where moist soil colloids can absorb the NH_3 released by urease enzymes. Penetration of N from surface applications to the plant root zone under normal growing conditions has not been a problem. Forages and grasslands are crops of choice for land application of nitrogenous wastes because of their accessibility and high capacity to assimilate N. Excessive fertilization with N can have negative environmental consequences, thus adding urgency to monitoring N use and estimating an N balance. Wise long-term land husbandry requires that records of past inputs be maintained so that future inputs are beneficial. Timing and controlling method of application are important aspects of the responsible use of N fertilizers (including nitrogenous wastes).

Nitrogen Source. Sources of N fertilizer include urea, 46.0% N; ammonium nitrate, 33.0% N; ammonium sulfate, 21.0% N; calcium nitrate, 15.5% N; sodium nitrate, 16.5% N; anhydrous ammonia, 82.0% N; and N solutions, 27.0%-53.0% N. Effectiveness varies with placement, soil, and environmental conditions. Ammonium nitrate often has the highest recovery efficiency. Losses of NH_3 can occur from surface-applied urea, resulting in a low N use efficiency. Effectiveness of surface-applied liquid fertilizers is also sometimes less than for solid fertilizers such as ammonium nitrate. Anhydrous NH_3 fertilizers are usually the lowest priced per unit of N and have low loss rates if properly injected into the soil. Slow release forms of N fertilizer (urea-formaldehydes) or nitrification inhibitors are sometimes used to avoid excessive N losses and to control rate of N availability to crops. Sometimes their use does not reduce losses so much as delay losses (Owens et. al 1992).

Nitrogen Fertilization and Botanical Composition. Nitrogen fertilization favors the grass component of a mixed grass-legume sward by increasing its competitiveness for light, nutrients, and water. Effects of light competition can be reduced greatly by frequent defoliation. Grasses, with their fibrous root systems, are more effective than most legumes in gathering nutrients and water from the soil. Thus, grasses are more competitive when N supply is not deficient. Legumes will often use available soil mineral N rather than fix N through the bacteria rhizobium. Replacement of legume species with grass species may also have implications for animal health and performance. (Dougherty and Rhykerd 1985). Nitrogen fertilization can also encourage one grass species over another (Belanger et al. 1989; Wilkinson et al. 1968).

Fertilization, Forage Quality, and Plant Composition. Fertilization generally increases nutrient concentrations as depicted in Figure 5.2. Low rates of N fertilization may increase yields but not N concentrations. As rates of N fertilization increase, N absorption by the roots exceeds assimilation of N into dry matter and causes N concentrations to increase. When NO_3^- is absorbed by roots, it must be reduced to NH_4^+ before the N is assimilated into organic (protein N) forms. Nitrate reduction is rate limited by the Fe- and Mo-based enzyme (nitrate reductase) and by the availability of reducing power in the form of reduced nicotinamide adenine dinucleotide (NADH). Nitrate absorbed by roots may be reduced in any plant cell containing cytoplasm, but reduction predominantly occurs in the chloroplasts of leaves and stems because energy and reducing power are available in these structures. When plants are fed NO_3^- in the root medium, the majority of N transported from roots to tops is in the NO_3^- form; while in plants fed solely on NH_4^+, the majority of transported N is in the form of amino acids (Lewis et al. 1982). The photosynthethic capacity of plants is highly correlated with N status, and this relationship appears whether comparing leaf N differences among plant species, leaf age, or varying N availabilities in the root medium during growth. For further discussion of metabolism of plant nutrients, the reader is referred to recent books by Mengel and Pilbeam (1992) and Mengel and Kirkby (1978).

As N concentrations increase in plants, either as a result of availability in the root medium or decreased ability to produce dry matter (assimilate C), the proportion of nonprotein N increases in such forms as amino acids, peptides (protein fragments), or NO_3^-. Free NH_4^+ does not accumulate because it is toxic to plants at relatively low concentrations. On the other hand, NO_3^- can accumulate in cell vacuoles to rather high concentrations without causing toxicity. These nonprotein sources may serve as reserves that can be recycled within plants to support continued growth if N becomes less available in the root medium. Nitrogen fertilization may decrease dry matter content (more succulent herbage) and water-soluble carbohydrates (fructosans) (Dougherty and Rhykerd 1985), or it may result in accumulation of N-rich secondary compounds such as amino acids, amides, and alkaloids (Bush et al. 1979). The effects are modified by plant growth stage and environmental stresses associated with temperature, water availability, and low light. For example, concentrations of the alkaloid perloline are increased when tall fescue growth is restricted by high temperatures and water stress.

Fertilization increases growth and may reduce or delay the number of days to maturity, depending upon species or cultural management. Changes in chemical composition may result from nutrient effects, and these changes in turn affect amount of forage consumed and its digestibility by animals. Generally fertilization with plant nutrients limiting plant growth increases quantity of herbage without greatly improving or adversely affecting the plant's quality for ruminant animal production. Nitrate accumulation to potentially toxic levels, an increase in alkaloid concentrations, and production of herbage having high K/Ca + mg equivalent ratios may be important exceptions (see Chap. 9).

PHOSPHORUS

Phosphorus deficiency occurs worldwide. Sanchez and Salinas (1981) estimate that P deficiency occurs in 82% of tropical soils in the western hemisphere. However, where fertilizer P has been applied in excess of crop removal, soil can accumulate large P reserves. Phosphorus occurs in the plant as inorganic P and in organic forms as enzymes, proteins, nucleic acids, adenosine triphosphate (ATP), lipids, and esters. In photosynthesis, light energy absorbed by chlorophyll reduces nicotinamide adenine dinucleotide phosphate (NADP) and synthesizes ATP. The NADP and ATP serve as energy donors to drive numerous biosynthetic reactions in the plant. Energy-transfer processes involving P include synthesis and degradation of starch and transport of nutrients from the soil through the roots to the plant tops. Phosphorus enhances cell division, fat formation, flowering, fruiting, seed formulation, and development of lateral and fibrous roots; it may improve disease resistance and forage quality.

The Phosphorus Cycle. Soil P supplies are from native inorganic and organic P compounds and return of plant residues, animal wastes, and fertilizers. Plant roots absorb P from the soil solution mainly as $H_2PO_4^-$. Soil solution P concentrations are very low (0.007-1.0 ppm) and are sustained by dissolution of inorganic P in soil and mineralization of organic P from soil organic matter, soil microorganisms, and plant and animal wastes. Solution P concentration depends on soil pH; presence and amounts of soluble or reactive Fe, Al, Mn, Ca, and other minerals in the soil;

form and amount of P in the original soil parent material; and amount and rate of mineralization of organic matter.

The preceding describes a P cycle (Fig. 5.7) in much the same manner as the N cycle is described. A favorable soil pH is essential in maintaining favorable levels of soil solution P; e.g., at soil pH below 6.0, Fe and Al phosphates of low solubility predominate, and the solubility increases until about pH 8.5. Calcium phosphates begin to form at about pH 6.0, and their solubility decreases with increasing pH in the alkaline range (pH > 7.0). The net effect of these changing solubilities is that maximum P solubility generally occurs in the soil pH range from 6.0 to 7.0. The unavailable pool of soil P consists of low-solubility P compounds and insoluble inorganic and organic P compounds. Uptake of solution P by bacteria and fungi is represented in Figure 5.7 as a revolving wheel. This is done to emphasize the central role of the microbial population in P cycling. The uptake of P by microorganisms is stimulated by addition of litter and other organic sources of C and release of both solution P and labile and stable organic P forms as the result of cell lysis or predation (Coleman et al. 1983; Tiessen et al. 1984). An important aspect of P cycling is that P only cycles through biological means because of virtual absence of

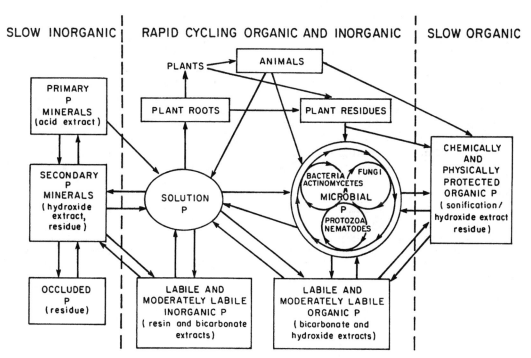

SLOW INORGANIC | RAPID CYCLING ORGANIC AND INORGANIC | SLOW ORGANIC

Fig. 5.7. The soil phosphorus cycle: its components and measurable fractions. (Adapted from Stewart and Sharpley 1987.)

a volatile form of P in most natural or improved forage or grasslands.

Phosphorus Availability in the Soil. Available P is frequently estimated by extraction with dilute acids or dilute solutions of sodium bicarbonate and is usually reported on a relative scale of low, medium, and high (deficient, marginal, and adequate supply, respectively). For maximum usefulness, these analyses must be calibrated to the soil and crop production situation. In general, critical external nutrient P concentration (soil test value) is between 17 and 37 ppm for dilute double acid extractant (humid regions) and 6 and 10 ppm for sodium bicarbonate extractions (less humid regions). Because of low soil solution P concentration, diffusion of P from the soil to the root is an important mechanism for P uptake by plants. Root extension (increasing root surface area) is also important in ensuring adequate P uptake by plants.

Adding P fertilizers to acid soils with high levels of reactive Fe and Al has limited usefulness because of rapid fixation of P. Liming acid soils to increase soil pH usually enhances P availability by reducing P fixation, which may vary among soil types. Fox and Searle (1978) identify soils by order of increasing P sorption as follows: soils high in quartz and organic materials, 2:1-type clays (montmorillonite), 1:1-type clays (kaolinitic), crystalline oxides, and desilicated volcanic ash. Most soils strongly absorb P.

Critical P concentration in forages may range from 0.14% to over 0.3% P, and high levels of available P are required for maximum N fixation by legumes (Follett and Wilkinson 1985). Also, high levels of solution P enhance uptake of both Mg and Ca in winter wheat seedlings (Reinbott and Blevins 1991). The amount of P fertilizer to apply must account for the amount of P supplied from the soil and the efficiency of P uptake from fertilizer, which is often low because of chemical fixation. Chemical fixation of P in soils may be reversible. However, the rate at which P is released from the "fixed" pool to the soil solution cannot yet be satisfactorily predicted. This rate depends on soil characteristics influencing the nature of P products formed during the fixation process. Management programs are needed that permit use of these P reserves without reducing yields.

Microbial Associations. Many genera and species of vesicular-arbuscular mycorrhizae form symbiotic associations with roots of higher plants. Fungal hyphae, in this associa-tion, extend the root system, resulting in a larger root surface area and increased P uptake. Yost and Fox (1979) found, at low P levels in soil solution, that P uptake averaged 25 times greater in mycorrhiza than in nonmycorrhiza-infected plants. Mycorrhizal associations with legumes in low-P soils have increased nodulation and N fixation. Soil microorganisms also can compete with plants for available P and immobilize it. Relatively large quantities of organic P can reside in soils. Under arid or semiarid conditions, slow decomposition of organic matter, including microbial biomass, reduces P availability (Cole et al. 1977).

Phosphorus Sources. Fertilizer P sources generally fall into four categories: (1) superphosphate (9% water-soluble P), made by treating ground rock phosphate with sulfuric acid; (2) triple superphosphate (20% water-soluble P), made by treating rock phosphate with phosphoric acid; (3) ammonium phosphates (7%-23% water-soluble P), made by reacting NH_3 and phosphoric acid; and (4) less soluble forms of P such as basic slag and ground rock phosphates. Finely ground rock phosphates may be effective in acid soils for supplying P to acid tolerant forages (Yost et al. 1982); however, their effectiveness per unit of P supplied is usually less than for normal superphosphate.

Phosphorus Placement. Immobility of P in soil causes concern about the effectiveness of surface applications of P for pastures and no-till forages. However, research with annual grasses and legumes seems to show no evidence of a problem. Mature forage plants with well-developed root systems may more effectively use P from low-P soils than young seedlings. In soils of low P availability, band-seeding and localized placement of fertilizer P just under or to the side of the immediate seedling root zone is desirable to ensure P availability. Legume seedlings are generally less able than grasses to obtain P from broadcast fertilizer applications. Band placement of P is beneficial for establishment of early vigor and health of forage grasses and legumes and may be more beneficial on early spring than on fall seedlings. Soil water affects plant response to P fertilization by increasing P availability for plant root absorption and by increasing root growth and extension.

Phosphorus Runoff and the Environment. Eutrophication of surface and coastal waters and estuaries leads to problems with their use for fisheries, recreation, industry, or drinking due to the increased growth of undesirable algae and aquatic weeds. Although N, C, and P are associated with accelerated eutrophication, most attention has focused on P. Phosphorus is often the limiting element in fresh water. Therefore, controlling its entry into surface water is of prime importance. A stream draining a watershed may receive unwanted P. The P received into a stream can result from an agricultural "point" source such as a confined livestock operation. Tremendous progress has been made in understanding and controlling agricultural point sources of nutrients, including development of guidelines and effective practices for utilizing animal wastes on crop- and pasturelands (Gilbertson et al. 1979).

There is little evidence for nutrient imbalance in forage plants until high P levels are reached; however, pollution hazards exist from erosion of surface soil or plant materials (Rechcigl et al. 1992) or on sandy soils from leaching losses. Factors determining loss of bioavailable P in agricultural runoff are described by Sharpley and Smith (1993). Bioavailable P is potentially available algae. "Nonpoint" sources of P to surface water and groundwater from agriculture may include cropland, pasture, forest, or rangeland. Although amount of direct P runoff from agricultural land is influenced by management, rainfall amount and timing and soil properties are important. Soils of low runoff potential have high infiltration rates even when wet. They often consist of deep, well to excessively drained sands or gravels. In contrast, soils with high runoff potential have one or more of the following properties: very slow infiltration rates when thoroughly wetted and containing high clay content, possibly of high swelling potential; high water tables; a claypan or clay layer at or near the surface; or shallow composition over nearly impervious material. A combination of soil conditions of high runoff potential with high precipitation is especially conducive to surface runoff. Probability of such occurrences increases in the southeastern US as opposed to portions of the more arid West and Great Plains (Stewart et al. 1976; USDA 1989).

POTASSIUM

The primary role of K in plant nutrition is metabolic. Although not part of any particular plant constituent, K is vital to many plant functions including formation and translocation of sugars and starch within the plant, protein synthesis, stomatal action, and the cation associated with organic anions (see Chap. 4, Vol. 1).

Soil K is found in three general forms: (1) soluble K, free to move with soil water; (2) exchangeable K, held on the soil colloids in equilibrium with soluble K and therefore considered available; and (3) nonexchangeable K, held within the clay lattice or in primary minerals and not readily available to plants. Soluble K and exchangeable K constitute a very low percentage of the total K content of most soils, with nonexchangeable K making up the large proportion. Exceptions are high-organic and acid-sandy soils, which are quite low in total K and contain little nonexchangeable K. Rate of exchange from nonexchangeable to exchangeable form determines the long-term K-supplying power of a soil.

A limited amount of total soil K is available annually from most soils. This amount may be adequate for relatively low-yielding forages without K additions, yet added K may be required as soon as yield potential is increased through introduction of more productive plants or addition of other fertilizers. Soils with low exchange capacities and low release rates of residual and nonexchangeable K require frequent, relatively small applications of K (i.e., sandy soils). Soils with high exchange capacity and high release rates of residual and nonexchangeable K can be fertilized less frequently.

Soluble soil K may be lost through leaching, which is important in use of K fertilizers, particularly on sandy soils. However, leaching losses of K are not high from soils supporting active plant growth and likely do not exceed 4-5 kg/ha/yr.

Potassium and the Plant. Demand for K by young seedlings is not large. Thus, on soils with good K-supplying power, a small amount of fertilizer banded or worked into the soil at seeding time is sufficient. However, band placement of K, if placed too close to the seed, may cause seedling injury. Potassium supply is more important to established stands. Highly productive, well-established forages remove 300-500 kg K/ha/yr (Table 5.2). Supply of K is especially critical for legumes used for hay because of the large quantities of K removed with harvest. Some soils may supply adequate amounts of K for seedling establishment but require substantial annual additions of K to maintain productivity.

LUXURY CONSUMPTION. The tendency for plants to absorb amounts of elements far in excess of actual requirements is called *luxury* *consumption* and can be a problem with K. For example, plant levels of around 2%-3% K are usually adequate for forages such as alfalfa (*Medicago sativa* L.), ladino clover (*Trifolium repens* L.), orchardgrass, and smooth bromegrass (*Bromus inermis* Leyss.), yet it is not unusual to find plants levels of 3.5%-4.5% K. The extra 1%-2% K is likely luxury consumption by the plant. Fertilization of established stands with K should bring soil test values to medium to high levels, depending on yield and desired productivity level and whether the stand is a grass-legume mixture or only grass.

COMPETITION. Requirements for K by grasses and legumes, adapted to the same environment, are usually about the same; however, grasses can frequently extract K from the soil more readily than the legumes growing with them. This difference becomes more pronounced as the K supply decreases. Therefore, higher levels of soil K may be needed for mixtures than for monoculture legume or grass stands. For example, in Table 5.3 K contents of grass and legume components of an alfalfa-orchardgrass mixture are compared under varying levels of applied K. Where K supply was plentiful, legume K content was about equal to that of the grass. However, as supply was reduced, legume K was deficient while grass K was still adequate. Thus, the problem in fertilizing legume-grass associations is that of applying enough K to maintain the health of the legume without unduly encouraging luxury consumption by companion grasses.

Maintenance of legume-grass mixtures requires both K and P. Legume yields are generally greater with a combined K and P topdressing than with only K. Fertilizing a legume-grass mixture with adequate K and P results in greater yields, higher-quality forage, and higher animal performance. Table 5.4 shows the effect of K in producing higher yields of four legume-grass mixtures. For mixtures, split applications of fertilizer, generally in the fall and after first grazing in the spring,

improve yield, the legume component, and quality (Griffith 1973). Fall application of fertilizer helps prevent heaving and legume stand losses over winter, and vigorous new tillers are developed by perennial grasses during the autumn-winter months.

TABLE 5.4. Effect of fertilizer potassium (K) on 3-yr average yield and botanical composition of four legume-grass mixtures

Mixture	Fertilizer rate	Forage yield	% legumes
		(kg/ha)	
Alfalfa-bromegrass	0	4461	73
	279	9146	90
Alfalfa-orchardgrass	0	4595	27
	279	6254	46
Ladino-orchardgrass	0	4192	7
	279	6792	12
Ladino-tall fescue	0	4573	9
	279	5963	28

Source: Griffith (1973).

Potassium Fixation. Most soil K occurs in minerals such as feldspars or mica and is unavailable until released by "weathering," a process where the outer surfaces are opened and linkages broken to make K available. The movement of K ions from the soil solution into montmorillonite, an expanding lattice clay, is called *K fixation*. The most important factor in K fixation is the type of clay mineral in the soil (Thompson and Troeh 1978). Application of fertilizer K can result in fixation. Fixed K is stored, rather than lost from the soil-plant system, and is more accessible for plant uptake than is K from unweathered minerals. However, fixed K is tightly held and may remain in place for years.

Potassium Sources. All K fertilizer salts (Table 5.5) are water soluble and have little or no effect on soil pH. Potassium chloride (40%-52% K) is an important source of K. Potassium sulfate (41.5% K) provides both K and S. Potassium magnesium sulfate (19%-25% K) is considered a highly desirable material where Mg is likely to be deficient. Potassium nitrate (37% K) is an excellent fertilizer that supplies both K and N; however, cost of production limits its general use.

LIME, CALCIUM, AND MAGNESIUM

Soil Acidity and Plant Growth. Soil acidity is detrimental to plant growth because of (1) increased solubility of toxic elements, (2) decreased availability of essential nutrients,

TABLE 5.3. Percentage potassium (K) in alfalfa and orchardgrass grown together

K (kg/ha)	K (% dry weight) Orchardgrass	K (% dry weight) Alfalfa	% K in alfalfa relative to grass
0.0	2.71	0.70	26
46.5	3.46	1.21	35
93.0	4.01	1.78	42
372.0	3.85	3.53	92

Source: Blaser and Kimbrough (1968).

TABLE 5.5. Fertilizer sources of potassium (K)

Name of fertilizer	Chemical formula	Grade and basis	
		Oxide N-P$_2$O$_5$-K$_2$O	Elemental N-P-K
Potassium chloride (muriate of potash)	KCl	0-0-60	0-0-49.8
Potassium sulfate	K$_2$SO$_4$	0-0-50	0-0-18.3 +17.6% S
Potassium-magnesium-sulfate	K$_2$SO$_4$-2MgSO$_4$	0-0-22	0-0-18.3 +22.7% S and 11.2% Mg
Potassium nitrate	KNO$_3$	13-0-44	13-0-36.5

and (3) repressed activity of desirable soil microorganisms. These relationships are extremely complex. For example, a combination of Al and Mn toxicity, Ca deficiency, Mo deficiency, and repressed *Rhizobium* activity for a legume forage grown on a strongly acid soil could simultaneously occur. The potential for Al toxicity is always present at low pH because of the Al in the clay mineral fraction of the soil. However, soils vary widely in Mn content; thus Mn toxicity may or may not occur.

ESSENTIAL ELEMENT AVAILABILITY. Lime has many beneficial effects on acid soils (Nelson 1980) including (1) Root expansion and penetration, water and nutrient use, crop yield, and increased net return are increased. (2) Soluble Al and Mn are decreased. Root growth and nutrient absorption are improved. High Al inhibits absorption of elements such as K, Ca, and Mg. (3) Cation exchange capacity (CEC) is increased by 2 to 3 meq 100/g in some soils due to pH-dependent exchange sites. Thus, potential K leaching on sandy soils may be decreased. (4) Availability of such elements as P, Mo, Mg, Ca, and B is improved. Soil pH plays a dominant role in P availability. Levels of available Mo are increased by liming while levels of available Cu, Fe, Mn, and Zn are decreased. (5) Rate of decomposition of organic matter is increased as is the rate of N release.

DESIRABLE SOIL MICROORGANISMS. Soil acidity generally results in low rates of mineralization of organic N by soil microorganisms and decreased capacity of *Rhizobium* for effective legume nodulation and N fixation. *Rhizobium* populations usually increase greatly after a soil is limed (Pearson and Hoveland 1974), thus favoring early and effective nodulation.

LIME REQUIREMENT AND SOURCES. Liming rates need to be based upon laboratory determinations for the location and soil in question (Follett and Follett 1980, 1983). The term

lime requirement (LR), although somewhat ambiguous, is simply the amount of lime that must be applied to the soil to grow a certain crop. It is most often considered to be the amount of CaCO$_3$, or its equivalent, that must be applied to increase soil pH to 6.5 or to some other desired value. Also, because of differences in soils, pH or LR measurements are primarily a diagnostic tool for predicting the need for lime; there is no single pH level below which plants can be expected to benefit from liming on all soils (Foy and Fleming 1978). The LR is affected by soil acidity and several other factors, including soil texture, type of clay mineral, amount of acidity, extractable Al, cation exchange capacity (CEC), and amount of organic matter. These factors influence the capacity of a soil to remain at a relatively constant pH level, known as its *buffering capacity*.

Fertilizers containing NH$_3$ reduce soil pH. Urea also hydrolyzes to NH$_4^+$ to acidify soils. Areas with permanent vegetation, such as those used for growing N-fertilized forages, are subject to soil acidification. In fact, the top few centimeters of soil in such areas may become acid more rapidly than in plowed fields because plowing returns lime back to the soil surface. Acidity in the top few centimeters can be detrimental to seedling establishment (Thompson and Troeh 1978). Lime can be applied and left on the soil surface when necessary for permanent vegetation. Reaction rates and effective downward movement of Ca and Mg are improved when a finer particle-sized liming material is used (Allen and Hosner 1991). The reaction rate of surface-applied lime is slower than when it is mixed with the soil. Thus, lower, but more frequent, lime applications may be required than for cultivated fields.

LEGUMES AND LEGUME-GRASS MIXTURES. Liming has long been recognized as needed for good forage production, particularly for legumes as discussed by Follett and Wilkinson (1985). Red (*T. pratense* L.) and white clovers (*T. repens* L.) and grasses are more tolerant of soil acidity than alfalfa. Satisfactory alfalfa production generally requires a pH of 6.5-7.0 (Table 5.6), whereas red and white clover growth is satisfactory at a pH of 6.0. In addition, an important effect of liming, coupled with proper fertilization, is that legume species can be added to pastures to obtain and maintain desirable botanical compositions. Tall fescue is tolerant of a soil pH range of 4.6-4.7, although soil pH should be maintained between 6.0 and 7.0 (Wilkinson and Mays 1979).

SUBSOIL ACIDITY. Besides causing topsoil acidity, N fertilizers can result in serious subsoil acidity where incorporation of limestone is not practical. Subsoil acidity limits full root zone exploitation by plants for water and nutrients. High N fertilizer requirements of forage grasses for high-yield management can result in serious subsoil acidity unless adequate lime is used concurrently with fertilization. Lime applied near the soil surface at the time of acidification from N fertilizers results in formation of $Ca(NO_3)_2$ and $Mg(NO_3)_2$, both residually basic. Leaching of Ca and Mg into subsoil horizons as NO_3^- salts is a practical means to help correct subsoil acidity and to increase the vertical distribution of exchangeable Ca and Mg in subsoil layers.

Calcium. Where needed to correct acid soil conditions, liming also provides Ca as a plant nutrient. Concentration of Ca in plant tissues ranges from 0.2 to several percent. These values are well in excess of minimal metabolic requirements, which have been shown to be only 2 ppm Ca in nutrient solution (Wallace et al. 1966). Calcium is commonly the major cation of the middle lamella of cell walls with Ca pectate as a principal cell wall constituent. Calcium is required for normal development of growing points in plants. General disorganization of cells and tissues with Ca deficiency suggests that Ca promotes membrane functions and most likely maintains membrane and intracellular organization (see Chap. 4, Vol. 1). Calcium may alleviate potentially toxic element effects in plants and animals, since their absorption is often inversely related to environmental Ca concentration.

Humid region soils are generally low in Ca while soils of dry areas are frequently rich in Ca. Genetic characteristics of plants help control Ca movement from the soil into the plant. Certain plant species always accumulate high concentrations of Ca and others low concentrations. Thus, Ca in livestock diets depends more on plant species in the diet than on supply of available soil Ca.

Magnesium. Magnesium is a constituent of chlorophyll, activates several enzymes, and has a major effect upon metabolism. Besides its role in chlorophyll and chloroplasts, Mg is contained in plastids and is a cofactor of nearly all enzymes that act on phosphorylated substrates; it therefore serves a major role in energy metabolism (see Chap. 4, Vol. 1).

Because of the importance of Mg to plant nutrition, crop quality, and animal health, dolomitic limestone (containing Mg carbonate) rather than calcic limestone (mostly Ca carbonate) is often preferred. After two or three rotations, soils treated with highly calcic limestone may develop Mg deficiency even with a soil pH near 7.0.

As with Ca, accumulation of Mg from the soil by plants is strongly affected by plant species. Legumes usually contain more Mg than grasses and other nonlegumes regardless of the level of available soil Mg. High levels of available soil K, heavy application of animal manure, or large application of fertilizer K may interfere with Mg uptake by plants. Magnesium fertilizers and dolomitic limestone are often effective for increasing crop yields and Mg concentrations in herbage on sandy or loamy soils but can be less effective on clayey soils, especially those having high levels of exchangeable K. See Wilkinson et al. 1990 for a fuller discussion of Mg in plants.

TABLE 5.6.   Effect of soil pH on seeding success, yield response, and stand survival for alfalfa

Soil pH	Crown counts		Survival	Average yield (2 yr)
	8/79	6/81		
	(numbers/m²)		(%)	(kg/ha)
4.9	52.74	1.08	2.0	2105.6
5.2	95.80	4.30	4.5	2307.2
5.5	119.48	11.84	9.9	3068.8
6.0	118.40	45.21	38.2	5376.0
6.9	128.09	35.52	27.7	5868.8
7.1	118.40	39.83	33.6	5891.2
LSD 0.05	40.90	8.61		380.8

Calcium and/or Magnesium Sources. Liming adjusts soil pH to a level where most plant toxic elements are rendered nontoxic and availability of most essential elements for plants is near maximum. Soil acidity develops gradually in humid regions as abundant precipitation percolates through soil and leaches Ca, Mg, and other basic cations from the topsoil. Increasingly, soil acidity is developing as a result of use of residually acidic forms of N fertilizer, especially NH_4^+ forms such as urea and ammonium sulfate. Neutralizing acid soils requires liming materials that have effective Ca and Mg compounds.

Limestone in nature is always impure calcite, impure dolomite, or a mixture of the two. Impurities are usually sand, silt, clay, and Fe. Effectiveness of a liming material in correcting acidity is determined by its neutralizing value or power. Pure calcium carbonate has a neutralizing power of 100; other liming materials can be compared on a percentage basis with pure calcium carbonate (Table 5.7).

SULFUR

The essentiality of S has been known for over 100 yr. However, until recently, scant attention was paid to S in plant nutrition be-

cause use of S-containing N and P fertilizers usually supplied adequate S. As higher-analysis fertilizers containing less or no S per unit of N or P became available, S deficiencies began appearing and have increased the need for S fertilization in many areas of North America and other parts of the world. However, S deficiencies are more or less offset in those regions where acid deposition is high (CAST 1984; Follett and Wilkinson 1985; Whitehead and Jones 1978).

Plant Requirements. All plant proteins apparently contain S amino acids (cystine, cysteine, and methionine). Deficiency of S affects forage yield and quality through the synthesis of these amino acids and through other biochemical processes including the formation of chlorophyll; synthesis of certain vitamins such as biotin, thiamine, and vitamin B; formation of disulfide linkages associated with the structural characteristics of protoplasm; and formation of ferrodoxin and a ferrodoxin-like compound that is involved in symbiotic N fixation (Follett and Wilkinson 1985; Stewart and Porter 1969).

Sulfur in Soils. In a manner similar to that

TABLE 5.7   Common forms of lime

Common name	Other names	Characteristics	Neutralizing power
			(%)
Limestone (calcic)	. . .	Almost entirely calcium carbonate	75-95
Limestone (dolomitic)	. . .	Contains up to 50% magnesium carbonate	95->100
Limestone (dolomite)	. . .	Almost entirely magnesium carbonate	100-120
Burned lime	Quicklime	Fast-acting, more expensive than limestone, caustic	150-185
Slaked lime	Hydrated or caustic lime	Fast-acting, more expensive than limestone, caustic	125-145
Marl	. . .	Soft, earthy material; localized use	90-95
Industrial lime	Includes many types and sources	Variable in composition; may contain toxic impurities	Highly variable
Wood ash	. . .	Hardwoods contain about one-third more calcium than do softwoods	30-70
Coal ash	. . .	May improve soil physical condition	Very little liming value
Ground or burned mollusk shells	. . .	Good liming material, localized use	Up to 95

Source: Schwartz and Follett (1979).

discussed for P, a conceptual S cycle is presented in Figure 5.8. The conceptual flow diagram describes the main form of S in soil, the main pathways of transformation, and the main sets of boundaries to the cycle (Stewart and Sharpley 1987). Total S content in soils varies widely; it is found as a variable mixture of primary minerals, soil solution sulfate (SO_4^{--}) ions, adsorbed SO_4^{--}, organically bound ester SO_4^{--}, and organic S compounds. Soil S exists in a wide variety of forms and oxidation states. Soil S is derived mainly from the weathering of rocks. With the weathering of primary minerals, sulfides are released and oxidized to SO_4^{--}. The SO_4^{--} is either organic material incorporated into soil by plants or organisms or relatively insoluble inorganic material.

Within the soil-plant system, most S is in organic combination but with several notable exceptions such as in saline and acid sulfate soils. Tropical and temperate soils differ in S content; in tropical soils adsorbed S is generally higher, organic S and ester SO_4^{--} are generally lower, and environmental conditions are generally conducive to rapid mineralization of organic matter and high S turnover rates (IFDC 1979). These generalizations are supported by research from Alaska where Laughlin et al. (1981) report that most S occurs in organic form, with the organic matter decomposition rate determining the rate of

conversion of organic S into plant-available SO_4^{--}.

Use of soil analyses to predict S deficiency has met with limited success (Follett and Wilkinson 1985; Murphy 1978; Laughlin et al. 1981) because of the variable mixture of S forms and their relative plant availability as well as the influence that soil properties have on S availability. Where sufficient information is known about soils and their S responses, soil testing is a useful guide to S fertilization for pastures dominated by shallow-rooted species. Plant analyses are desirable for deeper-rooted species. Deeper soil sampling and soil analysis may be impractical where plant roots are deeper that 1 m.

Nitrogen:Sulfur Ratios. Amount of S required by plants can be related to requirements for N and P. Plants generally require about as much S as P. The required S (% basis) for adequate plant nutrition depends upon stage of growth, whereas the %N to %S ratio (N:S) is more nearly constant at all stages of growth (Pumphrey and Moore 1965). The N:S ratio is, therefore, more useful to indicate S adequacy. When S is more limiting than N to plant growth, nearly all S is in proteins. Abundant nonprotein S can occur when N or other growth factors are limiting. With severe S deficiencies, applied S increases S content, dry matter yields, and N content (Wilkinson and

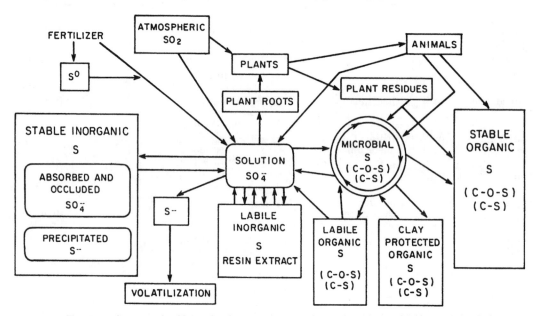

Fig. 5.8. Conceptual sulfur cycle: the main forms and transformations of sulfur in the soil-plant system. Both soil inorganic and organic sulfur are divided into labile and stable forms. *C-O-S* refers to ester sulfates, and *C-S* refers to carbon-bonded sulfur. (Stewart and Sharpley 1987)

Mays 1979). To ensure maximum production, according to Stewart and Porter (1969), an N:S ratio of 12:1 to 15:1 in plant tissue is needed. Generally, literature values indicate that the N:S ratio for optimum plant growth ranges from 14:1 to 16:1 when N content of the plant is adequate (Follett and Wilkinson 1985; IFDC 1979; Wilkinson and Mays 1979).

Sulfur Sources. Major sources of S fertilizer have been single superphosphate and sulfate of ammonia used primarily for supply N and P; S is applied incidentally. Gypsum and elemental S are two major sources of S applied to correct recognized S deficiencies. Soluble SO_4^{--} sources are immediately available to plants. Elemental S added to soil must undergo oxidation to form plant-available SO_4^{--}, a process causing increased soil acidity. In addition to fertilizers, gypsum, and elemental S, atmospheric deposition and waste products can also be important sources of S for forages. However, these latter sources may introduce environmental problems under some circumstances. A brief discussion of acid deposition is given in the following. Additionally, a discussion of two waste materials is included since agricultural lands are often considered as potential sites for their disposal. If the quality of the waste material is suitable, agricultural land disposal can be beneficial by helping to correct either nutrient deficiencies or low soil pH. However, a complete chemical analysis and quality control are important to prevent the introduction of harmful or toxic elements into the soil-plant-animal system.

Sulfur and the Environment

ACID DEPOSITION. The total addition of acids and acid-forming substances in all forms of precipitation as well as the dry deposition that occurs when plants, soils, and waters absorb acids and acid-forming substances from the atmosphere are referred to as *acid deposition*. The ill effects of acid deposition and polluted air on soils and vegetation near some sources that emit large amounts of sulfur dioxide are well-known. However, S and N that occur in acid deposition can be important to forages because both are needed for the synthesis of proteins and for building cells. Sulfur is the mineral nutrient added in greatest amount in acid deposition. Sulfur deficiencies in plants have been identified in 32 states, but none of them are in the Northeast (CAST 1984), where annual deposition of S from precipitation alone ranges from 10 to 12 kg S/ha. The corresponding rates from precip-

itation range from 2 to 5 kg S/ha for the remainder of the US. Importance of this source of S is illustrated by reports that up to one-half of the S required for maximum crop production on many of the soils of the Tennessee River Valley states is derived from atmospheric sources (CAST 1984).

FLUIDIZED BED COMBUSTION RESIDUE (FBCR). FBCR is a product of an industrial process by which S in high-S coals is removed by controlled combustion with calcitic or dolomitic limestones. The FBCR contains significant quantities of Ca (and/or Mg), sulfates, and oxides but may also contain certain heavy metals such as Pb, Cd, and Hg. The chemical composition of the FBCR and a limestone used as part of an experiment by Whitsel et al. (1988) is given in Table 5.8. Feeding trials conducted with rats (*Rattus rattus*) to examine the effect of soil application, or dietary inclusion, of FBCR versus lime on the composition and quality of foods indicated no major detrimental effects relating to the use of the FBCR as a soil amendment (Cahill et al. 1988). However, a similar trial using pigs (*Sus scrofa domestica*) showed a decrease in body weight gain; otherwise no other adverse effects on FBCR application to soils were apparent (Whitsel et al. 1988). Further studies, where FBCR was applied to hill pastures grazed with ewes through three lambing seasons, indicated no adverse effects on herbage growth or composition or on the health or productivity of animals consuming herbage produced on these FBCR-treated soils (Vona et al. 1992).

TABLE 5.8 Elemental composition of experimental fluidized bed combustion residue (FBCR) and limestone

Element	FBCR	Limestone
	(%)	
Calcium	32.0	20.4
Magnesium	0.5	11.9
Potassium	0.301	0.0165
Phosphorus	0.0451	0.005
Sodium	0.5	0.024
Sulfur	10.15	0.115
Copper	0.0056	0.0015
Manganese	0.0454	0.0312
Molybdenum	0.00072	0.0002
Zinc	0.0091	0.3058
Iron	1.85	0.245
Aluminum	1.55	0.0335
Nickel	0.0029	0.0015
Strontium	0.0487	0.0296
Chromium	0.0025	0.002
Cadmium	0.000013	0.00001
Lead	0.00022	0.00015
Cobalt	0.0031	0.0004

Source: Whitsel et al. (1988).

FLY ASH. Physical and chemical effects derived from use of fly ash can be attributed to its source and how it is produced. Coal, the source of fly ash, was formed from plant substances and then altered over long periods of time by various processes that included heat, pressure, and deposition of sedimentary materials. After the coal is burned in the boilers of large power-generating plants, the fly ash is collected as a powdery residue. It is composed of silicates, oxides, and sulfates, along with lesser amounts of carbonates and phosphates (Capp 1978). A number of other elements are also present including C, chiefly in the form of coke particles. Most fly ashes are mildly to moderately alkaline so that fly ash can substitute for limestone as well as a source of S. Results of greenhouse and field studies indicate that fly ash application can either completely or partially correct for B, Mo, P, K, and Zn deficiencies in plants. However, detrimental effects on plant growth may include B toxicity, soluble salt damage, and nutrient deficiencies, if soil pH is increased too much. Chemical analyses of fly ash are important for its safe and effective use on soils used to grow forages. See Table 5.9 for the range of elemental composition reported for a number of fly ashes used in research projects.

ESSENTIAL MICRONUTRIENTS

Even though required only in very small amounts, micronutrients are just as essential for plant growth as major or secondary nutrients. Transfer of essential micronutrients from soils into plants is a complicated process, perhaps more so than for macronutrients. This is also true for those mineral nutrients essential to animals that enter the food chain through plants but that are not essential to plants. Allaway (1975) describes the processes for nutrients that are essential to both plants and animals. Nickel has recently been demonstrated to be an essential element for higher plants (Brown et al. 1987a), but cases of deficiency in either plants or animals are apparently extremely rare. It is likely to be more important for grain or seed production. Cobalt (Co) is essential for microorganisms that live in the nodules on the roots of legumes. Its importance for forage production in field situations has yet to be determined. In Table 5.10 examples are given of general critical micronutrient ranges for plants. These values are associated with different forage crops; e.g., legumes have a relatively higher requirement for B than do grasses (Murphy and Walsh 1972).

TABLE 5.10   Generalized deficient, normal, and toxic levels in plants for eight micronutrients

Micronutrient	Deficient	Normal	Toxic
	(concentration, ppm)		
Boron (B)	<5	5-30	>75
Chloride	<2	50	20,000
Cobalt (Co)[a]	0.02	0.05-2.0	. . .
Copper (Cu)	4	4-15	>20
Manganese (Mn)	. . .	15-100	500
Molybdenum (Mo)	<0.1	1-100	. . .
Nickel (Ni)	<0.1	0.2-2	>30
Zinc (Zn)	<15	8-15	>200

Sources: Brown et al. (1987b), Follett and Wilkinson (1985), and Mengel and Kirkby (1978).

Removal of large yields, use of high-analysis fertilizers with fewer micronutrient-containing impurities, improper liming, and particular soil characteristics (such as inherently low B content) may result in micronutrient deficiencies. Dramatic cases of the importance of specific micronutrients do exist; e.g., Mo is unavailable in strongly acid soils. Instances have been noted in Australia and New Zealand where 40 g Mo/ha added to acid soil have given the same increase in legume yields as applications of several metric tons of lime (Brady 1974). Micronutrient deficiencies probably occur worldwide. Significant plant responses to micronutrient fertilization are reported from Armenia where yield increases and stimulation of legumes in grass-legume-forb stands are reported (Kasarian and Della-Rossa 1974).

Addition of B is frequently recommended for alfalfa and other legumes on light-colored sands and silt loams in humid regions (Murphy and Walsh 1972). Boron deficiency is readily corrected by addition of borax or B-containing fertilizers. Care is necessary when

TABLE 5.9   Range of elemental composition of seven experimental fly ashes

Element	Range (%)
Calcium	0.2 - 4.1
Magnesium	0.1 - 0.8
Potassium	1.7 - 3.2
Phosphorus	0.1 - 0.2
Carbon	1.0 - 38.1
Sodium	0.1 - 0.6
Sulfur	0.15 - 0.40
Boron	0.005 - 0.045
Copper	0.004 - 0.020
Manganese	0.000 - 0.030
Molybdenum	0.002 - 0.310
Zinc	0.008 - 0.020
Iron	5.1 - 16.0
Aluminum	10.3 - 19.7
Silicon	21.6 - 28.4

Source: Capp (1978).

fertilizing with B because of the narrow range of concentrations in soil where deficiency or toxicity may occur.

Animal manures can be particularly beneficial for correcting micronutrient deficiencies both as a micronutrient source and by solubilizing soil micronutrients to make them more plant-available. Correction of a micronutrient deficiency can be accomplished by addition of a salt of the deficient nutrient to the soil or addition of a chelate of the deficient nutrient sprayed in small quantities on plant leaves. Irrespectively, rate of application of micronutrients should be carefully regulated since an overdose can cause severe phytotoxicity problems or toxicity to consuming animals. Micronutrients should not be added unless the need for them is clearly established.

NONESSENTIAL ELEMENTS

Plants absorb nonessential mineral elements, such as heavy metals. These can be phytotoxic or, if consumed by animals or man, detrimental to their health. Phytotoxicity is the intoxication of living plants by substances present in the growth medium (soil) when these substances are taken up and accumulated in plant tissue (Chang et al. 1992). This aspect of mineral nutrition of plants has increased in importance because of the widespread application of municipal, industrial, and agricultural wastes to forage- and pastureland, as well as atmospheric contamination with various forms of these elements. Arsenic (As), Cd, chromium (Cr), Pb, Hg, and Se are presently elements of concern. Selenium is essential for animals and has a very narrow concentration range before toxicity to the animal occurs. Metals may induce phytotoxicity by (1) altering plant-water relations, thereby causing water stress and wilting; (2) increasing the permeability of root cell membranes, thereby causing roots to become leaky and less selective in extraction of constituents from soil; (3) inhibiting photosynthesis and respiration; and (4) adversely affecting the activity of metabolic enzymes. Chang et al. (1992) present a more complete discussion of heavy metal phytotoxicity.

INTERACTIONS AMONG PLANT NUTRIENTS

Documented plant nutrient interactions include (1) P and K fertilization is needed to obtain response to N for grasses and vice versa; (2) increased applications of N may increase K and P concentrations and uptake; (3) increased K applications may decrease amounts

of Mg uptake and cause Mg deficiency, while similar effects may occur less severely with Ca; (4) NH_4^+ forms of N may depress Mg uptake; (5) cultural x nutrient interactions such as species mixtures, irrigation, and grazing intensity may affect soil or crop nutrient requirements; (6) soil pH and nutrient interaction might involve balance between solubility of P, Ca, and Al; and (7) high soil P-pH interactions affect micronutrient availability. For further discussion of this topic see Allaway 1975.

QUESTIONS

1. Describe the importance of nutrient management planning for determining fertilizer requirements of forage-livestock production systems.
2. Explain the concept in soil fertility that is termed the "Principle of Limiting Factors" or "Liebig's Law of the Minimum."
3. Why do grasses and legumes differ in their response to soil fertility factors?
4. Describe the concepts and essential features of nutrient cycles in pasture ecosystems.
5. Describe the meanings of N requirement, N use efficiency, and N recovery as they relate to pasture fertilization.
6. What are the soil factors that affect P availability to plants?
7. Why is soil acidity detrimental to plant growth?
8. Is it usually desirable to include micronutrients in forage fertilizers? Explain.
9. Describe what plant nutrient interactions are. Give examples.
10. What roles do soil fauna and flora play in determining the fertilizer requirements of a pasture, hay, or silage crop?
11. What are some environmental considerations in developing nutrient management plans for forage—livestock production systems?

REFERENCES

Allaway, WH. 1975. The Effect of Soils and Fertilizers on Human and Animal Nutrition. USDA Agric. Inf. Bull. 378.

Allen, ER, and LR Hosner. 1991. Factors affecting the accumulation of surface-applied agricultural limestone in permanent pastures. Soil Sci. 151:240-48.

Arnold, GW, and ML Dudzinski. 1978. Ethology of Free Ranging Domestic Animals. Amsterdam: Elsevier, 68-86.

Bauder, JW, JS Jacobsen, and WT Lanier. 1992. Alfalfa emergence and survival response to irrigation water quality and soil series. Soil Sci. Soc. Am. J. 56:890-96.

Belanger, G, JE Richards, and RB Walton. 1989. Effects of 25 years of N, P, and K fertilization on yield, persistence, and nutrient value of a timothy sward. Can. J. Plant Sci. 69:501-12.

Blaser, RE, and EL Kimbrough. 1968. Potassium nutrition of forage crops with perennials. In VJ

Kilmer, SE Younts, and NC Brady (eds.), The Role of Potassium in Agriculture. Madison, Wis.: American Society of Agronomy, Crop Science Society of America, Soil Science Society of America, 423-45.

Brady NC. 1974. The Nature and Property of Soils 8th ed. New York: Macmillan.

Bristow, AW, and SC Jarvis. 1991. Effects of grazing and nitrogen fertilizer on the soil microbial biomass under permanent pastures. J. Sci. Food Agric. 54:9-21.

Brown, PH, and JT Madison. 1990. Effect of nickel deficiency on soluble anion, amino acid, and nitrogen levels in barley. Plant and Soil 125:19-27.

Brown, PH, RM Welch, and EE Cary. 1987a. Nickel: A micronutrient essential for higher plants. Plant Physiol. 85:801-3.

Brown, PH, EE Cary, and RT Checkai. 1987b. Beneficial effects of nickel on plant growth. J. Plant Nutr. 10:2125-35.

Brown, RH. 1978. A difference in N use efficiency in C_3 and C_4 plants and its implications in adaptation and evolution. Crop Sci. 18:93-98.

Bush, L, J Boling, and S Yates. 1979. Animal disorders. In RC Buckner and LP Bush (eds.), Tall Fescue, Am. Soc. Agron. Monogr. 20. Madison, Wis., 247-92.

Cahill, NJ, RL Reid, MK Head, JL Hern, and OL Bennett. 1988. Quality of diets with fluidized bed combustion residue treatment: I. Rat trials. J. Environ. Qual. 17:550-56.

Capp, JP. 1978. Power plant fly ash utilization for land reclamation in the eastern United States. In FW Schaller and P Sutton (eds.), Reclamation of Drastically Disturbed Lands. Madison, Wis.: American Society of Agronomy, Crop Science Society of America, Soil Conservation Society of America, 339-53.

Council of Agricultural Science and Technology (CAST). 1984. Acid Precipitation in Relation to Agriculture, Forestry, and Aquatic Biology. Counc. Agric. Sci. and Tech. Rep. 100. Ames, Iowa.

Chang, AC, TC Granato, and AL Page. 1992. A methodology for establishing phytotoxicity criteria for chromium, copper, nickel, and zinc in agricultural land application of municipal sewage sludges. J. Environ. Qual. 21:521-36.

Cole, CV, S Innis, and JWB Stewart. 1977. Simulation of phosphorus cycling in semiarid grasslands. Ecol. 58:1-15.

Coleman, DC, CPP Reid, and CV Cole. 1983. Biological strategies of nutrient cycling in soil systems. Adv. Ecol. Res. 13:1-56.

Devitt, DA. 1989. Bermudagrass response to leaching fractions, irrigation salinity, and soil types. Agron. J. 81:893-901.

Donahue, RL, RH Follett, and RW Tullock. 1983. Our Soils and Their Management. Danville, Ill.: Interstate Printers and Publishers.

Dougherty, CT, and CL Rhykerd. 1985. The role of nitrogen in forage-animal production. In ME Heath, RF Barnes, and DS Metcalfe (eds.), Forages: The Science of Grassland Agriculture, 4th ed. Ames: Iowa State Univ. Press, 318-25.

Dudzinsky, ML, SR Wilkinson, RN Dawson, and AP Barnett. 1983. Fate of NH_4NO_3 and broiler litter applied to Coastal bermudagrass. In RR

Lourance, RL Todd, LE Asmussen, and RA Leonard (eds.), Nutrient Cycling in Agricultural Ecosystems, Univ. Ga. Coll. Agric. Spec. Publ. 23, 373-88.

Fincher, GT, WG Monson, and GW Burton. 1981. Effects of cattle feces rapidly buried by dung beetles on yield and quality of Coastal bermudagrass. Agron. J. 73:775-79.

Follett, RH, and RF Follett. 1980. Strengths and weaknesses of soil testing in determining lime requirements for soils. In Proc. 1st Natl. Conf. Agric. Limestone, 16-18 Oct., Nashville, Tenn., 40-51.

———. 1983. Soil and lime requirement tests for the 50 states and Puerto Rico. J. Agron. Ed. 12:9-17.

Follett, RF, and DJ Walker. 1989. Ground water quality concerns about nitrogen. In RF Follett (ed.), Nitrogen Management and Ground Water Protection. Amsterdam: Elsevier, 1-22.

Follett, RF, and SR Wilkinson. 1985. Soil fertility and fertilization of forages. In ME Heath, RF Barnes, and DS Metcalfe (eds.), Forages: The Science of Grassland Agriculture, 4th ed. Ames: Iowa State Univ. Press, 304-17.

Follett, RF, SC Gupta, and PG Hunt. 1987. Conservation practices: Relation to the management of plant nutrients for crop production. In RF Follett, JWB Stewart, and CV Cole (eds.), Soil Fertility and Organic Matter as Critical Components of Production Systems, Spec. Publ. 19. Madison, Wis.: Soil Science Society of America, 19-51.

Fox, RL, and PGE Searle. 1978. Phosphate adsorption by soils of the tropics. IN Diversity of Soils in the Tropics. Spec. Publ. 34. Madison, Wis.: American Society of Agronomy, 97-119.

Foy, CD, and AL Fleming. 1978. The physiology of plant tolerance to excess available aluminum and manganese in acid soils. In GA June (ed.), Crop Tolerance to Suboptimal Land Conditions, Spec. Publ. 32. Madison, Wis.: American Society of Agronomy, 301-28.

Gerloff, GC. 1987. Intact plant screening for nutrient deficiency stress. Plant and Soil 99:3-16.

Gilbertson, CV, FA Norstadt, AC Mathers, RF Holt, AP Barnett, TM McCalla, CA Onstad, RA Young, LR Shapler, LA Christensen, and DL Van Dyne. 1979. Animal Waste Utilization on Cropland and Pastureland: A Manual for Evaluating Agronomic and Environmental Effects. USDA Util. Rep. 6—EPA-600/2-79-059.

Gough, LP, HT Shacklette, and AA Case. 1979. Element Concentrations Toxic to Plants, Animals, and Man. Geol. Surv. Bull. 1466, US Geological Survey. Washington, D.C.: US Gov. Print. Off.

Griffith, WK. 1973. Soil fertility and its relationship to yield and quality of forages. In Proc. 6th Am. Forage and Grassl. Counc., 1-12.

———. 1974. Satisfying the nutritional requirements of established legumes. In DA Mays (ed.), Forage Fertilization. Madison, Wis.: American Society of Agronomy, Crop Science Society of America, and Soil Science Society of America, 147-69.

Haynes, RJ, and PH Williams. 1993. Nutrient cycling and soil fertility in grazed pasture ecosystems. Adv. Agron. 49:119-99.

Hetrick, BAD, DG Kitt, and GT Wilson. 1988. Mycorrhizal dependence and growth habit of warm season and cool season tall grass prairie plants. Can. J. Bot. 66:1376-80.

Hoffman, GJ, RS Ayers, EJ Doering, and BL McNeal. 1980. Salinity in irrigated agriculture. In ME Jensen (ed.), Design and Operation of Farm Irrigation Systems, Am. Soc. Agric. Eng. Monogr. 3, 145-85.

International Fertilizer Development Center (IFDC). 1979. Sulfur in the Tropics. Ala. Tech. Bull. IFDC-T12. Muscle Shoals: IFDC.

Jarvis, SC, DJ Hatch, and DR Lockyer. 1989. Ammonia fluxes from grazed grassland: annual losses from cattle production systems and their relation to nitrogen inputs. J. Agric. Sci. 113:99-108.

Kasarian, Y, and R Della-Rossa. 1974. Availability of microelements in mountain meadows and the influence of microfertilizers on yields and quality of forage. IN Proc. 12th Intl. Grassl. Congr.: 260-64.

Laughlin, WM, GR Smith, and MA Peters. 1981. Sulfur, manganese, molybdenum, and magnesium influences on bromegrass forage yield and composition in southcentral Alaska. Commun. Soil Sci. Plant An. 12(4):299-317.

Lewis, OAM, DM James, and EJ Hewitt. 1982. Nitrogen assimilation in barley (Hordeum vulgare L. cv. Mazurha) in response to nitrate and ammonium nutrition. Ann. Bot. 49:39-49.

Maas, EV, and GJ Hoffman. 1977. Crop salt tolerance—current assessment. Am. Soc. Agric. Eng. Proc. J. Irrig. Drain. Div. 103(IR2):114-34.

McKimmie, T, and AK Dobrenz. 1991. Ionic concentrations and water relations of alfalfa seedlings differing in salt tolerance. Agron. J. 83:363-67.

Martin, DL, JR Gilley, and RW Skaggs. 1991. Soil water balance and management. In RF Follett, DR Keeney, and RM Cruse (eds.), Managing Nitrogen for Groundwater Quality and Farm Profitability. Madison, Wis.: Soil Science Society of America, 199-235.

Mengel, K, and EA Kirkby. 1978. Principles of Plant Nutrition. 4th ed. Worblaufen-Bern, Switzerland: International Potash Institute.

Mengel, K, and DJ Pilbeam (eds.). 1992. Nitrogen Metabolism of Plants. Proc. Phytochem. Soc. Eur. Oxford, England: 33 Clarendon Press.

Murphy, LS, and LM Walsh. 1972. Correction of micronutrient deficiencies with fertilizers. In JJ Mortvedt, PM Giordano, and WL Lindsay (eds.), Micronutrients in Agriculture. Madison, Wis.: Soil Science Society of America, 347-87.

Murphy, MD. 1978. Responses to sulphur in Irish grassland. In Symp. Proc.—Sulphur in Forages, 3-4 Oct., Wexford, Ireland, 95-109.

Nelson, WL. 1980. Agricultural liming: Its effect on soil fertility, plant nutrition, and yields. In Proc. 1st Natl. Conf. Agric. Limestone, 16-18 Oct., Nashville, Tenn., 34-39.

Nguyen, K, and M Goh. 1992. Nutrient cycling and losses based on a mass balance model in grazed pastures receiving long-term superphosphate applications in New Zealand: II. Sulphur. J. Agric. Sci. Camb. 119:107-22.

Olson, RA. 1985. Nitrogen problems. In Proc.: Plant Nutrient Use and Environ., Kansas City, Mo. Washington, D.C.: Fertilizer Institute, 115-37.

Olson, RA, and SR Wilkinson. 1992. Model evalua-

tion for perennial grasses in the southern United States. Agron. J. 84:523-29.

Overman, AR, and SR Wilkinson. 19992. Model evaluation for perennial grasses in the southern United States. Agron. J. 84:523-29.

Overman, AR, D Downey, and SR Wilkinson. 1989. Effect of N rate and split application on Bahiagrass production. Comm. Soil Sci. Plant Anal. 20:501-12.

Owens, LB, WM Edwards, and RW Van Keuren. 1992. Nitrate levels in shallow groundwater under pastures receiving ammonium nitrate, or slow-release nitrogen fertilizer. J. Environ. Qual. 21:607-13.

Pearson, RW, and CS Hoveland. 1974. Lime needs of forage crops. In DA Mays (ed.), Forage Fertilization. Madison, Wis.: American Society of Agronomy, Crop Science Society of America, Soil Science Society of America, 301-22.

Pumphrey, FV, and DP Moore. 1965. Sulfur and nitrogen content of alfalfa herbage during growth. Agron. J. 57:237-39.

Quin, BF. 1982. The influence of grazing animals on nitrogen balance. In Nitrogen Balances in New Zealand Ecosystems, Plant Physiol. Dir., DSIR. Palmerston North, New Zealand, 95-102.

Rechcigl, JE, GG Payne, AB Bottcher, and PS Porter. 1992. Reduced phosphorus application on bahaigrass and water quality. Agron. J. 84:463-68.

Reeder, JD, and B Sabey. 1987. Nitrogen. In RD Williams and GE Schuman (eds.), Reclaiming Mine Soils and Overburden in the Western United States: Analytical Parameters and Procedures. Ankeny, Iowa: Soil Water Conservation Society, 155-84.

Reinbott, TM, and DB Blevins. 1991. Phosphate interaction with uptake and leaf concentration of magnesium, calcium and potassium in winter wheat seedlings. Agron. J. 83:1043-46.

Rowarth, JS, RW Tillman, AG Gillingham, and PEH Gregg. 1992. Phosphorus balances in grazed hill-country pastures—The effect of slope and fertilizer input. N.Z. J. Agric. Res. 35(3):337-42.

Russelle, MP. 1992. Nitrogen cycling in pasture and range. J. Prod. Agric. 5:13-23.

Sanchez, PA, and JG Salinas. 1981. Low-input technology for managing oxisols and ultisols in tropical America. Adv. Agron. 34:279-406.

Schimel, DS. 1986. Carbon and nitrogen turnover in adjacent grassland and cropland ecosystems. Biogeochem. 2:345-57.

Schwartz, JW, and RF Follett. 1979. Liming Acid Soils. USDA Agric. Fact Sheet 4-5-4.

Soil Conservation Society of America (SCSA). 1982. Resource Conservation Glossary. Ankeny, Iowa: SCSA.

Sharpley, AN, and SJ Smith. 1993. Prediction of bioavailable phosphorus loss in agricultural runoff. J. Environ. Qual. 22:32-37.

Spain, JM, CA Francis, RH Howler, and F Calvo. 1975. Differential Species and Varietal Tolerance to Soil Acidity in Tropical Crops and Pastures. Soil Management in Tropical America. Univ. Consortium on Soils of the Tropics, North Carolina State Univ., Raleigh.

Stewart, BA, and LK Porter. 1969. Nitrogen-sulfur relationships in wheat (Triticum aestivum L.), corn (*Zea mays*), and beans (*Phaseolus vulgaris*). Agron. J. 61:267-71.

Stewart, BA, DA Woolhiser, WH Wischmeier, JH Caro, and MH Frere. 1976. Control of Water Pollution from Cropland. Joint ARS-EPA Rep. EPA-600/2-75-0266 or ARS-H-5-2. Washington, D.C.: US Gov. Print. Off.

Stewart, JWB, and AN Sharpley. 1987. Controls on dynamics of soil and fertilizer phosphorus and sulfur. In RF Follett, JWB Stewart, and CV Cole (eds.), Soil Fertility and Organic Matter as Critical Components of Production Systems, Spec. Publ. 19. Madison, Wis.: Soil Science Society of America, 101-21.

Thompson, LM, and FR Troeh. 1978. Soils and Soil Fertility. 4th Ed. New York: McGraw-Hill.

Tiessen, H, JWB Stewart, and HW Hunt. 1984. Concepts of soil organic matter transformations in relation to organo-mineral particle size fractions. Plant Soil 76:287-95.

Underwood, EJ. 1977. Trace Elements in Human and Animal Nutrition. 4th ed. London: Academic Press.

USDA. 1954. Diagnosis and Improvement of Saline and Alkali Soils. USDA Agric. Handb. 60, US Salinity Lab. Washington, D.C.: US Gov. Print. Off.

———. 1989. The Second RCA Appraisal: Soil, Water and Related Resources on Nonfederal Land in the United States—Analysis of Conditions and Trends. Washington, D.C.: US Gov. Print. Off.

Vona, LC, C Meredith, RL Reid, JC Hern, HD Perry, and OL Bennett. 1992. Effect of fluidized bed combustion residue on the health and performance of sheep grazing hill pastures. J. Environ. Qual. 21:335-40.

Wallace, A, E Frolich, and OR Lunt. 1966. Calcium requirements of higher plants. Nat. 209:634.

Walworth, JL, and ME Sumner. 1987. The diagnosis and recommendation integrated system (DRIS). Adv. Soil Sci. 6:149-88.

Wedin, DA, and D Tilman. 1990. Species effects on nitrogen cycling: A test with perennial grasses. Oecologia 84:433-41.

Welch, LF, AR Bertrand, and WE Adams. 1964. How nonuniform distribution of fertilizer affects crop yields. Ga. Agric. Res. 5:315-16.

West, CP, AP Mallarino, WF Wedin, and DB Mart. 1989. Spatial variability of soil chemical properties in grazed pastures. Soil Sci. Soc. Am. J. 53:784-89.

Whitehead, DC, and LHP Jones. 1978. Nitrogen/sulphur relationships in grass and legumes. In Symp. Proc.—Sulphur in Forages, 3-4 Oct., Wexford, Ireland, 126-41.

Whitsel, TJ, RL Reid, WL Stout, JL Hern, and OL Bennett. 1988. Quality of diets with fluidized bed combustion residue treatment: II. Swine trials. J. Environ. Qual. 17:556-62.

Wilkinson, SR, and RS Lowrey. 1973. Cycling of mineral nutrients in pasture ecosystems. In GW Butler and RW Baily (eds.), Chemistry and Biochemistry of Herbage, vol. 2. London: Academic Press, 248-315.

Wilkinson, SR, and DA Mays. 1979. Mineral nutrition. In RC Buckner and LP Bush (eds.), Tall Fescue, Am. Soc. Agron. Monogr. 20. Madison, Wis., 41-73.

Wilkinson, SR, LF Welch, and GA Hillsman. 1968. Compatibility of tall fescue and Coastal bermudagrass as affected by nitrogen fertilization and height of clip. Agron. J. 60:359-62.

Wilkinson, SR, JA Stuedemann, and DP Belesky. 1989. Distribution of soil potassium in grazed K-31 tall fescue pastures as affected by fertilization and endophytic fungus infection. Agron. J. 81:508-12.

Wilkinson, SR, RM Welch, HF Mayland, and DL Grunes. 1990. Magnesium in plants: Uptake distribution, function and utilization by animals and man. In H Sigel and A Sigel (eds.), Compendium on Magnesium and Its Role in Biology, Nutrition, and Physiology, vol. 26, Metal Ions in Biological Systems Series. New York: Marcel Dekker, 33-56.

Yost, RS, and RS Fox. 1979. Contribution of mycorrhizae to P nutrition of crops growing on an oxisol. Agron. J. 71:903-8.

Yost, RS, GC Naderman, EJ Klamprath, and E Lobato. 1982. Availability of rock phosphate as measured by an acid tolerant pasture grass and extractable phosphorus. Agron J. 74:462-68.

6

Quality-related Characteristics of Forages

DWAYNE R. BUXTON
Agricultural Research Service, USDA

DAVID R. MERTENS
Agricultural Research Service, USDA

ORAGES constitute the major portion of the diets of the world's livestock, especially ruminants. Ruminants, with their specially designed digestive systems and associated microbial populations, are well adapted for obtaining nutrients and energy from forages. Polysaccharides of plant cell walls are not degraded by mammalian enzymes but can be partially degraded by rumen microbes. Cell walls, composed of polysaccharides, lignin and phenolics, proteins, cutin, silica, and water, are major limits to forage digestibility. Cell contents, contained within cell walls, consist of organic acids, proteins, lipids, soluble minerals, and soluble carbohydrates; they are usually completely digestible and readily available.

Forage cell walls provide fiber to livestock, which is required for normal rumen function. Fiber stimulates the cardial region of the reticulum to induce regurgitation, rumination, and ruminal motility. This chapter focuses on the available energy and protein of

DWAYNE R. BUXTON is Research Leader of the Field Crops Research Unit, ARS, USDA; a cluster scientist of the US Dairy Forage Research Center; and Collaborator/Professor of Agronomy at Iowa State University. He holds an MS degree from Utah State University and a PhD from Iowa State University. With emphasis on limitations imposed by plant cell walls, his research is on physiological, morphological, and environmental factors that influence forage quality.

DAVID R. MERTENS is a ARS, USDA, Research Dairy Scientist with the US Dairy Forage Research Center in Madison, Wisconsin. He holds an MS degree from the University of Missouri and a PhD from Cornell University and conducts research on plant fiber utilization by dairy cows and computer modeling of digestion and passage in ruminants.

forages because they are often low in these attributes relative to animal requirements. A balance of nutrients including minerals and vitamins, however, is required for optimal animal production. Emphasis is placed on ruminants because of their importance as users of forages.

Forage quality can be defined as the relative performance of animals when fed herbage ad libitum, i.e., their consumption when fed excess forage. The quality of a forage is a function of nutrient concentration of the forage, its intake or rate of consumption, digestibility of the forage consumed, and partitioning of metabolized products within the animal. Forage quality is often estimated by in vitro or chemical means because of limitations of cost and time in using animals. Forage quality varies among plant species and plant parts and is influenced by herbage age and maturity, soil fertility, and the environment in which forages are grown. Each of these factors is presented and discussed in the following sections. As is emphasized, herbage maturity is the most important factor affecting forage quality within a species.

NUTRIENTS AND ENERGY IN FORAGES

Minerals and Vitamins. A balance of nutrients including minerals and vitamins is required for optimal animal production. Twenty-one mineral elements are known to be required by herbivores. Of these potassium (K), chlorine (Cl), iron (Fe), molybdenum (Mo), cadmium (Cd), lithium (Li), and nickel (Ni) are present in forages in sufficient quantities to meet much of the ruminant requirement. Amounts present in forages and their utilization by ruminants are discussed in detail by Minson (1990). Only calcium (Ca), P, magnesium (Mg), and S are considered here.

The Ca requirement of ruminants varies from a low of 2.0 g kg^{-1} dry matter (DM) for mature animals to 8.0 g kg^{-1} DM for lactating or rapidly growing young animals (NRC 1984, 1989), and the requirement for Ca of most moderately productive ruminants is met by forages. Legumes generally contain more Ca than grasses, concentrations being about 10.1 g kg^{-1} DM in cool-season legumes compared with 3.8 g kg^{-1} DM in cool- or warm-season grasses (Minson 1990). Average Ca concentration in leaves of cool-season grasses is about 7.2 g kg^{-1} compared with 2.6 g kg^{-1} DM in stems. Warm-season grasses have similar Ca concentrations in stems as cool-season grasses, but they have lower average leaf concentrations of 5.6 g kg^{-1} DM. As forages mature, Ca concentration of total herbage declines because the proportion of stems in herbage increases.

Ruminant requirements for P range from 2.0 to 5.0 g kg^{-1} DM (NRC 1984, 1989). When only mature forages are fed, P can become limiting. Minson (1990) found the mean P concentration of legumes was 3.2 g kg^{-1} DM and that of cool-season grasses was 2.7 g kg^{-1}. Phosphorus concentration declines as plants advance in maturity. Phosphorus fertilizer applied to low-P soils increases P concentration of herbage.

The Mg requirement for ruminants varies from 1.0 to 3.0 g kg^{-1} DM (NRC 1984, 1989). Legumes generally contain more Mg than cool-season grasses but less than warm-season grasses (on average about 2.6, 1.8, and 3.6 g kg^{-1} DM, respectively [Minson 1990]). At mature stages, leaves contain higher concentrations of Mg than stems. Normally, Mg is adequate in forages to meet animal requirements; however, only a low proportion may be available to the animal, resulting in grass tetany for animals grazing lush, rapidly growing herbage. These relationships are considered in Chapter 9.

Adequate S is required for efficient functioning of cell wall-degrading microorganisms in the rumen. These microorganisms synthesize proteins from ammonia and need a S:N ratio of approximately 1:10 to provide adequate S for the synthesis of the S-containing amino acids cysteine and methionine. The National Research Council (1984, 1989) recommends that ruminant diets contain 1-3 g kg^{-1} DM of S. Application of S fertilizer to soils that produce S-deficient forages raises the herbage concentration.

Plants do not require Na but accumulate small amounts, especially in saline soils. Animals require Na, and NaCl is usually a supplement for animals fed forage diets.

Vitamins are required in small quantities for normal body functions. Fat-soluble vitamins include A, D, E, and K, and water-soluble vitamins include thiamin (B$_1$), riboflavin (B$_2$), niacin, pantothenic acid, B$_6$, B$_{12}$, folacin, biotin, choline, and ascorbic acid (vitamin C). Microbial activity in the rumen normally results in enough synthesis to meet the requirement for water-soluble vitamins. In other herbivores, microbial synthesis in the colon or cecum can meet this need. Vitamin B$_{12}$ is not present in forages but is synthesized by rumen microorganisms in amounts influenced by the cobalt (Co) concentration in the diet (Minson 1990). Forages contain carotene, which is the precursor of vitamin A. Poor quality herbage, or herbage that has lost carotene through oxidation during prolonged storage, may be deficient in vitamin A. Herbivores exposed to sunlight or that eat sun-cured herbage usually have sufficient vitamin D.

Protein. Nitrogen (N) in forage can be divided into true protein and nonprotein N (NPN). True protein normally comprises 60%-80% of the herbage N. Most of the NPN is in nucleic acids, free amino acids, amides, and nitrate. When there is sufficient available carbohydrate to provide energy for microbial growth, NPN is normally converted to ammonia in the rumen and used for microbial protein synthesis. More than 90% of N in most cool-season forages is in cell solubles and is readily digestible. This value is lower in some warm-season grasses.

Protein requirements for ruminants usually are expressed as crude protein (CP), which is the sum of true protein and NPN expressed as CP equivalents (N × 6.25). Requirements range from 70 g CP kg^{-1} DM for mature beef cows to 190 g kg^{-1} DM for high-producing, lactating dairy cows (NRC 1984, 1989). Minson (1990) reports average CP concentrations of 100 g kg^{-1} DM for warm-season grasses, 129 g kg^{-1} DM for cool-season grasses, and 170 g kg^{-1} DM for cool-season legumes. Generally, concentrations of protein in leaf blades of forages are twice that of stems. Protein concentration is higher in immature than in more mature forage. For most species, the fastest rate of decline in CP concentration with advance in herbage age or maturity occurs in very immature forage, and the rate slows as the herbage becomes more mature (Fick and

Onstad 1988). The average decrease in CP concentration with advancing age for several species and experiments summarized by Minson (1990) was 2.2 g kg^{-1} DM d^{-1}.

When moderate to high rates of N fertilizer are used during spring growth in temperate areas, the protein concentration of immature grasses may be higher than the average values reported by Minson (1990), with an associated faster rate of decline with aging. For example in Iowa, CP concentration ranged from 305 g kg^{-1} DM in reed canarygrass (*Phalaris arundinacea* L.) to 264 g kg^{-1} in tall fescue (*Festuca arundinacea* Schreb.) in mid-May after being fertilized in early spring with 95 kg N ha^{-1} (Buxton and Marten 1989). Crude protein concentration declined at a rate of 4.2 g kg^{-1} DM d^{-1} in reed canarygrass and 3.8 g kg^{-1} d^{-1} in tall fescue. Concentrations of CP decreased in both leaves and stems, but the rapid decline occurred because stems made up a larger portion of total herbage as reproductive tillers matured. A slower rate of decline would be expected in aftermath growth.

The CP concentration of grass herbage is strongly influenced by available soil N level. Altering CP concentration by N fertilization represents the most important effect of fertilization on forage quality. Application of N fertilizer to grasses usually increases CP concentration as well as crop growth. The majority of the increased CP is NPN, however, in the form of nitrates and free amino acids. The increase in CP of N-fertilized herbage is usually accompanied by a decrease in water-soluble carbohydrate concentration. Legumes growing in association with grasses also transfer N to grasses. Smooth bromegrass (*Bromus inermus* Leyss.) grown with alfalfa (*Medicago sativa* L.) was found to have a total N concentration that was similar to that of smooth bromegrass fertilized with 125 kg of N ha^{-1} (Sanderson and Wedin 1989b).

Energy. Nutrients in plants that provide most of the energy for animals are carbohydrates, proteins, and lipids. Carbohydrates provide up to 80% of the energy for ruminants from herbage, whereas lipids contribute less than 5%. The gross energy (i.e., the caloric content obtained by burning) of forages has little nutritional importance because the portion that is digestible is more important in determining energy availability to animals than is the concentration of gross energy. The most obvious division of plant DM into energy-yielding components for animals is between cell walls and cell contents. Plant cell contents are almost completely available to livestock. The availability of plant cell walls to ruminant livestock varies greatly depending on their composition and structure (Van Soest 1982). Available energy, then, is inversely related to cell wall, or neutral detergent fiber (NDF), concentration in herbage.

The maximum cell wall concentration of the diet that will not hinder production may be as high as 700-750 g NDF kg^{-1} DM for mature beef cows and as low as 150-200 g NDF kg^{-1} for growing or fattening ruminants. The concentration of NDF in diets of high-producing dairy cows should be in the range of 270-290 g kg^{-1} DM to allow for adequate energy and also to maintain adequate fiber in the diet.

Stems have a greater concentration of cell walls than leaves in most forages. Thus, stems usually are of lower digestibility than leaves, and stem digestibility declines more rapidly with increased plant maturity than does that of leaves. Differences between leaf and stem digestibility, however, are normally less in grasses than in forage legumes.

Digestibility, expressed as DM, organic matter, energy, or total digestible nutrients (TDN), is the most commonly used term of energy availability. Total digestible nutrients is the sum of digestible protein, carbohydrates, and ether extract (lipids). Digestible ether extract is multiplied by 2.25 to account for the higher energy concentration of lipids compared with proteins and carbohydrates. The TDN requirement of ruminants varies from a low of 500 g kg^{-1} DM in diets for mature animals to 750-850 g kg^{-1} DM in diets for lactating females and rapidly growing young animals.

CONSUMPTION OF FORAGES

The amount of herbage consumed is the major determinant of herbivore production, yet it is one of the most difficult aspects of forage quality to measure or predict. Within a forage species, there is usually a positive relationship between digestibility and daily consumption or intake. However, many correlations between digestibility and intake are low, and deviations occur among plant species and plant parts. Determining the intake potential of a forage is difficult because the drive of animals to eat varies with physiological status. For example, when fed the same forage, lactating cows consume more than mature animals that are just maintaining their body condition. This explains why most tables of feed information do not include ad libitum intake. Evaluation of intake potential as a forage attribute requires that variability in intake due to animals be minimized or eliminated.

Intake potential of a forage can be understood better when the relative importance of animal and plant factors is known. Physical limitations, physiological control, and psychogenic factors are the major regulators of intake by herbivores. Physical limitations to intake, sometimes called *gut fill,* occur when animals eat until the rumen and/or lower intestines reach a constant fill. Physical fill limits intake when forages with high NDF concentration and low available energy are fed to animals with high energy demands. The filling effect of herbage, then, is related to its cell wall concentration, which is related to the bulk of the feed and total rate of disappearance (via digestion and passage) of digesta from the rumen. Passage from the rumen requires both particle size reduction and flow through the reticulo-omasal orifice. Plant cell walls must be chewed to reduce particle size enough for passage. Neutral detergent fiber concentration of forages is positively correlated with chewing activity (Beauchemin and Buchanan-Smith 1989). The increased chewing requirement of grasses with a high cell wall concentration may contribute to the decreased rate of passage and increased "fill effect" of grasses compared with legumes, as discussed in a following section.

An animal's energy demand differs with physiological status, i.e., whether it is growing, lactating, or simply maintaining body weight. Physiological control of feed intake occurs when animals regulate intake to meet their energy demand. Hence, physiological energy demand limits intake when animals with low energy requirements are fed diets high in available energy and low in NDF concentration. For example, when high-quality forages are fed to animals having only maintenance requirements, it is animal and not forage characteristics that determine intake.

Adjusting intake for animal differences by feeding a common diet to all animals can significantly improve relationships between plant characteristics and intake potential. By selecting a "standard animal unit," INRA scientists in France have developed an empirical system for evaluating intake potential of feeds (Jarrige 1989). With this system, 1 kg of reference grass pasture containing 150 g CP kg^{-1} DM, 250 g crude fiber kg^{-1} DM, and 770 g digestible organic matter kg^{-1} DM has a fill unit of 1.0. Intake of forages is adjusted for animal differences and expressed as fill units based on this pasture grass reference.

Because NDF concentration is positively correlated with fill and negatively correlated with energy availability, NDF concentration can be used to express both of these intake regulation mechanisms on a common basis (Fig. 6.1). Physical and physiological mechanisms provide a logical basis for understanding intake regulation and estimating intake at the extremes of forage quality. At intermediate levels of forage quality, however, interactions among the two mechanisms make prediction of intake more difficult. When animals of similar production potential are fed diets of different NDF concentrations, actual intake will be similar to what is shown by the dotted line in Figure 6.1.

Digestible NDF intake for cattle consuming grasses is generally greater than for ruminants consuming legumes such as alfalfa be-

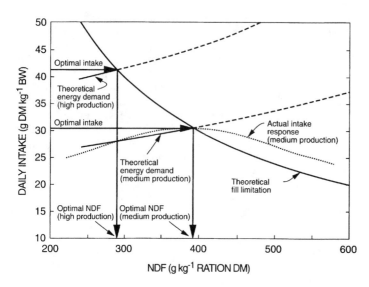

Fig. 6.1. Intake response to changes in neutral detergent fiber (NDF) concentration of rations fed to ruminants. Two regulators of intake by ruminants are fill limitation, based on the maximum daily dry matter intake, and energy demand. The NDF concentration of a ration is positively correlated with fill and negatively correlated with energy availability. The arrows show how this relationship can be used to estimate optimal NDF concentration for a targeted level of animal intake and production. BW = body weight of animal.

cause grasses generally contain more digestible NDF than legumes (Fig. 6.2). Weiss and Shockey (1991) found lactating dairy cows fed ad libitum diets of orchardgrass (*Dactylis glomerata* L.) or alfalfa silage consumed 43% more digestible NDF, 14% more NDF, and 8% less DM when fed orchardgrass rather than alfalfa. Digestible DM intake was similar for alfalfa and orchardgrass, and there was no difference in milk production. The expected smaller intake of orchardgrass compared with that of alfalfa at similar digestibilities has also been confirmed by other studies (Thomson et al. 1991). Animals may adapt to increased cell wall concentration in grass diets by increasing the maximum gut fill or rate of ruminal clearance.

As forages advance in maturity, voluntary intake generally declines. The decline is probably a function of fill limitations due to the increase in fill units or cell wall concentration with advance in maturity. Minson (1990) reports an average decline of 0.39 g kg^{-1} (body weight)$^{0.75}$ d^{-1} during primary spring growth. Leaf fractions are consumed in greater quantities than are stems in most species because leaf fractions usually have lower cell wall concentrations. Intake is also affected by psychogenic factors that influence an animal's intake in response to external stimuli such as

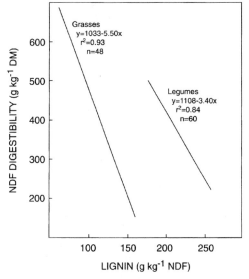

Fig. 6.2. In vitro digestibility of neutral detergent fiber (NDF) in stems of cool-season grasses and legumes as a function of the lignin concentration in the stem. At equivalent lignin concentrations, digestion of grass stems is more inhibited than in legume stems. (Adapted from Buxton and Russell 1988.)

palatability, feed flavor, feed pH, social interactions, stress, diseases, and management.

Palatability refers to the preference that animals have for a particular feed when offered a choice. The quantitative importance of palatability in controlling daily intake when animals are not allowed choice is difficult to determine.

AVAILABILITY OF NUTRIENTS AND ENERGY

Protein. Protein digestion by ruminants is complex and involves degradation and loss of protein in the rumen, transformation of forage protein to microbial protein, and ultimately digestion and absorption of amino acids from microorganisms and herbage that move out of the rumen. Under some circumstances excessive amounts of forage N are lost from the rumen and from the animal, and animal production is limited. The extent to which protein is degraded by rumen microbes is governed by the proteolytic rate and length of time plant residues are retained in the rumen. After proteins are degraded to amino acids and peptides, they can be assimilated by microorganisms and used to synthesize microbial protein, or if other sources of energy are limited, they can be deaminated and metabolized for energy. When amino acids are fermented and deaminated, ammonia is released into the rumen. If this ammonia is absorbed through the ruminal wall into the bloodstream, it is detoxified in the liver by conversion to urea. Some of the urea is recycled to the rumen via saliva and through the ruminal wall, but substantial amounts are excreted in the urine and lost to the animal. Feeding a source of energy, such as grain, with forages can improve the utilization of forage protein.

Under most situations, improved forage protein utilization occurs if a large portion of the forage protein passes from the rumen undegraded so that it can be digested in the intestines, where absorption is more efficient. Minson (1990) reports than on average, 75% of protein is degraded in the rumen, allowing only about 25% of protein in forages to escape ruminal fermentation and pass to the intestine of the animal ("escape or bypass protein"). There is evidence that the proportion of escape protein is greater in mature plants of some species. For example, Mullahey et al. (1992) found that the percentage of escape protein increased with maturity of switchgrass (*Panicum virgatum* L.) and reached 70% at the ripe seed stage but remained relatively constant with advance in maturity of smooth bromegrass. The average escape protein of smooth bromegrass was 20% compared to 51% for switchgrass.

Among legumes, those species with moderate levels of tannin, such as birdsfoot trefoil (*Lotus corniculatus* L.) and sainfoin (*Onobrychis viciifolia* Scop.), have a higher proportion of escape protein than those without tannin such as alfalfa. The amount of escape protein is higher for red clover (*Trifolium pratens* L.) than for alfalfa or white clover (*T. repens* L.) (Albrecht and Broderick 1993). The escape value may be higher for hays and artificially dried herbage than for silages and fresh forage. In addition, herbage preserved as low-moisture silage can undergo heating, which creates complexes between protein and carbohydrates that result in increased escape protein.

The amount of microbial protein produced in the rumen is determined predominately by availability of energy in the herbage for microbial growth. Protein synthesized by microbes often accounts for 50%-80% of N reaching the small intestine. When available energy is adequate, microbial protein can furnish animals with enough protein for maintenance and some growth. When limited by energy availability in forages, microbial protein may not be produced in enough quantity for optimum performance during rapid growth, late pregnancy, or early lactation.

Energy. Cell contents in feeds are digested at rates 3 to 10 times faster than rates of passage from the rumen. Thus under most situations, digestion of cell contents is almost complete. Conversely, much of the cell wall is indigestible (Fig. 6.2), and rate of degradation of the digestible portion of the cell wall by rumen microorganisms is similar to rate of passage. The large size and oblong shape of the rumen, and high location in the rumen of the reticula-omasal orifice, favor selective retention of large buoyant cell wall particles and support of microbial populations that degrade fibers.

Anaerobic microorganisms in the digestive tract produce carbon dioxide, methane, volatile fatty acids (acetic, propionic, butyric, and valeric), sometimes lactic acid, and microbial cells. Ruminants have evolved mechanisms that allow digestion, absorption, and metabolism of some of these end products. The loss of energy in methane, which is lost from the rumen as gas, and energy as heat during fermentation can be significant, how-

ever. In nonruminants, such as horses and rabbits, some cell wall may be digested by microorganisms in the colon or cecum.

Ruminal microbes normally gain access for attachment to cell walls through cut or sheared surfaces of herbage because epidermal layers of plants restrict microbe and enzyme access. Plant tissue types are colonized and degraded by rumen microorganisms at different rates (Akin 1989). Mesophyll and phloem cells are degraded more rapidly and before the outer bundle sheath and epidermal cells. The attachment of bacteria to sclerenchyma and vascular bundle cells is low, and these tissues are slowly and incompletely degraded. Cool-season grasses with the C_3 photosynthetic pathway have leaves with widely spaced vascular bundles and less distinct parenchyma bundle sheaths, whereas warm-season grasses with the C_4 photosynthetic pathway have leaves with Kranz anatomy, resulting in closely spaced vascular bundles and a distinct thick-walled parenchyma bundle sheath surrounding each bundle. The C_3 grass leaves have a higher proportion of loosely arranged mesophyll cells than C_4 grasses, which are usually the first to be digested. The proportions of epidermis and parenchyma bundle sheath are substantially higher in warm-season grass leaf blades than in those of cool-season grasses. These tissues are usually slowly or partially degraded in warm-season grasses but extensively degraded in cool-season grasses. In contrast to leaf anatomy, stem anatomy of C_3 and C_4 grasses is similar as is the extent of digestibility.

The cell wall fraction of forages can represent more than half of the organic matter in the herbage, even when forages are immature and cell walls are incompletely digested, depending on their chemical composition and structure. Thus, cell wall concentration can have a large influence on herbage digestibility. Buxton (1990) found that NDF concentration in stems of four cool-season grasses was near 700 g kg^{-1} DM when plants were at the flowering stage of development compared with about 500 g kg^{-1} DM in leaves. At this time, digestibility of leaves ranged from about 550 g kg^{-1} DM for lower leaves to 650 g kg^{-1} DM for upper leaves in the canopy, compared with 500 g kg^{-1} DM for bottom stems and 550 g kg^{-1} DM for top stems (Buxton and Marten 1989). By comparison, other studies found alfalfa at the midflower stage of maturity had leaves that contained only about 250 g NDF kg^{-1} DM with a digestibility of 750 g kg^{-1} DM (Buxton and Hornstein 1986; Buxton et al.

1985). Stems ranged in NDF concentration from 450 to 700 g kg^{-1} DM from top to bottom with digestibilities that ranged from 650 to 450 g kg^{-1} DM from top to bottom of the stems.

LIGNIN. Cell walls are fabricated from primary and secondary walls. Polysaccharides are the principal components, but lignin and associated phenolics, which add rigidity to the cell wall, are the chemical constituents most often identified as limiting cell wall polysaccharide digestibility by ruminants. Lignin is virtually indigestible and is present in forages in concentrations that usually range from 3% to 12% of DM. Lignin concentration is generally higher in legume cell walls than in those of grasses, but lignin in grasses is usually more inhibitory to digestion than that in legumes (Fig. 6.3). Lignin has been described as having both core and noncore components (Jung and Deetz 1993). Core lignin is a highly condensed, high-molecular weight polymer of three closely related phenylpropanoid monomers. These vary in methoxylation at the 3- and 5-carbon (C) positions of the aromatic ring: *p*-coumaryl (no methoxylation), coniferyl (monomethoxylation), and sinapyl (dimethoxylation) alcohols.

Noncore lignin consists primarily of *p*-coumaric and ferulic acids attached to either core lignin or to the hemicellulose arabinoxylan. Both acids are present in larger quantities in cell walls of grasses than of legumes, which may explain why grasses are more responsive to alkaline treatment than legumes. These phenolic acids occur as ester monomers or dimers, which are labile to alkali treatment. Grasses also contain some ether-bound phenolics that are not alkali labile.

Attachment of lignin polymers to polysaccharides may occur by cross-linking through ferulic and *p*-coumaric acids and other compounds. Most ether-linked *p*-coumaric acid is bound to the core lignin polymer and may be deposited in the cell during lignification of the secondary wall. Ferulic acid is thought to be present mostly in primary cell walls. Lignification begins in the middle lamella and primary wall regions and lags behind secondary wall thickening. Bidlack and Buxton (1992) found in stems of six cool-season grasses and legumes that maximum deposition rate of hemicellulose occurred first, followed by cellulose (1 to 6 d later) and then lignin (up to 13 d after maximum hemicellulose deposition).

CELL WALL CARBOHYDRATES. Structural polysaccharides often comprise more than 90% of cell walls of forages. Cell wall carbohydrates can be grouped into cellulosic, hemicellulosic, and pectic polysaccharides. Cellulose is a high-molecular weight linear polymer of beta 1,4 linked D-glucopyranose units. Many of the unique properties of cellulose arise from its secondary structure. Linear chains of glucose units combine to form microfibrils, which are extensively cross-linked by hydrogen (H) bonding (Hatfield 1993). Microfibrils are insoluble, hydrophobic, and more resistant to digestion than the glucan chains from which they are formed.

Hemicelluloses are a complex mixture of linear and branched chain polysaccharides containing polymers of xylose, arabinose, mannose, galactose, glucose, and uronic acids. Pectic polysaccharides include both polygalacturonans and neutral polysaccharides (primarily arabinans, galactans, and arabinogalactans). They are present only in the primary cell wall and middle lamella. The amount of pectic polysaccharides in mature grass cell walls is usually less than 1%, but they can make up 10%-20% of legume cell walls. Their concentration is greater in immature than in mature herbage. Pectic sub-

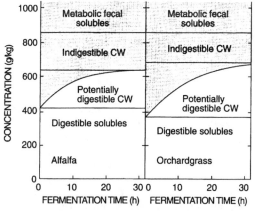

Fig. 6.3. In vivo digestion of alfalfa and orchardgrass organic matter. Unshaded area represents plant material that can be digested. Alfalfa contains more cell solubles that are very rapidly digested than orchardgrass, and orchardgrass has more cell wall material (CW) than alfalfa. A higher proportion of the cell wall in orchardgrass is potentially digestible, but the digestion rate is slower. The amount of indigestible cell wall is greater in alfalfa than in orchardgrass. The ordinate represents concentrations in feed plus metabolic fecal dry matter. Metabolic fecal solubles are produced by microbes during digestion. (Adapted from Waldo and Jorgensen 1981.)

stances, specifically those that are easily extracted, are usually rapidly and extensively digested (Titgemeyer et al. 1992).

The hemicellulose in most forages consists of over 90% xylans. Grasses have a higher proportion of hemicellulose in their cell walls than do legumes. Composition of xylans varies depending upon frequency of substitution with arabinose and glucuronic acids. The magnitude of branching or substitution on polysaccharide chains can govern the nature of interactions of these polysaccharides through H or ionic bonding.

Cellulose microfibrils are thought to be embedded in a hemicellulose and lignin matrix. Cross-linkage between lignin and cell wall polysaccharides can dramatically affect the digestibility of polysaccharides (Ralph and Helm 1993). Lignin may be bonded covalently through arabinose to hemicellulosic polysaccharides in the matrix, but there is no evidence of covalent bonding between lignin and cellulose. Galactose and arabinose in the cell wall are usually degraded more completely by rumen organisms than glucose or xylose (Buxton and Brasche 1991). Xylose generally is the least digestible of the sugars in cell walls.

Lignin may limit access of microbes and enzymes to polysaccharides in the cell wall rather than selectively protecting specific carbohydrate components. Differential digestibilities of cell wall polysaccharides and monosaccharides may be a reflection of the cell types being digested and their polysaccharide composition rather than differences in digestibilities of specific polysaccharides within a cell wall type. Evidence for this concept is discussed by Chesson (1993), who shows that individual cell types of ryegrass (*Lolium* L. sp.) differed in rate and extent of digestion, but within each cell type there was no observable difference in the rate of polysaccharide degradation. Likewise, Piwonka et al. (1991) found that digestibility of arabinose, xylose, and glucose did not differ within leaf tissues and cell types of caucasian bluestem (*Bothriochloa caucasia* [Trin.] C.E. Hubb.).

VARIATION AMONG FORAGES. Grasses develop lignified veins to provide mechanical support for leaves; the veins contribute to the high cell wall concentration of grass leaves. The NDF concentration of cool-season grass leaves is usually about twice that of legume leaves (Sanderson and Wedin 1989a). Consequently, legume leaves are often more digestible than those of grasses. But the difference in digestibility is not as great as suggested by the relative cell wall concentrations because the NDF of grasses is typically more digestible than that of legumes (Fig. 6.2). The total herbage of legumes generally has a lower cell wall concentration than grasses, and the digestion rate of the cell wall is usually faster in legumes than in grasses.

Minson (1990) reports that the average digestibility of cool-season grasses was 13 g kg^{-1} greater than that of warm-season grasses. Data comparing cool- and warm-season grasses are often confounded, however, because warm-season grasses are normally grown in warmer environments than cool-season grasses. Warm temperatures hasten maturity, which is usually associated with an increase in cell wall concentration and lower digestibility.

Only a few studies have compared cool-season and warm-season grasses in the same growth environment. Wilson et al. (1983) report that leaves of C$_3$ *Panicum* species averaged 70 g kg^{-1} DM less NDF than leaves of C$_4$ *Panicum* species. Similarly, Kephart and Buxton (1993) found C$_3$ grasses contained 45 g kg^{-1} DM less NDF in leaf blades than C$_4$ grasses. In the latter study there was no consistent difference between the C$_3$ and C$_4$ species for DM digestibility or for NDF lignin concentration. Cool-season grasses often have a lower proportion of stems in herbage than do warm-season grasses, which may account for much of the lower forage quality of warm-season grasses.

As plants advance in maturity, the leaf:stem ratio usually decreases (Fig. 6.4). Additionally, cell wall concentration within stems increases, usually also within leaves (especially in grasses), and the proportion of cell solubles decreases. For example, Sanderson and Wedin (1989a) report that DM digestibility of leaf blades and stems of timothy (*Phleum pratense* L.) and smooth bromegrass declined with increasing maturity, whereas those of alfalfa and red clover retained relatively high digestibility with advance in maturity.

As a result, digestibility of grass herbage usually declines faster with maturity than does that of legume herbage. Fastest rates of decline occur during primary growth of grasses when reproductive development is occurring. Buxton and Marten (1989) report that herbage digestibility of four cool-season grass species grown in Iowa decreased linearly with time during the spring. Plants with a moderate number of reproductive tillers had an average rate of decline in digestibility of 3.9 g

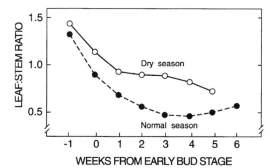

Fig. 6.4. Changes in leaf:stem ratio of alfalfa during spring growth. The leaf:stem ratio is normally high in immature herbage and low in more mature herbage, which is a major reason that mature forages are of lower quality than immature forages. The leaf:stem ratio is also often greater in plants grown under drought than under normal conditions. (Adapted from Albrecht et al. 1987.)

kg^{-1} d^{-1}, whereas the rate averaged 6.1 g kg^{-1} d^{-1} in plants with numerous reproductive tillers. By comparison, alfalfa herbage digestibility decreased 2.8 g kg^{-1} d^{-1} during spring growth (Buxton et al. 1985), which is similar to data reported by Fick and Onstad (1988).

Cell walls of leaves and stems of cool-season grasses may double in lignin concentration with advancing maturity, whereas most legumes show an increase of 20% or less in the amount of lignification of cell walls over the same period (Buxton and Hornstein 1986; Buxton 1990). As a result, cell wall digestibility decreases to a greater extent with maturity in grasses than in legumes. At the time of normal harvest or grazing, however, cell wall concentration increases to a greater extent with maturity in legume stems compared with those of grasses.

ENVIRONMENTAL FACTORS. Plant environment usually has a smaller effect on forage quality than on forage yield (Buxton and Casler 1993), and environmental effects on forage quality are generally much less than the effects of forage maturity or aging. Environmental effects are integrated through plant physiological processes and reflected in forage growth rate, developmental rate, yield, and herbage quality. Year to year and seasonal variation in environment alters herbage quality, however, even when forages are harvested at similar morphological stages. This can result in inconsistent performance of animals that consume the forage. Environment often exerts its greatest influence on forage

quality by altering leaf:stem ratios, but it also causes other morphological modifications and changes in chemical composition of plant parts. Environmental factors that have the greatest effect on nutritive value of forages are temperature, water deficits, solar radiation regime, and nutrient availability. Temperature usually has a greater influence on forage quality than other environmental factors.

Optimal growth temperatures are near 20°C for cool-season species compared with 30° to 35°C for warm-season species. At temperatures below the optimum for growth, soluble sugars accumulate because photosynthetic rate is less sensitive to low temperature than is growth. High temperatures normally increase rate of plant development and reduce plant leaf:stem ratio and digestibility. Herbage grown at a high temperature usually has lower forage quality than herbage grown at a low temperature, even when compared at the same morphological stage.

Wilson (1982) and Wilson and Minson (1983) conclude that an increase of 1°C in the range of normal plant growth decreases digestibility of legumes by 2-3 g kg^{-1} DM and that of warm-season grasses by about 6 g kg^{-1} DM. Fales (1986) likewise shows that elevated growth temperature decreased the digestibility of tall fescue leaves. The depressed digestibility associated with elevated temperatures is usually attributed to higher concentrations of cell wall constituents. Additionally, the cell wall is usually less digestible from plants grown in warm environments than from plants grown in cooler environments. This is one reason that forages produced at locations with cool temperatures, such as those closer to the earth's poles or at high elevations, tend to be of high quality.

Drought usually inhibits tillering and branching of forages and hastens death of established tillers. Leaf area is reduced by accelerating the rate of senescence of older leaves. Both N and soluble carbohydrates are mobilized and exported out of leaves as they go through senescence. Water stress typically slows development of forages (Halim et al. 1989). Hence, if the leaf loss associated with drought is not severe, water deficit may actually improve forage quality (Peterson et al. 1992; Sheaffer et al. 1992). Under severe, prolonged water stress, however, leaves are lost and some perennial species may go into dormancy, which causes most nutrients to be translocated from leaves to roots and results in low forage quality.

The effects of water stress on CP concentrations have been inconsistent, which probably depends (1) on whether or not the stress was great enough to cause leaf senescence and (2) on the distribution of N in the soil profile in relation to the limited available soil water. If both N and water are present in the same soil horizon, they may be taken up together and herbage CP concentration may be unaffected or may be increased if N is more available than soil water. If subsoil water is ample and most of the soil N is near the surface, however, growth may continue with reduced N uptake so that herbage CP concentration declines.

Forages frequently experience reduced solar radiation during periods of cloudiness; when shaded by neighboring plants, especially when grown in swards with a mixture of tall and short species; or when lower plant parts are shaded by upper plant parts. Photosynthetic rates of individual leaves decrease in a curvilinear manner with increasing shade, and growth rates and herbage yield likewise decrease. Shading also reduces tillering and stem development.

In some species, shading may also induce stem, petiole, or leaf elongation resulting in elevated leaves that are more likely to be exposed to high sunlight levels. Shaded grass leaves typically are longer, thinner, and sometimes narrower than when grown in full sunlight. Buxton and Lentz (1993) found that orchardgrass leaves of vegetative tillers were 22% longer when equally spaced plants were grown on 0.15-m centers, where interplant shading was greater, rather than on 0.60-m centers. Conversely, plant density had little effect on morphology of leaf blades from reproductive tillers.

Shading typically has a smaller effect on forage quality than on forage morphology or yield. Morphological changes that affect forage canopy height, however, can affect quality indirectly by altering bite size and intake potential of forages by grazing animals. Bite size, which is influenced by canopy height, is one of the most important determinants of intake rate of grazed forages. Crude protein concentration is much more responsive to shading of grasses than other quality parameters, and leaves are more responsive than stems (Kephart and Buxton 1993). The CP response for legumes, however, is generally smaller than for grasses. Increased concentration of CP from shading is usually at the expense of soluble carbohydrates. The effects of shading on DM digestibility have been small, usually less than 5%. Some studies have shown an increase in digestibility, whereas others have shown a decrease in digestibility.

Diurnal variation has been observed in forage digestibility and concentrations of nonstructural carbohydrates, with lowest values before sunrise and highest values in the afternoon. For example, Lechtenberg et al. (1971) found that digestibility of alfalfa was about 16 g kg^{-1} greater in the late afternoon than in the early morning before sunrise. Leaf starch concentration increased by 100 g kg^{-1} during the daylight, whereas that in stems did not change. These changes caused simultaneous shifts in the leaf:stem ratio from 1.1 to 1.5. Additionally, protein tends to be degraded at night through proteolysis and resynthesized the next day. The result is diurnal fluctuation in protein concentration. However, some of the advantages of high-quality forage in the late afternoon may be lost when it is harvested for hay because of high respiration rates associated with these soluble plant fractions.

Soil nutrients have only small effects on forage quality. Nitrogen fertilization has the greatest impact on plant growth and usually raises CP concentration of nonlegumes as discussed previously. Forage species with low N concentrations, such as warm-season grasses grown on soil with low N levels, may have improved digestibility as a result of N application because the additional N in the plant, resulting from fertilization, may stimulate rumen microbe activity.

During spring growth in temperate regions, the effect of warming temperatures interacts with that of advancing maturity to cause rapid declines in forage quality (Van Soest 1982). Temperatures do not change as markedly during summer regrowth, and the simple effect of advancing maturity on forage quality often results in a slower decline than during the spring.

KINETICS OF CELL WALL DIGESTION

Digestion is a time-dependent process. Rate of digestion relative to rate of passage is a critical dynamic property affecting digestibility in the digestive tract. To be utilized, cell walls must be retained in the digestive tract for 40-50 h. Long retention times occur because ruminants ingest a portion of their feed as large particles that will not pass the reticulo-omasal orifice. These particles must be regurgitated and chewed (ruminated) later. Large buoyant particles form a floating mat in the rumen that segregates them from the rumen liquid, slowing passage compared with small particles that sink in the rumen.

Extent and Rate of Cell Wall Digestion. Three parameters are needed to describe digestion kinetics of forage cell walls: indigestible fraction, fractional rate constant, and discrete lag time. A portion of the cell wall does not degrade during anaerobic fermentation (Fig. 6.3). The potentially digestible cell wall is degraded by first-order kinetics. Digestion does not begin instantaneously at zero time, and a discrete lag time is often added to models to quantify the lag phenomenon (Mertens 1993).

Long retention times of slowly digesting particles in the rumen result in rumen fill and reduced intake of additional forage. Simulations with mathematical models suggest that cell wall concentration of plants, extent of cell wall digestion, and rate of passage, including rate of particle size reduction, are the most important factors affecting intake (Mertens 1993). Rate of cell wall digestion and length of lag time before digestion begins have less effect on intake.

The indigestible fraction is an important kinetic parameter because it defines an intrinsic property of the cell wall and affects animal utilization (Mertens 1993). Measurement of the indigestible fraction is critical to the estimation of the potential digestibility fraction, fractional rate constant, and lag time. Ideally, indigestibility should be measured after an infinite fermentation time, but it can be reasonably estimated by the residue remaining after 72-96 h because digestion is typically >99% complete after these times. Indigestible NDF is greater in mature than in immature forage and can make up 20%-50% of stem DM in forages depending on species and maturity. A high positive correlation exists between lignin concentration and indigestible NDF in DM (Smith et al. 1972; Buxton 1989).

Fractional rates of digestion of NDF can vary from 0.02 to 0.19 h^{-1} (Smith et al. 1972), but variation in rate has not been related to chemical composition (Mertens 1993). The NDF concentration is often negatively related to the fractional rate constant of digestion (Smith et al. 1972; Fisher et al. 1989). Forages with high concentrations of NDF in DM may have walls that are thickened and, therefore, more resistant to mechanical disruption and microbial penetration and digestion. Available surface area for microbial attack may limit rate of digestion in these situations. Parenchyma cells usually are rapidly and extensively degraded in situ, which may occur in part because of their small size, thin cell walls, and large surface area exposed to rumen microorganisms after mastication. On the other hand, sclerenchyma cells are slowly

and incompletely degraded in situ, which may occur in part because of their large size, thick cell walls, and low surface area exposed to digestion after mastication (Grabber et al. 1992). Longer lag times have also been associated with the lower accessibility of forage cell walls to microorganisms. Attachment of bacteria to cells is a prerequisite to digestion that may be a determinant of lag time (Allen and Mertens 1988).

Two of the most important sources of variation in digestion kinetics are plant species and maturity. Legumes generally have more rapid rates of NDF degradation than grasses (Smith et al. 1972; Buxton 1989). Additionally, immature forages generally have a faster rate of digestion than more mature forages. Stems of a late maturity orchardgrass group had digestion rates that were 28% greater than those of an early-maturity group (Fig. 6.5).

Rate of Passage. In ruminants, rate of passage is limited by rate of particle size reduction (selective retention of large particles) and rate of flow from the rumen. Hence, horses and other nonruminant herbivores typically have a higher rate of passage and reduced digestion of cell walls than ruminants when fed similar forage (Rittenhouse 1986). Because

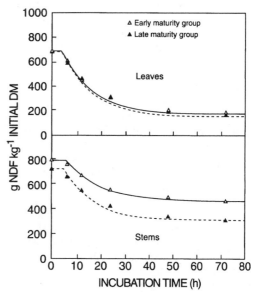

Fig. 6.5. In vitro digestion time course of neutral detergent fiber (NDF) in leaves and stems of two maturity groups of orchardgrass harvested on the same date in late spring. After an initial lag, the NDF in leaves was more digestible than that in stems, and the difference between maturity groups was greater in stems than in leaves. (Adapted from Lentz and Buxton 1992.)

they chew forage more thoroughly during eating, horses can digest high-quality forages nearly as well as cattle and sheep. Horses do not digest low-quality, high-fiber forages as well as cattle and sheep. Factors affecting rate of passage include cell wall concentration and composition, particle size, and density.

Digesta particles must be reduced to a size that will pass through a 1-mm screen before they can easily flow from the rumen (Minson 1990). Leaves are usually retained in the rumen for a shorter time than stems because of both higher rates of passage and faster rates of NDF digestion in leaves. Plant factors that affect rate and extent of cell wall digestion may also influence fragility (brittleness) and cell wall disintegration in the rumen. Cool-season grasses, with less leaf rigidity, are broken down to a lesser degree by chewing than warm-season grasses (Wilson 1993). Likewise, the number of times that ingested hay is chewed and the time spent masticating are greater for late-cut than for early-cut hays and greater for grasses than for legumes. Hence, the rate of disappearance of late-cut hay from the rumen can be up to five times greater for legumes than for grasses (Grenet 1989).

Anatomical structure of grass stems may also influence their rate of passage. Lignification of parenchyma in grass stems nearing maturity results in a compact, interlocked residue after digestion made up of epidermis, sclerenchyma ring, vascular tissues, and parenchyma. In contrast, the parenchyma of legume stems is digested even in mature plants, leaving an outer ring of lignified cells. The pattern of breakdown of legume stems into small particles probably provides an advantage in rate of passage for forage legumes over grasses. In general, legume fragments are more cuboidal, whereas grass particles are elongated and slender. The relatively long filamentous shape of grass particles may result in a greater resistance to escape from the rumen than the cuboidal fragments of legumes (Mertens et al. 1984; Wilson 1993).

In addition to particle size, particle density plays a role in passage of particles from the rumen. Specific gravity of ruminal contents is lowest after eating and approaches unity before the next meal. When particles begin to ferment rapidly, fermentation gasses can buoy them, keeping their specific gravity less than 1.0. Once digestion of particles slows as it nears completion, and particles are completely hydrated, they achieve a specific gravity of between 1.2 and 1.4, which is optimal

for passage from the rumen (Wattiaux et al. 1992).

METABOLISM AND PARTITIONING OF ABSORBED NUTRIENTS

Of the variation in energy intake among forages, approximately 65%-75% may be related to intake, 20%-30% to differences in digestibility, and only 5%-15% to differences in metabolic efficiency. Digestible energy obtained from forage is generally used less efficiently than equal digestible energy from grain. Additionally, when grasses and legumes are fed at similar levels of intake, energy from legumes may be used more efficiently for tissue gain than energy from grasses (Waldo et al. 1990). Although the difference between grasses and legumes in efficiency of energy use is not universally accepted, many studies suggest a greater efficiency in converting metabolizable energy of legumes to net energy for production. Some evidence indicates that the potential difference in efficiency relates to differences in protein absorption. Because legumes have higher protein concentrations, a higher proportion of the digestible energy absorbed is derived from amino acids. This difference in protein concentration may have its biggest effect on the performance of young rapidly growing animals. If the carbon skeletons from deaminated amino acids are used for glycolytic precursors to convert acetate to fat tissue, however, higher protein concentration may also improve energetic efficiency of acetate utilization.

An alternative explanation for less efficient use of energy from grasses than legumes is related to the fraction of digested organic matter that is derived from cell walls. Because grass cell walls normally are in greater concentration and the cell walls of grasses are more potentially digestible than those of legumes (Fig. 6.3), cell walls represent more of the digested organic matter in grasses than in legumes. Cell walls contribute about 30% of the digested organic matter of legumes, compared with 50% for grasses. Digestion of cell walls is thought to result in greater acetate and less propionate production in the rumen than digestion of cell solubles (Van Soest 1982). Similar to carbon skeletons from amino acids, propionate can provide glycolytic precursors for acetate conversion to fat tissue. Therefore, a greater proportion of propionate in volatile fatty acids produced during fermentation in the rumen may also contribute to greater efficiency of energy use for legumes than for grasses.

Because greater cell wall concentration of grasses compared with the concentration in legumes results in greater gut fill, it may also cause greater energy consumption by the gut. This difference, when combined with differences in the acetate:propionate ratio of absorbed volatile fatty acids, suggests that cell wall concentration in herbage may influence metabolic efficiency of energy utilization in herbage. Hence, immature grasses may be used more efficiently than mature grasses, and the cool-season grasses (having less cell wall) may have greater energetic efficiency than the warm-season grasses.

QUESTIONS

1. Which mineral required by animals is most limiting in forage-based diets?
2. Which fertilizer nutrients normally have the largest influence on forage quality? Which aspects of forage quality are affected?
3. Discuss the relationship between cell wall concentration of herbage and the amount of available energy for livestock.
4. Why is potential intake of a forage so difficult to predict?
5. Discuss physical limitations and physiological control as regulators of intake by herbivores and factors affecting the assessment of forage quality.
6. What factors determine the amount of protein N lost in the urine of ruminants and the amount of microbial protein produced in the rumen?
7. Which cell types in forages are degraded most rapidly, and which are degraded most slowly?
8. Describe the components of lignin and how lignin may inhibit the digestion of forage cell walls.
9. Explain why C_3 species are usually more digestible than C_4 species.
10. Explain why, with advancing age, the digestibility of grass stems and leaves usually declines at a faster rate than does that of legumes.
11. What are the most important sources of variation in digestion kinetics of plants in the rumen?
12. Why do horses have a faster rate of digesta passage than ruminants?
13. Explain why the rate of disappearance from the rumen is faster for legumes than for grasses.
14. Explain how particle density influences the rate of escape from the rumen.
15. Explain why legumes may be used more efficiently for tissue gain than grasses.

REFERENCES

Akin, DE. 1989. Histological and physical factors affecting digestibility of forages. Agron. J. 81:17-25.

Albrecht, KA, and GA Broderick. 1993. Ruminal in vitro degradation of protein from different forage legume species. In 1992 Research Summaries. Madison, Wis.: US Dairy Forage Research Center, 92-94.

Albrecht, KA, WF Wedin, and DR Buxton. 1987. Cell-wall composition and digestibility of alfalfa stems and leaves. Crop Sci. 27:735-41.

Allen, MS, and DR Mertens. 1988. Evaluating constraints on fiber digestion by rumen microbes. J. Nutr. 118:261-70.

Beauchemin, KA, and JG Buchanan-Smith. 1989. Effects of dietary neutral detergent fiber concentration and supplementary long hay on chewing activities and milk production of dairy cows. J. Dairy Sci. 72:2288-2300.

Bidlack, JE, and DR Buxton. 1992. Content and deposition rates of cellulose, hemicellulose, and lignin during regrowth of forage grasses and legumes. Can. J. Plant Sci. 72:809-18.

Buxton, DR. 1989. In vitro digestion kinetics of temperate perennial forage legume and grass stems. Crop Sci. 29:213-19.

———. 1990. Cell-wall components in divergent germplasm of four perennial forage grass species. Crop Sci. 30:402-8.

Buxton, DR, and MR Brasche. 1991. Digestibility of structural carbohydrates in cool-season grass and legume forages. Crop Sci. 31:1338-45.

Buxton, DR, and MD Casler. 1993. Environmental and genetic effects on cell wall composition and digestibility. In HG Jung et al. (eds.), Forage Cell Wall Structure and Digestibility. Madison, Wis.: American Society of Agronomy, 685-714.

Buxton, DR, and JS Hornstein. 1986. Cell-wall concentration and components in stratified canopies of alfalfa, birdsfoot trefoil, and red clover. Crop Sci. 26:180-84.

Buxton, DR, and EM Lentz. 1993. Performance of morphologically diverse orchardgrass clones in spaced and solid plantings. Grass and Forage Sci. 48:336-46.

Buxton, DR, and GC Marten. 1989. Forage quality of plant parts of perennial grasses and relationship to phenology. Crop Sci. 29:429-35.

Buxton, DR, and JR Russell. 1988. Lignian constituents and cell-wall digestibility of grass and legume stems. Crop Sci. 28:553-58.

Buxton, DR, JS Hornstein, WF Wedin, and GC Marten. 1985. Forage quality in stratified canopies of alfalfa, birdsfoot trefoil, and red clover. Crop Sci. 25:273-79.

Chesson, A. 1993. Mechanistic models of forage cell wall degradation. In HG Jung et al. (eds.), Forage Cell Wall Structure and Digestibility. Madison, Wis.: American Society of Agronomy, 337-76.

Fales, SL. 1986. Effects of temperature on fiber concentration, composition, and in vitro digestion kinetics of tall fescue. Agron. J. 78:963-66.

Fick, GW, and DW Onstad. 1988. Statistical models for predicting alfalfa herbage quality from morphological or weather data. J. Prod. Agric. 1:160-66.

Fisher, DS, JC Burns, and KR Pond. 1989. Kinetics of in vitro cell-wall disappearance and in vivo digestion. Agron. J. 81:25-33.

Grabber, JH, GA Jung, SM Abrams, and DB Howard. 1992. Digestion kinetics of parenchyma and sclerenchyma cell walls isolated from orchardgrass and switchgrass. Crop Sci. 32:806-10.

Grenet, E. 1989. A comparison of the digestion and reduction in particle size of lucerne hay (*Medicago sativa*) and Italian ryegrass hay (*Lolium italicum*) in the ovine digestive tract. Br. J. Nutr. 62:493-507.

Halim, RA, DR Buxton, MJ Hattendorf, and RE Carlson. 1989. Water-stress effects on alfalfa forage quality after adjustment for maturity differences. Agron. J. 81:189-94.

Hatfield, RD. 1993. Cell wall polysaccharide interactions and degradability. In HG Jung et al. (eds.), Forage Cell Wall Structure and Digestibility. Madison, Wis.: American Society of Agronomy, 285-313.

Jarrige, R (ed.). 1989. Ruminant Nutrition: Recommended Allowances and Feed Tables. Paris: Institut National de al Recherche Agronomique.

Jung, HG, and DA Deetz. 1993. Cell wall lignification and degradability. In HG Jung et al. (eds.), Forage Cell Wall Structure and Digestibility. Madison, Wis.: American Society of Agronomy, 315-46.

Kephart, KD, and DR Buxton. 1993. Forage quality responses of C_3 and C_4 perennial grasses to shade. Crop Sci. 33:831-37.

Lechtenberg, VL, DA Holt, and HW Youngberg. 1971. Diurnal variation in nonstructural carbohydrates, in vitro digestibility, and leaf to stem ratio of alfalfa. Agron. J. 63:719-24.

Lentz, EM, and DR Buxton. 1992. Digestion kinetics of orchardgrass as influenced by leaf morphology, fineness of grind, and maturity group. Crop Sci. 32:482-86.

Mertens, DR. 1993. Kinetics of cell wall digestion and passage in ruminants. In HG Jung et al. (eds.), Forage Cell Wall Structure and Digestibility. Madison, Wis.: American Society of Agronomy, 535-70.

Mertens, DR, TL Strawn, and RS Cardoza. 1984. Modelling ruminal particle size reduction: Its relationship to particle size description. In PM Kennedy (ed.), Techniques in Particle Size Analysis of Feed and Digesta in Ruminants, Can. Soc. Anim. Sci. Occas. Publ. 1. Edmonton, Alberta, Canada, 134-41.

Minson, DJ. 1990. Forage in Ruminant Nutrition. New York: Academic Press.

Mullahey, JJ, SS Waller, KJ Moore, LE Moser, and TJ Klopfenstein. 1992. In situ ruminal protein degradation of switchgrass and smooth bromegrass. Agron. J. 84:183-88.

National Research Council (NRC). 1984. Nutrient Requirements of Beef Cattle. 6th rev. ed. Washington, D.C.: National Academy Press.

———. 1989. Nutrient Requirements of Dairy Cattle. 6th rev. ed. Washington, D.C.: National Academy Press.

Peterson, PR, CC Sheaffer, and MH Hall. 1992. Drought effects on perennial forage legume yield and quality. Agron. J. 84:774-79.

Piwonka, EJ, JW MacAdam, MS Kerley, and JA Paterson. 1991. Composition and susceptibility to rumen microbial degradation of nonmesophyll cell walls isolated from caucasian bluestem (*Bothriochloa caucasica* [Trin.]) leaf tissue. J. Agric. Food Chem. 39:473-77.

Ralph, J, and RF Helm. 1993. Lignin/hydroxycinnamic acid/polysaccharide complexes: Synthectic

models for regiochemical characterization. In HG Jung et al. (eds.), Forage Cell Wall Structure and Digestibility. Madison, Wis.: American Society of Agronomy, 201-46.

Rittenhouse, R. 1986. The relative efficiency of rangeland use by ruminants. In Olafur Gudmundsson (ed.), Grazing Research at Northern Latitudes. New York: Plenum, 179-91.

Sanderson, MA, and WF Wedin. 1989a. Phenological stage and herbage quality relationships in temperate grasses and legumes. Agron. J. 81:864-69.

———. 1989b. Nitrogen in the detergent fibre fractions of temperate legumes and grasses. Grass and Forage Sci. 44:159-68.

Sheaffer, CC, PR Peterson, MH Hall, and JB Stordahl. 1992. Drought effects on yield and quality of perennial grasses in the north central states. J. Prod. Agric. 5:556-61.

Smith, LW, HK Goering, and CH Gordon. 1972. Relationships of forage compositions with rates of cell wall digestion and indigestibility of cell walls. J. Dairy Sci. 55:1140-47.

Thomson, DJ, DR Waldo, HK Goering, and HF Tyrrell. 1991. Voluntary intake, growth rate, and tissue retention by Holstein steers fed formaldehyde- and formic acid-treated alfalfa and orchardgrass silages. J. Anim. Sci. 69:4644-59.

Titgemeyer, EC, LD Bourquin, and GC Fahey, Jr. 1992. Disappearance of cell wall monomeric components from fractions chemically isolated from alfalfa leaves and stems following *in-situ* ruminal digestion. J. Sci. Food Agric. 58:451-63.

Van Soest, PJ. 1982. Nutritional Ecology of the Ruminant. Corvallis, Oreg.: O and B Books.

Waldo, DR, and NA Jorgensen. 1981. Forages for high animal production: Nutritional factors and effects of conservation. J. Dairy Sci. 64:1207-29.

Waldo, DR, GA Varga, GB Huntington, BP Glenn, and HF Tyrrell. 1990. Energy components of growth in Holstein steers fed formaldehyde- and formic acid-treated alfalfa or orchardgrass silages at equalized intakes of dry matter. J. Anim. Sci. 68:3792-3804.

Wattiaux, MA, LD Satter, and DR Mertens. 1992. Effect of microbial fermentation on functional specific gravity of small forage particles. J. Anim. Sci. 70:1262-70.

Weiss, WP, and WL Shockey. 1991. Value of orchardgrass and alfalfa silages fed with varying amounts of concentrates to dairy cows. J. Dairy Sci. 74:1933-43.

Wilson, JR. 1982. Environmental and nutritional factors affecting herbage quality. In JB Hacker (ed.), Nutritional Limits to Animal Production from Pastures. Farnham Royal, UK: CAB International, 111-31.

———. 1993. Organization of forage plant tissues. In HG Jung et al. (eds.), Forage Cell Wall Structure and Digestibility. Madison, Wis.: American Society of Agronomy, 1-32.

Wilson, JR, and DJ Minson. 1983. Influence of temperature on digestibility of tropical legume *Macroptilium atropurpureum*. Grass and Forage Sci. 38:39-44.

Wilson, JR, RH Brown, and WR Windham. 1983. Influence of leaf anatomy on the dry matter digestibility of C_3, C_4, and C_3/C_4 intermediate types of *Panicum* species. Crop Sci. 30:141-46.

Evaluating Forage Production and Quality

7

LYNN E. SOLLENBERGER
University of Florida

DEBBIE J. R. CHERNEY
Cornell University

MEASUREMENTS of forage production and quality are important in cultivar development programs and forage management research. Assessment of forage quality also is critical when formulating livestock rations and selling or purchasing forages. The value of forage is best understood in the context of its contribution to livestock production. Thus, quantifying forage production and quality in ways that relate meaningfully to livestock production is desirable.

CONCEPTS OF FORAGE PRODUCTION AND QUALITY

When forage is harvested mechanically, production generally is considered to be the amount of forage dry matter (DM) above the residual stubble height. When forage is grazed, its production can be described in a number of ways. Forage mass is total dry weight of forage per hectare above soil level or some reference height (Forage and Grazing Terminology Committee 1992). Forage accumulated is the increase in forage mass per

hectare over a specified period of time (FGTC 1992) and when expressed on a per day basis is a measure of growth rate. Because animals graze selectively and consume only a portion of the forage mass or accumulation, these measures may be of limited utility if they are not closely related to quantity of forage consumed by livestock. Quantifying the proportion of the forage mass or accumulation that is consumed, or available for consumption, allows useful comparisons of productivity among species or management practices and provides a basis for estimating an appropriate stocking rate.

Forage quality is best defined as production per animal. This definition is valid if animal potential is not limiting performance, no energy or protein supplements are fed, and forage is offered ad libitum (Mott and Moore 1970). Factors affecting forage quality include forage nutritive value and intake. Nutritive value is the chemical composition, digestibility, and nature of digested products of a forage. Intake is related positively to the accessibility and acceptability of forage and negatively to the amount of time that forage is retained in the rumen and reticulum. When it is not possible to measure production per animal, voluntary intake of DM, digestible DM, organic matter, or digestible organic matter can be used as measures of forage quality.

Animal factors and the determinants of forage quantity and quality affect livestock production per animal and per hectare (Fig. 7.1). Animal production per hectare often is used to characterize the potential of a production system or management strategy. It can be cal-

LYNN E. SOLLENBERGER is Professor of Agronomy, University of Florida. He received the MS degree from Pennsylvania State University and the PhD degree from the University of Florida. Since 1985 he has taught and conducted research in the areas of pasture and forage crop management and utilization.

DEBBIE J. R. CHERNEY is Senior Research Associate, Department of Animal Science, Cornell University. She received the MS degree from Louisiana State University and the PhD degree from the University of Florida. Her research interests include forage utilization and evaluation.

culated from measures of forage quantity (expressed as animals/hectare) and quality (expressed as product/animal), i.e.,

$$\text{product/animal} \times \text{animals/hectare}$$
$$= \text{animal product/hectare}.$$

FORAGE EVALUATION SCHEMES

Several schemes for evaluating forages have been outlined (Mott and Moore 1970; Mochrie et al. 1981; Wheeler 1981). They are useful for systematically identifying superior plants and providing management information to guide the use of these plants on a farm. Different schemes may be needed depending on whether the forage is to be used for grazing or mechanical harvest, whether resources for research are plentiful or limited, and whether the plant being introduced is a different species or is only a slight modification of a species in use in the region (Mochrie et al. 1981; Wheeler 1981). General consensus exists that testing with animals should occur early in the evaluation of a forage (Mochrie et

al. 1981; Matches 1992).

The Mott and Moore (1970) scheme has five phases: initial screening of introductions and breeders' lines, small plot clipping trials, studies of plant responses to grazing stress, animal response trials, and evaluation of forage-livestock systems. The first four phases of this scheme are used as an outline for discussing techniques to quantify forage production and quality.

Initial Screening. Often numerous entries are included in initial screening trials. Seed or vegetative planting materials generally are in limited supply, so spaced plants or a few rows of each entry may be used. Screening trials provide information on adaptation to the local environment, growth patterns, plant morphology, productivity, in vitro digestibility, and chemical composition. Preliminary evaluation of palatability and response to defoliation may be obtained in screening trials, but caution in interpretation must be exercised when numerous lines are included in the

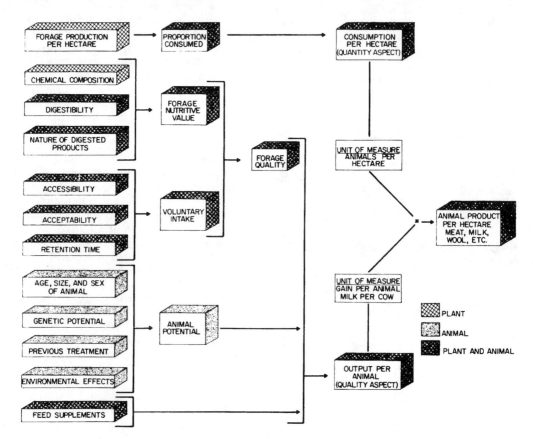

Fig. 7.1. The relationship of forage production and quality to animal production per head and per hectare (Mott and Moore 1985).

same grazing unit (Burns et al. 1989a). Palatable lines may be grazed more severely than unpalatable ones, making fair evaluation of persistence difficult. Initial screening should provide enough information that a relatively large proportion of entries can be discarded because of deficiencies. In later phases, cost per entry evaluated is greater, so severe culling is critical in the screening phase.

Clipping Trials. Clipping trials follow initial screening and use plots that often range in size from 5 to 30 m². Clipping trials are the most widely used technique for assessing forage production and nutritive value responses to fertilizer and defoliation management (Frame 1981). These trials are important for selecting forages that will be harvested mechanically on a farm.

In clipping trials, the most accurate way to quantify forage mass is to harvest the entire plot (minus the necessary border) at the soil surface (Burns et al. 1989b) or some consistent stubble height. Areas of the plot and unharvested border will vary with factors such as growth habit of the plant, type of treatment imposed, amount of seed or vegetative planting material available, and sampling method. Forage mass often can be determined with satisfactory precision from rectangular plots with areas of at least 5 m² (Frame 1981) and plot width of 1 m or more.

Small plots generally do not provide enough DM to determine forage quality through use of animals. Therefore, laboratory assays of chemical composition or digestibility are used to assess forage nutritive value and to predict forage quality. These assays are discussed later in the chapter.

Species and plant part composition are important determinants of forage quality. A commonly used method to determine composition is hand separation of harvested samples. This technique is time-consuming and costly, but it is precise and does not require expensive equipment. In recent years, other techniques have been developed. Near infrared reflectance spectroscopy (NIRS) has been used to quantify species composition (Coleman et al. 1990). The carbon ratio technique (Ludlow et al. 1976; Coates et al. 1991) has proven useful in determining the proportion of C_3 and C_4 species in a mixed sward, and its primary application is to tropical grass-legume mixtures.

Obtaining a representative sample for analysis of botanical composition or nutritive value is critical. In clipping studies, representative samples can be obtained by harvesting a narrow strip adjacent to the forage mass strip, subsampling from the forage mass sample, or using the entire forage mass sample. The approach chosen depends largely on heterogeneity of the plot area and the effect of sample size on cost of determining botanical composition or nutritive value.

Studies of Plant Responses to Grazing Stress. When forages are evaluated for use on grazed pastures, clipping trials should not be substituted for grazing evaluation. Factors such as selective grazing, treading, and nutrient return may cause responses of grazed plants to differ from those of clipped plants (Matches 1992). In addition, clipping instantaneously defoliates the plant, while grazing animals may remove only parts of a plant or may defoliate the plant over a period of days or weeks.

Pastures can be small in studies of plant responses to grazing stress. Areas in the range of 200-2000 m² are common. Grazing periods are short, from several hours up to 3 d, depending upon the treatments being imposed. Because there is not enough forage in pastures this size for animals to be stocked continuously, results are most applicable to rotational stocking. When pastures are continuously stocked, however, cattle do not continuously defoliate specific plants or areas of pasture. Thus, treatments defined by short rest periods (1-7 d), followed by short grazing periods (<1 d) to achieve a target stubble height or residual DM, may simulate conditions under continuous stocking.

Cost per treatment evaluated is much lower in studies of plant response to grazing stress than in animal performance trials on pasture. To identify treatments with limitations and to reduce the number of treatments for which measurement of animal production is warranted, studies of plant responses to grazing stress should almost always precede those measuring animal performance.

MEASURING FORAGE PRODUCTION. Measuring forage production is more difficult on grazed than on clipped swards. Reasons for this include large differences in forage mass from one site to another in grazed pastures and the requirement that a minimum of forage be removed by sampling so that most remains for grazing. Grazed swards are heterogenous in forage mass due to selective grazing by livestock and to nonuniform redistribution of nutrients in feces and urine.

Methods for measuring forage mass and

calculating forage accumulation on grazed pastures have been described in detail by other authors ('t Mannetje 1978; Frame 1981; Meijs et al. 1982). Clipping to soil level is the most accurate way to directly quantify forage mass at a given site. This direct approach has limitations because clipping to soil level may kill resident vegetation, and a large number of samples must be taken because forage mass is often highly variable within a grazed pasture. To avoid loss of plants due to clipping at soil level, sampling can occur at a taller stubble height. This reduces plant loss, and as long as the sampling height is below the grazing height, information about forage accumulation and consumption can be obtained. Use of indirect sampling methods that are correlated with forage mass can minimize the effects of sward heterogeneity. Indirect methods include sward height (Griggs and Stringer 1988), visual estimation (Sollenberger et al. 1987), rising plate meter (Gourley and McGowan 1991), disc meter (Bransby et al. 1977), and capacitance meter (Crosbie et al. 1987). The primary reason for using indirect measures is to increase the number of sampling units per pasture without excessive cost, time commitment, or removal of forage.

Calibration of the indirect method requires that both direct and indirect measurements (double sampling) of forage mass be made with a number of sampling units, and with a range of masses, to establish the relationship between them using regression. Then forage mass for a pasture can be predicted using this equation and the average of a relatively large number (often 20-50) of indirect measures made on that pasture. The relationship between direct and indirect measures may vary with species in the sward, season of the year, and pregraze versus postgraze sampling (Sollenberger et al. 1987). As a result, frequent and specific calibration is needed.

MEASURING BOTANICAL COMPOSITION. Botanical composition of grazed swards is indicative of stand persistence or competitiveness of species and may help explain differences in animal performance. Direct measurements can be used, but accurate determinations require a large number of samples. Calibration of visual estimation (indirect sampling) with values obtained by hand separation (direct sampling) can be a useful method of determining proportion of individual species in the DM for binary or simple mixtures (Sollenberger et al. 1987; Ortega-S. et al. 1992). Intensive training is required for this technique. Point-quadrant techniques can be used to estimate species or plant part composition of pastures (Tothill 1978; Grant 1981), but often they are too tedious to be practical. Analysis of fecal samples using the carbon ratio technique can quantify the proportion of C_4 and C_3 species in the diet (Jones and Lascano 1992). Other techniques to measure sward composition have been described (Tothill 1978; Grant 1981).

MEASURING NUTRITIVE VALUE. In studies of plant responses to grazing stress, nutritive value generally is assessed by analysis of chemical composition or digestibility of an appropriate sample. Samples taken to measure forage mass can be used, but it is difficult to adequately represent species or plant part composition of a heterogenous pasture. If forage mass samples contain forage from below the grazing height, they will underestimate nutritive value of the diet. If forage mass samples represent the pasture as a whole, then chemical composition or digestibility of forage accumulated or consumed (that disappeared during grazing) can be calculated by difference between pregraze and postgraze samples (Meijs et al. 1982).

Another approach is to collect 20-50 handplucked forage samples that represent the portion of the canopy being grazed. Samples are composited before grinding and analysis. This approach is most applicable when selective grazing is at a minimum, e.g., in single species swards that are rotationally stocked and moderately to heavily grazed.

Esophageally fistulated animals may be used to sample pastures and the extrusa analyzed for nutritive value. These "point in time" measures may not represent diet nutritive value over entire grazing periods, however, and fistulated animals moved to a pasture for this purpose may not always select a diet of the same composition as resident animals (Jones and Lascano 1992). This technique is most applicable when opportunities for selective grazing are great and when pastures are continuously stocked.

Animal Response Trials

PASTURE TRIALS. Characterizing pasture is essential for meaningful interpretation of animal production data. Forage mass, species composition, and nutritive value are important pasture attributes to measure (Burns et al. 1989b). Sampling strategies for these responses are described earlier, so this section focuses on methods for quantifying animal responses on pasture.

Experimental units in animal response trials are larger and more costly than in earlier phases of evaluation. Culling of less desirable entries or management treatments must be severe in early phases so that only the most promising remain to be compared in animal performance trials.

Measuring animal performance in grazing trials is complex because responses are affected by a large number of factors (Fig. 7.1). A thorough discussion of all pertinent factors is beyond the scope of this chapter, and readers are referred to Mott and Lucas 1952; Mott 1960; Matches 1970; Bransby 1989; and Stuedemann and Matches 1989. This discussion emphasizes experimental design and sources of variation, stocking decisions, number of levels of grazing intensity to impose, and measuring digestibility and intake in grazing trials.

Experimental Design and Sources of Variation. Generally, the pasture is the experimental unit in grazing trials. High cost per experimental unit limits grazing trials to few treatments and replicates. Simple experimental designs, most often completely randomized or randomized block designs, are used to maximize error degrees of freedom. Riewe (1961) and Bransby et al. (1988) propose evaluating forages at multiple levels of grazing intensity without replicates. This approach has not been used widely to date and involves considerable risk when the experimental site is not uniform (Matches 1970).

Forage quality and quantity often vary among pastures of a given forage that are managed the same. Likewise, performance varies among animals on the same pasture. Animal to animal variation is typically the greatest source of variation in grazing trials (Petersen and Lucas 1960). Mott and Lucas (1952) indicate that among-animal and among-pasture variation (expressed as coefficient of variation) are 10%-30% and approximately 5%, respectively, for measures of product per animal. For product per unit land area, coefficients are estimated to be 10% for both sources of variation. More replicates, greater number of animals per treatment, and increased length of experimental period reduce the magnitude of experimental errors and increase precision and the power of experimental tests for a particular experiment (Petersen and Lucas 1960; Stuedemann and Matches 1989). Weighing errors can be a large source of variation in trials that measure liveweight gain and are conducted over a short period of time (Matches 1970). Generally, animals are confined for 12-16 h (over-

night) without food or water before weighing to reduce variation.

If measurement of product per animal is of greatest priority, two to three larger pastures (i.e., capable of carrying more animals) per treatment are recommended, but if product per hectare is of greatest priority, then three or more smaller pastures (i.e., fewer animals per pasture) are best (Mott and Moore 1985). Stuedemann and Matches (1989) suggest that the optimal number of animals per treatment is the fewest required to obtain a desired degree of precision. In most cases, researchers are seeking a homogenous group of animals that represent the type of animal for which results will be applied; however, methods for assigning a heterogenous group of animals to pastures have been described (Mott and Lucas 1952).

Stocking Decisions. Choices to be made include whether to use rotational or continuous stocking and fixed or variable stocking rates. Factors affecting choice of rotational or continuous stocking are discussed in Chapter 13. Variable stocking involves periodic adjustment of stocking rate to maintain a similar grazing pressure (animal units per unit of forage DM) or sward state (e.g., height or mass) among treatments throughout the period of grazing, while fixed stocking implies that stocking rate is set at the beginning of an extended period of grazing and remains constant throughout.

Wheeler et al. (1973) propose that fixed stocking rates be used when (1) pasture growth is uniform and predictable, (2) quality and quantity of the forage can be maintained if left uneaten, (3) there is minimum flexibility in experimental and related production conditions, and (4) results are applied almost directly to farming practice. Likewise, variable stocking rates are recommended when the opposite conditions exist. The emphasis by Wheeler et al. (1973) on choosing the most appropriate technique for a particular set of experimental circumstances and objectives has provided a useful perspective to an issue that has produced considerable debate among grassland scientists.

Variable stocking rate studies include two categories of animals: testers and "put-and-takes." Tester animals remain on the pasture during the entire experiment, and the number of testers is based on the projected minimum stocking rate of the pasture during the experimental period. Put-and-take animals are added to or removed from the pasture as needed so that the target sward state or grazing pressure is maintained. Only data for tester animals are used to compute performance per animal. For determining average stocking rate and production per land area, both testers and put-and-take animals are considered. Methods for calculation of product per animal and per hectare are described by Mott and Lucas (1952).

Jones (1981) suggests that fixed stocking rate experiments historically have been favored in Australia because of their relevance to the grazing industry, the absence of researcher subjectivity once stocking rates have been chosen, and the relative ease with which such experiments can be carried out, even by minimally trained staff. In fixed rate studies, calculation of animal performance is straightforward because all animals remain on a given experimental unit during the entire experiment.

Data from variable stocking rate trials can be applied on a farm without requiring that animal number on the pasture or farm be altered frequently. During periods of rapid forage growth, stocking rate on the grazed area of a pasture can be increased by using electric fence to prohibit animal access to portions of the pasture. This allows excess forage to be stockpiled for grazing later in the season or harvested as hay to be fed during periods of shortfall.

Number of Levels of Grazing Intensity in Grazing Trials. Because stocking rate or grazing pressure have a profound effect on forage mass, application of multiple increments of either variable generates predictable and highly significant relationships with performance per animal or per unit land area (Burns et al. 1989b). Mott and Moore (1985) developed a model that illustrates the effect of grazing pressure on gain per animal (g) and gain per unit area (G) over a range of grazing pressures (n) (Fig. 7.2). Between n_u and n_o gain per animal is a function of forage quality. Variation in forage quality can be attributed to characteristics of the canopy (e.g., species composition, plant part composition, proportion of live forage, forage density, forage maturity, and chemical composition) and their effects on ingestive behavior and intake of grazing livestock (Chapter 6). From n_o to n_m opportunity for diet selection gradually decreases and gain per animal is determined primarily by forage mass. At n_m animals are able to consume only enough forage to meet their maintenance requirement, and gain is zero. Maximum gain per unit area occurs at a

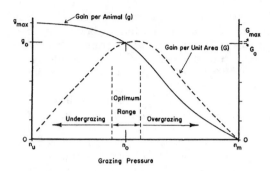

Fig. 7.2. The relationship of grazing pressure (n) with gain per animal (g) and gain per unit land area (G) (Mott and Moore 1985).

grazing pressure greater than that for maximum gain per animal.

In their model, Mott and Moore (1985) suggest that an optimum range in grazing pressure exists between conditions of undergrazing and overgrazing. This zone includes the grazing pressure at which gain per unit land area is maximum and extends to lower grazing pressures but not to the point where gain per animal is maximum. In an alternative model, Jones and Sandland (1974) propose that the relationship between stocking rate and gain per animal is linear across the range of possible stocking rates. A revised model by Jones (1981) is similar to that of Mott and Moore (1985) in that gain per animal plateaus when swards are understocked, but both models seem to suggest that the relationship tends to linearity in the range of grazing pressures or stocking rates that are used in practice.

Because of variation in forage mass among and within years due to environmental factors, fixed stocking rate experiments are best conducted with three or more stocking rates. As described by Mott (1960) and reiterated by Matches (1970), the variable stocking rate method attempts to adjust stocking rate so that comparisons among treatments are made at equal grazing pressures. These adjustments allow grazing pressure or some sward characteristic to be maintained relatively constant over a range of environmental conditions and forage growth rates.

Bransby (1989) suggests that regardless of the approach chosen, use of multiple levels of grazing intensity has advantages in that (1) treatment by grazing intensity interactions can be detected, (2) the relationship between animal production and forage mass can be determined for each treatment, and (3) data are more useful for economic analysis and modeling. The cost of including multiple levels of grazing intensity for each treatment is a major factor limiting use of this approach. If available resources do not permit more than one grazing intensity per treatment, then the variable stocking rate method would probably be the one of choice because it offers greater opportunity to graze at a near-optimum intensity (Bransby 1989).

Measuring Digestibility and Intake on Pasture. Measurements of digestibility and intake on pasture are laborious and often highly variable (Greenhalgh 1982). It has been suggested that such measurements be made only when (1) a definite requirement of them can be established and (2) they can be made with a degree of accuracy appropriate to the objectives of the experiment (Greenhalgh 1982).

There are no direct methods for measuring digestibility on pasture. Indirect techniques include ratio techniques, fecal index techniques, and in vitro-based procedures (Le Du and Penning 1982). Markers for ratio techniques must be unaltered in passage through the animal and be quantitatively recoverable, and the feed and feces must be sampled accurately (Le Du and Penning 1982). Lignin, because it is considered indigestible, has often been used as a marker. In actual practice, however, immature grasses and low-lignin forages may have apparent lignin digestibilities of 20% to 40% (Van Soest 1994). Use of in vitro indigestible neutral detergent fiber (NDF) or acid detergent fiber (ADF) as markers may alleviate this problem because of their relatively high concentrations in forages (Lippke et al. 1986).

The fecal index technique relates a fecal component to in vivo digestibility by regression. This relationship is developed using data from a confinement trial with forage similar to that being grazed. Component concentration in feces of grazing animals then can be used to estimate digestibility (Le Du and Penning 1982). An advantage of the fecal index technique is that forage samples are not needed and only small fecal samples are required. Because of difficulties in obtaining representative forage for feeding to confined animals, the fecal index method is not recommended when pastures are continuously stocked or when there is opportunity for selective grazing.

Procedures based on in vitro analyses require that the relationship between in vivo and in vitro digestibility of the forage be determined in a confinement study. Then representative pasture samples, obtained by hand plucking or by use of esophageally or ruminally fistulated animals, are analyzed for in vitro digestibility.

Indirect methods of estimating animal intake on pasture include sward difference (Meijs et al. 1982), animal gain, digestibility and fecal output equations, and grazing behavior. Of the indirect methods, digestibility and fecal output equations are used most often. Fecal output from grazing animals may be estimated by total collection or by the ratio of chromicoxide in the feed and feces. (Le Du and Penning 1982). Inaccurate estimates of fecal output may occur if fecal samples are not representative of total excretion. Diurnal variation in marker concentration in feces is a potential problem with pulse-dosed chromicoxide (Prigge et al. 1981). Continuous marker release methods, such as controlled release boluses or continuous infusion pumps, may eliminate this source of variation (Brandyberry et al. 1991).

Monitoring an animal's grazing behavior as a method of measuring intake (Hodgson 1982) is demanding in terms of time and labor. Data have greatest application to short-term measurements of intake and have been useful in identifying sward characteristics that affect intake. The telemetric grazing clock (Forwood and Hulse 1989) attempts to use advanced technology to refine this approach and reduce labor requirements.

More direct methods of measuring intake on pasture include the animal weight telemetry system, multiple impedance plethysmography, and conductivity transducing cannulas (Forwood and Hulse 1989). Use of electronic systems is limited by cost and size of equipment. Smaller affordable systems may become available with advancing technology.

CONFINEMENT STUDIES. Pasture and confinement trials that quantify animal performance are the best measures of forage quality, but high inputs of labor, time, and capital limit their use. Alternatively, forage quality can be estimated by determining forage intake and digestibility using confined animals. Advantages of confined feeding trials include (1) accurate and direct measurements of intake and digestibility can be made, (2) characteristics of consumed forage can be determined, (3) animal environment can be controlled, and (4) trials are not restricted to forage growing seasons (Minson 1981). Disadvantages include (1) feeding trials cannot duplicate forage selection by grazing animals, (2) forage must usually be preserved, and (3) chemical composition and nutritive value may be altered during preservation or storage (Minson 1981).

Voluntary intake must be determined at ad libitum levels of forage offered. Most researchers accomplish this by feeding 10%-15% excess above that consumed, but some researchers offer forage such that the excess remaining is the same for each animal (Minson 1981). Digestibility and intake are often determined simultaneously, but digestibility also may be determined at restricted levels of intake. Because of decreased opportunity for selection and less variation in retention time, measures of digestibility with restricted, relative to ad libitum, intakes should be affected less by animal differences and related more closely to intrinsic plant characteristics and to in vitro measures of plant digestibility (Cherney et al. 1990a). Digestibility measured at restricted intakes may not represent that obtained for grazing animals with ample opportunity for diet selection.

Because of the expense and amount of forage required to conduct animal trials in confinement, sheep often are used in place of cattle (Heaney 1980). While cattle and sheep generally have similar rankings for forages, important differences can occur (Miller et al. 1991). Caution always should be used when extrapolating sheep data to cattle. Generally at least three sheep are recommended for digestibility determinations, while six are recommended for intake determinations (Heaney 1980). Use of a standard forage can reduce the effects of animal variation on estimates of mean voluntary intake (Abrams et al. 1987).

LABORATORY METHODS FOR FORAGE ANALYSES

Laboratory analyses of forages (1) provide relative quality estimates in forage improvement research programs and (2) determine the nutritional value of forage to aid in management decisions, formulation of rations for livestock, diagnosis of nutrition-related herd health problems, and establishing equitable market values for forage crops. Effective laboratory methods of evaluating forages must be precise, economical, simple, rapid, and accepted by both the scientific community and the practitioner (Barnes 1981). Major categories of laboratory analyses include chemical methods and microbial or enzymatic methods.

Microbial and Enzymatic Methods. Microbial or enzymatic digestion methods are sensitive to most intrinsic factors that limit in vivo digestion of a forage (Van Soest 1994), and such methods usually are more highly correlated with in vivo digestibility than are chemical methods. Microbial and enzymatic methods, however, provide little information about the actual chemical constituents influencing digestion (Van Soest 1994).

Most microbial or enzymatic methods use in vitro or in situ systems to simulate digestive processes and estimate relative forage quality. In these systems, forage samples are incubated with ruminal microorganisms for 48 h at 39°C (Marten and Barnes 1980). Enzymes can be substituted for the ruminal microorganisms (Bughrara et al. 1992). Digestibility is determined by difference between residual and initial amounts of a component.

The value of these techniques depends on how accurately they reflect ruminal events (Van Soest 1994). Advantages are that a small amount of forage is required, animal digestion is simulated, several hundred forages may be evaluated simultaneously, and the procedure is relatively inexpensive. Major disadvantages are that (1) a ruminally fistulated animal must be maintained as a source of microbes, (2) inoculum varies with diet and time and method of collection, and (3) the systems may be susceptible to end-product inhibition.

IN VITRO METHODS. Of the laboratory methods available, in vitro DM disappearance (IVDMD) is most highly correlated with in vivo digestibility and is the best predictor of relative forage quality (Marten and Barnes 1980). Many factors can influence IVDMD (Marten and Barnes 1980; Cherney et al. 1993), but they do not preclude its use for routine evaluation of forage quality.

Many procedures for measurement of IVDMD have been described (Osbourne and Terry 1977; Marten and Barnes 1980), but most are modifications of the two-stage technique of Tilley and Terry (1963). The first stage, in which forages are incubated in ruminal fluid, simulates ruminal digestion of cell contents and structural carbohydrates. The second stage involves incubation with acid-pepsin or extraction with neutral detergent solution (Marten and Barnes 1980). Incubation with acid-pepsin simulates abomasal digestion, while extraction with neutral detergent solution simulates true digestion. In vitro organic matter disappearance, which requires one additional weighing after ashing undigested residue from IVDMD, may be used where soil contamination is a problem (Moore 1994).

IN SITU BAG METHODS. Indigestible synthetic bags containing forage are suspended in the rumen (Marten and Barnes 1980). Like in vitro techniques, rate and extent of digestion are measured as DM loss after specific incubation periods (Waldo and Glenn 1984; NRC 1985). In situ bag methods generally are subject to the same limitations as in vitro methods, although in situ results tend to be more variable (Van Soest 1994). In situ bag methods are not affected by product inhibition, as is possible with in vitro techniques, because the rumen is not a closed system. Standardized procedures to reduce error should include introduction of bags in reverse sequence and removing them all at once, specific bag pore sizes (40 to 60 μm), sample size-to-bag surface area ratios of 10 to 20 mg/cm² (Nocek 1988), and standardized bag-rinsing methods (Cherney et al. 1990b).

GAS PRODUCTION METHODS. The in vitro gas production procedure is similar to that of Tilley and Terry (1963), but gas production, rather than DM loss, quantifies fermentation (Blümmel and Ørskov 1993). Because this system focuses on fermentation products, both soluble and insoluble fractions of forages are considered (Pell and Schofield 1993). Use of a computerized pressure sensor system allows continuous monitoring of gas production (Pell and Schofield 1993). When measuring rate of digestion, this feature may provide an advantage over traditional systems, which generally use very few time points. In addition, because sample size for the entire time course of digestion is small (0.1 g), this technique may be particularly useful with tissue culture or synthetic compounds (Pell and Schofield 1993). Because of its lack of specificity and the small sample size used, gas production data need to be considered cautiously (Johnson 1963; Cherney et al. 1985). In addition, because gas production is strongly influenced by microbial population, ruminal fluid sampling times and techniques are critical, and direct comparison of forages varying widely in chemical composition may be difficult.

ENZYMATIC METHODS. Alternative methods for estimating forage digestibility include en-zyme solubility procedures. Advantages of these procedures over standard in vitro analyses include reduced cost and greater convenience and precision (Marten and Barnes 1980). Methods using cellulases alone, with proteolytic enzymes, or other chemical treatments have been developed (Minson 1981). Using commercial cellulases, Bughrara et al. (1992) report correlations with IVDMD ranging from 0.57 to 0.95. A major limitation to adoption of enzymatic techniques has been the inability to obtain cellulase preparations with consistent activity (Gabrielsen 1986). Also, purified cellulases are less efficient and digest cell walls less completely than do ruminal microorganisms (Van Soest 1994). Despite these possible limitations, a pepsin-cellulase method is rapidly replacing the rumen fluid-pepsin technique in Australia (Minson 1981). Adoption has been associated with (1) an inexpensive source of cellulase, (2) no need for a fistulated animal, (3) less odor, (4) smaller analytical error, (5) improved repeatability, and (6) suitability for small sample numbers (Minson 1981).

Chemical Methods
FIBER ANALYSES. Chemically, fiber can be characterized as consisting primarily of cellulose, hemicellulose, lignin, pectin, cutin, and silica. Several procedures have been developed to determine forage fiber concentration (Chap. 8, Vol. 1). Analysis of crude fiber is of limited value because forage fiber is not separated into readily digestible and less digestible fractions. In some instances the digestibility of crude fiber is higher than that of the supposedly digestible nitrogen-free (N-free) extract fraction. The Van Soest system of fiber analysis partitions forage carbohydrates into fractions based on nutritional availability and has become the primary standard for chemical evaluation of forages in the US (Marten 1981). Current procedures for NDF, ADF, and lignin are described by Van Soest et al. (1991). Van Soest et al. (1991) recommend that triethylene glycol replace the toxic 2-ethoxyethanol in the neutral detergent solution. Other research (Cherney et al. 1989) has shown that 2-ethoxyethanol is not necessary for the determination of NDF in forages.

Of the fiber fractions, NDF is best related to feed intake because it represents the total insoluble fiber matrix (Van Soest 1994). The ADF fraction has been correlated with digestibility, but the intended purpose of ADF is as a preparative residue for determination of cellulose, lignin, maillard products, silica, acid insoluble ash, and acid detergent insoluble N (ADIN). Although there may be a statistical association in some cases, there is no valid theoretical basis to link ADF to digestibility (Van Soest et al. 1991). Cellulose and hemicellulose have secondary effects on digestibility, which depend on their relationship with lignin (Chap. 6). The relationships of NDF and ADF with digestibility, therefore, are not cause and effect and may be negative, positive, or nonexistent. Lignin, however, has a direct role in reducing cell wall digestibility (Van Soest 1994).

PROTEIN AND RELATED ANALYSES. Crude protein (CP) and digestible CP have long been the standards for evaluating protein value of forages and are still acceptable for many situations (Vérité 1980). Crude protein is estimated by measuring total N in forages, usually by the Kjeldahl procedure, and multiplying $N \times 6.25$ (CP is usually about 16% N). Digestible protein is obtained from ADIN or pepsin insoluble N (Goering and Van Soest 1970). Total N is only 60%-80% true protein (Van Soest 1994), however, with the remainder being nonprotein N and unavailable N. The latter two tend to be higher in immature fresh forages and silages and when proteins are protected (Vérité 1980). Traditional measures of protein concentration may be inadequate to fully describe the quality of these forages. In addition, rations balanced for CP and digestible protein do not always account for dynamics of ruminal fermentation and ammonia production (Sniffen et al. 1992). New protein systems have been proposed that consider fractions of protein intake as degradable or undegradable (NRC 1985; Waldo and Glenn 1984). Major limitations to implementation of these systems are lack of sufficient protein fractionation data (NRC 1985) and standardized methods of protein fractionation. Roe and Sniffen (1990) have outlined procedures that partition N into biologically meaningful fractions, which include nonprotein N, soluble protein, and ADIN.

MOISTURE ANALYSES. Forage moisture and its determination are important because (1) drying method and moisture concentration affect nutrient availability and composition and (2) all other analyses are expressed on a DM basis. On-farm methods for determining moisture concentration include microwave oven, Koster tester, electronic moisture me-

ters for silage, and the Delmhorst moisture meter for hay (Brusewitz et al. 1993). In the laboratory, methods include drying at 105°C in conventional forced air ovens, Karl Fischer, toluene distillation (corrected for volatiles), and NIRS (Brusewitz et al. 1993). Freeze-drying forage results in material that is more similar to fresh tissue than what is produced by other drying methods (Smith 1981), but it is time-consuming, and only a relatively small number of samples can be processed at a time. Heat drying is used more often because large numbers of samples may be processed quickly and inexpensively. Samples generally are dried at 60°C or lower because at higher temperatures tissue damage may result from nonenzymatic-browning reactions and other heat thermodegradations (Smith 1981).

Other Analyses. Forage mineral concentration can be determined by atomic absorption, emission spectroscopy, and NIRS. With most minerals no attempt is made to determine bioavailability (Minson 1981). Other laboratory analyses are used by forage researchers, but because many are laborious and expensive, their potential for routine use in forage evaluation is limited. The value of these analyses is that they can improve understanding of factors limiting forage quality to an extent not possible with standard analyses. For example, differences in forage microanatomy have explained differences in digestibility between forages (Flores et al. 1993). Leaf-to-stem ratio influenced both intake and digestibility (Poppi et al. 1985). Particle size, grinding energy, specific gravity, and buffering capacity are other forage characteristics that have been associated with forage quality (Van Soest 1994). Sometimes antiquality components such as alkaloids, tannins, prussic acid, etc., limit animal production (Van Soest 1994). Laboratory determination of these compounds then becomes crucial to forage quality evaluation.

Near Infrared Reflectance Spectroscopy. Near infrared reflectance spectroscopy is a technique that can estimate major chemical constituents in forages rapidly and reproducibly (Shenk et al. 1981; Chap. 8). Because it generates results more quickly than conventional analyses and is relatively inexpensive per sample, NIRS is becoming the preferred method of analysis for routine evaluation of DM, CP, ADF, NDF, and ADIN concentration of forages, and it has official

Association of Official Analytical Chemists approval for determination of DM, CP, and ADF (Linn and Martin 1991a). In research applications, NIRS is routinely used for IVD-MD (Marten 1989). Constituent concentrations are not measured directly by NIRS but are calculated from prediction equations developed from calibration samples (Barnes 1981).

Spectral reflectance properties of plant tissues can be measured by NIRS. Observed NIR spectra are the summation of the reflectance spectra of the functional organic constituents of a sample. Major forage organic constituents include combinations of CH, OH, and NH chemical bonds. Because major mineral constituents tend to be associated with the major organic constituents, NIRS has been used to predict concentration of certain minerals (Smith et al. 1991).

The main advantages of NIRS over conventional systems of analysis include speed (<3 min per sample), ease of sample preparation, and performance of multiple analyses with one operation (Norris 1989). These result in reduced costs of analyses in large experiments and allow some experiments to be conducted that could not be done otherwise. Disadvantages include relatively expensive start-up costs, complex data treatments, dependence on calibration procedures, and a lack of sensitivity for minor constituents (Norris 1989). Accuracy of NIRS methods depends on (1) sampling errors, (2) accuracy of wet chemistry analyses used to calibrate the instrument, (3) calibration samples that reflect the population being tested, (4) selection of wavelengths and mathematical treatments for equation calibration, and (5) technician competency (Martin and Linn 1989).

APPLICATION OF FORAGE ANALYSES

Farmers request analysis of forages primarily to establish a market price or to provide estimates of forage quality that aid in formulating livestock rations. To accomplish these objectives the farmer generally requires more than a simple reporting of forage nutritive value data. Accurate prediction of forage quality or of indices of forage quality (from the nutritive value data) is needed. The remainder of this chapter provides a brief discussion (1) of indices of quality that may be useful in marketing forage and (2) of assessment of forage quality for ration formulation. A thorough review of forage quality indices

and prediction of forage quality is provided by Moore (1994).

Hay Marketing. An index that reflects relative differences in forage quality across a wide spectrum of forages would simplify hay pricing. Moore (1994) defines an index of forage quality as a single number that represents the combination of a forage's potential for voluntary intake and nutritive value and allows for relative comparisons among forages differing in genotype, season of growth, and maturity. Relative feed value (RFV) is an index that is intended to achieve this objective. Other indices include the nutritive value index, which is similar to the RFV, and the quality index, which is defined as voluntary total digestible nutrient intake of the forage as a multiple of the total digestible nutrient requirement of the animal for maintenance (Moore 1994).

The RFV index is based on predictions of intake and digestibility from assays of forage NDF and ADF concentrations, respectively (Linn and Martin 1991b). The Hay Marketing Task Force of the American Forage and Grassland Council has proposed standards for RFV (Linn and Martin 1991b), with a value of 100 representing a hay with 410 g ADF/kg and 530 g NDF/kg. The RFV is routinely used in several states and has increased awareness among producers of the importance of forage quality and how to produce higher-quality forage (Moore 1994).

The RFV does have limitations. Because ADF is not always negatively correlated with digestibility, the RFV may not always reflect true forage quality potential, particularly for late-season cuttings. In addition, the RFV assumes that NDF of grass and legume has similar intake potential, and that may not be true (Chapter 6).

Ration Formulation. Dynamic models, such as the Cornell Net Carbohydrate and Protein System (Fox et al. 1990), are being used increasingly to formulate rations for specific groups of animals. These models rely on accurate descriptions of animal, environment, feed composition, and management conditions on a particular farm. Accurate and rapid analyses of representative samples of forages and feeds on the farm are essential for use of models in making management decisions. Near infrared reflectance spectroscopy has made it possible to obtain nutritive value data quickly. If appropriate equations are available to predict forage quality from nutritive value, potential exists for successful application of models.

QUESTIONS

1. Outline the phases of the Mott and Moore scheme for forage evaluation, and list plant or animal responses that may be quantified in each phase.
2. Why is it important to study the actual effect of the grazing animal upon a pasture rather than to simulate the effect of grazing with clipping methods?
3. Why is it more difficult to accurately measure forage mass and botanical composition in grazed than in clipped swards?
4. Describe the difference between direct and indirect sampling methods to determine forage mass, and explain how indirect methods are used.
5. Discuss the differences between fixed and variable stocking rate experiments, and indicate when it may be appropriate to use each of these techniques.
6. Describe the methods used to measure intake and digestibility on pasture and the reasons for choosing these methods.
7. Laboratory analyses of forages are conducted for what purposes?
8. Assume you are opening a commercial forage-testing laboratory. What laboratory procedures will your lab conduct? What are your reasons for choosing these analyses?
9. List advantages and disadvantages of using NIRS for forage quality evaluation.

REFERENCES

Abrams, SM, HW Harpster, PJ Wangsness, JS Shenk, E Keck, and JL Rosenberger. 1987. Use of a standard forage to reduce effects of animal variation on estimates of mean voluntary intake. J. Dairy Sci. 70:1235-40.

Barnes, RF. 1981. Infrared reflectance spectroscopy for evaluating forages. In Forage Evaluation: Concepts and Techniques. Lexington, Ky.: American Forage and Grassland Council, 89-102.

Blümmel, M, and ER Ørskov. 1993. Comparison of in vitro gas production and nylon bag degradability of roughages in predicting feed intake in cattle. Anim. Food Sci. Tech. 40:109-19.

Brandyberry, SD, RC Cochran, ES Vanzant, and DL Harmon. 1991. Technical note: Effectiveness of different methods for continuous marker administration for estimating fecal output. J. Anim. Sci. 69:4611-16.

Bransby, DI. 1989. Compromises in the design and conduct of grazing experiments. In Grazing Research: Design, Methodology, and Analysis. Madison, Wis.: Crop Science Society of America, 53-67.

Bransby, DI, AG Matches, and GF Krause. 1977. Disk meter for rapid estimation of herbage yield in grazing trials. Agron. J. 69:393-96.

Bransby, DI, BE Conrad, HM Dicks, and JW Drane. 1988. Justification for grazing intensity experiments: Analysing and interpreting grazing data. J. Range Manage. 41:274-79.

Brusewitz, GH, LE Chase, M Collins, SR Delwiche, JW Garthe, RE Muck, and RE Pitt. 1993. Forage Moisture Determination. NRAES-59. Ithaca, N.Y.: Northeast Regional Agricultural Engineering Service.

Bughrara, SS, DA Sleper, and PR Beuselinck. 1992. Comparison of cellulase solutions for use in digesting forage samples. Agron. J. 84:631-36.

Burns, JC, DS Fisher, and KR Pond. 1989a. Evaluating plant responses to defoliation: Importance, objectives, and approaches. In Proc. 45th South. Pasture and Forage Crop Improv. Conf., 87-96.

Burns, JC, H Lippke, and DS Fisher. 1989b. The relationship of herbage mass and characteristics to animal responses in grazing experiments. In Grazing Research: Design, Methodology, and Analysis. Madison, Wis.: Crop Science Society of America, 7-19.

Cherney, DJR, JA Patterson, and JH Cherney. 1989. Use of 2-ethoxyethanol and a-amylase in the neutral detergent fiber method of feed analysis. J. Dairy Sci. 72:3079-84.

Cherney, DJR, DR Mertens, and JE Moore. 1990a. Intake and digestibility by wethers as influenced by forage morphology at three levels of forage offering. J. Anim. Sci. 68:4387-99.

Cherney, DJR, JA Patterson, and RP Lemenager. 1990b. Influence of in situ bag rinsing technique on determination of dry matter disappearance. J. Dairy Sci. 73:391-97.

Cherney, DJR, J Siciliano-Jones, and AN Pell. 1993. Technical note: Forage in vitro dry matter digestibility as influenced by fiber source in the donor cow diet. J. Anim. Sci. 71:1335-38.

Cherney, JH, JJ Volenec, and WE Nyquist. 1985. Sequential fiber analysis of forage as influenced by sample weight. Crop Sci. 25:1113-15.

Coates, DB, APA Van Der Weide, and JD Kerr. 1991. Changes in fecal ^{13}C in response to changing proportions of legume (C_3) and grass (C_4) in the diet of sheep and cattle. J. Agric. Sci. Camb. 116:287-95.

Coleman, SW, S Christiansen, and JS Shenk. 1990. Prediction of botanical composition using NIRS calibrations developed from botanically pure samples. Crop Sci. 30:202-7.

Crosbie, SF, BM Smallfield, H Hawker, MJS Floate, JM Keoghan, PD Enright, and RJ Abernethy. 1987. Exploiting the pasture capacitance probe in agricultural research. J. Agric. Sci. 108:155-63.

Flores, JA, JE Moore, and LE Sollenberger. 1993. Determinants of forage quality in Pensacola bahiagrass and Mott elephantgrass. J. Anim. Sci. 71:1606-14.

Forage and Grazing Terminology Committee (FGTC). 1992. Terminology for grazing lands and grazing animals. J. Prod. Agric. 5:191-210.

Forwood, JR, and MM Hulse. 1989. Electronic measurement of grazing time and intake in free roaming livestock. Proc. 16th Int. Grassl. Congr., 4-11 Oct., Nice, France. Vol. 2, 799-800.

Fox, DG, CJ Sniffen, JD O'Connor, JB Russell, and PJ Van Soest. 1990. The Cornell Net Carbohydrate and Protein System for Evaluating Cattle Diets. Search: Agriculture (no. 34). Cornell Univ., Ithaca, N.Y.

Frame, J. 1981. Herbage mass. In Sward Measurement Handbook. Hurley, England: British Grassland Society, 39-69.

Gabrielsen, BC. 1986. Evaluation of marketed cellulases for activity and capacity to degrade forage. Agron. J. 78:838-42.

Goering, HK, and PJ Van Soest. 1970. Forage Fiber Analysis: Apparatus, Reagents, Procedures and Some Applications. USDA-ARS Agric. Handb. 379. Washington, D.C.: US Gov. Print. Off.

Gourley, CJP, and AA McGowan. 1991. Assessing differences in pasture mass with an automated rising plate meter and a direct harvesting technique. Aust. J. Exp. Agric. 31:337-39.

Grant, SA. 1981. Sward components. In Sward Measurement Handbook. Hurley, England: British Grassland Society, 71-92.

Greenhalgh, JFD. 1982. An introduction to herbage intake measurements. In Herbage Intake Handbook. Hurley, England: British Grassland Society, 1-10.

Griggs, TC, and WC Stringer. 1988. Prediction of alfalfa herbage mass using sward height, ground cover, and disk technique. Agron. J. 80:204-8.

Heaney, DP. 1980. Sheep as pilot animals. In Pigden, WJ et al. (eds.), Standardization of Analytical Methodology for Feeds. Ottawa, Canada: International Development Research Centre, 44-48.

Hodgson, J. 1982. Ingestive behaviour. In Herbage Intake Handbook. Hurley, England: British Grassland Society, 113-18.

Johnson, RR. 1963. Symposium of microbial digestion in ruminants: In vitro rumen fermentation techniques. J. Anim. Sci. 22:792-800.

Jones, RJ. 1981. Interpreting fixed stocking rate experiments. In Forage Evaluation: Concepts and Techniques. Lexington, Ky.: American Forage and Grassland Council, 419-30.

Jones, RJ, and CE Lascano. 1992. Oesophageal fistulated cattle can give unreliable estimates of the proportion of legume in diets of resident animals grazing tropical pastures. Grass and Forage Sci. 47:128-32.

Jones, RJ, and RL Sandland. 1974. The relation between animal gain and stocking rate. J. Agric. Sci. Camb. 83:335-42.

Le Du, YLP, and PD Penning. 1982. Animal based techniques for estimating herbage intake. In Herbage Intake Handbook. Hurley, England: British Grassland Society, 37-75.

Linn, JG, and NP Martin. 1991a. Forage quality analyses and interpretation. Vet. Clin. North Am.: Food Anim. Pract. 7:509-23.

———. 1991b. Forage quality tests: Tests predicting animal performance. In Proc. 21st Natl. Alfalfa Symp., Rochester, Minn., 145-51.

Lippke, H, WC Ellis, and BF Jacobs. 1986. Recovery of indigestible fiber from feces of sheep and cattle on forage diets. J. Dairy Sci. 69:402-12.

Ludlow, MM, JH Troughton, and RJ Jones. 1976. A technique for determining the proportion of C_3 and C_4 species in plant samples using stable natural isotopes of carbon. J. Agric. Sci. Camb. 87:625-32.

't Mannetje, L. 1978. Measuring quantity of grassland vegetation. In Measurement of Grassland Vegetation and Animal Production. Hurley, England: Commonwealth Agriculture Bureaux, 63-95.

Marten, GC. 1981. Chemical, in vitro, and nylon bag procedures for evaluating forage in the USA. In Forage Evaluation: Concepts and Techniques. Lexington, Ky.: American Forage and Grassland Council, 39-55.

———. 1989. Current applications of NIRS technology in forage research. In GC Marten et al. (eds.), Near Infrared Reflectance Spectroscopy (NIRS): Analysis of Forage Quality, USDA-ARS Agric. Handb. 643. Washington, D.C.: US Gov. Print. Off., 45-48.

Marten, GC, and RF Barnes. 1980. Prediction of energy digestibility of forages with in vitro rumen fermentation and fungal enzyme systems. In Standardization of Analytical Methodology for Feeds. Ottawa, Canada: International Development Research Centre, 61-71.

Martin, NP, and JG Linn. 1989. Extension applications in NIRS technology transfer. In GC Marten et al. (eds.), Near Infrared Reflectance Spectroscopy (NIRS): Analysis of Forage Quality, USDA-ARS Agric. Handb. 643. Washington, D.C.: US Gov. Print. Off., 48-53.

Matches, AG. 1970. Pasture research methods. In Proc. Natl. Conf. Forage Qual. Eval. and Util. Lincoln: Nebraska Center for Continuing Education, I1-I32.

———. 1992. Plant response to grazing: A review. J. Prod. Agric. 5:1-7.

Meijs, JAC, RJK Walters, and A Keen. 1982. Sward methods. In Herbage Intake Handbook. Hurley, England: British Grassland Society, 11-36.

Miller, PS, WN Garrett, and N Hinman. 1991. Effects of alfalfa maturity on energy utilization by cattle and nutrient digestibility by cattle and sheep. J. Anim. Sci. 69:2591-2600.

Minson, DJ. 1981. An Australian view of laboratory techniques for forage evaluation. In Forage Evaluation: Concepts and Techniques. Lexington, Ky.: American Forage and Grassland Council, 57-71.

Mochrie, RD, JC Burns, and DH Timothy. 1981. Recommended protocol for evaluating new forages for ruminants. In Forage Evaluation: Concepts and Techniques. Lexington, Ky.: American Forage and Grassland Council, 553-59.

Moore, JE. 1994. Forage quality indices: Development and application. In GC Fahey, Jr., et al. (eds.), Natl. Conf. Forage Qual., Eval., and Util., 967-98.

Mott, GO. 1960. Grazing pressure and the measurement of pasture production. In Proc. 8th Int. Grassl. Congr., 606-11.

Mott, GO, and HL Lucas. 1952. The design, conduct, and interpretation of grazing trials on cultivated and improved pastures. In Proc. 6th Int. Grassl. Congr., 1380-85.

Mott, GO, and JE Moore. 1970. Forage evaluation techniques in perspective. In Proc. Natl. Conf. Forage Qual. Eval. and Util. Lincoln: Nebraska Center of Continuing Education, L1-L10.

———. 1985. Evaluating forage production. In ME Heath, RF Barnes, and DS Metcalfe, Forages: The Science of Grassland Agriculture, 4th ed. Ames: Iowa State Univ. Press, 422-29.

National Research Council (NRC). 1985. Ruminant Nitrogen Usage. ISGNO-309-359-K, Subcommittee on Nitrogen Usage in Ruminants. Washington, D.C.: National Academy of Science.

Nocek, JE. 1988. In situ and other methods to estimate ruminal protein and energy digestibility: A review. J. Dairy Sci. 71:2051-69.

Norris, KH. 1989. NIRS instrumentation. In GC Marten et al. (eds.), Near Infrared Reflectance Spectroscopy (NIRS): Analysis of Forage Quality, USDA-ARS Agric. Handb. 643. Washington, D.C.: US Gov. Print. Off., 12-17.

Ortega-S., JA, LE Sollenberger, KH Quesenberry, JA Cornell, and CS Jones, Jr. 1992. Productivity and persistence of rhizoma peanut pastures under different grazing managements. Agron. J. 84:799-804.

Osbourne, DF, and RA Terry. 1977. In vitro techniques for the evaluation of ruminant feeds. In Proc. Nutr. Soc. 36:219-25.

Pell, AN, and P Schofield. 1993. Computerized monitoring of gas production to measure forage digestion in vitro. J. Dairy Sci. 76:1063-73.

Petersen, RG, and HL Lucas. 1960. Experimental errors in grazing trials. In Proc. 8th Int. Grassl. Congr., 747-50.

Poppi, DP, RE Hendricksen, and DJ Minson. 1985. The relative resistance to escape of leaf and stem particles from the rumen of cattle and sheep. J. Agric. Sci. Camb. 105:9-14.

Prigge, EC, GA Varga, JL Vincini, and RL Reid. 1981. Comparison of ytterbium chloride and chromium sesquioxide as fecal indicators. J. Anim. Sci. 53:1629-33.

Riewe, ME. 1961. Use of the relationship of stocking rate to gain of cattle in an experimental design for grazing trials. Agron. J. 53:309-13.

Roe, MB, and CJ Sniffen. 1990. Techniques for measuring protein fractions in feedstuffs. In Proc. Cornell Nutr. Conf. for Feed Manuf., Cornell Univ., Ithaca, N.Y., 81-88.

Shenk, JS, I Landa, MR Hoover, and MO Westerhaus. 1981. Description and evaluation of near infrared reflectance spectro-computer for forage and grain analysis. Crop Sci. 21:355-58.

Smith, D. 1981. Removing and Analyzing Total Nonstructural Carbohydrates from Plant Tissue. Univ. Wis. Coll. Agric. Life Sci. Bull. 2107. Madison.

Smith, KF, SE Willis, and PC Flinn. 1991. Measurement of the magnesium concentration in perennial ryegrass (Lolium perenne) using near infrared reflectance spectroscopy. Aust. J. Agric. Res. 42:1399-1404.

Sniffen, CJ, JD O'Connor, PJ Van Soest, DG Fox, and JB Russell. 1992. A net carbohydrate and protein system for evaluating cattle diets: II. Carbohydrate and protein availability. J. Anim. Sci. 70:3562-77.

Sollenberger, LE, KH Quesenberry, and JE Moore. 1987. Effects of grazing management on establishment and productivity of aeschynomene overseeded in limpograss pastures. Agron. J. 79:78-82.

Stuedemann, JA, and AG Matches. 1989. Measurement of animal response in grazing research. In GC Marten et al. (eds.), Grazing Research: Design, Methodology, and Analysis. Madison, Wis.: Crop Science Society of America, 21-35.

Tilley, JMA, and RA Terry. 1963. A two stage technique for in vitro digestion of forage crops. J. Br. Grassl. Soc. 18:104-11.

Tothill, JC. 1978. Measuring botanical composition of grasslands. In Measurement of Grassland Vegetation and Animal Production. Hurley, England: Commonwealth Agriculture Bureaux, 22-62.

Van Soest, PJ. 1994. Nutritional Ecology of the Ruminant. Ithaca, N.Y.: Cornell Univ. Press.

Van Soest, PJ, JB Robertson, and BA Lewis. 1991. Methods for dietary fiber, neutral detergent fiber, and nonstarch polysaccharides in relation to animal nutrition. J. Dairy Sci. 74:3583-97.

Vérité, R. 1980. Appreciation of the nitrogen value of feeds for ruminants. In Standardization of Analytical Methodology for Feeds. Ottawa, Canada: International Development Research Centre, 87-96.

Waldo, DR, and BP Glenn. 1984. Comparison of new protein systems for lactating dairy cows. J. Dairy Sci. 67:1115-33.

Wheeler, JL. 1981. Forage evaluation in developing countries. In Forage Evaluation: Concepts and Techniques. Lexington, Ky.: American Forage and Grassland Council, 561-69.

Wheeler, JL, JC Burns, RD Mochrie, and HD Gross. 1973. The choice of fixed or variable stocking rates in grazing experiments. Exp. Agric. 9:19-302.

Forage Analysis by Near Infrared Spectroscopy

8

JOHN S. SHENK
Pennsylvania State University

MARK O. WESTERHAUS
Pennsylvania State University

NEAR infrared reflectance spectroscopy (NIRS) had its beginnings in the 1930s (Gordy and Martin 1939; Kaye 1954; Whetzel 1968). It was first used in agriculture in the late 1960s by Karl Norris for measuring moisture in wheat (*Triticum aestivum* L. emend. Thell.). The first publication of its use in forage analysis is Norris et al. 1976. Since then, more than 100 papers have been published.

The method has many advantages. It is rapid, requiring less than a minute for analysis, and it can predict multiple constituents from the spectra collected. It is easy to use, requiring little or no sample preparation. Generally, the forage sample is oven dried and ground for the measurement. It is safe to the environment, requiring no hazardous chemicals. Its primary disadvantage is its role as a secondary method of analysis, requiring calibrations to primary methods. A major goal of NIRS analysis is to predict chemical and physical properties of a forage sample that relate to animal performance. Although animal response can be predicted directly from the spectra with an accuracy equal to or better than that of the current laboratory reference

analyses (Eckman et al. 1983; Harpster et al. 1982; Holecheck et al. 1982; Ward et al. 1982), NIRS analysis still is used primarily to predict the outcomes of current laboratory analyses.

The precision, i.e., the ability to give the same reading repeatedly, of NIRS analyses is more repeatable than that of the reference method. When both accuracy, i.e., the ability to give the correct value, and precision are acceptable, there is no better analytical procedure presently available for the forage industry.

Details of sample presentation methods (Blosser 1985; Blosser et al. 1986; Shenk and Westerhaus 1992; Shenk and Westerhaus 1993a), sampling methods (Abrams 1989), methods of drying and grinding (Shenk and Westerhaus 1992), and laboratory reference methods used in calibration (Goering and Van Soest 1970; Albrecht et al. 1987; Aufrere and Michalet-Doreau 1988; Clark et al. 1987, 1989; Coelho et al. 1988; Windham et al. 1989; Windham and Barton 1991) have been discussed in Shenk and Westerhaus 1992. The new methods of analyzing undried and unground forages in their natural form have been described by Shenk (1993).

FORAGE SPECTRA

The reflected spectra of forages are very similar to those of other agricultural commodities like grains, oilseeds, and feeds. Molecular absorptions, i.e., wavelengths not reflected, in the near infrared region are primarily due to bending and stretching of x-hydrogen (H) bonds, where x is carbon (C), nitrogen (N), or oxygen (O). Stretching vibra-

JOHN S. SHENK is Professor of Agronomy, Pennsylvania State University. He received the MS and PhD degrees from Michigan State University. He developed the use of near infrared reflectance (NIR) spectroscopy for the analysis of forage and feed.

MARK O. WESTERHAUS is Near Infrared Statistical Analyst, Pennsylvania State University. He received the MS degree from Case Western Reserve University. He has worked with Dr. Shenk since 1977 to develop statistical techniques that improve the accuracy of NIR analyses.

tions occur at lower wavelengths than bending vibrations. An NIR absorption band occurs whenever the frequency of NIR radiation matches the frequency of a bending or stretching vibration of a chemical bond in the sample (Shenk et al. 1992).

The NIR absorptions are called *bands* because monochromators do not measure radiant energy at single wavelengths but as bands about 10 nm wide. An NIR spectrum is obtained by exposing forage to near infrared radiation and measuring the reflected radiation. The resulting spectrum contains information about the chemical and physical properties of the sample. The large amount of information present in a spectrum can be seen by decomposing a spectrum by frequency space deconvolution and a curve-fitting method as described by Nadler (1988) and Nadler et al. (1989). Figure 8.1 illustrates the complexity of a decomposed spectrum of hay that contained 12% moisture, 23% protein, and 46% cell wall fiber.

The estimation of height and width of individual absorption bands is difficult because of the large number of overlapping bands. Theoretical shape of an absorption band is Lorentzian in the frequency (wave number) scale. The slit shape of the instrument widens the band with a smoothing function that approaches a normal distribution. The final shape of an absorption band is a combination of Lorentzian and normal distributions.

CALIBRATION

When used as a secondary method, NIRS must be calibrated to primary reference

methods. If the constituent to be measured is related to the concentration of C-H, O-H, or N-H bonds, such as fiber constituents or crude protein, the calibration is usually straightforward and accurate. If the constituent to be measured is not related to the concentration of C-H, O-H, or N-H bonds, such as minerals, the calibration depends more on the calibration samples and is usually accurate.

Product Libraries. Near infrared spectroscopy relies heavily on the definition of a target population of samples, the collection of representative samples with accurate laboratory reference values, and the use of advanced chemometric procedures to obtain the most accurate calibration. Sample populations for calibration must be broad enough to encompass most samples encountered during routine analysis yet narrow enough to maintain analytical accuracy. Often, many samples are available, but obtaining laboratory reference values for all samples would be very expensive.

Two procedures, CENTER and SELECT, have been developed to define a population on the basis of spectra and to select samples for calibration. The procedure CENTER (Shenk and Westerhaus 1991a) was developed to rank spectra in a file, using principal component scores, according to the Mahalanobis distance of spectra from the average spectra of the file. The procedure SELECT (Shenk and Westerhaus 1991b) was developed to remove redundant samples from the population and to identify new samples that "fill in the gaps" in the product library.

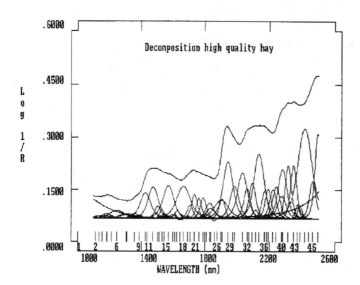

Fig. 8.1. Decomposition of the spectrum from high-quality hay. Individual components cannot be resolved clearly but contribute to the final shape.

To demonstrate the advantages of CENTER and SELECT, three natural (undried and unground) product libraries of forage samples were analyzed. The first library file consisted of 128 pasture samples. The population boundaries were established with CENTER. The library was first reduced with SELECT by eliminating redundant samples, then enlarged with SELECT by adding samples from a larger population of 250 samples that filled in the gaps in the library. This process was repeated to obtain 200 samples from 500 haylage samples and 150 samples from over 250 hay samples obtained with a core sampler. These natural product libraries were accumulated over 2 yr.

The spectral relationships among these libraries were displayed on a three-dimensional graph. To construct the graph, spectra from each library were reduced into the first three principal component scores. These scores represented the three largest patterns of variation among the library spectra. In order to display all three libraries on the same axes, one set of principal components was selected. Since haylage was found to have the largest spectral variation, the first three principal components from the haylage spectra were used to obtain all the scores shown in Figure 8.2. The pasture and haylage spectra were separated along axis three but were more similar to each other than to the hay samples, largely because haylage and pasture samples contained high moisture (45%-75%). The scores of the hay samples were clustered near the bottom of the plot and had much less variation than the other two groups of scores.

This method of compressing the spectra into three values and displaying them in three dimensions has several benefits. First, new spectra can be added to the display to show how well new samples are represented by the library spectra. Second, if the library has large gaps, the probable chemical analysis of samples needed to fill these neighborhoods can be ascertained. Finally, by combining qualitative information (location, maturity, etc.) with the display, the structure of the library can be better understood. By using the display to understand and improve the structure of the library, more accurate calibrations can be developed.

Calibration Concepts and Methods. For the purposes of calibration, NIR spectra consist of overlapping absorption bands, scatter coefficient effects, and other nonlinear effects, especially at high absorption levels. A successful calibration avoids wavelengths with high absorption levels, corrects or minimizes scatter coefficient effects, and selects wavelengths at absorption bands that relate to laboratory data plus wavelengths at absorption bands needed to correct for overlapping bands.

If NIR spectra were not affected by particle size, scatter coefficient, and path length, calibrations could be performed directly on the absorbance data. When these interferences are present, spectral data transformation can help the calibration. None of these derivative treatments compensates fully for the multiplicative changes resulting from scatter and path length changes, but derivatizing spectra minimizes the effect of scatter by emphasizing local changes.

There has been some discussion in the literature concerning which derivative is best. No single derivative gives the best predictions for all constituents and products. In general, calibrations based on first and second derivative with De-trend (Barnes et al. 1989) are among the most accurate. Currently, however, the only way to obtain the lowest prediction error is by trial and error.

CALIBRATION PROCEDURES. Several calibration procedures have been used to maximize the relationships between sample spectra and reference method determinations. These would include multiple regression, principal component regression, partial least squares (PLS) regression, and neural networks (Westerhaus and Reeves 1992).

Multiple regression is the simplest statisti-

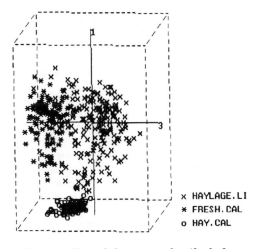

x HAYLAGE.LI
* FRESH.CAL
o HAY.CAL

Fig. 8.2. Natural forage samples (fresh forage, hay, and haylage) displayed against the haylage PCA file.

cal procedure used to predict one variable from one or more other variables. The mathematical model fit by multiple regression is $Y = B_0 + B_1X_1 + B_2X_2 + \ldots + B_pX_p$. Y is the variable to be predicted, B_0 is the equation intercept, and B_i is the scaling factor that relates changes in variable X_i to Y. In this example, Y is the laboratory reference value and X_i is the absorbance values at a given wavelength.

When data from all wavelengths are not needed in the regression model, some form of wavelength selection is needed. The simplest form of wavelength selection is step-up regression, in which variables are added to the model one at a time. An external validation file is recommended to detect overfitting. Overfitting occurs when too many wavelengths are added to the regression model in an attempt of fit individual samples in the calibration set. Stepwise regression is similar to step-up regression except wavelengths are not forced to remain in the model once selected.

The principal component procedure is a data reduction technique. It is designed to reduce the intercorrelated spectral data points (wavelengths) to a smaller set of independent variables called *principal components,* or *eigenvectors.* Patterns of variation are identified, and a score is assigned to the spectra for each pattern. Samples that exhibit a large amount of a pattern are given a high score, and samples that exhibit a small amount of a pattern are given a low score. Principal component regression (PCR) is simply multiple regression on the scores.

Partial least squares is similar to PCR but differs in that the patterns of variation (factors) are identified with the help of the laboratory reference values. Partial least squares calibrations are usually better than PCR calibrations for predicting forage quality because the laboratory reference values are used to form "relevant" factors, i.e., factors that correlate highly with the laboratory values. Modified PLS is a procedure where the NIRS residuals at each wavelength, obtained after each factor is calculated, are standardized (divided by the standard deviations of the residuals at a wavelength) before calculating the next factor.

VALIDATION PROCEDURES. Both PCR and PLS use external validation or cross validation to prevent overfitting. Cross validation is a procedure of partitioning the calibration set into several groups. The calibration is performed once for each group, reserving that group for validation and calibrating on the remaining groups, until every sample has been predicted once. Cross validation uses samples more efficiently than separate calibration and validation sample sets because all samples are used for both calibration and validation.

Experience shows that validations performed on small sets randomly selected from the calibration set can be quite variable. Changing the validation set changes the validation error and often the recommended number of factors. Another advantage of cross validation is the availability of prediction residuals identifying samples that are not consistent with the population.

Cross validation works best for the full spectrum calibration methods because full spectrum calibration models do not change drastically when a few samples are changed. Calibration methods based on wavelength selection, however, often select different wavelengths when different samples are included in the calibration set.

REPEATABILITY FILE. A repeatability file should be used with any of the regression methods described (Westerhaus 1990). This file contains spectra from one or more samples scanned at different sample temperatures, at different room temperatures, on different instruments, etc. The average scan for each sample is subtracted from the individual scans, leaving a file of temperature effects, instrument effects, etc. By accounting for these sources of variation in the calibration procedure, the resulting calibration is developed that is less sensitive to these types of variation. Adding this file to the calibration process substantially improves the performance (robustness) of the calibration in routine operation.

Accuracy of NIRS Analysis Equations. The accuracy and precision of NIRS analyses are always of great concern. Accuracy describes how well the prediction of an unknown sample agrees with the laboratory reference method analysis. There has been some confusion concerning what level of accuracy is required for NIRS analyses to be useful. Many would conclude that the agreement between NIRS analyses and laboratory analyses depends on the precision of the laboratory analyses. While the laboratory precision is the minimum attainable error for NIRS analyses, it does not indicate the level of accuracy required for a measurement to be useful. For many applications, NIRS analyses are used to keep products within certain tolerance limits. This requires accuracy that is better than the tolerance limits. Even in these cases, the error of using a small sample to represent a larger amount of the product (bale, truckload, field, silo, etc.) must be considered.

Accuracy is the measure of agreement between the NIRS-predicted values and laboratory reference method values for samples excluded from the calibration and can be divided into three categories. Single chemical entities, such as crude protein and moisture in most forages, are measured with high precision. Combinations of chemical entities, such as fiber or digestibility measurements in forages, are measured with less accuracy. For these constituents, samples with completely different chemical compositions can receive the same laboratory analysis. The lowest accuracy occurs with constituents, such as minerals, that are present in small quantities or that do not directly absorb energy in the NIR region. Calibrations for these constituents usually rely on secondary correlations (correlations between the constituent to be predicted and another component of the sample more easily measured by NIRS). Calibrations for the last two categories require that the samples to be predicted are represented in the calibration set. NIRS analysis in the third group is no substitute for the laboratory analysis but can be a useful guide if the alternative is to assume an average value from a feed composition table.

Analyzing a sample that represents the largest quantity is essential; i.e., the accuracy of the NIRS analysis is acceptable only if it is equal to or less than the error of subsampling. If this error is too large, both the sampling error and the NIRS analysis error must be improved. Even if NIRS accuracy is low by this criterion, there may be no analytical alternative, and the NIRS analysis is better than no analysis.

Population Diversity and Calibration Method. In general, NIRS calibrations are more accurate for homogeneous products than for heterogeneous products. In heterogeneous products, the chemical entities measured by most laboratory methods vary qualitatively as well as quantitatively. The more diverse the spectra in the population, the larger the standard deviation of the reference values, and the larger the number of eigenvectors required to describe the population. Shenk and Westerhaus (1991a, 1991b) use

these concepts and mathematical procedures, to show that the lowest prediction errors were obtained using a standardized "H" (Mahalanobis distance between two spectra divided by the average Mahalanobis distance to the mean) limit of 0.6 to select samples for calibration. When comparing the performance of stepwise and modified partial least squares (MPLS) regression, they found that MPLS had similar or better prediction accuracy than stepwise regression across constituents and populations. Frank (1987) found this also, using Monte Carlo simulation. In the study by Shenk and Westerhaus (1991b), the average improvement across populations and constituents for MPLS over stepwise regression was 18%.

Calibration sample sets should contain samples with special features. These features may result from different processing methods, an unusual growing season, differing harvest techniques, contamination, or mixture with other products. Spectra can acquire additional special characteristics during sample preparation, so processing and presentation should be standardized. However, small differences in grinding procedures, drying methods, and sample handling inevitably will occur and should be included in the calibration set.

Traditionally, NIRS calibrations for forage have been limited to single products. Specific calibrations are developed for silage, sun-cured hay, and fresh forage. Silage is sometimes subdivided into grass/legume haylage and small grain silage of wheat, barley (*Hordeum vulgare* L.), rye (*Secale cereale* L.), and oat (*Avena sativa* L.). These products require different processing before NIRS analysis. Silage and fresh forage are dried by microwave or forced draft oven before grinding with a cyclone mill. Field-cured hay is ground with a cyclone mill. Hay too moist to be ground is first dried with a forced draft or microwave oven. The classification of forage samples into these different products has been arbitrary, confusing, and often misleading.

There are two advantages of developing and using multiproduct calibrations. In a laboratory situation, the instrument operator is responsible for correctly classifying a sample and selecting the appropriate equation. Improper identification usually results in an incorrect analysis. By combining related products into broad categories, selecting the correct calibration should be easier. Furthermore, the cost of obtaining reference laboratory values should be lower because fewer samples are required for calibration. If only a few samples are available for a product, it may be better to expand a similar product file by adding the new samples and recalibrating rather than obtaining a complete set of new samples for the product (Shenk and Westerhaus 1993a).

CALIBRATION SCOPE. A local calibration equation is designed to analyze all reasonable samples in a small well-defined population. Samples for a local calibration are selected from a group of samples to be analyzed. Laboratory analyses are obtained on the calibration samples. The calibration is developed and used to predict the remaining samples. Developed properly, a local calibration can be more accurate that a global calibration for samples in the target population. The local calibration method is most appropriate in a research service laboratory where samples come in groups such as harvests, years, and experiments (Shenk and Westerhaus 1993b).

A global or broad-based calibration is designed to analyze 90%-95% of all samples of a single product or multiple ones. The goal in deriving a global calibration is to cover a broad range of samples while maintaining acceptable accuracy (Abrams et al. 1987). The accuracy of a global calibration on new samples must be carefully monitored to detect when new samples are no longer represented by the calibration samples. When this occurs, new samples can be added to the calibration samples, and a new calibration developed. If adding new samples to the calibration raises the error of analysis, a separate calibration should be considered.

Adjustments to calibrations for slope and bias have often been used to improve the performance of a calibration on a particular set of samples. After slope and bias adjustment, however, the calibration no longer performs accurately on the original calibration set. If a calibration requires adjustment because new samples are not represented in the calibration set, the calibration should be expanded with some new samples. Recently, calibration expansion has been made easier by the introduction of expandable equations. Expandable equations include all the information necessary to reconstruct the NIR data used by the calibration program, to apply the spectral transformation used with the original calibration data, and to recalibrate with the old and new data using the original calibration technique and options. This method of calibration

adjustment is superior to slope and bias adjustment.

Outlier Tests. Four different outlier tests are available during routine analysis. They are the global "H" test, the neighborhood "H" test, a "T" test, and an "RMS" test. The global H is a measure of the distance between the sample just scanned and the average spectrum of the calibration samples. The neighborhood H is a measure of the distance between the sample just scanned and the closest sample in the calibration set. The T test indicates if the analytical values of the sample just scanned are extreme compared with the laboratory values of the calibration samples. The RMS test ensures subsamples have similar spectra before they are averaged. These outlier tests are very useful for providing confidence in the accuracy of the NIRS analysis.

INSTRUMENT NETWORKS

An instrument network is a group of instruments that share calibrations. Instrument networks have many benefits. Developing NIRS calibrations for forage products is time-consuming and expensive. Instrument networks permit calibrations to be developed on a "master" instrument and shared among compatible instruments on the network, saving time and money. Another benefit of instrument networks is consistency of analyses. The highest level of NIRS analysis consistency can only be achieved by using a single calibration.

Standardized Instruments. Even if instruments meet the manufacturers' specifications, their spectra may not be alike enough to allow consistent predictions across instruments without adjustments to the spectra or equations. A standardization procedure has been developed to adjust the spectra of any monochromator, tilting filter, or discrete filter instrument to match those produced on a "master" instrument. This allows repeatability of NIRS analyses across instruments (Shenk and Westerhaus 1987, 1991c, 1992; Shenk 1990, 1992). Repeatability of NIRS analysis of the same sample across instruments is acceptable when it is equal to or lower than the repeatability of subsampling from the same container of material on a single instrument.

Instrument standardization is accomplished by scanning characterization samples on the master instrument and the instrument to be standardized. The spectra of the instrument to be standardized are adjusted to match those produced by the master instrument. The challenge in instrument standardization is selecting samples that represent the products to be routinely analyzed by the network as characterization samples. Monochromators are standardized by making a wavelength and photometric correction using 30 characterization samples. This procedure can be simplified to a photometric offset correction using a single sample (Shenk and Westerhaus 1991c) for NIRS systems instruments. Filter instruments (discrete or tilting) are standardized to a virtual master instrument using 2 block PLS (Shenk 1990). A virtual master instrument is the "master" monochromator with its spectra transformed to look like a particular filter instrument.

Monitoring Network Performance. Monitoring a network of instruments is similar to monitoring a single instrument. Samples should be routinely analyzed by all instruments on the network and by a laboratory. Three criteria can be monitored: consistency of spectra across instruments, consistency of calibration performance across instruments, and accuracy of NIRS analyses. As with single sample monitoring, agreement between NIRS analyses and the laboratory should be equal to or less than sampling error from the material represented by the sample. Agreement among NIRS analyses should be equal to or less than subsampling error from a container of material.

Many factors can cause NIRS analysis error to exceed the level computed when the calibration was developed. This increase in error can only be identified by routine monitoring of calibration performance. A simple monitoring procedure using confidence limits based on calibration errors has been presented in the 1989 USDA Handbook 643 (Shenk et al. 1989).

ROUTINE FORAGE ANALYSIS

Routine forage analysis requires a precalibrated instrument to perform fast and accurate analyses. The instrument can be configured to analyze up to 50 finely ground samples automatically or to analyze undried and unground samples with a sample transport attachment. Undried and unground samples should be analyzed in duplicate, checking the agreement between subsamples and omitting extreme subsample scans. Instrument performance should be checked daily with a check cell. Instrument diagnostics should be performed at least once a month. If the cali-

brations were developed on spectra from another instrument, the agreement between instruments must be checked before the calibration is used.

The computer software should be designed to identify samples with extreme analytical values (high T values) and unusual spectra (high global and neighborhood H values). Samples that have unusual spectra (large H values) should be saved for reference method analysis. After 10 or more samples have been identified in routine analysis, the calibration should be expanded to include these samples.

Recalibration and Equation Expansion. Recalibration is necessary whenever (1) the instrument changes, (2) samples that should be represented by the calibration samples exhibit large H values or, (3) monitoring the NIRS analyses against laboratory determinations results in values outside the control limits.

Samples for recalibration should be selected by their spectra. The neighborhood H statistic is used to "fill out" the calibration space. The neighborhood H distance between a new sample and each sample in the calibration set is computed. If a small neighborhood H exists, the new sample is already represented by a sample in the calibration set. If the smallest neighborhood H is large, however, the new sample is not represented by the calibration samples and should be added.

After the selected samples are analyzed by the laboratory reference methods and added to the calibration samples, recalibration should be performed. As with the initial calibration, the new calibration must be monitored to maintain analytical accuracy. If the product being analyzed is highly variable, the calibration monitoring and recalibration procedure may need to be repeated several times.

Expandable calibration equations provide a new way for users to update a calibration. An expandable equation contains all the information necessary to reproduce the calibration calculations. When new spectra and laboratory values are added to this, an expanded equation can be derived that performs well on both the new and original calibration samples. This method of equation update is superior to slope and bias adjustments.

To ensure continued accuracy of NIRS analysis, instrument diagnostics should be performed routinely, spectra checked against those in the calibration set, NIRS analyses monitored, and recalibration should be performed when necessary. Participating in a lab certification program, such as the National

Forage Testing Association program sponsored by the National Hay Association and the American Forage and Grassland Council, can provide additional insurance of continued calibration accuracy.

Examples of Successful Use of the Technology. The Universities of Minnesota and Wisconsin have set up a consortium of 17 laboratories in the Midwest to monitor the forage sample analyses by chemistry laboratories and NIRS instruments. This project was designed to certify the accuracy of analytical technology for farmers in this part of the US. This program boosts the creditability of the analytical technology and provides training sessions to the laboratories. It is a good example of how NIRS technology is properly supported.

Plant breeding is another area where NIRS analysis is popular. Many plant-breeding companies are using NIRS to evaluate germplasm and to analyze cultivar trials for quality constituents. One of the first papers published about the potential of NIRS as an analytical method was written by a plant breeder (Marum et al. 1979). Plant-breeding companies are beginning to credit NIRS for their quality analyses in cultivar development.

Hay marketing was one of the first suggested applications of NIRS. Today, most of the hay marketed and sold in the upper Midwest is first analyzed by NIRS. NIRS analysis of undried and unground forages plus a green-brown color index have recently been introduced.

THE POTENTIAL AND FUTURE OF NIRS ANALYSIS

NIRS analysis of agricultural products has only a 30-yr history. Karl Norris began to explore its potential in the mid-1960s. In this short 30-yr period, NIRS analysis has progressed from the simple measurement of moisture in ground wheat using fixed filter instruments to the analysis of fresh pasture grass using a visible near infrared monochromator with sophisticated software. NIRS analysis has kept pace with the rapidly developing electronic industry, and the forage industry has benefited greatly from its development.

At present, a major disadvantage of NIRS in the forage industry is its dependence on primary reference methods for calibrations. The potential of using NIRS to formulate rations for livestock has not been realized. During the next 30 yr, NIRS should be exploited to help predict animal performance, much like current laboratory procedures and feed composition tables did when they were developed.

Another challenge for the next 30 yr will be cost reduction for instrumentation and software. Also, robust sensors need to be developed for hay balers, forage choppers, and other harvesting equipment to assist the farmer with real time decisions. NIR sensors must be miniaturized into small devices, perhaps even hand-held, for the hay-marketing industry.

Research on NIRS technology is being conducted at Pennsylvania State University in three major areas: making instruments more alike by improving instrument standardization, improving population structuring, and developing new calibration methods for more accurate NIRS analyses. Researchers at the University of Nebraska have shown NIRS can be used to determine ploidy in forage grasses. South Dakota State researchers have investigated C^{15} uptake in forage material. Researchers at the Universities of Illinois and Missouri have derived calibrations to detect mold in hay. Many laboratories are developing in vitro digestion calibrations for forage and maize (*Zea mays* L.) silage. These are only a few of the new and interesting research areas currently under investigation.

The potential and future of NIRS in forage evaluation rest in the hands of the instrument manufacturers, the academic research community, and the livestock-feeding industry. NIRS technology will benefit from advances in the electronics industry and remains the most environmentally safe method of analysis.

QUESTIONS

1. Briefly explain the principles upon which the use of near infrared reflectance spectroscopy is based for the analysis of forages.
2. What are the advantages of using NIRS as an analytical technique of the quality of forages and feedstuffs?
3. What are the major limitations for the use of NIRS as an analytical tool?
4. Speculate as to the impact of NIRS analysis on future research and extension programs.
5. Consult with an agricultural extension service specialist or crop analytical consultant, and determine the impact of the forage-testing and ration formulation service provided in your area.

REFERENCES

Abrams, SM. 1989. Sampling. In GC Marten et al. (eds.), Near Infrared Reflectance Spectroscopy

(NIRS): Analysis of Forage Quality, USDA-ARS Agric. Handb. 643. Washington, D.C.: US Gov. Print. Off., 22.

Abrams, SM, JS Shenk, MO Westerhaus, and FE Barton II. 1987. Determination of forage quality by near infrared reflectance spectroscopy: Efficacy of broad based calibration equations. J. Dairy Sci. 70:806-13.

Albrecht, KA, GC Marten, JL Halgerson, and WF Wedin. 1987. Analysis of cell-wall carbohydrates and starch in alfalfa by near infrared reflectance spectroscopy. Crop Sci. 27:586.

Aufrere, J, and B Michalet-Doreau. 1988. Comparison of methods for predicting digestibility of feeds. Anim. Feed Sci. and Tech. 20:203-18.

Barnes, RJ, MS Dhanoa, and SJ Lister. 1989. Standard normal variate transformation and detrending of near-infrared diffuse reflectance spectra. Appl. Spectros. 43:772-77.

Blosser, TH. 1989. High-moisture feedstuffs, including silage. In GC Marten et al. (eds.), USDA-ARS Agric. Handb. 643. Washington, D.C.: US Gov. Print. Off., 55-57.

Blosser, TH, JB Reeves III, and VF Colenbrander. 1986. Near infrared reflectance spectroscopy (NIRS) for predicting chemical composition of undried feedstuff. J. Dairy Sci. 69:abstr. suppl. 136.

Clark, DH, HF Mayland, and RC Lamb. 1987. Mineral analysis of forages with near infrared reflectance spectroscopy. Agron. J. 79:485-90.

Clark, DH, EE Cary, and HF Mayland. 1989. Analysis of trace elements in forages by near infrared reflectance spectroscopy. Agron. J. 81:91-95.

Coelho, M, FG Hembry, FE Barton, and AM Saxton. 1988. A comparison of microbial, enzymatic, chemical and near-infrared reflectance spectroscopy. Methods in forage evaluation. Anim. Sci. and Tech. 20:219-31.

Eckman, DD, JS Shenk, PJ Wangsness, and MO Westerhaus. 1983. Prediction of sheep responses by near infrared reflectance spectroscopy. J. Dairy Sci. 66:1983-87.

Frank, IE. 1987. Intermediate least squares regression method. Chemometrics and Intelligent Lab. Syst. 1:233-42.

Goering, HK, and PJ Van Soest. 1970. Forage fiber analysis. In USDA Agric. Handb. 379. Washington, D.C.: US Gov. Print. Off., 8-15.

Gordy, W, and PC Martin. 1939. The infrared absorption of HCl in solution. J. Chem. Phys. 7:99.

Harpster, HW, E Keck, PJ Wangsness, SM Abrams, and JS Shenk. 1982. Predicting the feeding value of hay from infrared spectroscopy and chemical lab analysis. J. Anim. Sci. 55(suppl. 1):310-11.

Holecheck, JL, JS Shenk, M Vavra, and D Arthur. 1982. Prediction of forage quality using near infrared reflectance spectroscopy on esophageal fistula samples from cattle on mountain range. 55:971-75.

Kaye, W. 1954. Near infrared spectroscopy. 1. Spectral identification and analytical application. Spectrochim. Acta 6:257-87.

Marum, P, AW Hovin, GC Marten, and JS Shenk. 1979. Genetic variability for cell wall constituents and associated quality traits in reed canarygrass. Crop Sci. 19:355-60.

Nadler, TK. 1988. Identifying resonances in the near infrared absorption spectrum. MS thesis, Pennsylvania State Univ., University Park.

Nadler, TK, ST McDaniel, MO Westerhaus, and JS Shenk. 1989. Deconvolution of near-infrared spectra. Appl. Spectros. 43:1354.

Norris, KH, RF Barnes, JE Moore, and JS Shenk. 1976. Predicting forage quality by infrared reflectance spectroscopy. J. Anim. Sci. 43:889.

Shenk, JS. 1990. Standardizing NIRS instruments. In R Biston and N Bartiaux-Thill (eds.), Proc. 3d Int. Conf. Near Infrared Spectros. Gembloux, Belgium: Agriculture Research Center Publishing, 649-54.

———. 1992. Networking and calibration transfer. In I Murray and IA Cowe (eds.), Making Light Work: Advances in Near Infrared Spectroscopy. New York: VCH Publishers, 223-28.

———. 1993. Analysis of undried, unground forage with a visible near infrared monochromator. In Proc. 17th Int. Grassl. Congr., Palmerston North, New Zealand, and Australia, 591-93.

Shenk, JS, and MO Westerhaus. 1987. Software standardization of NIR instrumentation. 14th FACSS. 129.

———. 1991a. Population definition, sample selection, and calibration procedures for near infrared reflectance spectroscopy. Crop Sci. 31:469-74.

———. 1991b. Population structuring of near infrared spectra and modified partial least squares regression. Crop Sci. 31:1548-55.

———. 1991c. New standardization and calibration procedure for NIRS analytical systems. Crop Sci. 31:1694-96.

———. 1992. NIRS analysis of agriculture products. Near infrared spectroscopy. Bridging the gap between data analysis and NIR application. In KI Hildrum, T Isaksson, T Naes, and A Tandberg (eds.), Norwegian Food Research Institute, 235-40.

———. 1993a. Near infrared reflectance analysis with single- and multiproduct calibrations. Crop Sci. 33:582-84.

———. 1993b. Using near infrared reflectance product library files to improve prediction accuracy and reduce calibration costs. Crop Sci. 33:578-81.

Shenk, JS, MO Westerhaus, and SA Abrams. 1989. Protocol for NIRS calibration: Monitoring analysis results and recalibration. In GC Marten et al. (eds.), Near Infrared Reflectance Spectroscopy (NIRS): Analysis of Forage Quality, USDA-ARS Agric. Handb. 643. Washington, D.C.: US Gov. Print. Off., 104-10.

Shenk, JS, JJ Workman, Jr., and MO Westerhaus. 1992. Application of NIR spectroscopy to agriculture products. In DA Burns and EW Ciurczak (eds.), Handbook of Near-infrared Analysis, New York: Marcel Dekker, 383-431.

Ward, RG, GS Smith, JD Wallace, NS Urguhart, and JS Shenk. 1982. Estimates of intake and quality of grazed range forage by near infrared reflectance spectroscopy. J. Anim. Sci. 54:399-402.

Westerhaus, MO. 1990. Improving repeatability of NIR calibrations across instruments. In R Biston and N Bartiaux-Thill (eds.), Proc. 3d Int. Conf. Near Infrared Spectrosc. Gembloux, Belgium: Agriculture Research Center Publishing, 671-74.

Westerhaus, MO, and JB Reeves III. 1992. NIR calibrations of agricultural products with neural networks. Near infra-red spectroscopy. Bridging the gap between data analysis and NIR applications. In KI Hildrum, T Isaksson, T Naes, and A Tandberg (eds.), Norwegian Food Research Institute, Ellis Horwood, West Sussex, England, 79-84.

Whetzel, KB. 1968. Near-infrared spectroscopy. Appl. Spectrosc. Res. 2(1):1-67.

Windham, WR, and FE Barton II. 1991. Moisture analysis in forage by near infrared reflectance spectra: Collaborative study of calibration methodology. J. Assoc. Off. Anal. Chem. 74:324-31.

Windham, WR, DR Mertens, and FE Barton II. 1989. Protocol for NIRS calibration: Sample selection and equation development and validation. In GC Marten et al. (eds.), Near Infrared Reflectance Spectroscopy (NIRS): Analysis of Forage Quality, USDA-ARS Agric. Handb. 643. Washington, D.C.: US Gov. Print. Off., 96-103.

9

Forage-induced Animal Disorders

HENRY F. MAYLAND
Agricultural Research Service, USDA

PETER R. CHEEKE
Oregon State University

THIS chapter presents antiquality factors of forage crops that affect animal performance. A summary is given of some herbage minerals that occur in concentrations deficient or toxic to animals. Significant attention is given to organic compounds that are secondary products (1) of photosynthesis or (2) of associated organisms that alter forage quality and reduce animal performance.

MINERAL DISORDERS

Mineral element concentrations in herbage may affect animal health because low levels may lead to deficiency, because high levels may lead to toxicity, or because one mineral element interacts with another to reduce the relative availability. Herbage levels and ruminant diet requirements are given in Table 9.1.

Grass Tetany (Hypomagnesemia). Grass tetany is a magnesium (Mg) deficiency of ruminants usually associated with their grazing of cool-season grasses during spring. Next to

HENRY (HANK) F. MAYLAND is Research Soil Scientist with the Agricultural Research Service, USDA, Kimberly, Idaho. His studies center on mineral cycling in the soil-plant-animal system, with emphasis on magnesium, selenium, silicon, sulfur, and zinc.

PETER R. CHEEKE is Professor of Comparative Nutrition in the Department of Animal Sciences, Oregon State University. His research focus is the study of natural toxicants in feeds and poisonous plants. He has also extensively studied the nutrition of small herbivores.

bloat, this is probably the most important nutritional problem in grazing livestock. Annual losses in the US are estimated at $50 million to $150 million (Mayland and Sleper 1993). Grass tetany occurs in all classes of cattle (*Bos taurus* and *Bos indicus*) and sheep (*Ovis aries*) but is most prevalent among older females early in their lactation. Magnesium must be supplied daily because it is excreted in urine and milk.

Plasma Mg levels are normally 18-32 mg/L. Blood plasma or serum and urine values less than 18 mg/L are considered indicative of hypomagnesemia. Increasing concern is expressed when blood plasma or urine levels fall below 15 mg/L (Mayland 1988; Vogel et al. 1993), and the dramatic signs of tetany are evidenced at levels below 10 mg/L. Physical symptoms proceed from reduced appetite, dull appearance, and staggering gait to signs of increased nervousness, frequent urination and defecation, muscular tremors, and excitability followed by collapse, paddling of feet, coma, and death.

Grass tetany is a complex disorder. First, Mg requirements are greater for lactating than for nonlactating animals and greater for older than for younger animals. Second, there are differences among bovine breeds in susceptibility to grass tetany with Brahman (*Bos indicus*) and Brahman crossbreeds being more tolerant and European breeds being least tolerant (Greene et al. 1989). Third, many factors influence Mg concentration and availability in the herbage. The principal factor is a high level of potassium (K), which negatively affects soil Mg uptake by plants and availability of herbage Mg to animals. Also,

TABLE 9.1.   Spectrum of nutrient–element concentrations found in grass and legume herbage and a guide to herbage concentrations required by sheep and cattle

Element		Concentrations in		Herbage concentrations for	
		Grasses	Legumes	Sheep	Cattle
		(mg/g)			
Calcium	Ca	2–5	2–14	3–4	
Chlorine[a]	Cl	0.1-20.0		1	2
Magnesium	Mg	1–3	2–5	1.2	1.9[c]
Nitrogen	N	10–40	10–50	10–15	
Phosphorus	P	2–4	3–5	2.5	3.0
Potassium	K	10–30	20–37	2–6	
Silicon[a]	Si	10–40	0.5–1.5	not established	
Sodium[a]	Na	0.1–3.0	0.1–2.0	1	2
Sulfur	S	1–4	2–5	1–2	
		(µg)			
Boron	B	3–40	30–80	. . . .	. . . .
Cobalt	Co	0.1–0.2	0.2–0.3	0.11	0.08
Copper	Cu	3–15	3–30	5–6	7–10[d]
Fluorine[a]	F	2-20		1–2	
Iodine[a]	I	0.004–0.8		0.5	0.5[e]
Iron[b]	Fe	50–250	50–250	30	40
Manganese	Mn	20–100	20–200	25	25
Molybdenum	Mo	1–5	1–10	not established	
Selenium[a]	Se	0.01–1.0		0.03–0.7	
Zinc	Zn	15–50	15–70	25–40	

Sources: Herbage data are from the author's files, Gough et al. (1979), and Mayland (1983). Animal data are adapted from Grace (1983), Grace and Clark (1991), and NRC (1984).
[a]Not required for plant growth.
[b]Values in excess of 100–150 µg/g are often reflective of soil contamination.
[c]2 mg/g if K and N are high.
[d]Influenced by Mo and S.
[e]2 µg/g in presence of goitrogens.

low levels of readily available energy, calcium (Ca), and phosphorus (P) and high levels of organic acids, higher fatty acids, and nitrogen (N) in the herbage reduce the absorption of or retention of Mg by the animal. Forage Mg levels greater than 2.0 mg/g dry matter and a milliequivalent 0 ratio of less than 2.2 for K/(Ca + Mg) are considered safe (Grunes and Welch 1989).

Agronomic practices to reduce risk potential of tetany include splitting applications of N and K fertilizers, liming acid soils with dolomitic rather than calcitic limestone, or spraying Mg on herbage (Robinson et al. 1989). One might also use cultivars or species having high Mg and Ca and low K concentrations (Sleper et al. 1989; Moseley and Baker 1991; Mayland and Sleper 1993). Animal husbandry practices include assigning breeds and classes of livestock to pastures that are less susceptible to tetany (Greene et al. 1989). Magnesium could be supplemented through additions to herbage, drinking water, stock salt, molasses licks, or other energy sources (Robinson et al. 1989).

Nitrate Poisoning. Nitrate (NO_3^-) accumulates in plant tissue because of luxuriant uptake of soil N when plant metabolism of N is slow or even stopped. This condition is promoted by cool temperature, drought, or other physiological stress that slows growth (Wright and Davison 1964). Plant NO_3^- itself is not normally toxic to animals, but it is usually reduced to nitrite (NO_2^-) in the rumen. If the NO_2^- is not reduced further, the accumulated NO_2^- causes various intensities of methemoglobinemia that may cause acute poisoning of the animal (Wright and Davison 1964; Singer 1972; Deeb and Sloan 1975).

Intensive applications of N fertilizer to silage crops often result in appreciable amounts of NO_3^- in the herbage. During ensiling, the NO_3^- is completely or partially degraded (Spoelstra 1985). The end products are ammonia (NH_3) and nitrous oxide, with NO_2^- and nitric oxide occurring as intermediates. Silage crops containing high NO_3^- levels should be checked for NO_3^- before the ensiled crops are fed to livestock.

Silofiller's disease is an illness of farm workers caused by inhalation of nitric and nitrous oxides from fermenting forages containing high N concentrations. Good air ventilation will reduce the health hazard of these noxious gases.

The most common effect of NO_3^- poisoning is the formation of methemoglobin that occurs when NO_2^- oxidizes the ferrous iron of blood hemoglobin to ferric iron. This produces a chocolate-brown methemoglobin that cannot release oxygen (O) to body tissue. As the toxicity intensifies, the brownish-colored blood casts a brownish discoloration to the nonpigmented areas of skin and the mucous membranes of the nose, mouth, and vulva. Clinical signs progress with staggering, rapid pulse, frequent urinating, and labored breathing, followed by collapse, coma, and death. Sublethal toxicity may be evidenced by abortion of pregnant females.

The rate and degree of NO_3^- reduction in the rumen depend on the microflora present and the energy available for continued reduction of NO_2^- to NH_3. This interaction of available energy and NO_3^- reduction results in mixed responses. Furthermore, cattle and sheep can adapt to some extent to prolonged high levels of NO_3^- feeding (Wright and Davison 1964). Sheep have been poisoned by as little as 500 µg N/g as NO_3^-, whereas in other experiments 8000 µg N/g as NO_3^- had no effect (Singer 1972). Other data show that 2000 µg N/g as NO_3^- in forage was the threshold limit for cattle (Bush et al. 1979) while no problem occurred with feeding forage containing 4600 µg N/g as NO_3^- (Davison et al. 1964).

Forages containing 3400 to 4500 µg N/g as NO_3^- should be considered potentially toxic and should be mixed with safer feeds prior to use (Wright and Davison 1964). Energy supplements will help in the complete reduction of inorganic N. Suspect forages should be tested for NO_3^- levels. Although risky, feeding additional forages low in NO_3^- would dilute the NO_3^- intake. Be aware that these test levels are reported on an elemental N basis.

Sudangrass (*Sorghum bicolor* [L.] Moench), sorghums (*Sorghum* spp.), corn (*Zea mays* L.), and small grains are often indicted in NO_3^- poisoning. These crops are often heavily fertilized and when subject to drought, frost, or other stress may accumulate large concentrations of NO_3^-. Perennial grasses generally are less of a problem because they usually receive lower levels of N fertilizer and have greater tolerance of drought and frost. Legumes fix their own N into reduced forms and do not contribute to NO_3^- poisoning.

Copper, Molybdenum, and Sulfur Interaction.

Copper (Cu) deficiency significantly affects ruminant livestock production in large areas of North America (Gooneratne et al. 1989), mainly in areas where soils are naturally high in molybdenum (Mo) and sulfur (S)

(Kubota and Allaway 1972). In ruminant nutrition, the two-way and three-way interactions between these three elements are unique in their complex effects on animal health. Different clinical terms describe each primary factor affecting Cu nutrition.

Sheep consuming a complete diet, low in S and Mo and with modest Cu (12-20 mg/kg dry matter), can succumb to Cu toxicity. Sheep grazing another pasture of similar Cu concentration, but high in Mo and S, will produce Cu-deficient lambs showing clinical signs of swayback disease (Suttle 1991). Copper-deficient cattle and sheep appear unthrifty and exhibit poor growth and reproduction. Copper deficiency reduces the level of melanin pigments in hair and wool so that normally dark-colored fibers will be white or grey (Grace 1983; Grace and Clark 1991). Low dietary Cu levels may limit effectiveness of the immune system in animals.

Forage plants growing on soil or peat that contain high levels of Mo will produce a scouring disease known as *molybdenosis*. In the presence of S, high intake of Mo can induce a Cu deficiency due to formation of insoluble Cu-Mo-S complexes in the digestive tract that reduce the absorption of Cu. Several pathways exist by which Cu × Mo × S interactions mediate their effect on ruminants (Suttle 1991). Most clinical signs attributed to the three-way interaction are the same as those produced by simple Cu deficiency and probably arise from impaired Cu metabolism. The tolerable risk threshold of Cu:Mo is not fixed at 3:1 but declines from 5:1 to 2:1 as pasture Mo concentrations increase from 2 to 10 µg Mo/g. Risk assessment has been only partially successful because the interactions are not yet fully understood (Suttle 1991).

Cattle are more sensitive than sheep to molybdenosis; however, sheep are more susceptible to Cu toxicity (Suttle 1991). Sheep should not be allowed to graze pastures that recently received poultry or swine manure. This manure may contain high Cu concentrations originating from the Cu salts fed to poultry or swine to control worms.

Copper-depleted animals should respond equally well to dietary Cu supplements, oral Cu boluses or pellets, and Cu injections. Copper fertilization of Cu-deficient pastures should be done carefully because the range between sufficiency and toxicity is quite small.

Selenium Deficiency and Toxicity.

Herbage selenium (Se) concentrations are marginal to

severely deficient for herbivores in many areas of the world. These include the Pacific Northwest and the eastern one-third of the US (Kubota and Allaway 1972; Mayland et al. 1989). Herbage Se concentrations of 0.03 µg/g are generally adequate. However, 0.1 µg/g may be necessary when high S reduces Se availability. Climate conditions and management practices that favor large increases in forage yield may dilute Se concentrations to critical levels in herbage.

Selenium deficiency causes white muscle disease in lambs and calves. The young may be born dead or die suddenly within a few days of birth because Se intake by the gestating dam was inadequate. A delayed form of white muscle disease occurs in young animals, while a third form is identified as illthrift in animals of all ages. Injectable Se, often with vitamin E, or oral supplementation (selenized salt or Se boluses) may meet animal requirements. Selenium fertilization of soils is legal only in New Zealand and Finland.

In many semiarid areas of the world, grasses and forbs contain adequate (0.03-0.1 µg/g) to toxic (>5 µg/g) levels of Se for grazing animal requirements. These areas include desert, prairie, and plains regions of North America where Se toxicity is observed in grazing animals. Some plants growing in these areas will accumulate 100-1000 µg/g Se. Animals eating these generally unpalatable plants will likely die. Grasses, small grains, and some legumes growing on the Se-rich Cretaceous geological materials may contain 5-20 µg/g. Some animals eating this herbage may die, but most are likely to develop chronic selenosis called *alkali disease*. There is hair loss, and hoof tissues become brittle. In these instances, some animals may develop a tolerance for Se as high as 25 µg/g.

A second chronic disorder in ruminants, called "blind staggers," or polioencephalomalacia, also occurs in these areas. This disorder, while historically attributed to Se, is likely caused by excess S (Mayland, unpublished data). High sulfate levels in the drinking water and ingested herbage have led to the occurrence of blind staggers (Beke and Hironaka 1990). Changing to high-quality, low-sulfate water and forage has reduced the risk.

Under marginal levels of Se, increased S reduces the uptake of Se by plants and the bioavailability of dietary Se to animals. However, the S antagonist has not been effective against Se toxicosis. Replacing high-Se forage with low-Se forage is the most effective way of countering Se toxicity.

Cobalt, Iodine, and Zinc. Cobalt (Co) deficiencies in herbivores have been identified in the southeastern US and Atlantic seaboard states and along the Wasatch Front in Utah (Kubota and Allaway 1972). Cobalt is a metal cofactor in vitamin B_{12}, which is required in energy metabolism in ruminants. The signs of Co deficiency include a transient unthriftiness and anemia leading to extreme loss of appetite and eventually death. Two other conditions attributed to Co deficiency are ovine (sheep) white liver disease and phalaris staggers (Graham 1991). The mechanism by which oral Co supplementation prevents staggers is not understood. Pasture herbage levels of at least 0.11 and 0.08 µg Co/g will provide adequate Co for sheep and cattle, respectively. Cobalt injections or oral supplements can be given to animals. Pastures may also be fertilized with cobalt sulfate. Manganese (Mn) and iron (Fe) are antagonists to Co absorption (Grace 1983).

Plants do not require iodine (I); nevertheless, herbage in the northern half of the US is generally deficient for animal requirements. These deficiencies are noted by the occurrence of goiter in animals (Kubota and Allaway 1972). The use of iodized salt easily meets iodine needs of animals on pasture.

Zinc (Zn) concentration in pasture plants ranges from 10 to 70 µg/g but is most often in the 10-30 µg/g range. In one study, cattle grazing forage having 15-20 µg/g gained weight faster when supplemented with additional Zn (Mayland et al. 1980). Blood Zn levels were statistically higher in supplemented than in control animals, but this difference was very small and would not be useful as a diagnostic tool. There was no other obvious difference between the control and Zn-supplemented groups. More information on Zn requirements and diagnostic tests is needed.

Fluorosis and Silicosis. Plants do not require fluorine (F), but herbage generally contains 1-2 µg F/g, which is adequate for bone and tooth development in animals. At higher levels, the development of fluorosis is influenced by the age, species, dietary form, and length of exposure. Fetuses and young animals are most susceptible to excess F. Otherwise, long-term intakes of 30 µg F/g may be tolerated by ruminants before bone abnormalities appear (Underwood 1977). In areas of endemic fluorosis, plants may be contaminated by naturally fluoridated dust from rock phosphate or other smelters. During manufacture of super-phosphate and dicalcium phosphate, 25%-50% of the original F is lost. Excess F may also be absorbed by plants that are sprinkler irrigated with thermal groundwater. Rock phosphate supplement and naturally fluoridated drinking water are the primary dietary sources of excess F.

Grasses contain more silicon (Si) than do legumes and account for the large amount of Si ingested by grazing animals. Silicon may be needed in trace amounts by animals but is not required by herbage plants. Silicon adversely affects forage quality and may affect animal preference of plants. Silicon serves as a varnish on the cell walls, complexes microelements and reduces their availability to rumen flora, and inhibits the activity of cellulases and other digestive enzymes (Shewmaker et al. 1989). The net effect of forage Si is to reduce dry matter digestibility by 3 percentage units for each percentage unit of Si present (Van Soest and Jones 1968).

Silicon is also responsible for urolithiasis (urinary calculi or range waterbelly) in steers. Waterbelly is correlated negatively with urine volume and water intake and only weakly related to herbage Si. Management strategies in high-Si areas include stocking only heifers and providing adequate drinking water. If feasible, the Ca:P ratio in the diet should be reduced and urine acidified by supplementing animals with ammonium chloride (Stewart et al. 1991).

Soil Ingestion. Most grazing animals and some humans eat soil. The ingestion may be indirect because herbage is dusty. Nevertheless, animals may actively eat soil for unknown reasons. Ingested soil may serve as a source of minerals. In addition, soil may contain residual chemicals applied to pasture or derived from atmospheric fallout. Some soil particles are harder than tooth enamel and will cause excessive abrasion and premature loss of teeth. Ingested soil, as a possible contaminant, must also be considered in experimental pasture studies (Mayland et al. 1977).

NATURAL TOXICANTS IN FORAGES

Plants are protected against herbivory (being eaten by vertebrate and invertebrate herbivores) by many diverse physical and chemical means. Physical defenses include leaf hairs, spines, thorns, highly lignified tissue (e.g., wood), and growth habitat (e.g., prostrate form). Chemical defenses of plants include a bewildering array of often complex chemicals that are toxic or poisonous to various herbivores. These chemicals may be synthesized by the plant itself or may be produced by symbiotic or mutualistic fungi growing with the plant. The defensive chemicals produced by the plant are considered secondary compounds, as distinct from the primary compounds that are essential in plant metabolism. Secondary compounds (e.g., alkaloids) do not function directly in cellular metabolism but apparently are synthesized to serve as the plant's defensive arsenal.

Chemicals synthesized by fungi are known as *mycotoxins,* which may be produced by fungi living on or in forage plants. Mycotoxins are responsible for many disorders of grazing animals. For example, the endophytic fungi of tall fescue (*Festuca arundinacea* Schreb.) produce ergot alkaloids that cause fescue foot, summer fescue toxicosis, and reproductive disorders, while endophytes in perennial ryegrass (*Lolium perenne* L.) produce lolitrems, which cause ryegrass staggers. In the US, total livestock-related losses attributed to the tall fescue endophyte are estimated between $500 million and $1 billion a year (Ball et al. 1993). The economic impact of poisonous plants on livestock production has been reported by James et al. (1988).

Toxins of plant origin can be classified into several major categories, including alkaloids, glycosides, proteins and amino acids, and phenolics (tannins). Alkaloids are bitter substances containing N in a heterocyclic ring structure. There are hundreds of different al-

Fig. 9.1. An example of ryegrass staggers, a neurological disorder of livestock grazing endophyte-infected perennial ryegrass pastures. Lolitrem B is the major toxin involved. *Courtesy of R. E. G. Keogh.*

kaloids, which are classified according to the chemical structure of the N-containing ring(s). For example, the pyrrolizidine alkaloids in common pasture weeds (*Senecio* spp.) have a pyrrolizidine nucleus of two five-membered rings, while ergot alkaloids have an indole ring structure. Lupins (*Lupinus* spp.) contain quinolizidine alkaloids, which are based on two six-membered rings.

Glycosides are composed of a carbohydrate (sugar) portion linked to a noncarbohydrate group (the aglycone) by an ether bond. Examples are cyanogenic glycosides, glucosinolates, saponins, and coumarin glycosides. Their toxicity is associated with the aglycone, such as cyanide in cyanogenic glycosides. Glycosides are hydrolyzed by enzymatic action, releasing the aglycone. This often occurs when the plant tissue is damaged by wilting, freezing, mastication, or trampling. A good example of this is the production of toxic cyanide when forage sorghums such as sudangrass are frosted. The breakdown of cell structure releases the glycoside from storage vacuoles, al-

lowing it to be hydrolyzed by enzymes in the cytosol and to release free cyanide.

Many toxic amino acids occur in plants. One of the best known is mimosine, a toxic amino acid in the tropical forage legume *Leucaena leucocephala* de Wit. Others include lathyrogenic amino acids in *Lathyrus* spp., indospecine in *Indigofera* spp., and the brassica anemia factor, which is caused by S-methylcysteine sulfoxide, a metabolic product of forage brassicas (*Brassica* spp.).

Phenolic compounds, which include the condensed and hydrolyzable tannins, are substances containing aromatic rings with one or more hydroxyl groups. The hydroxyl groups are chemically reactive and can react with functional groups of proteins to form indigestible complexes. The tannin-protein complexes are astringent and adversely affect feed intake. All plants contain phenolic compounds. In some cases, their type or concentration may cause negative animal responses. These include reduced feed intake and protein digestibility of birdsfoot trefoil (*Lotus cor-*

Fig. 9.2. (*a*) A steer in Australia that has been grazing on *Leucaena leucocephala* for several months. Note the loss of hair and unthrifty appearance due to the effects of mimosine, a toxic amino acid that occurs in *Leucaena*. (*b*) The same steer a few months after receiving a dose of rumen bacteria that detoxifies mimosine. These photographs illustrate quite dramatically the potential of rumen microbes to detoxify some of the natural toxins in plants. *Courtesy of R. J. Jones.*

niculatus L.) and sericea lespedeza (*Lespedeza cuneata* [Dum.-Cours.] G. Don). Oak (*Quercus* spp.) poisoning is caused by tannins in oak browse. Many tree legumes used in tropical agroforestry can contain sufficient levels of tannins to impair animal performance.

Toxins and Animal Disorders Associated with Forage Legumes. Some animal disorders are associated with forage legumes. For example, pasture bloat is a very common problem with ruminants grazing many forage legumes, such as alfalfa (*Medicago sativa* L.) and clovers (*Trifolium* spp.). Pasture bloat is caused by rapid release of cell contents of succulent, immature legume forage during rumen fermentation. The succulent material has a rapid rate of cell rupture, releasing the soluble proteins and fermentable carbohydrates. Soluble proteins are foaming agents, causing the formation of a stable foam in the rumen, which prevents eructation (belching) of rumen gases formed by the fermentation. High sugar content of lush legume forage supports a vigorous microbial population, producing large amounts of gas. The combination of vigorous gas production and presence of foaming agents produces a stable frothy foam, which blocks the esophagus, culminating in rumen distention and respiratory paralysis. Bloat can be minimized by maintaining at least 50% grass in the pasture or by giving animals access to blocks containing an antifoaming agent (e.g., poloxalene).

Some legume forages, such as birdsfoot trefoil, cicer milkvetch (*Astragalus cicer* L.), sericea lespedeza, and sainfoin (*Onobrychis viciifolia* Scop.), do not cause bloat. These plants contain tannins, which at low concentrations complex with the cytoplasmic proteins and prevent their ability to foam. However, *Lespedeza* spp. and trefoils (*Lotus* spp.) can contain sufficiently high tannin levels to inhibit animal performance by reducing feed intake and protein digestibility. Although saponins are foaming agents that occur in most bloat-producing legumes, they do not seem to have a significant role in bloat (Majak et al. 1980).

Phytoestrogens occur in many forage legumes, including alfalfa, red clover (*T. pratense* L.), subterranean clover (*T. subterraneum* L.), and other *Trifolium* spp. In clovers, the estrogenic compounds are mainly isoflavones such as formononetin, genistein, biochanin A, and diadzein, while coumestrol and other coumestans are the principal estrogens in alfalfa and medics, like black medic (*Medicago lupulina* L.).

Historically, phytoestrogens have had their greatest impact on reproduction of sheep grazing subterranean clover in Australia. "Clover disease" is characterized by a dramatic decrease in fertility of ewes, along with abnormalities of the genitalia. Since the identification of phytoestrogens as the causative agents of clover disease, Australian plant breeders have selected for and developed low-estrogen cultivars of subterranean clover. This has largely eliminated the problem of clover-induced infertility in Australia.

The phytoestrogens owe their physiological activity to their structural resemblance to endogenous estrogens, allowing them to bind to estrogen receptors and elicit an estrogen response (Adams 1989). The activity of phytoestrogens is altered by rumen fermentation. Biochanin A and genistein are converted to inactive compounds in the rumen, while formononetin is bioactivated to the more potent compounds daidzein and equol (Adams 1989).

Toxins associated with specific forage legume species will be briefly described. Further detail is provided by Cheeke and Shull (1985).

RED CLOVER (*Trifolium pratense* L.) When infected with the black patch fungus (*Rhizoctonia leguminicola* Gough & ES Elliott), red clover hay may contain the indolizidine alkaloid slaframine. Slaframine is a cholinergic agent that causes excessive salivation (clover slobbers), eye discharge, bloat, frequent urination, and watery diarrhea. These effects are due to stimulation of the autonomic nervous system. The fungal infection and potential toxicity develop most rapidly in periods of high humidity. Prompt removal of the toxic forage from livestock generally alleviates all signs of intoxication.

WHITE CLOVER (*Trifolium repens* L.). White clover may contain cyanogenic glycosides. These may have some importance in conferring resistance to slugs and other pests. The concentration of cyanogens in white clover is not sufficiently high to cause livestock poisoning but may contribute to lower midsummer palatability of the clover.

ALSIKE CLOVER (*Trifolium hybridum* L.). Poisoning from alsike clover has been reported in Canada and the northern US. It has not been conclusively proven to be caused by the clover, but circumstantial evidence strongly suggests that it is clover caused (Nation 1991). Toxicity signs include photosensitization, neurological effects such as depression and stupor, liver damage, and contradictory effects on liver size. Sometimes, the liver is extremely enlarged, in others, it is shrunken and fibrotic.

SWEETCLOVER (Melilotus spp.). Sweetclover poisoning has been an important problem in North America. Sweetclover contains coumarin glycosides, which are converted by mold growth to dicoumarol. Dicoumarol is an inhibitor of vitamin K metabolism, thus causing an induced vitamin K deficiency. Sweetclover poisoning causes a pronounced susceptibility to prolonged bleeding and hemorrhaging because of the essential role of vitamin K in blood clotting. Wet, humid weather that favors mold growth during curing of sweetclover hay increases the likelihood of poisoning. Cattle are the main livestock affected.

Moldy sweetclover hay should not be fed or should be used with caution. Ammoniation of stacked hay with anhydrous ammonia reduces dicoumarol levels. Animals with signs of sweetclover poisoning are treated with injections of vitamin K. Low-coumarin cultivars of sweetclover have been developed and should be used in areas where sweetclover poisoning is a problem. Coumarin has a vanillalike odor and is responsible for the characteristic smell of sweetclover.

OTHERS. Additional forage legumes contain various toxins. As already mentioned, birdsfoot trefoil and lespedeza contain tannins. Crownvetch (*Coronilla varia* L.) contains glycosides of 3-nitropropionic acid, which are metabolized in ruminants to yield NO_2^-. Concentrations are rarely sufficient to cause NO_2^- poisoning, but the glycosides contribute to a low palatability of crownvetch. Cicer milkvetch, a minor forage legume in the northern US, has caused photosensitization in cattle and sheep (Marten et al. 1987, 1990).

The seeds of common (*Vicia sativa* L.) and hairy vetch (*V. villosa* Roth) are poisonous. They contain toxic lathyrogenic amino acids, which cause damage to the nervous system, with signs such as convulsions and paralysis. This occurs primarily in nonruminants that consume seeds as a contaminant of grain. Hairy vetch poisoning of ruminants has been reported in the US (Kerr and Edwards 1982) and South Africa (Kellerman et al. 1988). Signs include severe dermatitis, skin edema, conjunctivitis, corneal ulcers, and diarrhea. About 50% of affected animals die. The toxic

agent in hairy vetch has not been identified.

Many *Lathyrus* spp. contain toxic amino acids that cause neurological problems and skeletal defects known as lathyrism. Flatpea (*L. sylvestris* L.) has potential as a forage crop for degraded soils such as reclaimed strip-mined areas. Flatpea is nearly free of toxicity, but Foster (1990) notes that "the question of flatpea toxicity must be answered conclusively before this plant can be recommended for use by livestock producers." Toxicity of flatpea hay to sheep, with typical signs of neurolathyrism, has been reported by Rasmussen et al. (1993).

Lupinus spp. contain a variety of alkaloids of the quinolizidine class. The sweet lupines such as *L. albus* L. and *L. angustifolius* L. contain low levels of various alkaloids (e.g., cytisine, sparteine, lupinine, lupanine). These cause feed refusal and neurological effects. Sheep are frequently poisoned by wild lupines on rangelands, as they avidly consume the seedpods. On rangelands in western North America, there are many species of wild lupines that are toxic to livestock. Species such as *L. sericeus* Pursh., *L. caudatus* Kellogg, and *L. argenteus* var. *tenellus* (Dougl. ex G.Don) D.Dunn (also known as *L. laxiflorus*) contain anagyrine, an alkaloid that is teratogenic in cattle. It causes crooked calf disease, if consumed by pregnant cows during days 40-70 of gestation. Severe skeletal deformations in the fetuses may occur. This teratogenic alkaloid does not occur in domesticated *Lupinus* spp.

In Australia, sweet lupines are extensively grown as a grain crop; sheep are grazed on the lupine stubble after harvest. Often the stems of lupines are infected with a fungus (*Phomopsis leptostromiformis* [Kühn] Bubak in Kab. & Bubak) that produces toxic phomopsins. These mycotoxins cause liver damage, including fatty liver and necrosis, eventually leading to liver failure and death. This condition is referred to as *lupinosis*.

Leucaena contains a toxic amino acid, mimosine. In the rumen, mimosine is converted to various metabolites, including 3,4-dihydroxypyridine (DHP). Both mimosine and DHP are toxic to ruminants, causing dermatitis and poor growth (mimosine) and goitrogenic (thyroid-inhibitory) effects. Australian researchers (Jones and Megarrity 1986) learned that Hawaiian ruminants, adapted to a leucaena diet, had mimosine-degrading rumen bacteria that eliminated the toxicity. These bacteria have now been introduced into cattle in Australia, allowing leucaena to be used as a productive source of high-protein forage (Quirk et al. 1988). Hammond et al. (1989) in Florida also reports detoxification of mimosine by rumen microbes.

Various other tropical legumes contain toxic factors. Many *Indigofera* spp. contain the toxic amino acid indospecine (Aylward et al. 1987). The jackbean (*Canavalia ensiformis* [L.] DC) contains canavanine, an amino acid analog of arginine (Cheeke and Shull 1985). Generally, the grazing diet contains other species that dilute the effects of the toxins.

Toxins and Animal Disorders Associated with Grasses. In contrast to herbaceous plants, grasses are generally not well defended chemically. Most grasses have coevolved with grazing animals and by growth habit survive frequent defoliation. Hence, there are few intrinsic toxins in common forage grasses. More frequent are mycotoxins produced by fungi living in or on grasses. Fungi living within plant tissues or tissue spaces are called *endophytes*. Many livestock syndromes are attributed to endophyte toxins. Examples of toxins intrinsically present in grasses are the alkaloids of *Phalaris* spp., cyanogens in forage sorghums (e.g., sudangrass), and oxalates in many tropical grasses. These have been reviewed by Cheeke (1995).

PHALARIS POISONING. This poisoning produces two syndromes in livestock: a neural disorder (phalaris staggers) and a sudden death syndrome. These are disorders of both sheep and cattle consuming various *Phalaris* spp., either by grazing or as hay or other conserved feed. Phalaris poisoning has particularly been reported from Australia and New Zealand, primarily in animals grazing pastures containing *P. aquatica* L. (formerly *P. tuberosa* L.). In the US, phalaris poisoning of sheep *P. caroliniana* Walt. is associated with *P. caroliniana* Walt. (Nicholson et al. 1989) and *P. aquatica* L. (East and Higgins 1988). Reed canarygrass (*P. arundinacea* L.) poisoning of sheep has been observed in New Zealand (Simpson et al. 1969). Contamination of a sheep feedlot diet with a source of phalaris alkaloids led to an outbreak of staggers in California (Lean et al. 1989).

Phalaris staggers is characterized by convulsions and other neurological signs due to brain damage, culminating in mortality. The syndrome is caused by tryptamine alkaloids in *Phalaris* spp., which are believed to inhibit serotonin receptors in specific brain and spinal cord nuclei (Bourke et al. 1990). These tryptamine alkaloids are responsible for the low palatability of the grass and poor performance of animals on reed canarygrass pastures (Marten et al. 1976). Low-alkaloid cultivars of reed canarygrass have been developed that give improved animal productivity (Marten et al. 1981; Wittenberg et al. 1992). The cause of the phalaris sudden death syndrome has not been conclusively identified but appears not to be the tryptamine alkaloids (Bourke et al. 1988). As many as four different factors, including a cardiorespiratory toxin, thiaminase and amine cosubstrate, cyanogenic compounds, and NO_3^- compounds, have been implicated (Bourke and Carrigan 1992).

HYDROCYANIC ACID (HCN) POISONING. Forage sorghums such as sudangrass contain cyanogenic glycosides, from which free cyanide can be released by enzymatic action. The glycosides occur in epithelial cells of the plant, while the glycoside-hydrolyzing enzymes are in mesophyll cells. Damage to the plant from wilting, trampling, frost, drought stress, etc., results in the breakdown of the plant structure, causing exposure of the glycosides to the hydrolyzing enzymes and formation of free cyanide (Fig. 9.3). Cyanide causes acute respiratory inhibition by inhibiting the enzyme cytochrome oxidase. Signs of poisoning include labored breathing, excitement, gasping, convulsions, paralysis, and death.

The cyanide potential and risk of poisoning decrease as forage sorghums mature (Wheeler et al. 1990). The likelihood of acute cyanide poisoning may be greater when feeding sorghum hay than when grazing the fresh plant because of the more rapid dry matter intake. Ground and pelleted sorghum hay may be especially toxic because of the rapid rate of intake and cyanide release (Wheeler and Mulcahy 1989). Ensiling markedly reduces the cyanide risk.

Nitrogen fertilization of forage sorghums enhances the cyanide risk. Special caution is needed when livestock is grazing lush, immature forage sorghums and when frost has occurred. Laboratory analysis of the cyanide potential of sorghum hay is advisable (Wheeler and Mulcahy 1989). Cattle deaths on sorghum pastures occur most frequently in late summer or early fall, following an overnight frost.

OXALATE POISONING. Many tropical grasses contain high levels of oxalate. Upon ingestion by ruminants, the oxalate complexes dietary Ca and forms insoluble calcium oxalate. This leads to disturbances in Ca and P metabolism

Fig. 9.3. Forage sorghums like sudangrass are excellent feeds for grazing animals. However, they must be managed carefully to avoid cyanide toxicity. *Courtesy of C. Mulcahy.*

involving excessive mobilization of bone mineral. The demineralized bones become fibrotic and misshapen, causing lameness and "bighead" in horses. Ruminants are less affected, but prolonged grazing by cattle and sheep of some tropical grass species can result in severe hypocalcemia, i.e., Ca deposits in the kidneys and kidney failure. Tropical grasses that have high oxalate levels include *Setaria* spp., *Brachiaria* spp., buffelgrass (*Cenchrus ciliaris* L.), 'Pangola' digitgrass (*Digiteria eriantha* Steud.), and kikuyugrass (*Pennisetum clandestinum* Hochst. ex Chiov.). Provision of mineral supplements high in Ca to grazing animals overcomes the adverse effect of oxalates in grasses.

FACIAL ECZEMA. Many livestock poisons are caused by fungi growing on or in grass. Other organisms like nematodes, insects, and bacteria are sometimes involved. Facial eczema of grazing ruminants is a classic example of secondary or hepatogenous photosensitization, due to liver damage. Facial eczema is a major problem of sheep and cattle on perennial ryegrass pastures in New Zealand and has been reported sporadically in other countries.

The fungus *Pithomyces chartarum* (Berk. & M.A. Curtin) M.B. Ellis grows on the dead litter in ryegrass pastures and produces large numbers of spores. The spores contain a hepatotoxin, sporidesmin, which is only slowly broken down in the liver. Spores are consumed during grazing of infected pastures, which leads to sporidesmin-induced liver damage. The damaged liver is unable to metabolize phylloerythrin, a metabolite of chlorophyll, which then accumulates in the blood. Phylloerythrin is a photodynamic agent that reacts with sunlight, causing severe dermatitis of the face, udder, and other exposed areas.

There are species differences in susceptibility to sporidesmin: for example, goats are much more resistant to facial eczema than sheep, probably because of a faster rate of sporidesmin detoxification in the liver (Smith and Embling 1991).

ERGOT. Ergotism is another mycotoxicosis associated with grasses. Seed heads of many grasses are susceptible to infection with *Claviceps paspali* F Stevens & JG Hall and other *Claviceps* spp. In the US, dallisgrass (*Paspalum dilatatum* Poir.) poisoning is the major *Claviceps*-caused ergotism. Ergot alkaloids cause vasoconstriction and reduced blood supply to the extremities, resulting in sloughing of ear tips, tail, and hooves. There are also neurological effects, including hyperexcitability, incoordination, and convulsions. Ergotism can be avoided by not allowing grasses to set seed.

TALL FESCUE TOXICOSIS. Tall fescue infected with the endophytic fungus (*Acremonium* *coenophialum* Morgan-Jones & Gams) is responsible for three types of livestock disorders when animals consume forage or seed. These include fescue foot, summer fescue toxicosis, and fat necrosis. Animal performance is reduced, and reproduction is impaired.

Tall fescue toxicity is caused by ergot alkaloids such as ergovaline, which is produced by the endophyte growing systemically within the plant tissues. One effect of ergovaline is the inhibition of prolactin from the pituitary gland. Physiologically, tall fescue toxicosis and its associated reproductive effects are explained by reduced prolactin levels that, via effects on brain neurotransmitters, inhibit smooth muscle contraction. Because prolactin also stimulates lactation, agalactia, i.e., lack of milk production, occurs with consumption of endophyte-infected tall fescue. This is particularly severe in horses (Porter and Thompson 1992). The problems associated with endophyte-infected tall fescue are further described in Chapter 28, Volume 1.

RYEGRASS TOXICITY. There are several livestock disorders associated with consumption of ryegrass. Besides facial eczema (already described), two other major syndromes are perennial ryegrass staggers and annual ryegrass (*L. multiflorum* Lam.) toxicity.

Perennial ryegrass staggers is caused by compounds called *tremorgens*. Affected animals exhibit various degrees of incoordination and other neurological signs (head shaking, stumbling and collapse, and severe muscle spasms), particularly when disturbed or forced to run. Even in severe cases, there are no pathological signs in nervous tissue. Affected animals usually spontaneously recover. With sheep, ryegrass staggers is primarily a problem in animal management, as affected animals are difficult to move from one pasture to another. Growth rate of the animal is also reduced (Fletcher and Barrell 1984). In Australia and New Zealand, ryegrass staggers occurs in sheep, cattle, horses, and deer. It has been reported in sheep and cattle in California (Galey et al. 1991). It also occurs in sheep grazing winter forage and stubble residue of endophyte-enhanced turf-type ryegrasses in Oregon.

The main causative agents of ryegrass staggers are a group of potent tremorgens called *lolitrems,* the most important of which is lolitrem B (Gallagher et al. 1984). Lolitrem B is a potent inhibitor of neurotransmitters in the brain. The lolitrems are produced by an endophytic fungus, *Acremonium lolii,* which

is often present in perennial ryegrass. Turf cultivars of both tall fescue and perennial ryegrass are often deliberately infected with endophytes because the endophyte increases plant vigor, in part by producing ergot alkaloids (*A. coenophialum*) and tremorgens (*A. lolii*). While the presence of the endophyte is advantageous when the grass is used for turf purposes, it has negative effects on animal performance when the grass is consumed by livestock.

In Australia and South Africa, annual ryegrass toxicity is a significant disorder of livestock. It has an interesting etiology, involving annual ryegrass, a nematode, and bacteria. Although the neurological signs are superficially similar, annual ryegrass toxicity and ryegrass staggers are totally different disorders. In contrast to the temporary incoordination seen with ryegrass staggers, there is brain damage with annual ryegrass toxicity. The neurological damage is evidenced by convulsions of increasing severity that terminate in death.

Annual ryegrass toxicity is caused by corynetoxins, which are chemically similar in structure to the tunicamycin antibiotics. Corynetoxins are produced by a *Clavibacter* spp. (formerly designated *Corynebacterium* spp.). This bacterium parasitizes a nematode (*Anguina agrostis*) that infects annual ryegrass. Ryegrass is toxic only when infected with the bacteria-containing nematode. The nematodes infect the seedling shortly after germination, and the larvae are passively carried up the plant as the plant stem elongates. They invade the florets, producing a nematode gall instead of seed. When consumed by animals, corynetoxins inhibit an enzyme involved in glycoprotein synthesis, leading to defective formation of various blood components of the reticulo-endothelial system. This impairs cardiovascular function and vascular integrity, causing inadequate blood supply to the brain.

Corynetoxins have been identified in other grasses besides annual ryegrass, including *Polypogon* and *Agrostis* spp. (Finnie 1991; Bourke et al. 1992). Annual ryegrass toxicity can be avoided by not allowing animals to graze mature grass containing seed heads or by clipping pastures to prevent seed head development. In Australia, these measures are often impractical because of the extensive land areas involved.

OTHER GRASS POISONS. Kikuyu grass is a common tropical forage that is occasionally toxic to livestock (Peet et al. 1990). Clinical signs include depression, drooling, muscle twitching, convulsions, and sham drinking (Newsholme et al. 1983). There is a loss of rumen motility and severe damage to the mucosa of the rumen and omasum. In many but not all cases, kikuyu poisoning occurs when the pasture is invaded by armyworm (*Spodoptera exempta* [Walker] [Lepidoptera, Noactuidae]). The causative agent has not been identified, and it is not conclusively known if the armyworm has a role in the toxicity.

Photosensitization of grazing animals often occurs with *Panicum* and *Brachiaria* spp. (Bridges et al. 1987; Cornick et al. 1988; Graydon et al. 1991). The condition is usually accompanied by crystals in and around the bile ducts in the liver. Miles et al. (1991) have shown that the crystals are metabolites of saponins, which are common constituents of *Panicum* spp. The crystals impair biliary excretion, leading to elevated phylloerythrin levels in the blood, causing secondary (hepatic) photosensitization, as previously described.

Toxins in Other Forages. Most common forages are legumes or grasses. A few others are sometimes used, including buckwheat (*Fagopyrum esculentum* Moench.), spineless cactus (*Opuntia* spp.), saltbush (*Atriplex* spp.), and forage brassicas such as kale (*B. oleracea* L.), rape (*B. napus* L.), cabbage (*B. oleracea* L.), and turnips (*B. rapa* L.).

Brassica spp. contain glucosinolates (goitrogens) and the brassica anemia factor. Glucosinolates are primarily of concern in the brassicas grown for seed, such as rapeseed and mustard. Forage brassicas contain a toxic amino acid, S-methylcysteine sulfoxide (SMCO), the brassica anemia factor. Ruminants often develop severe hemolytic anemia on kale or rape pastures, and growth is markedly reduced.

S-methylcysteine sulfoxide is metabolized in the rumen to dimethyl disulfide, an oxidant that destroys the red blood cell membrane. This leads to anemia, hemoglobinuria (red urine), and liver and kidney damage. Mortality frequently occurs. Because the SMCO content of brassicas increases with plant maturity, it is not advisable to graze mature brassica or to use these crops for late winter pasture in temperate areas. Avoiding the use of S and high N in fertilizer reduces SMCO levels and toxicity. Brassica anemia is reviewed by Cheeke and Shull (1985) and Smith (1980).

Buckwheat is a fast-growing, broad-leaved annual sometimes grown as a temporary pasture. Buckwheat seed and forage contain a photosensitizing agent, fagopyrin. This compound is absorbed by the body and moved to the surface of the skin, where it reacts with sunlight, causing photodermatitis (photosensitization). Light-skinned animals are particularly susceptible.

Acute bovine pulmonary emphysema may occur when cattle are moved from sparse dry pasture to lush grass, legume, or brassica pasture. The abrupt change in pasture type results in a disturbance in the rumen microbes, leading to excessive conversion of the amino acid tryptophan to a metabolite, 3-methyl indole (3-MI). The 3-MI is absorbed and is toxic to the lung tissue, causing pulmonary edema and emphysema (Carlson and Breeze 1984). The condition, also called *summer pneumonia,* may be fatal. Provision of supplementary feed before moving cattle onto lush meadows is helpful in preventing the disorder.

Animal Metabolism of Plant Toxins. Plants and animals have coevolved. As plants have developed the enzymatic means to synthesize defensive chemicals, animals have evolved detoxification mechanisms to overcome the plant defenses. The most fundamental of these are the drug-metabolizing enzyme systems of the liver, such as the cytochrome P450 system. This enzyme system (also called the *mixed function oxidase system*) oxidizes hydrophobic, nonpolar substances such as plant toxins and introduces a hydroxyl group. The hydroxyl group increases the water solubility of the compound, mainly by providing a site to react (conjugate) with other water soluble compounds such as amino acids (e.g., glycine), peptides (glutathione), and sugars (e.g., glucuronic acid). These conjugated compounds, such as glucuronides, are much less toxic and can be excreted in the urine or bile (e.g., saponin glucuronides in the bile of animals consuming *Panicum* spp.). Most differences in susceptibility among livestock species to plant toxins are due to differences in liver metabolism. Some toxins are bioactivated, or made more toxic, because of liver metabolism (e.g., aflatoxin or slaframine). The relative rates at which the active metabolites are formed and detoxified determine the extent of cellular damage.

On an evolutionary basis, browsing animals such as sheep and goats have been exposed to greater concentrations of plant toxins than have grazing animals such as horses and cattle. As a result, these browsing species generally are more resistant to many plant toxins than are the strict grazers, and they find plants containing toxins more palatable than do cattle and horses (Cheeke 1991). Sometimes, as with pyrrolizidine alkaloids in *Senecio* spp., the resistance of sheep and goats is due to a lower rate of bioactivation of the compounds in the liver to the toxic metabolites (Cheeke 1988).

Browsing animals are better able than grazers to resist adverse effects of dietary tannins and phenolic compounds, which are common constituents of shrubs, trees, and other browse plants. For example, deer, which are browsers, have salivary tannin-binding proteins that counteract the astringent effects of tannins (Austin et al. 1989). These salivary proteins are absent in sheep and cattle. Mehansho et al. (1987) reviews the roles of salivary tannin-binding proteins as animal defenses against plant toxins. Resistance to tannin astringency would result in tannin-containing plants being more palatable to browsers than to grazers, which lack the tannin-binding proteins.

In ruminants, metabolism of toxins by rumen microbes is an important factor in altering sensitivity to plant toxins. In some cases, e.g., cyanogenic glycosides and the brassica anemia factor, the toxicity is increased by rumen fermentation. Sometimes, e.g., mimosine or oxalate toxicity, the compounds are detoxified by microbial metabolism. The toxic amino acid mimosine in *Leucaena* spp. has been of particular interest in this regard. As discussed earlier, the successful use of leucaena as a high-protein forage was not possible in Australia and many other areas until ruminants were dosed with mimosine-degrading bacteria.

QUESTIONS

1. What is the genetic potential of various forage species for increased palatability, digestibility, and bioavailability of elements like Mg, Ca, and Se?
2. What is the empirical relationship of bioavailable Cu to forage Cu, Mo, and S concentrations?
3. What affect does Si and soil ingestion have on digestive processes and bioavailability of trace elements?
4. What is the toxicity of excess S in herbage and its role in animal nutrition and health, specifically polioencephalomalacea?
5. How does the element requirement for nutrition (production functions) differ from that required for adequate immune response in animals?
6. Why do many plants contain toxic substances?
7. Why does pasture bloat occur more often with legumes than with grasses? Why are some legumes nonbloating?
8. How have the problems associated with *Leucaena leucocephala* toxicity been overcome? Does that technique have potential application with other plant toxins?
9. What are endophytes? What is the role of endophytes in perennial ryegrass staggers?
10. Under what conditions is sudangrass most toxic? How can toxicity be prevented?
11. Why do cattle, sheep, and goats differ in their susceptibility to plant toxins?

REFERENCES

Adams, NR. 1989. Phytoestrogens. In PR Cheeke (ed.), Toxicants of Plant Origin, vol. 4. Boca Raton, Fla.: CRC Press, 23-51.

Austin, PJ, LA Suchar, CT Robbins, and AE Hagerman. 1989. Tannin-binding proteins in saliva of sheep and cattle. J. Chem. Ecol. 15:1335-47.

Aylward, JH, RD Court, KP Haydock, RW Strickland, and MP Hegarty. 1987. *Indigofera* species with agronomic potential in the tropics. Rat toxicity studies. Aust. J. Agric. Res. 38:177-86.

Ball, DM, JF Pedersen, and GD Lacefield. 1993. The tall-fescue endophyte. Am. Sci. 81:370-79.

Beke, GJ, and R Hironaka. 1990. Toxicity to beef cattle of sulfur in saline well water: A case study. Sci. Total Environ. 101:281-90.

Bourke, CA, and MJ Carrigan. 1992. Mechanisms underlying *Phalaris aquatica* "sudden death" syndrome in sheep. Aust. Vet. J. 69:165-67.

Bourke, CA, MJ Carrigan, and RJ Dixon. 1988. Experimental evidence that tryptamine alkaloids do not cause *Phalaris aquatica* sudden death syndrome in sheep. Aust. Vet. J. 65:218-20.

———. 1990. The pathogenesis of the nervous syndrome of *Phalaris aquatica* toxicity in sheep. Aust. Vet. J. 67:356-58.

Bourke, CA, MJ Carrigan, and SCJ Love. 1992. Flood plain staggers, a tunicaminyluracil toxicosis of cattle in northern New South Wales. Aust. Vet. J. 69:228-29.

Bridges, CH, BJ Camp, CW Livingston, and EM Bailey. 1987. Kleingrass (*Panicum coloratum* L.) poisoning in sheep. Vet. Pathol. 24:525-31.

Bush, L, J Boling, and S Yates. 1979. Animal disorders. In RC Buckner and LP Bush (eds.), Tall Fescue, Am. Soc. Agron. Monogr. 20. Madison, Wis., 247-92.

Carlson, JR, and RG Breeze. 1984. Ruminal metabolism of plant toxins with emphasis on indolic compounds. J. Anim. Sci. 58:1040-49.

Cheeke, PR. 1988. Toxicity and metabolism of pyrrolizidine alkaloids. J. Anim. Sci. 66:2343-50.

———. 1991. Applied Animal Nutrition: Feeds and Feeding. New York: Macmillan.

———. 1995. Endogenous toxins and mycotoxins in forage grasses and their effects on livestock. J. Anim. Sci. 73:909-18.

Cheeke, PR, and LR Shull. 1985. Natural Toxicants in Feeds and Poisonous Plants. Westport, Conn.: AVI Publishing.

Cornick, JL, GK Carter, and CH Bridges. 1988. Kleingrass-associated hepatotoxicosis in horses. J. Am. Vet. Med. Assoc. 193:932-35.

Davison, KL, W Hansel, L Krook, K McEntee, and MJ Wright. 1964. Nitrate toxicity in dairy heifers. I. Effects on reproduction, growth, lactation and vitamin A nutrition. J. Dairy Sci. 47:1065-73.

Deeb, BS, and KW Sloan. 1975. Nitrates, Nitrites, and Health. Ill. Agric. Exp. Stn. Bull. 750.

East, NE, and RJ Higgins. 1988. Canary grass (*Phalaris* sp) toxicosis in sheep in California. J. Am. Vet. Med. Assoc. 192:667-69.

Finnie, JW. 1991. Corynetoxin poisoning in sheep in the south-east of South Australia associated with annual beard grass (*Polypogon monspeliensis*). Aust. Vet. J. 68:370.

Fletcher, LR, and GK Barrell. 1984. Reduced liveweight gains and serum prolactin levels in hoggets grazing ryegrasses containing *Lolium* endophyte. N.Z. Vet. J. 32:139-40.

Foster, JG. 1990. Flatpea (*Lathyrus sylvestris* L.): A new forage species? A comprehensive review. Adv. Agron. 43:241-313.

Galey, FD, ML Tracy, AL Craigmill, BC Barr, G Markegard, R Peterson, and M O'Connor. 1991. Staggers induced by consumption of perennial ryegrass in cattle and sheep from northern California. J. Am. Vet. Med. Assoc. 199:466-70.

Gallagher, RT, AD Hawkes, PS Steyn, and R Vleggaar. 1984. Tremorgenic neurotoxins from perennial ryegrass causing ryegrass staggers disorder of livestock: Structure elucidation of lolitrem B. J. Chem. Soc., Chem. Commun. 614-16.

Gooneratne, SR, WT Buckley, and DA Christensen. 1989. Review of copper deficiency and metabolism in ruminants. Can. J. Anim. Sci. 69:819-45.

Gough, LP, HT Shacklette, and AA Case. 1979. Element Concentrations Toxic to Plants, Animals, and Man. Geol. Surv. Bull. 1466. Washington, D.C.: US Gov. Print. Off.

Grace, ND. 1983. The Mineral Requirements of Grazing Ruminants. N.Z. Soc. Anim. Prod., Occas. Publ. 9. Palmerston North, New Zealand.

Grace, ND, and RG Clark. 1991. Trace element requirements, diagnosis and prevention of deficiencies in sheep and cattle. In Physiological Aspects of Digestion and Metabolism in Ruminants, Proc. 7th Int. Symp. Ruminant Physiol., New York: Academic Press, 321-46.

Graham, TW. 1991. Trace element deficiencies in cattle. Vet. Clin. North Am.: Food Anim. Pract. 7:153-215.

Graydon, RJ, H Hamid, P Zahari, and C Gardiner. 1991. Photosensitisation and crystal-associated cholangiohepatopathy in sheep grazing *Brachiaria decumbens*. Aust. Vet. J. 68:234-36.

Greene, LW, JF Baker, and PF Hardt. 1989. Use of animal breeds and breeding to overcome the incidence of grass tetany: A review. J. Anim. Sci. 67:3463-69.

Grunes, DL, and RM Welch. 1989. Plant contents of magnesium, calcium, and potassium in relation to ruminant nutrition. J. Anim. Sci. 67:3486-94.

Hammond, AC, MJ Allison, MJ Williams, GM Prine, and DB Bates. 1989. Prevention of leucaena toxicosis of cattle in Florida by ruminal inoculation with 3-hydroxy-4-(1H)-pyridone-degrading bacteria. Am. J. Vet. Res. 50:2176-80.

James, LF, MH Ralphs, and DB Nielsen. 1988. The Ecology and Economic Impact of Poisonous Plants on Livestock Production. Boulder, Colo.: Westview.

Jones, RJ, and RG Megarrity. 1986. Successful transfer of DHP-degrading bacteria from Hawaiian goats to Australian ruminants to overcome the toxicity of leucaena. Aust. Vet. J. 63:259-62.

Kellerman, TS, JAW Coetzer, and TW Naude. 1988. Plant Poisonings and Mycotoxicoses of Livestock in Southern Africa. Cape Town, South Africa: Oxford University Press.

Kerr, LA, and WC Edwards. 1982. Hairy vetch poisoning in cattle. Vet. Med. Small Anim. Clin. 77:257-61.

Kubota, J, and WH Allaway. 1972. Geographic distribution of trace element problems. In JJ Mortvedt, PM Giordano, and WL Lindsay (eds.), Micronutrients in Agriculture. Madison, Wis.: American Society of Agronomy, 525-54.

Lean, IJ, M Anderson, MG Kerfoot, and GC Marten. 1989. Tryptamine alkaloid toxicosis in feedlot sheep. J. Am. Vet. Med. Assoc. 195:768-71.

Majak, W, RE Howarth, AC Fesser, BP Goplen, and MW Pedersen. 1980. Relationships between ruminant bloat and the composition of alfalfa herbage. II. Saponins. Can. J. Anim. Sci. 60:699-708.

Marten, GC, RM Jordan, and AW Hovin. 1976. Biological significance of reed canarygrass alkaloids and associated palatability to grazing sheep and cattle. Agron. J. 68:909-14.

———. 1981. Improved lamb performance associated with breeding for alkaloid reduction in reed canarygrass. Crop Sci. 21:295-98.

Marten, GC, FR Ehle, and EA Ristau. 1987. Performance and photosensitization of cattle related to forage quality of four legumes. Crop Sci. 27:138-45.

Marten, GC, RM Jordan, and EA Ristau. 1990. Performance and adverse response of sheep during grazing of four legumes. Crop Sci. 30:860-66.

Mayland, HF. 1983. Assessing nutrient cycling in the soil/plant/animal system of semi-arid pasture lands. In Nuclear Techniques in Improving Pasture Management. Vienna, Austria: International Atomic Energy Agency, 109-17.

———. 1988. Grass tetany. In DC Church (ed.), The Ruminant Animal: Its Physiology and Nutrition. Englewood Cliffs, N.J.: Prentice-Hall, 511-23 and 530-31.

Mayland, HF, and DA Sleper. 1993. Developing a tall fescue for reduced grass tetany risk. In Proc. 17th Int. Grassl. Congr., Palmerston North, New Zealand, and Australia, 1095-96.

Mayland, HF, GE Shewmaker, and RC Bull. 1977. Soil ingestion by cattle grazing crested wheatgrass. J. Range Manage. 30:264-65.

Mayland, HF, RC Rosenau, and AR Florence. 1980. Grazing cow and calf responses to zinc supplementation. J. Anim. Sci. 51:966-74.

Mayland, HF, LF James, KE Panter, and JL Sonderegger. 1989. Selenium in seleniferous environments. In LW Jacobs (ed.), Selenium in Agriculture and the Environment, Spec. Publ. 23. Madison, Wis.: Soil Science Society of America, 15-50.

Mehansho, H, LG Butler, and DM Carlson. 1987. Dietary tannins and salivary proline-rich proteins: Interactions, induction, and defense mechanisms. Annu. Rev. Nutr. 7:423-40.

Miles, CO, SC Munday, PT Holland, BL Smith, PP Embling, and AL Wilkins. 1991. Identification of a sapogenin glucuronide in the bile of sheep affected by *Panicum dichotomiflorum* toxicosis. N.Z. Vet. J. 39:150-52.

Moseley, G, and DH Baker. 1991. The efficacy of a high magnesium grass cultivar in controlling hypomagnesemia in grazing animals. Grass and Forage Sci. 46:375-80.

Nation, PN. 1991. Hepatic disease in Alberta horses: A retrospective study of "alsike clover poisoning" (1973-1988). Can. Vet. J. 32:602-7.

National Research Council (NRC). 1984. Nutrient Requirements of Beef Cattle. 6th rev. ed. Washington, D.C.: National Academy Press.

Newsholme, SJ, TS Kellerman, GCA Van Der Westhuizen, and JT Soley. 1983. Intoxication of cattle on kikuyu grass following army worm (*Spodoptera exempta*) invasion. Onderstepoort J. Vet. Res. 50:157-67.

Nicholson, S, BM Olcott, EA Usenik, HW Casey, CC Brown, LE Urbatsch, SE Turnquist, and SC Moore. 1989. Delayed phalaris grass toxicosis in sheep and cattle. J. Am. Vet. Med. Assoc. 195:345-46.

Peet, RL, J Dickson, and M Hare. 1990. Kikuyu poisoning in goats and sheep. Aust. Vet. J. 67:229-30.

Porter, JK, and FN Thompson, Jr. 1992. Effects of fescue toxicosis on reproduction in livestock. J. Anim. Sci. 70:1594-1603.

Quirk, MF, JJ Bushell, RJ Jones, RG Megarrity, and KL Butler. 1988. Live-weight gains on leucaena and native grass pastures after dosing cattle with rumen bacteria capable of degrading DHP, a ruminal metabolite from leucaena. J. Agric. Sci. 111:165-70.

Rasmussen, MA, MJ Allison, and JG Foster. 1993. Flatpea intoxication in sheep and indications of ruminal adaptation. Vet. Human Toxicol. 35:123-27.

Robinson, DL, LC Kappel, and JA Boling. 1989. Management practices to overcome the incidence of grass tetany. J. Anim. Sci. 67:3470-84.

Shewmaker, GE, HF Mayland, RC Rosenau, and KH Asay. 1989. Silicon in C-3 grasses: Effects on forage quality and sheep preference. J. Range Manage. 42:122-27.

Simpson, BH, RD Jolly, and SHM Thomas. 1969. *Phalaris arundinacea* as a cause of deaths and incoordination in sheep. N.Z. Vet. J. 17:240-44.

Singer, RH. 1972. The nitrate poisoning complex. In Proc. US Anim. Health Assoc., 310-322.

Sleper, DA, KP Vogel, KH Asay, and HF Mayland. 1989. Using plant breeding and genetics to overcome the incidence of grass tetany. J. Anim. Sci. 67:3456-62.

Smith, BL, and PP Embling. 1991. Facial eczema in goats: The toxicity of sporidesmin in goats and its pathology. N.Z. Vet. J. 39:18-22.

Smith, RH. 1980. Kale poisoning: The brassica anemia factor. Vet. Rec. 107:12-15.

Spoelstra, SF. 1985. Nitrate in silage. Grass and Forage Sci. 40:1-11.

Stewart, SR, RJ Emerick, and RH Pritchard. 1991. Effects of dietary ammonium chloride and variations in calcium to phosphorus ratio on silica urolithiasis in sheep. J. Anim. Sci. 69:2225-29.

Suttle, NF. 1991. The interactions between copper, molybdenum, and sulphur in ruminant nutrition. Annu. Rev. Nutr. 11:121-40.

Underwood, EJ. 1977. Trace Elements in Human and Animal Nutrition. 4th ed. New York: Academic Press.

Van Soest, PJ, and LHP Jones. 1968. Effect of silica in forages upon digestibility. J. Dairy Sci. 51:1-5.

Vogel, KP, BC Gabrielsen, JK Ward, BE Anderson, HF Mayland, and RA Masters. 1993. Forage quality, mineral constituents, and performance of beef yearling grazing two crested wheatgrasses. Agron. J. 85:584-90.

Wheeler, JL, and C Mulcahy. 1989. Consequences for animal production of cyanogenesis in sorghum forage and hay—a review. Trop. Grassl. 23:193-202.

Wheeler, JL, C Mulcahy, JJ Walcott, and GG Rapp. 1990. Factors affecting the hydrogen cyanide potential of forage sorghum. Aust. J. Agric. Res. 41:1093-1100.

Wittenberg, KM, GW Duynisveld, and HR Tosi. 1992. Comparison of alkaloid content and nutritive value for tryptamine- and β-carboline-free cultivars of reed canarygrass (*Phalaris arundinacea* L.). Can. J. Anim. Sci. 72:903-9.

Wright, MJ, and KL Davison. 1964. Nitrate accumulation in crops and nitrate poisoning in animals. Adv. Agron. 16:197-247.

10

Emerging Technologies: Forage Harvesting Systems

RICHARD G. KOEGEL

Agricultural Research Service, USDA,
and University of Wisconsin

FORAGE crops, especially perennial legumes, can produce impressive amounts of both energy and protein per hectare with minimum negative environmental impact. They suffer from two major drawbacks, however: (1) harvesting losses, especially those due to rain, are higher than for most other crops and (2) valuable nutrients in the herbage are intimately associated with large quantities of fiber, which generally limits their use, in significant quantities, to ruminant rations.

Two harvesting systems under development to partially overcome these problems include quick-drying forage mats and wet fractionation. Both systems attempt to reduce the probability of rain damage by decreasing the time required between mowing and removal from the field. Both attempt to increase the value of the nutrients retained in the forage. Artificial dehydration of forages is a further means of alleviating rain damage and improving retention of forage quality. Forage/livestock farmers are continually searching for systems that will reduce harvesting and storage costs, and researchers and educators must develop and test emerg-

RICHARD G. KOEGEL is a Research Agricultural Engineer, ARS, USDA, at the US Dairy Forage Research Center, Madison, Wisconsin, and has a joint appointment as Professor of Agricultural Engineering at the University of Wisconsin, Madison. He received the MS degree from Utah State University and the PhD from the University of Wisconsin, Madison. His major areas of research, since 1968, have been harvesting, processing, and storage of forage crops.

ing technologies to improve forage harvesting systems.

QUICK-DRYING FORAGE MATS

This process consists of four steps: (1) mowing; (2) severe conditioning referred to as *maceration, fiberization,* or *superconditioning;* (3) pressing the macerated forage into thin cohesive mats; and (4) returning the mats to the stubble for rapid drying. The four steps are carried out concurrently in a single machine that takes the place of the conventional mower-conditioner (Koegel et al. 1988; Shinners et al. 1992). With properly formed forage mats no thicker than 8 mm, drying time may be reduced to one-third the time required to dry conventionally harvested forage. This means that under good field-drying conditions (Shinners et al. 1987; Rotz et al. 1990) forage may be mowed and removed from the field as dry hay in one daylight period. The wilting period to make silage is 2-3 h under good conditions. Under less favorable conditions the difference in drying times between forage mats and conventionally harvested forage may be less pronounced. If forage mats are rained on, leaching losses (losses of the most soluble carbohydrates) may be 3-5 times those of conventionally harvested forage (Rotz et al. 1991). The probability of having mats rained on, however, is drastically reduced because of the short drying period.

Two different types of macerating devices have been used. The first type passes the forage between successive, opposing, roughened cylindrical surfaces rotating at different peripheral speeds. The crushing and scuffing action causes the stems to be split into ribbonlike splinters while the less fibrous upper plant parts are mashed. One version uses a large roughened drum (40 cm diameter) surrounded by five to seven roughened rolls of 10 cm diameter. Another version uses five to seven roughened rolls in a staggered configuration so that each roll is opposed by two other rolls. Opposing rolls typically run with about 30% difference in peripheral speeds. The second macerator type has one pair of steel crushing rolls that flatten the stems to create longitudinal cracks. A third high-speed roll with ridges impacts the crushed forage to open up the cracks and fiberize the stems (Kraus et al. 1992).

Because the maceration process creates many small fragments that would be lost during the subsequent harvesting process, it is necessary to form the forage into a dense cohesive structure, referred to as a *mat,* to prevent excessive losses. This is accomplished by means of a continuous press that accepts the output of the macerator, presses the macerated forage into mats, and deposits the mats on to the stubble at exactly ground speed. Presses have been made in a number of configurations, with forage being pressed between rollers or belts or both.

There are at least three reasons why macerated forage mats have high drying rates relative to conventionally harvested forage: (1) stems are opened up creating a large specific surface area exposing the internal moisture to air; (2) bruising of the plant tissue darkens it, causing it to absorb more solar energy than conventionally harvested forage (Ajibola et al. 1982); and (3) properly formed forage mats have a high dry matter density, which gives them high conductivity of both heat and moisture. This allows heat absorbed at the upper surface to flow throughout the mat and moisture to flow from moister to dryer locations.

If macerated forage is not compacted into mats, the upper surface will dry, creating an insulation layer, and the remainder will then dry slowly. Also, if forage mats are made too thick, the top and bottom will dry, leaving a moist layer in the middle. At thicknesses of 12 mm or more, forage mats may not dry faster than conventionally harvested forage.

Both in the laboratory and in feeding trials, forage mats have averaged about 15% higher neutral detergent fiber (NDF) digestibility than control forage (Hong et al. 1988b; Mertens and Hintz 1990). This is thought to be related to the increased surface area and tissue damage that provides many more colonization sites for rumen microorganisms (Hong et al. 1988a). In addition, mat-harvested hay has been found to have approximately 20% more "bypass" protein (that protein that escapes the rumen undegraded but that can be digested in the intestine for more efficient utilization) than conventional hay (Yang et al. 1993). This is thought to be due to the maillard reaction taking place between protein and sugars in juice spread over the surface of the herbage and exposed to the warmth of the sun.

Feeding trials with ruminants fed forage mat materials compared with conventionally harvested forage have been conducted at the US Dairy Forage Research Center, Madison, Wisconsin (Hong et al. 1988b; Mertens and Hintz 1990). In two sheep trials, the increased energy available from the NDF fraction of the forage mat ration was 19.6% and 23.6% greater than that of conventionally harvested forage, as estimated by multiplying dry matter intake times NDF digestibility (Table 10.1). Since the NDF fraction was approximately 47% of the total forage, the increase in energy from the forage as a whole was approximately 10%. The amount of 4% fat-corrected milk (FCM) produced by lactating goats was increased by 12% and protein by 5% (Table 10.2).

Mertens and Hintz (1990) show further evidence of increased forage quality values derived from mat-harvested alfalfa (*Medicago sativa* L.) silage compared with conventional alfalfa silage in feeding trials with sheep (Table 10.3) and with lactating cows (Table 10.4). With sheep, differences among various

TABLE 10.1.   Comparison of processing methods on quality of alfalfa hay

	Control	Mat	% Difference
Trial 1. Eight wethers for 12 wk			
Dry matter intake, kg/d	1.15	1.22[a]	6.1
Apparent NDF digestibility, %	43.0	48.5[b]	12.8
Trial 2. Four wethers for 4 wk			
Dry matter intake, kg/d	1.22	1.28	4.9
Apparent NDF digestibility, %	35.3	41.6[b]	17.8

Source: Hong et al. (1986).
[a]Mean is higher than control at $p < .10$.
[b]Mean is higher than control at $p < .05$.

TABLE 10.2. Digestibility trials with lactating goats fed on 60% alfalfa hay and 40% grain (10 goats for 4.5 wk)

	Control	Mat	% Difference
Dry matter intake, kg/d	2.44	2.58[a]	5.7
Milk, 4% fat corrected, kg/d	3.3	3.7[a]	12.1
Protein, kg/d	0.1026	0.1080[a]	5.3

Source: Hong et al (1988b).
[a]Mean is higher than control at $p < .10$.

nutritional value parameters averaged 14% to 31% units greater for mat-harvested material (Table 10.3). With lactating cows, the calculated apparent energy was 11% greater from mat-harvested alfalfa silage compared with conventional alfalfa silage.

Apparent energy was calculated for each type of forage, taking into account production of FCM, body weight change, maintenance requirement, and energy derived from concentrates. Based on the increased amount of energy derived from macerated forage, the potential may exist for maintaining milk production at the genetic potential with rations having a higher forage:concentrate ratio. This potential remains to be verified, however, and the challenge is to learn to use these forages to the best advantage.

Forage mats may be stored either as silage or hay. A belt and tine pickup, similar to those used on combines to pick up swathed grain, has been successfully used on both balers and forage harvesters. Because of its soft, compliant nature, macerated forage can be made into very dense bales and has excellent packing characteristics for silos.

Other potential advantages of mats include (1) less damage to the forage stand by reducing trips over the field to two per cutting and (2) the capability of maceration to partially reverse the effects of maturity on digestibility; this increases the period during which harvesting may take place. Increasing the length of the harvesting period, in turn, would permit a given size of machine to harvest more forage. It could also improve the feasibility of forage harvesting as a custom operation.

As of 1993, a number of European equipment manufacturers had developed prototype forage mat machines that they were evaluating for commercial potential. No North American manufacturers were known to be working on commercialization of this harvest technology.

WET FRACTIONATION

Wet fractionation consists of expressing juice from freshly cut forage to yield two fractions: (1) a high-fiber fraction that may be ensiled, dehydrated, or fed fresh and (2) a juice fraction that is low in fiber. The juice fraction

TABLE 10.3. Digestibility trials with alfalfa silage fed ad libitum to sheep.

	Control	Mat	% Difference
Trial 1. Eight sheep for 6 wk (digestion stalls)			
Average dry matter digestibility, %	59.68	69.18	15.9
Daily dry matter intake % body weight	3.42	4.49[a]	31.3
Trial 2. Twenty-six sheep for 2 wk (pen of 13)			
Daily dry matter intake, % body weight	2.65	3.02	14.0
Average weight gain, kg	2.98	3.68[b]	23.5
kg gain/kg dry matter	0.149	0.177	18.8

Source: Mertens and Hintz (1990)
[a]Mean is higher than control at $p < .01$.
[b]Mean is higher than control at $p < .06$.

TABLE 10.4. Digestibility trial with lactating cows fed on 65% alfalfa silage and 35% concentrate (12 cows for 8 wk).

	Control	Mat	% Difference
Milk production, kg/d	24.5	24.2	n.s.d.[b]
Fat, %	3.7	3.5	n.s.d.
Body weight increase, kg/d	0.08	0.44[a]	450
Dry matter intake, kg/d	19.9	19.6	n.s.d.
Calculated energy from forage, MJ/kg	4.61	5.11	0.8

Source: Mertens and Hintz (1990).
[a]Mean is higher than control at $p < .01$.
[b]n.s.d. indicates means are not significantly different.

is frequently separated into a protein concentrate and a deproteinized liquid containing sugars, dissolved minerals, and nonprotein nitrogen (N). Wet fractionation can result in an essentially "weather independent" harvesting system when the fibrous fraction is ensiled immediately. It can also greatly reduce the energy requirement for dehydration and consequently the cost. The low-fiber, high-protein concentrate obtained from the juice can be used for feeding mongastrics, including humans, as well as ruminants. It constitutes a "value-added" product and allows greater flexibility in utilization of the forage.

While a number of variations in the wet fractionation process exist, one processing pathway, with relative amounts of the various fractions, is shown in Figure 10.1.

The first step in the wet fractionation process is the disruption of a large fraction of the cells of freshly cut herbage. This process has been called *maceration* or *pulping* and

has been carried out in a number of ways including crushing between rolls, impacting, abrasion/shearing, and extrusion through orifices.

The second step is the expression of juice usually by means of some type of continuous press such as screw presses, belt presses, and cone presses. With properly functioning equipment, the weight of juice is typically slightly more than half of the fresh crop weight. With a juice dry matter content of 8%-10% and a fibrous fraction dry matter content of 30%-35% this results in approximately 25% of the total dry matter and 30%-35% of the protein being in the juice fraction.

At this point, the fibrous fraction may be fed fresh, ensiled, or dehydrated. Although the fibrous fraction has a higher concentration of cell wall constituents (fiber) than the original forage, several feeding trials have shown its feeding value to be equivalent per unit of dry matter to the original forage in

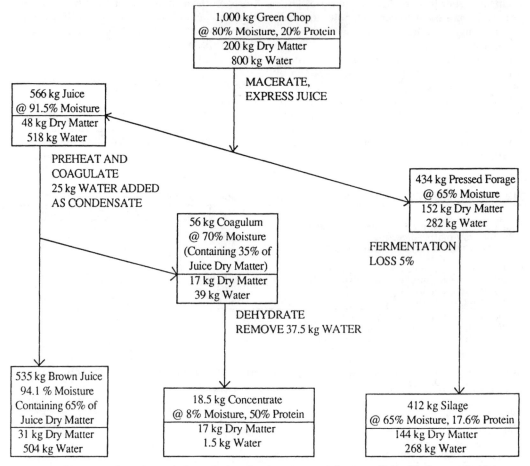

Fig. 10.1. Typical mass balance for wet fractionation carried out on alfalfa (*Medicago sativa* L.).

feeding value (Lu et al. 1979; Connell and Cramp 1975). This is usually attributed to the fact that the fiber has been made more digestible by the severe mechanical disruption it has undergone. The fibrous fraction, however, contains only about 75% as much dry matter as the original crop.

While the juice can be used fresh in monogastric or ruminant rations, be mixed with grain and ensiled, or be preserved by acidification, it is more frequently processed to obtain concentrates of protein and other constituents. The protein in the juice is of two general types: (1) particulate or chloroplastic and (2) soluble or cytoplasmic. The chloroplastic protein is deep green and contains carotenoids including beta carotene and xanthophylls. The soluble protein is cream colored when concentrated.

The simplest method of removing both types of protein from the juice is to heat the juice to 80°C (175°F). This causes both types of protein to coagulate so that the proteins can be skimmed, filtered, or centrifuged from the juice. Alternately, the chloroplastic protein can be coagulated at 55°-60°C and removed followed by heating the remaining liquid to 80°C to coagulate the cytoplasmic protein, which can then be removed. If it is desired to maximize the yield of the more valuable soluble protein, this process has two disadvantages: (1) approximately 20% of the soluble protein coagulates together with the particulate protein, and (2) the soluble protein is irreversibly coagulated, which limits its usefulness.

The protein can also be precipitated by lowering the pH to approximately 4.5. This process does not allow the separate recovery of the particulate protein and the soluble protein, but the soluble protein can be resolubilized by adjusting the pH to below 3.0 or above 10.0. The protein precipitate recovered by acidification is very fine grained compared with heat-coagulated protein and is thus harder to separate from the liquid. While the pH can be lowered by the addition of acid, it is also possible to have the necessary acid generated by microorganisms during anaerobic fermentation of the juice. Ajibola et al. (1982) show that separation could consistently be achieved in 24 h by adding to fresh juice 5% v/v (volume basis) of supernatant from juice fermented the previous day as a "starter." This technique is necessary for quick and reliable fermentations since the naturally occurring microflora in the juice varies greatly depending on the conditions under which the forage crop was grown and harvested.

The whole juice coagulum from alfalfa averages about 45% protein on a dry matter basis. In addition to being used fresh, it can be preserved by drying, acidifying, or salting. It can also be mixed with ground grain and either ensiled or dehydrated to yield high-protein and high-energy concentrate. It has been fed to swine and poultry as well as to ruminants. When fed as a very high percentage of total protein in the ration, it has caused digestive problems in both swine and poultry. This intolerance has usually been attributed to saponins, detergentlike substances, in the juice. If fed at lower levels in the ration or if the juice has undergone fermentation, saponins do not appear to be a problem.

The coagulum has been used as a major constituent in a milk replacer for calves. During the first 6 wk of life, calves do not perform well because they lack the proper enzymes for digesting the coagulum. After 6 wk of age, calves fed on coagulum-based milk replacer do equally well as those fed on commercially available milk replacer.

Biotechnologists have pointed out and demonstrated the possibility of inserting genes into plants that would cause them to produce industrially valuable substances, such as enzymes. Plants, such as forage crops, could thus serve as "factories" for value-added products useful for biopulping, environmental cleanup, food processing, or domestic cleaning. Very large quantities of enzymes would be needed for these applications, and farm-scale production would thus be appropriate and attractive. Being able to do preliminary separation and concentration of valuable fractions at the production site is necessary to avoid excessive transport costs and unmanageably large waste streams concentrated at a centralized processing site.

The target enzyme(s) would be separated out of the soluble protein fraction. While the chloroplastic protein is frequently considered to be feed grade protein, the soluble protein is a higher-value, more versatile protein. There is thus an incentive to separate these two types of protein while attempting to maximize the yield of soluble proteins. Heat coagulation and acidification are not suitable processes because they generally deactivate target enzymes, irreversibly change protein, and fail to maximize the yield of the soluble protein. An important goal, therefore, is to develop a low-temperature-processing pathway to clarify the juice by removal of the particulate green material. Then the remaining soluble protein can be concentrated by a factor of at least five to make its transport and subsequent isolation of the target enzyme(s) economically feasible.

A proposed processing pathway consists of four steps: (1) heating the juice to a temperature between 35° and 45°C, (2) holding the juice at this temperature long enough to allow aggregation of particulate material, (3) steady flow centrifugation at a force of 10,000 g to remove particulates, and (4) ultrafiltration through a nominal 10,000 molecular weight filter to obtain a soluble protein concentrate.

One method of heating the juice is to pass AC electricity directly through it. The passage of AC electricity through the juice, in addition to quickly and uniformly warming it, has at least two functions according to the literature: (1) chloroplast membranes are disrupted, freeing soluble protein from within the chloroplasts, and (2) surface charges on chloroplasts and chloroplast fragments are significantly reduced, allowing these particulates to aggregate, which greatly facilitates their removal by centrifugation. Aggregation was found to require certain minimum times, which were inversely related to juice temperature. For example, at 50°C aggregation is essentially instantaneous whereas at 35° C the hold time for maximum particulate removal is approximately 60 min. After removal of the particulates by centrifugation, the resulting concentrate is ultrafiltered to concentrate the soluble protein 5- to 10-fold. This concentrate could be considered food grade protein after any target enzymes are extracted from it. The green particulate fraction is frequently considered feed grade protein although valuable constituents such as beta carotene can be extracted from it.

Wet fractionation has also been proposed as a means of "bioremediation," or removal of toxic substances from the environment, including organic, inorganic, or radioactive substances. In this method, a plant species would be chosen that had exhibited luxury uptake and concentration of the target pollutant in its tissue. The maximum possible amount of the pollutant would be concentrated in one of the fractions. It would then be destroyed, isolated, or utilized depending on the nature of the pollutant.

ARTIFICIAL DEHYDRATION OF FORAGES

The difficulty of field-drying forages during periods of bad weather leads to speculation on whether or not forages can be profitably dehydrated by artificial means. Shown in Table 10.5 are some estimates of the amount of energy and the costs involved in dehydrating forage from various moisture contents to 10% moisture content per metric ton of end product. These cost estimates do not include the cost of additional handling necessitated by the dehydration process or the cost of transporting extra moisture from the field to the dehydration site.

The figures clearly show that, given typical prices of hay, it is not economically feasible to artificially reduce the moisture of a standing crop to a suitable level for stable storage. For this reason, most artificial drying operations aim to field dry the crop to the lowest possible moisture content, say 25%-40% before finish drying. Dehydration equipment may take a number of forms. Commercial dehydrating operations in the US have generally used rotary drum dehydrators with combustion gases from a gas or oil flame providing the neces-

TABLE 10.5. Moisture removed, energy required, and estimated costs of dehydrating forages from various initial moisture percentages (wet weight basis)

Initial moisture (% w.b.)	Water removed mt (water/mt product) final moisture (% w.b.) 20%	10%	Thermal efficiency[a] (%)	Energy for evaporation (kWh/mt)	Fuel oil required (L/mt)	Energy cost[b] ($/mt)	Total cost[c] Low capital	High capital
85	4.33	5.00	63	5642	535	185	190	200
80	3.00	3.50	61	4079	387	133	138	148
70	1.87	2.00	55	2585	245	85	90	100
60	1.00	1.25	49	1813	172	59	64	74
50	0.60	0.80	41	1387	132	46	51	61
40	0.33	0.50	32	1111	105	36	41	51
30	0.14	0.29	23	896	85	29	34	44
25	0.06	0.20	18	790	75	26	31	41
20	0.0	0.13	13	711	67	23	28	38

Note: Figures are for a final moisture content of 10% unless otherwise noted.
[a]Based on Butler and Hellwig (1970).
[b]Cost assumptions: 1.5 kWh of electricity is used for each liter of oil; oil price = $0.22/L; electricity price = $0.083/kWh.
[c]Capital costs of $5/mt and $15/mt are added to the energy costs for the low- and high-capital systems, respectively.

sary heat. Three concentric cylinders within the dehydrator are arranged to cause the crop to traverse the length of the dehydrator three times before exiting. Such dehydrators may remove 2700-3600 kg of water per hour from high-moisture material (Butler 1965).

In certain areas of southern Germany, Italy, and Switzerland, cheese factories pay a premium for milk produced without the use of silage. This has led to the use of barn installations for finish drying hay. In these installations, hay mows with slatted floors are uniformly filled by blowers capable of handling long loose hay. Waste heat from an internal combustion engine, which also drives the aeration blower, is used to raise the temperature of the air that is forced through the hay. Successive layers of hay are placed over the original layer and are dried by blowing air from the bottom thorough the entire accumulated depth.

In North America and Europe, considerable work has been done on finish drying large round bales by forcing air through them. This is frequently accomplished by setting the bales on end over openings in ducts pressurized by axial flow fans. The air may be at ambient temperature or may be heated by solar energy or by fossil fuels. While the use of ambient air may appear attractive, it results in slower drying, typically requiring 1-2 wk. This, in turn, increases the energy input to the fan(s) and leads to a larger drying installation, which increases capital costs. During periods of bad weather, quality of the aerated forage may suffer if drying proceeds too slowly.

The erection of solar collectors can largely eliminate the use of fossil fuels but increases capital costs. It may also require some supplementation by fossil fuel late in the season or during extended periods of bad weather. Economics are improved if the drying structure is used for drying other crops or if it can have multiple uses such as feed or machinery storage (Bledsoe et al. 1981).

Bales et al. (1992) used solar collectors that were an integral part of the drying structure roof and south wall. These consisted of a clear plastic cover, a corrugated sheet metal absorbing surface painted black that had air passing over both sides, and an insulated back wall. These collectors had overall efficiencies approaching 60%. Drying times were 3-4 d.

If it is necessary to complete harvesting of the crop in a 10-d period for near-optimum maturity and it takes 2 d to finish dry the crop, then it is necessary for the drying installation to accommodate 20% of the total crop on any given day. At this capacity, any delays due to bad weather would extend the harvesting period beyond 10 d.

Periodically, proposals or claims for machines that dehydrate in the field, as part of the harvesting process, receive publicity. In 1990, US Patent No. 4,912,914, was issued to S. G. Wingard of Columbia, South Carolina. His proposal was for a machine that would mow, dehydrate, and bale the crop in a single pass. Dehydration was to be by means of electrically heated rolls with the electricity presumably being generated by the tractor engine. Assuming the electrical generating unit and the heated rolls functioned with 100% efficiency, producing 1 mt per hour of 20% moisture hay from 80% moisture content crop would require the evaporation of 3 mt of water per hour, which requires the power equivalent of over 2100 kilowatts. A throughput of 5 mt/h would thus require over 10,000 kilowatts! The capital and operating cost of such a machine would, of course, be unthinkable.

At approximately the same time the state of Minnesota issued a grant for the development of a field-going machine that would use LP gas to finish dry forage from as much as 40% moisture content and concurrently bale it. At a realistic dehydrator efficiency of 30%, this would require over 700 kilowatts (2.5 million kJ) for each metric ton of product. At 50,000 kJ per kilogram of gas, this would require around 50 kg of gas per metric ton of hay. To harvest 5 mt per hour of product would require the removal of about 1650 kg of water. At a starting moisture of 40%, this is approximately the moisture removal rate of a large stationary triple pass dehydrator (Butler 1965). Thus, the required machine would have to be so large as to make it impractical as a field-going machine, not to mention the great capital cost involved.

REDUCING HARVESTING AND STORAGE COSTS

The cost of harvesting and storing forages makes up a large fraction of total forage costs. Important among these costs are the capital costs of both structures and harvesting equipment, including power units.

Many forage/livestock farmers have been penalized financially by having to purchase and maintain dual sets of forage harvesting equipment: one set for making silage and the other set for baling dry hay. In attempting to reduce harvesting and storage costs several options are being tried: (1) a silage-only system, (2) a hay-only system, (3) a common set of equipment for making both hay and silage, and (4) seasonal pasturing to reduce the quantity of harvested and stored forage.

The silage-only system has some obvious advantages in being highly mechanized and having the lowest harvest losses, especially under difficult weather conditions. Its disadvantages include (1) higher storage losses, especially of true protein, (2) higher equipment and storage costs, (3) higher energy requirements, (4) difficulty in segregating and feeding higher- and lower-quality materials to best advantage, and (5) poor adaptation to off-farm marketing.

The all-hay system almost totally reverses the advantages and disadvantages of the all-silage system and has been shown to frequently be the more profitable system of the two despite its general lack of popularity and the high level of frustration under difficult weather conditions.

Large bales, both cylindrical and rectangular, offer the possibility of harvesting both dry hay and silage with the same machine. Their popularity for silage making has grown very rapidly in Europe and is frequently carried out by custom operators. An airtight covering is, of course, required to make bale silage. The most reliable and popular covering so far appears to be plastic stretch wrapping. Bale stacks with a common airtight cover are more economical of time and plastic, but excellent technique and management are needed to avoid large losses. These can result if the cover is damaged by wind, rodents, or birds.

Plastic for covering bales raises questions since it represents the use of a nonrenewable resource and its disposal or recycling is problematic. A much better long-range solution would appear to be sealing the bale surfaces by spraying on an organic polymer derived from agricultural products (starches, sugars, proteins, and/or oils). Such a polymer would presumably be fed along with the forage and, due to its biodegradability/solubility, would probably have to be stored under a roof.

A least-cost strategy for many regions of the US might be to make hay whenever weather permitted and bale silage the remainder of the time. The advantage of being able to segregate and select higher- and lower-quality materials for most advantageous use would apply to both hay and silage bales. Use of total mixed rations (TMR) would undoubtedly require comminution of the forage prior to ration mixing. Harvesting corn Zea mays L. silage using this system remains speculative. A large round baler manufactured in Scandinavia is equipped with a flail mower to allow making direct-cut bale silage. It is possible that such a system could be adapted for corn silage. Alternatively, corn silage harvesting could be carried out as a custom operation since its harvest timing is less crucial than that for other forage crops.

Seasonal pasturing of forage crops may be used to reduce the scale and cost of forage harvesting and storage. Since forage production and grazing needs rarely match throughout the year, the harvesting of excess production and the feeding of stored forage to make up deficits can result in good overall forage utilization.

QUESTIONS

1. Why do macerated forage mats dry faster than conventionally harvested forage?

2. How much moisture is expressed from forage as mats are formed?

3. In addition to quick drying, what are some advantages and disadvantages of forage mats?

4. What are some advantages of the wet fractionation process?

5. What two major types of protein are found in forage juice?

6. Name some potential large-scale uses of enzymes that may some day be extracted from transgenic forage crops?

7. A friend of yours is considering investing money in the development of a machine that will mow, dehydrate, and bale forage in a single pass. What is your advice to the friend?

REFERENCES

Ajibola, OO, RJ Straub, RG Koegel, and HD Bruhn. 1982. Fermentation of Plant Juice as a Protein Separation Technique. Proc. Int. Conf. Leaf Protein Res, Aurangabad, India.

Bales, BM, BL Bledsoe, and RS Freeland. 1992. Solar dryer performance during hay drying cycle. Am. Soc. Agric. Eng. Pap. 921578. St. Joseph, Mich.

Bledsoe, BL, RL Reid, K Pierce, BA McGraw, and ZA Henry. 1981. A multi-use modular dryer for large round hay bales using solar heated air. Am. Soc. Agric. Eng. Pap. 81-4553. St. Joseph, Mich.

Butler, JL. 1965. Energy comparisons in processing bermuda grass and alfalfa. Trans. Am. Soc. Agric. Eng. 8(2):175-76.

Butler, JL, and RE Hellwig. 1970. Effect of partially field drying on energy requirements for processing and pelleting Coastal bermudagrass. Trans. Am. Soc. Agric. Eng. 13:315-19.

Connell, J, and DG Cramp. 1975. The nutritive value for dairy cows of artificially dried lucerne wafers made from fresh crop or after dewatering.

Proc. Br. Soc. Anim. Prod. 4:112-13.

Hong, BJ, GA Broderick, MP Panciera, RG Koegel, and KJ Shinners. 1988a. Effect of shredding alfalfa stems on fiber digestion determined by in vitro procedures and scanning electron microscopy. J. Dairy Sci. 71:1536-45.

Hong, BJ, GA Broderick, RG Koegel, KJ Shinners, and RJ Straub. 1988b. Effect of shredding alfalfa on cellulolytic activity, digestibility, rate of passage, and milk production. J. Dairy Sci. 71:1546-55.

Koegel, RG, KJ Shinners, FJ Fronczak, and RJ Straub. 1988. Prototype for production of fast-drying forage mats. Appl. Eng. Agric. 4(2):126-29.

Kraus, TJ, RG Koegel, KJ Shinners, and RJ Straub. 1992. Evaluation of a crushing-impact macerator. Am. Soc. Agric. Eng. Pap. 921005. St. Joseph, Mich.

Lu, CD, NA Jorgensen, and GP Barrington. 1979. Wet fractionation process: Preservation and utilization of pressed alfalfa silage. J. Dairy Sci. 62:1399-1407.

Mertens, DR, and R Hintz. 1990. Personal communication. USDA Agricultural Research Services, Dairy Forage Research Center, Madison, Wis.

Rotz, CA, RG Koegel, KJ Shinners, and RJ Straub. 1990. Economics of maceration and mat drying of alfalfa on dairy farms. Appl. Eng. Agric. 6(3):248-56.

Rotz, CA, RJ Davis, and SM Abrams. 1991. Influence of rain and crop characteristics on alfalfa damage. Trans. Am. Soc. Agric. Eng. 34(suppl. 4):1583-91.

Shinners, KJ, RG Koegel, and RJ Straub. 1987. Drying rates of macerated alfalfa mats. Trans. Am. Soc. Agric. Eng. 30(4):909-12.

Shinners, KJ, TJ Kraus, RG Koegel, and RJ Straub. 1992. A Crushing-Impact Macerator, Beltless Press Forage Mat Formation Machine. Proc. AG ENG 1992, Uppsala, Sweden.

Yang, JH, GA Broderick, and RG Koegel. 1993. Effect of heat treating alfalfa hay on chemical composition and ruminal in vitro protein degradation. J. Dairy Sci. 76:154-64.

11

Postharvest Processing of Forages

MICHAEL COLLINS
University of Kentucky

KENNETH J. MOORE
Iowa State University

STORED forages provide most of the feed needed for livestock during winter and periods of pasture shortage. This forage is referred to as *hay, silage, haylage,* or *dehy,* depending upon moisture levels and harvesting practices. Silage is preserved by maintenance of anaerobic conditions, whereas hay and dehy are preserved by reducing moisture concentrations below those that support microbial growth (Pitt 1990). Haylage is produced by ensiling forage between 400 and 600 g kg^{-1} moisture and undergoes less fermentation than silage. Field-cured hay, generally below 200 g kg^{-1} moisture, and barn-dried or preservative-treated hay, between 200 and 350 g kg^{-1} moisture, are the primary emphasis of this chapter.

Silage (see Chap. 12) is fermented and stored at higher moisture concentrations than hay and usually has lower harvesting losses; however, hay typically has lower storage losses (Fig. 11.1). In areas with high rainfall and poor herbage drying conditions, ensiling is frequently the predominant method of feed storage. Hay, due to its low water content, is more suitable for long-distance transportation and is an important cash crop in many areas. Round-baled or stacked hay may be stored outside, thus avoiding the costs associated with specialized storage structures.

HAY QUALITY

Maturity Stage at Harvest. Maturity is the most important single factor determining feeding value of hay, regardless of species. As forage plants mature, their DDM [digestible dry matter concentration, in vivo, or DDM estimated by dry matter disappearance, in vitro (IVDMD)] decreases sharply. First-growth, cool-season grasses and legumes often contain 800-850 g kg^{-1} DDM during the first 2-3 wk of spring growth (Collins 1991). The DDM declines by 3-5 g kg^{-1} daily thereafter as plants develop, flower, and produce seed, eventually reaching final levels below 500 g kg^{-1} at the ripe seed stage.

Maturity at harvest not only affects digestibility but also influences hay consumption by animals. Typical DDM and CP (crude protein) concentrations for hays harvested at different stages of maturity are shown in Figure 11.2. Hay intake decreases dramatically as maturity advances due to increasing fiber concentrations. Aftermath hay crops, especially cool-season grasses and legumes grown in the northern states, usually average about 600 g kg^{-1} and commonly range from 570 to 640 g kg^{-1} DDM.

Leaf:Stem Ratio. Leaf tissue of grasses and legumes is much higher in nutrient content than stem tissue except at very early maturity stages. Figure 11.3 illustrates typical dif-

MICHAEL COLLINS is Professor of Agronomy, University of Kentucky, Lexington. He received the MS degree from West Virginia University and the PhD from the University of Kentucky. His research has emphasized postharvest physiology of forage crops and management effects on forage quality.

KENNETH J. MOORE is Professor of Agronomy, Iowa State University, Ames. He received the MS and PhD degrees from Purdue University. His research has emphasized cell wall carbohydrate chemistry and postharvest physiology of forage crops.

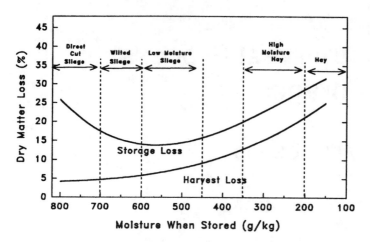

Fig. 11.1. Estimated total field and harvest loss and storage loss when legume-grass forages are placed in storage at varying moisture levels. (Adapted from Hoglund 1964.)

ferences in leaf and stem composition of alfalfa (*Medicago sativa* L.) herbage and hay. Leaf tissue is higher in CP and DDM and lower in neutral detergent fiber (NDF) than stem tissue. Similarly, nonstructural carbohydrate concentrations as high as 200 g kg^{-1} have been reported in alfalfa leaves (Lechtenberg et al. 1971). Stems are lower in digestible energy (DE) than are leaves, and leaf percentage is positively correlated with intake by animals (Minson 1977).

Leaf and stem differences in quality of herbage persist in the hay produced from it (Fig. 11.3). Consequently, hays that contain large amounts of leaves are likely to be of high quality. Hay-making systems aimed at producing high-quality hay emphasize harvesting and conserving as much of the leaf component as possible. Harvesting losses of DM, nitrogen (N), ash, and DDM are general-

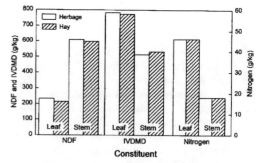

Fig. 11.3. In vitro dry matter disappearance, detergent fiber (NDF), and total nitrogen concentrations of leaf and stem fractions of fresh alfalfa herbage immediately after cutting and of hay produced by field curing. At early bloom maturity, the leaf fraction contains most of the digestible dry matter and nitrogen in the crop (Collins 1991).

ly greater from the leaf component than from the stem component, especially for legume forages. Thus, leaf generally makes up a smaller fraction of the total DM in dry hay than in the fresh herbage.

HAY PRODUCTION

The Drying Process. Moisture concentrations above 750 g kg^{-1} are common in standing forages cut for hay and are highest in immature forage (Savoie 1988). The removal of this moisture prior to storage is limited by plant resistances to drying including internal tissue resistance, stomatal, cuticular, and boundary layer resistances (Jones and Harris 1980; Harris and Tullberg 1980). A layer of still air near the plant surface called the *boundary layer* slows the exchange of water vapor between the crop and atmosphere (Monteith 1973). Although stomata occupy only about 1% of the surface area and close soon after

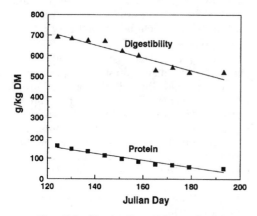

Fig. 11.2. Forage digestibility and crude protein concentration of perennial forages decrease during spring as plants advance in maturity, making the timing of cutting an important factor affecting the quality of the harvested crop.

cutting, one-third of the crop moisture content may be lost by that route (Harris and Tullberg 1980; Jones and Harris 1980).

KINETICS OF DRYING. Moisture concentration declines exponentially during drying (Macdonald and Clark 1987). Two distinct phases are apparent during field hay curing, a rapid one just after cutting when stomata are open and a protracted phase encompassing the majority of the drying period when stomatal resistance is high (Jones 1979). Drying rates can be calculated as described by Rotz and Sprott (1984) using the following equation:

$$\text{Drying rate } (k) = (-1/t)\,[\ln(M - M_e/M_0 - M_e)]$$

where

k = drying rate ([kg water lost] [kg plant water^{-1}] [h^{-1}]).

t = length of drying interval in hours.

M = moisture concentration (dry basis) at the end of time t.

M_0 = moisture (dry basis) at the beginning of time t.

M_e = equilibrium moisture concentration (dry basis).

ENVIRONMENTAL FACTORS. The environment sets the potential limits of the rate of water loss during hay drying, and other factors such as species and hay management practices determine the degree to which that potential is achieved. Vapor pressure deficit, total radiation, wind speed, and soil moisture are important environmental factors affecting field-drying rates of hay (Hill et al. 1977; Rotz and Chen 1985; Gupta et al. 1989). Grass-drying rates are increased by increasing radiation level, increasing vapor pressure deficit, decreasing initial moisture content of the forage, and decreasing density of the windrow (Savoie and Mailhot 1986). It appears that wind speed is not limiting to drying rates when it is above 2.2 m sec^{-1} at the hay surface (Klinner and Shepperson 1975; Gupta et al. 1989).

Under constant relative humidity (RH) and temperature conditions, hay reaches a moisture equilibrium with the atmosphere and remains stable as long as these conditions persist (Hill et al. 1977). Figure 11.4 illustrates RH and temperature effects on equilibrium moisture of alfalfa hay. Above 70%-75% humidity, hay may not dry sufficiently for safe storage even with extended curing times. Un-

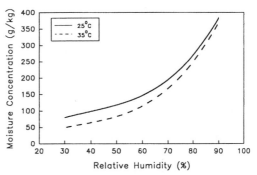

Fig. 11.4. Equilibrium moisture is the moisture content at which hay would stabilize after being exposed to uniform environmental conditions for an extended time period. Higher relative humidity levels lead to higher equilibrium moistures, and higher temperatures reduce equilibrium moisture. (Adapted from Hill et al. 1977.)

der high solar radiation conditions, leading to higher hay temperatures, equilibrium moistures may approach zero (Rotz and Chen 1985).

During the day, the windrow surface reflects about 20% of the solar radiation back into the atmosphere; the remainder is absorbed and may contribute to water evaporation (Jones and Harris 1980). Radiation is dissipated by heating the tissue, by evaporation of water, or by photochemical reactions (Monteith 1973). Only 50% of the solar radiation penetrates to a depth of 2 cm into the swath or windrow. Thus, increasing swath surface area is generally advantageous to hay drying because it utilizes a greater proportion of the available solar radiation.

Plant Factors Affecting Drying. Legume leaflets and grass leaf laminae generally dry faster than their respective stem or pseudostem components. Stomatal densities are greater for leaves than for stems, and leaves have lower internal resistances to drying because they are thinner. Detached leaf blades of annual ryegrass (*Lolium multiflorum* Lam.) dried three times faster than true stems from the same plant during the early portion of the drying cycle and seven times faster during the last phase (Jones 1979). Drying agent and mechanical-conditioning treatments are directed primarily toward increasing stem drying rates.

SPECIES DIFFERENCES IN DRYING RATE. Legume forage usually dries less rapidly than grass, and red clover (*Trifolium pratense* L.)

dries more slowly than alfalfa, with or without mechanical conditioning (Clark et al. 1985). Hay-drying time for several species was correlated ($r = 0.69$, $p < 0.05$) with stem diameter. Tetlow and Fenlon (1978) report drying rate coefficients for annual ryegrass, perennial ryegrass (*L. perenne* L.), tall fescue (*Festuca arundinacea* Schreb.), and alfalfa of 0.111, 0.079, 0.131, and 0.057 h^{-1}, respectively. Inclusion of 8% smooth bromegrass (*Bromus inermis* Leyss.) in red clover swards hastened hay drying (Collins 1985).

It might be possible to modify cuticle characteristics of forage species through plant breeding or management. Considerable, highly heritable, variation exists in leaf and stem cuticular wax content within red and white clover (*T. repens* L.) populations (Moseley 1983) compared with alfalfa (Galeano et al. 1986). Alfalfa grown under dry conditions had 4% more epicuticular wax than with irrigation.

Early work (Willard 1926) suggests that some moisture lost from legume stems during drying exits via the leaf. Drying of grass pseudostems and red clover stems is slowed when leaf laminae are removed (Jones 1979; Harris and Tullberg 1980). Transpirational flux in leaves increases tension on xylem water in the stem and provides a gradient for water movement from surrounding tissues to the xylem.

Mechanical and Chemical Treatment

MECHANICAL CONDITIONING. Forage crushing or crimping during the mowing operation generally increases drying rates, as much as twofold under favorable environmental conditions (Jones 1990). Crimpers pass the forage between intermeshing corrugated rollers to bend stems at intervals along their length whereas crushing usually involves smooth or ribbed rollers of different patterns. Flail conditioners, which increase drying rate by abrading forage plant leaves and stems as they pass through the harvester, are utilized in harvesting grasses but are considered too harsh for legumes.

Mechanical conditioning disrupts the cuticle to reduce cuticular resistance to movement of water vapor and may also reduce internal tissue resistance. The finding that drying rates of orchardgrass (*Dactylis glomerata* L.) stems were doubled by rubbing to disrupt the waxy layer illustrates the importance of the cuticle in limiting hay drying (Firth and Leshem 1976). Reducing internal tissue resistance to drying by splitting the stems also increased drying rate from 0.08 h^{-1} for rubbed stems to 0.44 h^{-1} for split stems. Treatment to reduce cuticular resistance has more impact on stem than on leaf-drying rates (Harris et al. 1974).

Conditioning can negatively affect drying in the event of rain (Gupta et al. 1989; Fig. 11.4) by making rewetting easier at high RH. Severe conditioning treatment of legume forage may also make it difficult to maintain optimum swath structure for hay curing by reducing the structural integrity of stems needed to provide support and increase air space within the swath (Dernedde 1980).

CHEMICAL CONDITIONING. Potassium carbonate and other drying agents may be applied as water solutions to hasten drying, especially of legume hay. Typical drying responses of alfalfa to application of potassium carbonate are illustrated in Figure 11.5. Physical modification of the epicuticular waxes, chemical alteration of the cuticle, and interference with stomatal closure are possible mechanisms. Increasing cation radius from 0.097 to 0.167 nm using carbonates of sodium (Na), potassium (K), and cesium (Cs) increased effectiveness in drying enhancement for alfalfa hay (Panciera et al. 1989). Drying agents usually do not enhance drying of cool-season grasses possibly because leaf sheaths prevent application of the drying agent to stem tissue (Harris and Dhanoa 1984; Jones 1990).

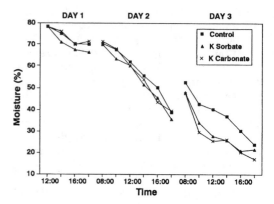

Fig. 11.5. Field-drying rates of alfalfa hay can be increased by application of chemical drying agents. Potassium carbonate or potassium sorbate, both at 4 g kg^{-1} forage weight, increased drying rate equally. Hay treated with either compound could be baled several hours earlier than untreated hay. (Adapted from Jaster and Moore 1992.)

TEDDING AND RAKING. Initial drying occurs primarily from leaves at swath and windrow surfaces exposed to solar radiation (Jones and Harris 1980). Internal swath humidity is initially high due to the high rate of water loss but declines as drying proceeds. Windrows 1.8 m wide gave faster drying rates of timothy (*Phleum pratense* L.) (0.14 h^{-1}) than did windrows 1.1 m wide (0.11 h^{-1}) just after cutting (Pattey et al. 1988). Tedding after cutting maximizes surface area and thus maximizes solar radiation interception by the hay crop. Tedding partially dried hay may double drying rates down to 670 g kg^{-1} moisture by exposing wetter material from beneath the swath surface (Savoie 1988).

Combining windrows to increase depth may improve later phases of drying because wind speed increases linearly with height just above the soil surface (Jones and Harris 1980; Wilman and Owen 1982). Reducing surface area by raking and by combining multiple windrows also reduces the area exposed to repeated wetting by dew and minimizes the resulting color loss in the hay.

The presence of the drying swath for extended periods generally reduces sward regrowth. In one study, delaying the removal of cut grass from a sward dominated by perennial ryegrass for 5 d reduced future yields by 9%, and a 10-d delay reduced yields by 16% (Owen and Wilman 1983). Tiller densities after harvest were reduced by 26% and 50% for swards with 5- and 10-d delays in removal, respectively.

FIELD LOSSES

Hay drying may require several days under average weather conditions. Field-dried hay must generally be reduced to 200 g kg^{-1} or below to prevent spoilage during storage and to even lower moistures (160 g kg^{-1}), if it is desirable to completely eliminate mold.

Harvest losses include mechanical losses such as mowing, conditioning, chopping, raking, packaging, and handling as well as respiration and leaching (Pitt 1982; Rees 1982; Collins 1991). A general relationship between field and storage losses over a wide range of moistures for an alfalfa-grass mixture is shown in Figure 11.1 (Hoglund 1964). High storage losses for wet silage are due to effluent loss and fermentation pattern. Storage losses decline with decreasing moisture level at the time of storage, but field losses increase, due mainly to increased physical losses, especially when forage must be handled or processed when below 400 g kg^{-1} moisture.

Physical Loss. Mechanical-harvesting losses are often higher for legume than for grass forage at a given moisture level due to differences in shoot morphology (Savoie 1988; McGechan 1990). The major factor is that leaflets of legumes are attached by a brittle petiole, whereas grass leaf laminae are attached firmly to the leaf sheath at the collar. Combined mowing and tedding and raking losses of alfalfa DM were found to be about twice that of perennial ryegrass (Klinner 1975). Physical losses of alfalfa may be elevated to 10% to 30% with severe treatments such as flail mowing or flail conditioning.

Raking should be completed when moistures are above 400 g kg^{-1} to minimize physical losses (Fig. 11.6). Tedding at moisture concentrations below 400 g kg^{-1} increases DM losses substantially (Savoie 1988). Tedding losses are also generally greater for legume than for grass forage due to differences in shoot morphology (McGechan 1990). Tedding losses for timothy are only 1%-2% above 400 g kg^{-1} moisture but increase to as much as 7% below 100 g kg^{-1}. Alfalfa DM losses of 4% resulted from tedding at 600 g kg^{-1} moisture compared with 8% when tedded at 400 g kg^{-1} moisture.

Respiration Loss. Respiration continues after cutting due to the combined activity of natural plant enzymes and microbes on the surfaces of the plants. Respiration in cut grass may consume 2% of the DM d^{-1} under good conditions and 5% d^{-1} under poor conditions

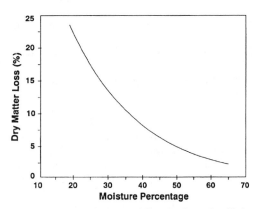

Fig. 11.6. Raking of legume hay should be completed before moisture concentration falls below 400 g kg^{-1} to avoid excessive dry matter losses. Losses during raking of alfalfa are below 5% if the operation is completed when crop moisture is 500 g kg^{-1} or more but increase dramatically as the crop dries further. (Adapted from Moser 1980.)

(Rucker and Knabe 1980). Rates are higher at high temperatures than at low temperatures. Respiration rates decline as the crop dries and are near zero below 400 g kg^{-1} moisture. Figure 11.7 illustrates the rate of DM loss to respiration for drying grass over a range of moisture contents (Honig 1980). Forage quality is reduced directly by respiration because nonstructural carbohydrates consumed during respiration are also readily digestible by livestock.

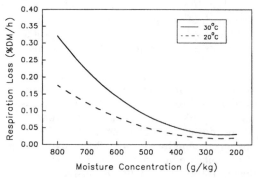

Fig. 11.7. Respiration contributes to DM losses of cut forage during field drying. Losses due to respiration in grass decline as the crop dries and are minimal below 400 g kg^{-1} moisture. Higher temperatures increase respiration rate. (Adapted from Honig 1980.)

Harvesting forage for silage does not eliminate respiration but redirects it from an aerobic to anaerobic type. The products of microbial growth in silage include organic acids that lower the pH to preserve the product. Thus, these losses can be minimized by rapid drying or fermentation but cannot be eliminated (see Chap. 12).

Rain Damage. Rainfall during hay curing reduces both yield and quality by delaying dry down through the leaching effects of water and by increasing physical losses (Collins 1983). Loss of alfalfa DM over 54 hay harvests averaged 17% without rain damage and 22% with rain damage (Collins 1990). The extent of loss due to a rain event becomes greater as crop moisture declines (Rucker and Knabe 1980). The combined effects of maturity stage and harvesting losses on digestibility of first-harvest alfalfa hay are shown in Figure 11.8. The IVDMD of cured alfalfa hay averaged 50 g kg^{-1} less than the initial with no rain and more than twice that amount when rain damage occurred (Collins 1990). Attempts to avoid the negative effects of rain on hay quality by

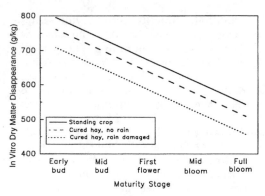

Fig. 11.8. In vitro dry matter disappearance (IVDMD) declines as alfalfa shoots mature. Respiration and leaf shatter decrease hay IVDMD by about 50 g kg^{-1} compared with the standing crop, and exposure to rain damage during curing decreases it by an additional 80 g kg^{-1}. (Adapted from Collins 1990.)

delaying harvest usually prove unsuccessful because advancing maturity during the period of delay continues to reduce hay quality. Grasses are affected somewhat less by rain during curing than are legumes.

The impact of rain on hay quality is exerted primarily on the leaf fraction (Collins 1991; Fig. 11.9). In terms of DM, N, ash, and IVDMD of alfalfa, it was found that more than 60% of the total losses of these constituents due to leaching came from the leaf fraction and 40% came from the stem. Three-fourths of the total combined respiration and leaf shatter losses of the same constituents came from the leaf fraction and only one-fourth from the stem. In vitro dry matter disappearance is affected more by rain than total

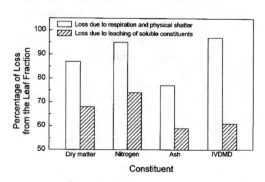

Fig. 11.9. Physical shattering losses, respiration losses, and leaching losses from a rainfall event during field curing of alfalfa hay come disproportionately from the leaf fraction. Since most of the losses are from the leaf fraction, stem percentage is increased in the hay compared with that in the crop just after cutting. (Adapted from Collins 1991.)

N in both timothy (Flipot et al. 1984) and alfalfa (Collins 1990).

Baling Loss. Baling loss, i.e., the change from the windrow to the bale, for rectangular bales of alfalfa generally ranges from 2% to 5% (Friesen 1978; Collins et al. 1987). Baling losses for round balers range from about 1% for grass hay or moist legume hay to as much as 15% for legume hay below 140 g kg^{-1} moisture. In dry environments, night baling may reduce losses to less than 1% by allowing leaves to remoisten.

Except for very dry hay, studies indicate that moisture content at baling has little effect on hay composition (Collins et al. 1987; Wittenberg and Moshtaghi-Nia 1990). Round-baled alfalfa was higher in N and lower in NDF and ADF (acid detergent fiber) when baled at 130 g kg^{-1} than when baled at 380 g kg^{-1} moisture.

CHANGES IN HAY YIELD AND QUALITY DURING STORAGE

Microbial Activity. Changes in hay composition during storage result from crop and microbial respiration and from nonenzymatic chemical reactions (Hlodversson and Kaspersson 1986). Plant enzymatic activity is markedly reduced but does cause measurable respiration during hay storage. However, the occurrence of significant respiration and heating of hay during storage is a result of microbial activity (Wood and Parker 1971). Counts of 13 different fungal genera peaked after 9 d of storage. Humidities above 70% and temperatures above 20°C are necessary for meaningful fungal growth in hay (Rees 1982).

FATE OF ANTIQUALITY COMPOUNDS DURING STORAGE

Concentrations of a number of antiquality compounds of importance in livestock feeding are influenced by forage preservation method, with concentrations frequently being reduced by hay curing or silage fermentation. For example, *Sorghum* spp. contain cyanogenic glycosides that, upon hydrolysis, release HCN, which is toxic to livestock. Hay drying denatures the enzyme responsible for the hydrolysis and generally reduces HCN levels, but toxic levels may still remain, in part because hay intake rates may be very rapid relative to those of grazed herbage (Wheeler and Mulcahy 1989). Ensiling, on the other hand, reduces HCN potential below toxic levels provided 6 to 8 wk are allowed between ensiling and feeding.

Hay produced from plants with toxic levels of nitrate (NO$_3^-$) should be used cautiously because this compound is stable in hay. Hay produced from grasses heavily fertilized with N or produced under drought conditions may have toxic levels of NO$_3$ and should be tested prior to feeding. Fermentation of wilted silage normally reduces NO$_3$ concentrations below toxic levels but contributes to the production of toxic "silo gases." Reductions in NO$_3$ concentration are smaller for low-moisture silage.

Tannins are phenolic compounds present in significant concentrations in birdsfoot trefoil (*Lotus corniculatus* L.), sericea lespedeza (*Lespedeza cuneata* [Dum.-Cours.] G. Don), and some other species. Field drying of fresh sericea lespedeza reduced tannin levels by 50% and increased in vivo DM digestibility of the forage by sheep (*Ovis aries*) by 43 g kg^{-1} (Terrill et al. 1989).

The problem of blister beetle (*Epicauta* sp.) poisoning in horses (*Equus caballus*) increased after the introduction of mechanical conditioning as a routine operation during hay harvesting. Blister beetles, which may be found in large concentrations in isolated areas of hayfields, contain cantharidin, a compound that is toxic to horses. Beetles were not killed in large numbers when hay was harvested using a sickle mower and swaths were placed so as to avoid wheel traffic during mowing. Cantharidin is very stable in hay.

Red clover hay from plants infected with the black patch fungus (*Rhizoctonia leguminicola* Gough & ES Elliott) contains the alkaloid slaframine. Symptoms include lachrymation, salivation, diarrhea, and excessive urination. Slaframine levels decline gradually during hay storage.

Tall fescue infected with the endophytic fungus *Acremonium coenophialum* Morgan-Jones & Gams contains a number of alkaloids with negative effects on livestock referred to as *fescue toxicosis* or *summer slump*. Limited research on alkaloid behavior during hay curing and storage and during silage fermentation indicate that N-formyl- and N-acetyl loline alkaloids are reduced by about 20% during hay curing but are very stable in dry, stored hay (Bush et al. 1993). Concentrations of these alkaloids change little during silage fermentation.

Heat Damage. Storage DM loss for untreated hay is highest at high moisture concentrations. Increased DM loss is generally associated with the increased heating that occurs during storage of moist hay (Rotz et al. 1988). Kjelgaard et al. (1983) found that losses of DM increased by 1% of the initial hay dry weight for each 10 g kg^{-1} increase in moisture above 100 g kg^{-1}. Fiber concentrations generally increase and water-soluble carbohydrates decrease during storage (r = 0.66 to 0.85) (Hlodversson and Kaspersson 1986). Heat generated by respiration and chemical reactions in moist hay evaporates water, which aids in cooling the hay mass and reduces hay moisture concentration, provided the water vapor can escape the hay mass (Nash and Easson 1977) (Fig. 11.10).

Green color is replaced by various shades of brown when hay undergoes significant heating during storage. The extent of the color change is proportional to the extent of maillard reaction, which involves a condensation of sugars and amino acids that renders both reactants indigestible (Moser 1980). At one time, heat-damaged hay, with its characteristic tobaccolike odor, was thought to indicate improved quality because the resulting forage is often highly palatable to livestock (Nash 1985). However, it is now clear that the nutritive value is reduced by spontaneous heating of hay and silage because CP and DM digestibility are reduced and fiber concentrations are increased.

The extent of heat damage can be quantified by determining acid detergent insoluble N (ADIN). Poststorage concentrations of

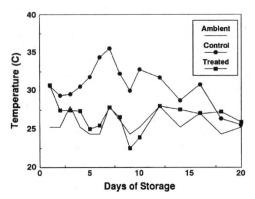

Fig. 11.10. Afalfa hay, at 300 g kg^{-1} moisture at baling, receiving 30 kg Mg^{-1} of hay weight of an 80:20 propionic:acetic acid mixture had lower storage temperatures than untreated hay at the same moisture. Hay is frequently well above ambient temperature immediately after baling due to the effects of solar radiation on the exposed windrow. In hay of moderate moisture, peak temperatures occur after a few days of storage, and temperatures then decline to near ambient levels within 3 wk. (Adapted from Lawrence et al. 1987.)

ADIN increased linearly with increasing moisture at baling between 150 and 250 g kg^{-1} for both round bales and small rectangular bales; however, the rate of increase with increasing moisture was 1.7 times greater for the round bales (Collins et al. 1987). This effect probably results from the larger mass of round bales and large rectangular bales, measuring 1 m or more in each dimension, which restricts heat and moisture loss during storage.

SPONTANEOUS COMBUSTION. Spontaneous combustion temperatures may occur during storage of moist hay (Miller et al. 1967). With adequate moisture to maintain the relative humidity of the air in the hay mass at 95% to 97%, heat generated by plant enzymatic activity and microbial growth may elevate hay temperatures to 70°C within a few days (Currie and Festenstein 1971). This progression may take several weeks with drier air. Above 70°C, oxidative chemical reactions are responsible for additional heat generation, greatly increasing the potential for further rapid increase to combustion temperatures (Festenstein 1971). Mixed timothy-meadow fescue (*F. pratensis* Huds.) grass hay at 440 g kg^{-1} moisture took 1 mo to reach a temperature of 90°C but then increased to 165°C in only 3 more days. Soluble carbohydrates were almost completely eliminated in hay that reached 165°C.

Spontaneous combustion requires large quantities of oxygen (O_2) (Currie and Festenstein 1971). Thermal conductivity of dry hay is lower than that of moist hay, so heat transfer to the outside of the stack becomes progressively less effective as the hay dries. Thus, hay temperatures may increase rapidly after much of the moisture has been removed. Spontaneous combustion occurs near the outside of the stack because O_2 levels in the interior are reduced by microbial activity.

Health Effects of Moldy Hay. Exposure to moldy hay and bedding contributes to a number of respiratory and digestive problems in horses (Webster et al. 1987; Madelin et al. 1991). Chronic obstructive pulmonary disease, or heaves, is associated with moldy hay and straw. Mold spores also contribute to colic in horses. Breathing spores of the fungus *Aspergillus fumigatus* and actinomycetes such as *Micropolyspora faeni* and *Thermoactinomyces vulgaris* during the handling of moldy hay can also cause farmer's lung, a sometimes debilitating disease in which the fungus grows in lung tissue (Lacey et al. 1978). Cattle (*Bos taurus*) are generally less affected by moldy hay than horses (Lacey and Lord 1977; Moser 1980); however, they are subject to mycotic abortion and aspergillosis in response to the presence of certain fungi in moldy hay.

HAY PRESERVATIVES

Risk of Rain Damage. The occurrence of rain during field curing of hay results in reduced yield and quality due to leaching, shattering, and respiration (Rees 1982; Collins 1985). Pitt (1982) models forage harvesting and concludes that shortened field-drying times would lead to increased hay yields by reducing these harvesting losses. Using the Markov chain probability model with weather data from Illinois, Feyerherm et al. (1966) calculated the probability of rain-free periods of various lengths. Assuming that a rain-free day is selected for mowing, the probability of having a 2-d rain-free period during May is 73%. However, the probability of having a 5-d period without rain to complete hay curing is only 28%. The reduction in risk of rain damage by shortening the field-curing period is an important part of the rationale for baling hay above 200 g kg⁻¹ moisture. Baling at 200-300 g kg⁻¹ also has small positive effects on harvested yield and forage quality (Wilkins 1988).

Types of Compounds and Modes of Action. Microbial growth in moist hay can be controlled by effective preservatives (Tetlow 1983; Henning et al. 1990), by barn drying (Tetlow 1983; Parker et al. 1992), by dehydration, or by gamma irradiation (Conning 1983). Sodium diacetate, propionic acid, ammonium propionate, urea, anhydrous ammonia, and some others are effective hay preservatives at the proper rates and with adequate distribution (Wood and Parker 1971; Lacey et al. 1978; Tetlow 1983). Ammonia-type additives may also improve forage quality by increasing fiber digestibility and adding nonprotein N. These compounds are usually applied during the baling process but may also be applied after baling (Khalilian et al. 1990).

ORGANIC ACIDS. Acetic and propionic acids are the most commonly organic acid-type hay preservatives used to prevent growth of fungi and actinomycetes (Lacey et al. 1978). The acids have the inherent problems of corrosiveness to machinery and of high volatility. In one study only about 15% of the propionic acid

applied to mixed grass-clover hay remained in the forage after storage periods ranging from 2 to 10 mo (Davies and Warboys 1978). At pH 5.0, acidity itself inhibits growth of actinomycetes; however, organic acid salts, such as ammonium propionate and sodium diacetate, are effective at pH 6.0 (Woolford 1984). These compounds avoid the corrosiveness and volatility associated with the acids. A mixture of acids and salts provides some acidifying action that aids in inhibition of fungal growth.

Round baled hay can also be preserved using organic acids and their salts; however, care must be taken to ensure adequate distribution of preservative in the windrow. Although distribution might be improved by application to the windrow or swath, hay preservatives are generally applied during the baling process due to the volatile nature of many of the materials used.

Effective preservatives can reduce peak storage temperatures of moist hay by up to 25°C or more (Sheaffer and Clark 1975; Khalilian et al. 1990). It was found that moist, untreated alfalfa-timothy hay had lower IVDMD and higher ADF and ADIN than similar hay receiving propionic acid or ammonium isobutyrate at 30 or more g kg⁻¹. In another study, concentrations of NDF and ADIN in grass round bales were reduced by treatment at an adequate rate (Wood and Parker 1971). Horses consumed moist alfalfa hay treated with 80:20 propionic:acetic acid as well as they consumed control hay that was field dried (Battle et al. 1988).

AMMONIA COMPOUNDS. Ammonia and ammonia-producing compounds, such as urea, have been used successfully in the preservation of both moist hay (Moore et al. 1985a, 1985b; Hlodversson and Kaspersson 1986; Henning et al. 1990) and crop residues at rates of approximately 30 g kg⁻¹ of crop weight. Ammonia or ammonia-producing compounds can reduce microbial growth during storage of high-moisture hay to as little as one-sixth that of control hay (Woolford and Tetlow 1984; Henning et al. 1990). Woolford and Tetlow (1984) found reduced levels of total viable organisms, molds, thermophilic actinomycetes, and bacteria in ammonia-treated grass hay.

Total N concentration and forage digestibility are frequently increased following ammonia treatment of grass hay (Table 11.1) (Moore et al. 1983; Woolford and Tetlow 1984; Verma et al. 1985; Moore and Lechtenberg 1987). Treatment with ammonia increased NDF digestibility of control grass hay from 422 up to 682 g kg^{-1} for treated hay. Ammonia treatment increases fiber digestibility by cleaving ester bonds between lignin and cell wall hemicellulose without actually removing lignin from the treated forage. The effect of urea treatment on fiber digestibility is greatest in grass hay of 230 g kg^{-1} moisture or more (Belanger et al. 1987).

Forage quality improvement by ammonia treatment is typically smaller for legume hay than for grass hay. Fiber digestibility of legume hay is less affected than that of grass hay, and legumes have much lower fiber concentrations (Knapp et al. 1975; Wittenberg and Moshtaghi-Nia 1990). In Indiana, tall fescue treated with ammonia had 110 g kg^{-1} higher NDF digestibility than untreated hay, about twice the improvement seen for tall fescue-ladino clover (*T. repens* L.) hay.

Care must be taken in handling and applying anhydrous ammonia. Anhydrous ammonia is extremely volatile, and stacks must be covered prior to treatment to prevent loss. Treatment is accomplished by releasing the required weight of ammonia into the covered stack. Covers should be removed 3 d prior to feeding to allow excess ammonia to dissipate.

Urea is more easily handled and applied than anhydrous ammonia and, upon hydrolysis, releases ammonia and carbon dioxide. Urease is a naturally occurring enzyme that catalyzes the hydrolysis of added urea (Belanger et al. 1987). Urease activity is generally sufficient in grass hay above 200 g kg^{-1} to rapidly hydrolize added urea to ammonia (Tetlow 1983). About 70% of added urea on perennial ryegrass hay was hydrolized within 2 d with or without added urease. Adding urea to tall fescue hay increased IVDMD and

in vitro cell wall disappearance, even for samples collected on the day of baling (Henning et al. 1990). Cattle consuming hay treated with rates above 10 g kg^{-1} of ammonia sometimes exhibit hyperexcitability that can result in injury to the animal and/or in reduced forage intake (Wilkins 1988). A reaction between ammonia and sugars present in the forage produces 4-methyl imidazol, which is thought to be the active compound. Lower-quality grass hay and crop residues, products for which the non-protein nitrogen and improved fiber utilization provided by ammonia or urea treatment would be most useful, generally have low levels of these sugars and thus are not likely to exhibit this problem.

MICROBIAL ADDITIVES. The possible mode of action of bacterial inoculants in aiding hay preservation under aerobic conditions has not been clearly elucidated. The principle is that bacteria might inhibit fungal growth by reducing the pH or by utilizing limited substrate supplies. Wittenberg and Moshtaghi-Nia (1991) treated moist alfalfa with commercial products containing viable organisms capable of producing lactic acid and found no effect on fungal species present in the hay or on plate counts. Bacteria-treated and untreated hays received similar visual ratings for mold development in another study conducted under controlled conditions (Rotz et al. 1988). There are some indications of positive effects of inoculation on hay preservation (Wittenberg and Moshtaghi-Nia 1990; Tomes et al. 1990), but taken as a whole, they do not exhibit the pattern of changes in hay temperature, composition, and dustiness typical of effective preservatives.

PACKAGE SIZE AND STORAGE METHOD

Much of the hay produced for on-farm livestock feeding is stored in large round bales weighing 0.3 mt or more. Package type, size, and storage method interact with moisture concentration to impact storage losses in yield and quality. Safe storage moisture for large round bales appears to be near 180 g kg^{-1}, slightly lower than the 200 g kg^{-1} considered acceptable for small rectangular bales (Russell and Buxton 1985).

Round bales are frequently stored outside without protection from the effects of precipitation and other weather conditions between baling and feeding. The surface layer of round hay bales stored outside loses any green color present at baling and becomes brown. The af-

TABLE 11.1 Anhydrous ammonia effects on forage quality of annual ryegrass hay in round bales treated with ammonia under plastic for 5 mo or left untreated and uncovered

Moisture concentration and quality constituent	No ammonia	30 g kg^{-1} ammonia
	(g kg^{-1})	
400 g kg^{-1} moisture		
Acid detergent fiber	510	444
Crude protein	171	205
IVDMD	537	651
200 g kg^{-1} moisture		
Acid detergent fiber	403	414
Crude protein	146	155
IVDMD	537	589

Source: Verma et al. (1985).
Note: Ammoniation effects on quality are greatest at moisture contents above about 230 g kg^{-1}.
IVDMD = In vitro dry matter disappearance.

fected portion of the package is referred to as being *weathered* and has reduced forage quality, yield, and palatability to livestock compared with the unweathered hay. Forage quality of the unweathered portion of the package is comparable to that of similar bales protected by inside storage.

Table 11.2 shows typical effects of outside storage on hay quality (Lechtenberg et al. 1979). Quality is reduced in weathered hay since DDM is greatly reduced and ADF and acid detergent lignin (ADL) are increased. Nitrogen concentration is frequently unchanged or increased because losses of soluble sugars, some minerals such as K, and some other compounds exceed those of N-containing compounds. Concentrations of NDF are typically higher in this weathered hay. For example, Collins et al. (1987) report 658 g kg^{-1} NDF in weathered alfalfa from round bales compared with 541 g kg^{-1} NDF in whole-bale samples. The magnitude of weathering effects can be very large since approximately one-third of the total volume of a typical round bale would be contained in a weathered layer 15-cm thick over the exposed surface.

Dry hay stored in round bales is not inherently subject to greater storage losses than hay in small rectangular bales (Lechtenberg et al. 1974). Round bales stored inside have DM losses comparable to those for small rectangular bales, in the range of 4% to 6%. However, due to storage under more adverse conditions, losses between 9% and 15% of the DM are common for round bales (Belyea et al. 1985; Collins et al. 1987). Forage cut at late maturity produces lower-density round bales that might be less effective in shedding water during outside storage (Friesen 1978).

Outside storage losses of round bales can be reduced by solid or perforated plastic covers or sleeves (Wood and Parker 1971; Scales et al. 1978; Russell and Buxton 1985). It was found that individual covers reduced the percentage of total bale volume weathered during storage from 23% to 11% (Wood and Parker 1971). Another study found plastic mesh covering mixed alfalfa-smooth bromegrass hay resulted in greater DDM and lower NDF in the surface layer compared with control bales, indicating less weathering (Russell et al. 1990).

Research indicates that much of the loss in DM and quality occurs at the point of bale contact with the soil. Low-density bales generally "settle," resulting in greater areas of the bale in contact with the soil surface, leading to greater storage losses. Treatments to prevent direct contact between soil and bale, such as crushed rock or wooden pallets, reduce outside storage losses by up to 50% (Verma and Nelson 1983).

Dry matter and quality losses for grass round bales are generally lower than for legume hay. Under extremely wet conditions, alfalfa round bales lost four times as much DM as ryegrass round bales even though both were covered with plastic and elevated above the ground (Verma and Nelson 1983). Weathering affects quality of legume hay more than that of grass hay (Table 11.2) (Lechtenberg et al. 1979).

Barn Drying. Heated or unheated forced air can be used to remove moisture from baled hay prior to storage (Miller 1946; Parker et al. 1992). Electric fans (610-mm diameter, 3.7-5.2 kW) maintained static pressure near 600 Pa in the duct for alfalfa at 160 kg m^3 DM density (Parker et al. 1992). In 8 of 13 trials drying alfalfa hay from near 350 g kg^{-1} moisture to final moistures of 60-120 g kg^{-1}, pre- and postdrying concentrations of CP, NDF, ADF, and DDM were not significantly different, and differences were very small in the remaining five trials. Solar-heated air 15°-20°C above ambient temperatures hastened drying considerably. Bales ranging in DM density between 80 and 166 kg m^3 were successfully dried except when large variation existed be-

TABLE 11.2. Losses of DM and quality during outside storage of round-baled grass and mixed alfalfa-grass hay

Hay type	Fraction	IVDMD	Nitrogen	Acid detergent fiber	Acid detergent lignin
			(g kg^{-1})		
Grass	Unweathered	588	21.6	444	97
	Weathered	425	26.2	496	158
Alfalfa-grass	Unweathered	565	22.8	450	137
	Weathered	342	27.0	487	189

Source: Lechtenberg et al. (1979).

Note: Bales were stored outside, on the ground for 8 mo between May and February in Indiana. Weathered hay exhibits discoloration from exposure to precipitation, radiation, and other weather conditions. Quality is drastically reduced in weathered material, but hay from the unweathered bale core is well preserved.

tween bales within a batch.

Earlier work indicates that the pressure required to force air through bales increased with increasing density and that less pressure was required for bales stacked on edge (Davis and Baker 1951). In the typical rectangular baler design, bales measuring 0.36 × 0.46 × 0.80 m are produced as mechanical fingers move hay into the bale chamber, a knife cuts the hay at one end, and the "flakes" thus produced are compressed into the forming bale. The cut edge is the one adjacent to the knife and tends to have many stems cut more or less perpendicular to their long axes. Stacking on edge thus has the effect of placing these groups of parallel stems vertically in the stack and would appear to reduce the pressure required to move air upward through the stack. Many hay producers routinely stack small rectangular bales on edge based on the observation that fewer storage-related problems such as dustiness and heating are observed compared with bales stacked flat.

HAYLAGE

Haylage is as suitable to mechanical handling and feeding as silage but undergoes less fermentation in the silo and thus has lower storage losses than silage. Haylage also requires an anaerobic environment for proper preservation and is often stored in sealed, upright silos. Larsen et al. (1971) found that ad libitum DM intake of mixed alfalfa-grass haylage increased from 11 kg head^{-1} d^{-1} for haylage at 600 g kg^{-1} moisture to 14 kg head^{-1} d^{-1} for haylage at 350 g kg^{-1} moisture. In practice, haylage moisture contents below 400 g kg^{-1} are not recommended because of the increased likelihood of heat damage during storage of very dry haylage.

An average of 14% of the total N in alfalfa haylage was in the ADIN fraction compared with only 11% ADIN in silages of the same species, indicating that more heating occurred in haylage (Janicki and Stallings 1987). Proper chop length, around 8 mm, is critical in haylage production to overcome inherently greater difficulty in O$_2$ exclusion from the lighter material compared with direct-cut or wilted silages. Haylage temperature is increased more by a given amount of heat production than is the temperature of silage because of its lower water content.

DEHYDRATION

Forage may also be preserved by rapid drying of chopped forage for 2-10 min at high temperature, often followed by grinding and pelleting or by cubing. Very rapid evaporation of moisture from the forage keeps it cool to preserve quality while temperatures of 300° to 1000°C are used. Field wilting prior to dehydration reduces the energy cost of drying. Wilting for just a few hours can reduce moisture to around 500 g kg^{-1} without greatly affecting forage quality (Hellwig 1965; Livingston et al. 1977).

Dehydrated forage may be pelleted after drying and grinding, or it may be pressed into cubes. Cubing or pelleting the dehydrated forage eliminates dustiness and eases mechanical handling. Grinding and pelleting of forages frequently increases DM intake by ruminants compared with the same crop in long form. Uden (1988) found chopped grass hay was more digestible (730 g kg^{-1}) than ground and pelleted hay (670 g kg^{-1}), but solids' retention time in the digestive tract was 25% less for the pelleted forage, a reflection of a faster rate of passage for finer particles.

Dehydration preserves vitamins A, E, and K, protein, and pigments such as xanthophyll, which is used in poultry feeding, better than field hay curing or silage fermentation because dehydration minimizes exposure to sunlight, plant enzymatic activity, and microbial activity (Moser 1980). Controlled atmosphere storage using CO$_2$ has been used to minimize oxidative losses of quality constituents in dehydrated forages, which are frequently used as feed supplements.

QUESTIONS

1. Why are we able to store hay successfully under aerobic conditions?
2. What are the sources of DM and quality loss during hay curing under field conditions?
3. What environmental factors influence hay-drying rates in the field?
4. What plant factors limit hay-drying rate, and what management practices can be used to partially overcome these limitations?
5. Why are legumes more susceptible to leaf loss during hay production than are grasses?
6. Why do large round bales commonly suffer greater losses of DM and quality than small rectangular bales?
7. Why are elevated temperatures observed when hay is baled above 200 g kg^{-1} moisture?
8. What influence does ensiling have on antiquality constituents such as HCN and NO$_3^-$ in forages?
9. What plant compounds are involved in the maillard reaction, and what effect does the reaction have on their nutritive value?
10. Why is dehydration more effective in preserving vitamins and pigments in forage than field hay curing?
11. Why does grinding and pelleting forages reduce their digestibility?

REFERENCES

Battle, GH, SG Jackson, and JP Baker. 1988. Acceptability and digestibility of preservative-treated hay by horses. Nutr. Rep. Int. 37:83-99.

Belanger, G, AM St. Laurent, CA Esau, JWG Nicholson, and RE McQueen. 1987. Urea for the preservation of moist hay in big round bales. Can. J. Anim. Sci. 67:1043-53.

Belyea, RL, FA Martz, and S Bell. 1985. Storage and feeding losses of large round bales. J. Dairy Sci. 68:3371-75.

Bush, LP, FF Fannin, MR Siegel, DL Dahlman, and HR Burton. 1993. Chemistry, occurrence and biological effects of saturated pyrrolizidine alkaloids associated with endophyte-grass interactions. Agric. Ecosystems and Environ. 44:81-102.

Clark, EA, SV Crump, and S Wijnheijmer. 1985. Morphological determinants of drying rate in forage legumes. In Proc. Am. Forage and Grassl. Conf., Hershey, Pa., 137-41.

Collins, M. 1983. Wetting and maturity effects on the yield and quality of legume hay. Agron. J. 75:523-27.

———. 1985. Wetting effects on the yield and quality of legume and legume-grass hays. Agron. J. 77:936-41.

———. 1990. Composition and yields of alfalfa fresh forage, field cured hay, and pressed forage. Agron. J. 82:91-95.

———. 1991. Hay curing and water soaking: Effects on composition and digestion of alfalfa leaf and stem components. Crop Sci. 31:219-23.

Collins, M, WH Paulson, MF Finner, NA Jorgensen, and CR Keuler. 1987. Moisture and storage effects on dry matter and quality losses of alfalfa in round bales. Trans. Am. Soc. Agric. Eng. 30:913-17.

Conning, DM. 1983. Evaluation of the irradiation of animal feedstuffs. In PS Elias and AJ Cohen (eds.), Recent Advances in Food Irradiation. Amsterdam: Elsevier, 247-83.

Currie, JA, and GN Festenstein. 1971. Factors defining spontaneous heating and ignition of hay. J. Sci. Food Agric. 2:223-30.

Davies, MH, and IB Warboys. 1978. The effect of propionic acid on the storage losses of hay. J. Br. Grassl. Soc. 33:75-82.

Davis, RB, Jr., and VH Baker. 1951. Fundamentals of drying baled hay. Agric. Eng. 32:21-25.

Dernedde, W. 1980. Treatments to increase the drying rate of cut forage. In Proc. Forage Conserv. in the '80's, Occasional Symp. 11. Brighton, UK: British Grassland Society, 61-66.

Festenstein, GN. 1971. Carbohydrates in hay on self-heating to ignition. J. Sci. Food Agric. 22:231-34.

Feyerherm, AM, LD Bark, and WC Burrows. 1966. Probabilities of Sequences of Wet and Dry Days in Illinois. Kansas Agric. Exp. Sta. Tech. Bull. 139k.

Firth, DR, and Y Leshem. 1976. Water loss from cut herbage in the windrow and from isolated leaves and stems. Agric. Meteorol. 17:261-69.

Flipot, P, W Mason, and G Lalande. 1984. Chemical composition and animal performance of grass forage of varying maturity stored as hay or silage. Anim. Feed Sci. and Tech. 11:35-44.

Friesen, O. 1978. Evaluation of hay and forage harvesting methods. In Grain and Forage Harvesting. St. Joseph, Mich.: American Society of Agricultural Engineering, 317-22.

Galeano, R, MD Rumbaugh, DA Johnson, and JL Bushnell. 1986. Variation in epicuticular wax content of alfalfa cultivars and clones. Crop Sci. 26:703-6.

Gupta, ML, RH Macmillan, TA McMahon, and DW Bennett. 1989. A simulation model to predict the drying time for pasture hay. Grass and Forage Sci. 44:1-10.

Harris, CE, and MS Dhanoa. 1984. The drying of component parts of inflorescence-bearing tillers of Italian ryegrass. Grass and Forage Sci. 9:271-75.

Harris, CE, and JN Tullberg. 1980. Pathways of water loss from legumes and grasses cut for conservation. Grass and Forage Sci. 35:1-11.

Harris, CE, R Thaine, and HI Marjatta Sarisalo. 1974. Effectiveness of some mechanical, thermal and chemical laboratory treatments on the drying rates of leaves and stem internodes of grass. J. Agric. Sci. Camb. 83:353-58.

Hellwig, RE. 1965. Effect of physical form on drying rate of Coastal bermudagrass. Trans. Am. Soc. Agric. Eng. 8:253-55.

Henning, JC, CT Dougherty, J O'Leary, and M Collins. 1990. Urea for preservation of moist hay. Anim. Feed Sci. and Tech. 31:193-204.

Hill, JD, IJ Ross, and BJ Barfield. 1977. The use of vapor pressure deficit to predict drying time for alfalfa hay. Trans. Am. Soc. Agric. Eng. 20:372-74.

Hlodversson, R, and A Kaspersson. 1986. Nutrient losses during deterioration of hay in relation to changes in biochemical composition and microbial growth. Anim. Feed Sci. and Tech. 15:149-65.

Hoglund, CR. 1964. Comparative storage losses and feeding values of alfalfa and corn silage crops when harvested at different moisture levels and stored in gas-tight and conventional tower silos: An appraisal of research results. Mich. State Univ., Dept. Agric. Econ. Mimeogr. 946.

Honig, H. 1980. Mechanical and respiration losses during pre-wilting of grass. In Proc. Forage Conserv. in the '80's, Occasional Symp. 11. Brighton, UK: British Grassland Society, 201-4.

Janicki, FJ, and CC Stallings. 1987. Nitrogen fractions of alfalfa silage from oxygen-limiting and conventional upright silos. J. Dairy Sci. 70:116-22.

Jaster, EH, and KJ Moore. 1992. Hay desiccation and preservation with potassium sorbate, potassium carbonate, sorbic acid and propionic acid. Anim. Feed Sci. and Tech. 38:175-86.

Jones, L. 1979. The effect of stage of growth on the rate of drying of cut grass at 20°C. Grass and Forage Sci. 34:139-44.

———. 1990. The effect of formic acid on the rate of water loss from cut grass and lucerne. Grass and Forage Sci. 45:83-90.

Jones, L, and CE Harris. 1980. Plant and swath limits to drying. Occasional Symp. 11. Brighton, UK: British Grassland Society, 53-60.

Khalilian, A, MA Worrell, and DL Cross. 1990. A device to inject propionic acid into baled forages. Trans. Am. Soc. Agric. Eng. 33:36-40.

Kjelgaard, WL, PM Anderson, LD Hoftman, LL Wilson, and HW Harpster. 1983. Round baling from field practices through storage and feeding. In JA Smith and VW Hays (eds.), Proc. 14th Int. Grassl. Congr., Lexington, Ky. Boulder, Colo.: Westview, 657-60.

Klinner, WE. 1975. Design and performance characteristics of an experimental crop conditioning system for difficult climates. J. Agric. Eng. Res. 20:149-65.

Klinner, WE, and G Shepperson. 1975. The state of haymaking technology—a review. J. Br. Grassl. Soc. 30:259-66.

Knapp, WR, DA Holt, and VL Lechtenberg. 1975. Hay preservation and quality improvement by anhydrous ammonia treatment. Agron. J. 67:766-69.

Lacey, J, and KA Lord. 1977. Methods for testing chemical additives to prevent moulding of hay. Ann. Appl. Biol. 87:327-35.

Lacey, J, KA Lord, HGC King, and R Manlove. 1978. Preservation of baled hay with propionic and formic acids and a proprietary additive. Ann. Appl. Biol. 88:65-73.

Larsen, HJ, EL Jensen, and RF Johannes. 1971. A Comparison of Haylage and Wilted Grass Silage plus Hay in the Dairy Cow Diet. Wis. Agric. Ext. Stn. Res. Rep. R2251.

Lawrence, LM, KJ Moore, HF Hintz, EH Jaster, and L Wischover. 1987. Acceptability of alfalfa hay treated with an organic acid preservative for horses. Can. J. Anim. Sci. 67:217-20.

Lechtenberg, VL, DA Holt, and HW Youngberg. 1971. Diurnal variation in nonstructural carbohydrates, in vitro digestibility, and leaf to stem ratio of alfalfa. Agron. J. 63:719-24.

Lechtenberg, VL, WH Smith, SD Parsons, and DC Petritz. 1974. Storage and feeding of large hay packages for beef cows. J. Anim. Sci. 39:1011-15.

Lechtenberg, VL, KS Hendrix, DC Petritz, and SD Parsons. 1979. Compositional changes and losses in large hay bales during outside storage. In Proc. Purdue Cow-Calf Res. Day, West Lafayette, Ind., 11-14.

Livingston, AL, RE Knowles, and GO Kohler. 1977. Nutrient changes during alfalfa wilting and dehydration. J. Agric. Food Chem. 25:779-83.

Macdonald, AM, and EA Clark. 1987. Water and quality loss during field drying of hay. Adv. Agron. 41:407-37.

McGechan, MB. 1990. A review of losses arising during conservation of grass forage: Part 2, storage losses. J. Agric. Eng. Res. 45:1-30.

Madelin, TM, AF Clarke, and TS Mair. 1991. Prevalence of serum precipitating antibodies in horses to fungal and thermophilic actinomycete antigens: Effects of environmental challenge. Equine Vet. J. 23:247-52.

Miller, LG, DC Clanton, LF Nelson, and OE Hoehne. 1967. Nutritive value of hay baled at various moisture contents. J. Anim. Sci. 26:1369-73.

Miller, RC. 1946. Air flow in drying baled hay with forced ventilation. Agric. Eng. 27:203-8.

Minson, DJ. 1977. Predicting forage intake by laboratory methods. In Proc. 13th Int. Grassl. Congr., Leipzig, Germany, 516-21.

Monteith, JL. 1973. Principles of Environmental Physics. In EJW Barrington and AJ Willis (eds.).

London: Edward Arnold, 79-99.

Moore, KJ, and VL Lechtenberg. 1987. Chemical composition and digestion in vitro of orchardgrass hay ammoniated by different techniques. Anim. Feed Sci. and Tech. 17:109-19.

Moore, KJ, VL Lechtenberg, KS Hendrix, and JM Hertel. 1983. Improving hay quality by ammoniation. In JA Smith and VW Hays (eds.), Proc. 14th Int. Grassl. Congr., Lexington, Ky. Boulder, Colo.: Westview, 626-29.

Moore, KJ, VL Lechtenberg, and KS Hendrix. 1985a. Quality of orchardgrass hay ammoniated at different rates, moisture concentrations, and treatment durations. Agron. J. 77:67-71.

Moore, KJ, VL Lechtenberg, RP Lemenager, JA Patterson, and KS Hendrix. 1985b. In vitro digestion, chemical composition and fermentation of ammoniated grass and grass-legume silage. Agron. J. 77:758-63.

Moseley, G. 1983. Variation in the epicuticular wax content of white and red clover leaves. Grass and Forage Sci. 38:201-4.

Moser, LE. 1980. Quality of forage as affected by post-harvest storage and processing. In CS Hoveland (ed.), Crop Quality, Storage, and Utilization. Madison, Wis.: American Society of Agronomy, 227-60.

Nash, MJ. 1985. Crop Conservation and Storage in Cool Temperate Climates. Oxford, England: Pergamon Press.

Nash, MJ, and DL Easson. 1977. Preservation of moist hay with propionic acid. J. Stored Prod. Res. 13:65-75.

Owen, IG, and D Wilman. 1983. Differences between grass species and varieties in rate of drying at 25°C. J. Agric. Sci. Camb. 100:629-36.

Panciera, MT, RP Walgenbach, and RJ Bula. 1989. Cation radius and pH of drying agent solutions influence on alfalfa drying rates. Agron. J. 81:174-78.

Parker, BF, GM White, MR Lindley, RS Gates, M Collins, S Lowry, and TC Bridges. 1992. Forced-air drying of baled alfalfa hay. Trans. Am. Soc. Agric. Eng. 35:607-15.

Pattey, E, P Savoie, and PA Dube. 1988. The effect of a hay tedder on the field drying rate. Can. Agric. Eng. 30:43-50.

Pitt, RE. 1982. A probability model for forage harvesting systems. Trans. Am. Soc. Agric. Eng. 25:549-62.

———. 1990. Silage and Hay Preservation. Cornell Univ. Coop. Ext. Bull. NRAES-5. Ithaca, N.Y.: Northeast Regional Agricultural Engineering Service.

Rees, DVH. 1982. A discussion of sources of dry matter loss during the process of haymaking. J. Agric. Eng. Res. 27:469-79.

Rotz, CA, and DJ Sprott. 1984. Drying rates, losses and fuel requirements for mowing and conditioning alfalfa. Trans. Am. Soc. Agric. Eng. 27:715-20.

Rotz, CA, and Y Chen. 1985. Alfalfa drying model for the field environment. Trans. Am. Soc. Agric. Eng. 28:1686-91.

Rotz, CA, RJ Davis, DR Buckmaster, and JW Thomas. 1988. Bacterial inoculants for preservation of alfalfa hay. J. Prod. Agric. 1:362-67.

Rucker, G, and O Knabe. 1980. Non-mechanical field losses in wilting grasses as influenced by different factors. In Proc. 13th Int. Grassl. Congr., Leipzig, Germany, 1379-81.

Russell, JR, and DR Buxton. 1985. Storage of large round bales of hay harvested at different moisture concentrations and treated with sodium diacetate and/or covered with plastic. Anim. Feed Sci. and Tech. 13:69-81.

Russell, JR, SJ Yoder, and SJ Marley. 1990. The effects of bale density, type of binding and storage surface on the chemical composition, nutrient recovery and digestibility of large round hay bales. Anim. Feed Sci. and Tech. 29:131-45.

Savoie, P. 1988. Hay tedding losses. Can. Agric. Eng. 30:39-42.

Savoie, P, and A Mailhot. 1986. Influence of eight factors on the drying rate of timothy hay. Can. J. Agric. Eng. 28:145-48.

Scales, GH, RA Moss, and BF Quin. 1978. Nutritive value of round hay bales. N.Z. J. Agric. 137:52-53.

Sheaffer, CC, and NA Clark. 1975. Effects of organic preservatives on the quality of aerobically stored high moisture baled hay. Agron. J. 67:660-62.

Terrill, TH, WR Windham, CS Hoveland, and HE Amos. 1989. Forage preservation method influences on tannin concentration, intake, and digestibility of sericea lespedeza by sheep. Agron. J. 81:435-39.

Tetlow, RM. 1983. The effect of urea on the preservation and digestibility in vitro of perennial ryegrass. Anim. Feed Sci. and Tech. 10:49-63.

Tetlow, RM, and JS Fenlon. 1978. Pre-harvest desiccation of crops for conservation. 1. Effect of steam and formic acid on the moisture concentration of lucerne, ryegrass and tall fescue before and after cutting. J. Br. Grassl. Soc. 33:213-22.

Tomes, NJ, S Soderlund, J Lamptey, S Croak-Brossman, and G Dana. 1990. Preservation of alfalfa hay by microbial inoculation at baling. J. Prod. Agric. 3:491-97.

Uden, P. 1988. The effect of grinding and pelleting hay on digestibility, fermentation rate, digesta passage and rumen and faecal particle size in cows. Anim. Feed Sci. and Tech. 19:145-57.

Verma, LR, and BD Nelson. 1983. Changes in round

bales during storage. Trans. Am. Soc. Agric. Eng. 26:328-32.

Verma, LR, BD Nelson, and CR Montgomery. 1985. High moisture ryegrass preservation with ammonia. In Proc. 15th Int. Grassl. Congr., Kyoto, Japan, 992-94.

Webster, AJF, AF Clarke, TM Madelin, and CM Wathes. 1987. Air hygiene in stables 1: Effects of stable design, ventilation and management on the concentrations of respirable dust. Equine Vet. J. 19:448-53.

Wheeler, JL, and C Mulcahy. 1989. Consequences for animal production of cyanogenesis in sorghum forage and hay—a review. Trop. Grassl. 23:193-201.

Wilkins, RJ. 1988. The preservation of forages. In ER Orskov (ed.), Feed Science. New York: Elsevier, 231-55.

Willard, CJ. 1926. Do legume leaves hasten the curing process by pumping moisture from the stems? J. Am. Soc. Agron. 18:369-75.

Wilman, D, and IG Owen. 1982. Effects of stage of maturity, nitrogen application and swath thickness on the field drying of herbage to the hay stage. J. Agric. Sci. Camb. 99:577-86.

Wittenberg, KM, and SA Moshtaghi-Nia. 1990. Influence of anhydrous ammonia and bacterial preparations on alfalfa forage baled at various moisture levels. I. Nutrient composition and utilization. Anim. Feed Sci. and Tech. 28:333-44.

———. 1991. Influence of anhydrous ammonia and bacterial preparations on alfalfa forage baled at various moisture levels. II. Fungal invasion during storage. Anim. Feed Sci. and Tech. 34:67-74.

Wood, JGM, and J Parker. 1971. Respiration during the drying of hay. J. Agric. Eng. Res. 16:179-91.

Woolford, MK. 1984. The antimicrobial spectra of organic compounds with respect to their potential as hay preservatives. Grass and Forage Sci. 39:75-79.

Woolford, MK, and RM Tetlow. 1984. The effect of anhydrous ammonia and moisture content on the preservation and chemical composition of perennial ryegrass hay. Anim. Feed Sci. and Tech. 11:159-66.

12

Silage: Basic Principles

KEITH K. BOLSEN
Kansas State University

SILAGE is the feedstuff produced by the fermentation of a crop (or green forage) of high moisture content, generally greater than 50%. Material ensiled below 50% moisture is commonly referred to as *haylage*. *Ensiling* is the name given to the process, and the container (if used) is called a *silo*. The historical development of silage dates back to 1500-2000 B.C. In the Old Testament (Isaiah 30:24) it is said, "the oxen and young asses ate salted, seasoned, green fodder." The Roman historian Cato, in about 100 A.D., wrote, "The Teutons stored green fodder in a pit and covered it with dung." The modern era of silage began in 1877, when a French farmer, A. Goffart, published the first book on silage, which was based upon his own experiences of ensiling maize (*Zea mays* L.). About a year later, an English translation of his book was printed in the US, and this new preservation technique was quickly taken up by American farmers.

Since the early 1950s, both the total quantity of forage produced and the percentage of it preserved as silage have increased steadily in the US. In most of North America, silage making is much less weather dependent than hay making, and silage is more easily mechanized, is better suited to large-scale livestock production, and is adapted to a wider range of crops—i.e., corn, sorghums (*Sorghum bicolor* [L.] Moench), and winter or spring cereals. Knowledge of the biochemistry and microbiology of silage fermentation has increased tremendously in the second half of the twentieth century, as evidenced by several comprehensive reviews (McCullough 1978; McCullough and Bolsen 1984; Woolford 1984; McDonald et al. 1991).

Successful ensiling maximizes nutrient preservation. Good silage is achieved by harvesting the crop at the proper stage; minimizing the activities of plant enzymes and undesirable, epiphytic microorganisms (i.e., those naturally present on forage crops); and encouraging the dominance of lactic acid bacteria (LAB). In the initial stages, plant respiratory enzymes use the oxygen (O_2) entrapped in the ensiled mass to oxidize water-soluble carbohydrates (WSC), producing heat and using up sugars that would otherwise be available for fermentation. Plant proteases hydrolyze proteins to nonprotein nitrogen (NPN), amino acids, peptides, and ammonia (NH_3). The undesirable microorganisms, primarily those in the family Enterobacteriaceae, clostridial spores, and yeasts and molds, compete with LAB for WSC, and many of their end products have no preservative action. Clostridia can cause secondary fermentation, which converts lactic acid to butyric acid and degrades amino acids to amines and NH_3. Yeasts and molds, especially lactic acid-assimilating yeasts, are also linked to aerobic deterioration, particularly when the silage is fed. The following section examines, in depth, the basic principles of producing good silage.

KEITH K. BOLSEN is Professor of Forage Preservation and Cattle Nutrition, Kansas State University. He received his MS degree from the University of Illinois and his PhD from the University of Nebraska. His research includes silage production, fermentation, preservation, and nutritional value studies, with special emphasis on agronomic performance of silage crops, additives, microbiology, and in-silo losses.

PRINCIPLES OF SILAGE PRESERVATION

The major processes for every silage can be divided into four phases: (1) aerobic, (2) fermentation, (3) stable, and (4) feedout. Each phase has distinctive characteristics that must be considered to maintain forage quality throughout the harvesting, silo-filling, storing, and feeding periods.

Aerobic Phase. As the forage enters the silo, two principal processes are occurring: respiration and proteolysis, which are both attributed to the activities of plant enzymes. Respiration is the complete breakdown of plant sugars to carbon dioxide and water, the breakdown using O_2 and releasing heat, as shown by the following reaction:

$$C_6 H_{12} O_6 + 6 O_2 \rightarrow 6 CO_2 + 6 H_2O + \text{heat}.$$
$$\text{(sugar)} \quad \text{(oxygen)} \quad \text{(carbon dioxide)} \quad \text{(water)}$$

During the wilting process after cutting, respiratory activity of the plant continues. Once in the silo, aerobic and facultative anaerobic microorganisms, such as yeasts, molds, and certain bacteria that occur naturally on the forage can reach high enough populations to be significant sources of respiration. The objective for good silage is to use up the O_2 quickly so the anaerobic process can continue.

Harvesting crushes and chops the forage, damaging the cells and releasing plant enzymes. Some of these enzymes, amylases and hemicellulases, break down starch and hemicellulose, respectively, increasing the level of sugars in the ensiled material. Simultaneously, plant protease enzymes degrade proteins into peptides, amino acids, amides, and NH_3 (the process of proteolysis).

The loss of sugar through respiration is crucial from the standpoints of both silage preservation and nutritional value. Sugars are the principal substrates for the LAB that produce the acids to lower the pH and preserve the crop. The loss of sugar also reduces the energy value of the forage and indirectly increases the concentration of fiber constituents. Prolonged aerobic conditions also allow yeasts and molds to grow to high levels, which can predispose the silage to heating when the silo is opened for feedout. Finally, the heat produced by respiration raises the temperature of the ensiled mass.

Elevated temperatures increase the rate and extent of protein breakdown to soluble NPN components. Excessive temperatures (above 42°-44°C) can result in maillard or browning reactions. In the maillard reaction, sugars and free amino acids are formed into polymers that are slow to digest and are measured as acid detergent fiber (ADF) and acid detergent insoluble N. In all silages, excessive heating will significantly reduce digestibility of protein, fiber, and other nutrient components. In low-moisture silages, there is the additional risk of fire in the silo. Proteolysis is a negative activity, particularly for high-producing dairy cows, because excess soluble NPN can result in lower milk production.

The negative effects of the aerobic phase on silage quality can be minimized by rapid filling, proper packing, and effective sealing of the ensiled material to exclude air and water. The main aerobic losses come from exposure to air before a given layer of forage is covered by a sufficient quantity of additional forage to effectively separate it from the atmosphere or by a relatively impermeable cover, such as polyethylene.

Fermentation Phase. Once anaerobic conditions are reached in the ensiled material, several processes begin. The intact plant cells start to lyse or break down under anaerobic conditions. In wet forages, this process occurs within several hours, whereas in drier forages, it can extend to over a day or more. This lysis of the plant cells has both positive and negative aspects. It provides sugars for the LAB to ferment to organic acids. It also releases numerous plant enzymes like those released by the chopping process. The enzymes that degrade polysaccharides such as starch are beneficial and provide additional sugars for fermentation. However, the proteolytic enzymes that are released break down protein to soluble NPN components.

The other negative aspect is the production of effluent. If forages are ensiled at 30%-35% dry matter (DM) or more, effluent is rarely a problem, except in very large tower or bunker silos. Below this DM range, the amount of effluent produced is related to the DM content of the forage and the degree of packing or pressure (Pitt and Parlange 1987). The effluent contains soluble sugars, organic acids, minerals, proteins, and other NPN components. If the effluent is not collected and used, it represents a loss of readily digestible nutrients from the silage. If the effluent drains into a lake or stream, "fish kills" and other environmental damage can result. After the effluent is collected, it can be either fed to cattle or spread on fields as a fertilizer.

With the onset of anaerobic conditions,

anaerobic microorganisms begin to grow and reproduce rapidly. The microorganisms of greatest concern in silage preservation are LAB, Enterobacteriaceae, yeasts and molds, and clostridial spores (McDonald et al. 1991). The populations of epiphytic microorganisms on forage crops are quite variable and are affected by forage species, stage of maturity, weather, mowing, field wilting, and the chopping process. Alfalfa (*Medicago sativa* L.) and corn are the two major silage crops in the US. Although their chemical compositions are fairly well-known, scant information is available about the microflora dwelling on them.

Recent studies identify and enumerate the microflora on alfalfa and whole-plant corn and investigate the factors that influenced them. Lin et al. (1992) studied four cuttings of alfalfa, each harvested at three stages of maturity, and three corn hybrids (Table 12.1). The numbers of epiphytic microorganisms were higher on standing corn than on alfalfa, and enterobacteria were predominant on both crops. Yeasts and molds also were major epiphytes on whole-plant corn. The LAB constituted only a small proportion of the total population (<0.5%) on both crops. Clostridial spores were not detected on standing alfalfa, and occurrences of these spores on standing corn plants were due to soil contamination from rainfall prior to harvest.

The chopping process tended to increase the microflora numbers compared with those on the standing crops, and the LAB population was most enhanced. This phenomenon was explained earlier as inoculation from the harvesting machine and as microbial multiplication in the plant juices liberated during harvest (McDonald et al. 1991). However, recent findings of Pahlow (1991) demonstrate that these large increases in microflora numbers were impossible to achieve (1) by microbial proliferation and growth because the time involved was too short or (2) by contamination from harvesting equipment, which could occur in the first load but not in later loads. A new "somnicell" hypothesis proposes that bacteria assume a viable stage on the surface of intact plants, but that stage is not culturable in the laboratory. The chopping process activates the previously dormant population by releasing plant enzymes (i.e., catalase and superoxide dismutase) and manganese compounds (Pahlow 1991).

Because forages are preserved by lactic acid, the most important microorganisms to the ensiling process are LAB. They ferment sugars to lactic acid primarily but also produce some acetic acid, ethanol, carbon dioxide, and other minor products. This is a rather large group of bacteria, which includes species in six genera (Table 12.2). They are divided into two categories: the homofermentative LAB produce only lactic acid from fermenting glucose and other six-carbon (6-C) sugars, whereas the heterofermentative LAB produce ethanol, acetic acid, and carbon dioxide in addition to lactic acid.

Because lactic acid is a much stronger acid than acetic acid and reduces the pH faster, homofermentative LAB are more desirable than heterofermentative. With a natural fermentation, competition between strains of LAB determines how homofermentative the fermentation will be. Use of bacterial inoculants, as discussed later in this chapter, can help develop a homofermentative fermentation, even under some crop and environmental conditions when the natural LAB population is high.

TABLE 12.1.   Epiphytic microflora on alfalfa and corn

Crop and microflora	Standing	Windrow	Chopped
		(log$_{10}$ cfu/g of crop)[c]	
Alfalfa			
Lactic acid bacteria (LAB)[a]	3.76	3.35	5.10
Enterobacteriaceae	6.06[d]	6.34[d,e]	6.53[e]
Yeasts and molds	5.07	5.26	5.35
Lactate-assimilating yeasts	4.19	3.78	4.37
Clostridial spores	ND[b]	1.32	1.93
Corn			
Lactic acid bacteria (LAB)[a]	4.22	. . .	6.31
Enterobacteriaceae	6.87	. . .	7.49
Yeasts and molds	6.85	. . .	7.12
Lactate-assimilating yeasts	6.36	. . .	6.65
Clostridial spores	1.97	. . .	2.88

Source: Lin et al. (1992).
[a]Included only LAB from the *Lactobacillus, Pediococcus,* and *Leuconostoc* genera.
[b]ND = not detected.
[c]cfu = colony-forming unit, i.e., a group of microorganisms on solid medium arising from a single cell.
[d,e]Means in the same row with different superscripts differ (*p* < .05).

TABLE 12.2. Lactic acid bacteria of importance in the ensiling process and their fermentation products

Genus	Species	Glucose fermentation
Lactobacillus	acidophilus casei cornyiformis curvatus plantarum salivarus	Homofermentative[a]
	brevis buchneri fermentum viridescens	Heterofermentative[b]
Pediococcus	acidilactici cerevisiae pentosaceus	Homofermentative
Enterococcus	faecalis faecium	Homofermentative
Lactococcus	lactis	Homofermentative
Streptococcus	bovis	Homofermentative
Leuconostoc	mesenteroides	

Source: McDonald et al. (1991).

[a]Microorganisms that ferment sugars to predominantly lactic acid.

[b]Microorganisms that ferment sugars to a variety of organic acidds, ethanol, and carbon dioxide.

Homofermentative fermentation of sugars results in little or no loss of DM and only small losses of energy (Table 12.3). Even a totally heterofermentative fermentation will rarely result in more than a 5% loss of DM, and gross energy loss is negligible.

Indirectly, the fermentation process metabolizes a small portion of the ensiled material to preserve other nutritional components in

TABLE 12.3. Pathways of importance in the ensiling process

Pathway: Substrate	End products	Substrate recovery, % Dry matter	Substrate recovery, % Energy
Homofermentative			
1 glucose	→ 2 lactic acid	100	97
Heterofermentative			
1 glucose	→ 1 lactic acid + 1 ethanol + 1 CO_2	76	97
3 fructose	→ 1 lactic acid + 2 mannitol + 1 acetic acid + 1 CO_2	95	98
Clostridial			
2 lactic acid	→ 1 butyric acid + 2 CO_2 + 2 H_2	49	81
2 alanine	→ 2 propionic acid + 1 acetic acid + 2 NH_3 + 1 CO_2	78	81
Yeast			
1 glucose	→ 2 ethanol + 2 CO_2	51	97

Source: McDonald (1981).

the silage. Quickly lowering the pH reduces the activity of proteolytic enzymes, sparing proteins. The low pH also stops the growth of other anaerobic bacteria, such as enterobacteria, *Bacillus* spp., clostridia, and *Listeria*. Finally, low pH also increases the rate of chemical hydrolysis of some polysaccharides, such as hemicellulose, which can lower the fiber content of the ensiled forage.

However, two groups of bacteria, enterobacteria and clostridia, can have a significant negative impact on silage quality. Enterobacteria affect silage during the first few days in the silo, whereas clostridial spores typically affect silage after the LAB have used up the sugars in the ensiled forage, but the pH is not low enough to arrest their activity.

The enterobacteria include a number of genera: *Escherichia, Klebsiella,* and *Erwinia*. These bacteria are anaerobic but can grow in the presence of O_2 like most LAB. They also ferment sugars and produce lactic acid, acetic acid, and ethanol, as do the LAB. However, their main product of fermentation is acetic acid, which slows the rate of pH decline in the ensiled material and increases the loss of DM during the fermentation phase. In addition, high levels of acetic acid can reduce silage DM intake in beef and dairy cattle.

Enterobacteria can be minimized by a rapid fermentation by LAB because the optimum pH for their growth is pH 6.0-7.0, and most strains of Enterobacteriaceae will not grow at a pH below 5.0. Consequently, the population of enterobacteria, which is usually high in the ensiled crop, is active during the first 12-24 h of ensiling. Then their numbers decline rapidly, so they are not a factor after the first few days of the fermentation phase.

Growth of clostridial spores can have a pronounced effect on silage quality. Because these bacteria require anaerobic conditions to grow, they are not normally active in the presence of O_2. Clostridia can be divided into two groups: (1) those that ferment sugars and organic acids (like lactic acid) and (2) those that ferment free amino acids (Table 12.3). Both groups affect silage quality negatively. The main end product of sugar and lactic acid fermentation is butyric acid, and there is a 50% loss of ensiled DM and almost a 20% loss of ensiled energy from this fermentation. Thus, substantial losses can occur in the quantity and quality (reduced digestibility and increased fiber components) of the silage.

The proteolytic clostridia ferment amino acids to a variety of products, including NH_3, amines, and volatile organic acids, which are of poor nutritional value. Some of the fermentation pathways result in significant losses of DM and energy, whereas other pathways result in little DM loss. The most significant effect of these clostridia is the production of amines, which are known to decrease silage DM intake.

Like the enterobacteria, clostridial spores are sensitive to low pH, and clostridia require wet conditions for active development. In general, clostridial growth is rare in crops ensiled with less than 65% moisture because there are usually sufficient sugars to quickly reduce the pH below 4.6-4.8, at which point clostridia will not grow. For wetter silages (70% moisture or more), reducing the pH to less than 4.6 is the only practical means of preventing the growth of these bacteria with today's technology.

The period of active fermentation lasts from 7 to 30 d. Forages ensiled at wetter than 65% moisture ferment rapidly, whereas fermentation is quite slow when the moisture content is below 50%, such as in wilted haylages. For forages ensiled in the normal moisture range (55%-65%), active fermentation is over within 7 to 21 d. At this point, fermentation of sugars by LAB has ceased, either because the low pH (below 4.0-4.2) stopped their growth or there was a lack of sugars for fermentation.

Stable Phase. Following the active growth of LAB, the ensiled material enters the stable phase. If the silo is properly sealed, little biological activity occurs in this phase. Very slow rates of chemical breakdown of hemicellulose can occur, however, releasing some sugars. If active fermentation ceased because of a lack of WSC, the LAB might ferment the sugars released by hemicellulose breakdown, causing a slow rate of pH decline during the stable phase.

The major factor affecting silage quality during the stable phase is the permeability of the silo to air. Oxygen entering the silo is used by aerobic microorganisms (via microbial respiration), causing increases in yeast and mold populations, losses of silage DM, and heating of the ensiled mass. As in the silo-filling period, this respiration results in reduced nutrient digestibilities and increased fiber content.

The amount of aerobic losses in this phase is related to the permeability of the silo and the density of the silage. If the silage is left unsealed, substantial DM losses can occur at the exposed surface. These losses can be reduced by covering the surface of the ensiled material with polyethylene sheeting, whether in vertical tower or horizontal bunker, trench, or stack silos. Oxygen can pass through polyethylene but at a very slow rate. Cracks in the silo wall or holes in the polyethylene seal obviously increase the rate at which O_2 can penetrate the silage mass.

Feedout Phase. When the silo is opened for feedout, O_2 usually has unrestricted access to the silage at the face. During this phase, the largest losses of DM and nutrients can occur because of aerobic microorganisms consuming sugars, fermentation products such as lactic and acetic acids, and other soluble nutrients in the silage. These soluble components are respired to carbon dioxide and water, producing heat. Yeasts and molds are the most common microorganisms involved in the aerobic deterioration of the silage, but bacteria such as enterobacteria and *Bacillus* spp. have also been shown to be important in some circumstances. Besides the loss of highly digestible nutrients in the silage, some species of molds can produce mycotoxins and/or other toxic compounds that can affect animal health.

This microbial activity is the same as that occurring because of O_2 infiltration during the stable phase. The major difference is the amount of O_2 available to the microorganisms. At feedout, the microorganisms at the silage face have unlimited quantities of O_2, allowing them to grow rapidly. Once yeasts or bacteria reach a population of 10^7-10^8 colony-forming units (cfu, i.e., a group of microorganisms on solid medium arising from a single cell)/g of silage, or once molds reach 10^6-10^7 cfu/g, the silage will begin to heat, and digestible components, like sugars and fermentation products, will be quickly lost. The time required for heating to occur is dependent on several factors including (1) numbers of aerobic microorganisms in the silage, (2) time exposed to O_2 prior to feeding, (3) silage fermentation characteristics, and (4) ambient temperature.

NUMBERS OF AEROBIC MICROORGANISMS. The higher the aerobic microorganism populations in the silage at feedout, the more quickly the silage will begin to heat. Filling and sealing the silo quickly provide the maximum level of sugars for fermentation by LAB, creating an environment unfavorable for the proliferation of aerobic microorganisms. Good sealing is also important to reduce the amount of O_2 diffusing into the silo during storage, which also minimizes aerobic microbe numbers.

TIME EXPOSED TO OXYGEN. The length of time the silage is exposed to O_2 prior to feedout is determined primarily by two factors: the density of the silage mass and the feedout rate. The less dense the silage, the greater the distance from the face that O_2 can penetrate. Porosity is determined by the amount of packing and the moisture content of the ensiled crop. Drier forages are more difficult to pack, leaving the silage more porous for a given silo pressure or degree of packing. Higher feeding rates are required with more porous material to reduce the potential of heating and substantial aerobic losses.

FERMENTATION CHARACTERISTICS. The most important of the silage characteristics is acid content. The volatile fatty acids (acetic, propionic, and butyric) suppress the growth of yeasts and molds. Butyric acid is the most inhibitory of the three; acetic acid, the least inhibitory. The strong antifungal characteristics of butyric acid mean that clostridial silages, while not as desirable, are usually very stable and might not heat even with prolonged exposure to air. Lactic acid can also inhibit the growth of yeasts and molds but is not as effective as butyric or acetic acid, because some yeasts can utilize lactic acid and reduce its effectiveness.

Overall, the greater the level of desirable fermentation products (lactic and acetic acids), the more aerobically stable the silage should be. This partly explains why a highly buffered crop like alfalfa, which requires the production of large quantities of fermentation products, is less prone to heating than corn silage, which ferments readily with lower levels of fatty acids such as lactic and acetic.

Low silage pH also slows the rate of heating. Yeasts and molds are relatively insensitive to pH, and many species will grow even at pH 4.0 and below. However, as concentrations of lactic or volatile fatty acids increase, the pH is lower and the acids are more inhibitory for yeasts and molds. Thus, fermentation to a low pH is also beneficial for providing aerobic stability to a silage during feedout.

Residual sugar content of the silage can also be important during the feedout phase. Yeasts and molds grow approximately twice as fast on sugars as on fermentation products, such as lactic and acetic acids, and ethanol. This means that utilizing all the sugars during the fermentation phase has some benefit, as long as sufficient fermentation has occurred to preserve the ensiled material. Alfalfa silages in the 60%-70% moisture range generally use up the available sugars, which contributes to their aerobic stability during feedout, whereas corn silages often have excess sugars, which partly cause their aerobic instability.

Moisture content also affects heating of the silage. First, if the crop is wetter, a greater mass (DM plus water) per unit of DM lost during deterioration will be there to absorb the heat produced by microbial respiration. Second, the heat capacity of water is 2.2 times higher than that of the dry forage. Therefore, the wetter the ensiled crop, the more heat will be required to raise the silage temperature, and the slower the rate of DM loss once heating begins.

AMBIENT TEMPERATURE. Microbial growth rates increase exponentially with temperature up to approximately 55°C. This means that heating occurs much more quickly during summer than during winter, and higher feeding rates are needed during the summer to prevent feedout losses.

Under farm conditions, feedout losses are largely a function of silage management practices. There are few data to quantitate feedout losses in farm silos, but laboratory studies indicate that DM losses are about 1.5%-3.0% per day for each 8°-12°C rise in the silage temperature above ambient (Woolford 1984). A fast filling rate and tight sealing of the silo minimize the buildup of aerobic microorganisms in the silage and maximize the production of fermentation products that will inhibit their growth. Adequate packing of the ensiled material reduces the distance that O_2 can penetrate the silage face at feedout. Finally, feeding rate and silage density determine the length of time the silage is exposed to O_2 prior to feedout, and the shorter the exposure time, the less likely a silage is to heat during the feedout phase.

FACTORS AFFECTING PRESERVATION EFFICIENCY AND SILAGE QUALITY

The ensiling process is influenced by numerous biological and technological factors (Table 12.4). Because many of these factors are interrelated, it is difficult to discuss their significance individually. However, there are two dominant features to be considered for all silages: (1) the crop and its stage of maturity (Table 12.5) and (2) the technology (management and know-how) imposed by the silage maker.

TABLE 12.4.   Factors influencing the ensiling process and silage quality

Biological (genotype and ecological)		Technological (management and "know-how")	
Crop characteristics	Epiphytic microflora	Improving suitability	Storage conditions
Hybrid or cultivar	Substrate	Field wilting	Silo construction
DM content	Climate/weather	Additives	Rate of filling
WSC content	Soiling	Mechanical treatment	Compaction and density
Buffering capacity	Additives		Sealing method
Plant structure			Air and oxygen access
Stage of maturity			Temperature and insulation
Weather			Mechanical treatment
Effective harvest time			
Additives			

Source: Bolsen (1985).

TABLE 12.5.   Harvest recommendations for various silage crops

Crop	Stage of maturity[a]
Corn	Kernels at two-thirds milk line to black layer
Sorghum, grain and forage type	Kernels at mid- to late dough
Cereal grains	Late boot to early-flowering or kernels at mid- to late dough
Alfalfa	Late bud to early bloom
Red clover	Early bloom
Summer annual grasses	Late vegetative to late boot
Bermudagrass	38-46 cm plant height
Orchardgrass, smooth bromegrass, and timothy	Boot to early heading
Grass mixtures	Boot to early heading
Legume-grass mixtures	Grasses at boot to early heading

Sources: Noller and Thomas (1985) and Holland and Kezar (1990).
[a]Applies to the first harvest. Most grasses do not head after the first harvest.

Silage Crops. Whole-plant corn is recognized as the "nearly perfect" silage crop! Its nutrient digestibility and DM yield per hectare plateau near maximum between the 80% milk line and 7-10 d post-black layer stages. Maturity of corn silage can best be determined by the location of the milk line. This line is the interface between the liquid endosperm (milk) and solid endosperm (starch) portions of the kernel, and it will not appear until the corn is at the dent stage of maturity. Black layer formation is a visible black demarkation at the base of the kernel that indicates transport of nutrients to the kernel has stopped. The DM content of corn is usually in an acceptable range (28%-42%) for about 2 wk during the harvest season, it has an adequate WSC content and a low buffering capacity, and its high proportion of grain contributes to a high-density silage.

Sorghums are becoming increasingly more important silage crops in the US, and many grain and forage sorghums have silage yields comparable to those of corn. Under dryland conditions, it was found that grain sorghum is equivalent to corn for silage production, and

grain sorghum silages are of similar nutritional quality to corn silages (Suazo et al. 1993). However, mid- to late-season forage sorghums are often harvested and ensiled at a low DM content, which results in high DM losses during storage and low silage intakes by beef or dairy cattle (White et al. 1988; Sonon et al. 1991). Some forage sorghum cultivars have ADF contents above 40% of the DM, and nutritive values are well below those of corn or grain sorghum silages, perhaps because some forage types have very low grain production, which detracts from the feeding value.

The following are grown for silage in the US: numerous legumes, i.e., alfalfa, red clover (*Trifolium pratense* L.), white clover (*T. repens* L.), and other perennial clovers; grasses, i.e., bermudagrass (*Cynodon dactylon* [L.] Pers.), orchardgrass (*Dactylis glomerata* L.), smooth bromegrass (*Bromus inermis* Leyss.), summer annuals, tall fescue (*Festuca arundinacea* Schreb.), and timothy (*Phleum pratense* L.); and cereal grains, i.e., barley (*Hordeum vulgare* L.), oat (*Avena sativa* L.), rye (*Secale cereale* L.), triticale (X *Triticosecale* Wittmack), and wheat (*Triticum aestivum* L. emend. Thell.) (Noller and Thomas 1985). Of these, alfalfa is the most widely used, but it is also the most difficult to ensile.

Alfalfa has a very high buffering capacity due to its high concentrations of organic acids, minerals, and protein, and it has a low WSC content compared to other hay crops. It must be field wilted to 35%-50% DM before ensiling. If alfalfa is ensiled too wet, it can be predisposed to a clostridial fermentation with high losses of DM and energy and protein quality in addition to extensive production of effluent. If alfalfa is ensiled too dry, it can undergo an extended aerobic phase that utilizes the WSC before the fermentation phase. This results in excessive heating and molding and reduced energy and protein digestibilities.

Stressed Crops. The yield and quality of drought- and frost-damaged crops are usually

Fig. 12.1.   Harvesting whole-plant grain sorghum for silage in the late dough stage of kernel maturity. *Kansas State Univ photo.*

maximized when harvested as silage. This is also true for crops that are immature because of late planting or unfavorable growing conditions (Holland and Kezar 1990). Although the DM yields of drought-damaged corn and sorghum silages can be quite low, their feeding values are 75%-95% of normal silages.

NITRATE TOXICITY. The potential for high nitrate concentrations occurs when crops such as corn, sorghum, and some grasses are exposed to drought, hail, frost, cloudy weather, or soil fertility imbalance. Nitrates accumulate in the lower portion of the plant when stresses reduce the crop yield to less than that expected based on the supplied N fertility level. Nitrates are responsible for the production of a lethal gas in the silo and nitrogen dioxide. When fed to livestock, nitrates interfere with the ability of the blood to carry O_2.

Plants in the field that recover from the stress situation will convert nitrates to a nontoxic form. When harvesting stressed crops, a 20-25 cm stubble should be left, and if it rains, 3 d should be allowed before resuming the harvest. A nitrate analysis prior to feeding is advised for all silages made from stressed forages. Silages should be analyzed before and after ensiling because the fermentation process converts only about 50% of the nitrates to a nontoxic form.

Livestock feeding programs should be modified if the silage contains more than 1000 ppm of nitrate N. Silages with more than 2000 ppm of nitrate N should not be fed to pregnant animals, and silages containing over 4000 ppm should not be fed. The stressed silage should be diluted with low-nitrate forage or grain, and use of nonprotein N supplements such as urea or ammonia should be avoided.

PRUSSIC ACID. Prussic acid can accumulate in sorghum and sudangrass (*Sorghum bicolor* [L.] Moench.) that grow rapidly following a stress situation. Prussic acid is degraded by the animal to release hydrogen cyanide (HCN), which affects the animal. Poisoning occurs when livestock consume young plants, drought-stressed plants, or damaged plants that are high in prussic acid. Minimum plant height before harvesting is 45-50 cm for sudangrass and 75-80 cm for sorghum-sudangrass hybrids. Harvest should be delayed until the late boot to early-heading stage for forage sorghum. The fermentation process does not decrease the prussic acid concentration in the silage; however, field wilting or field-drying before ensiling will release about 50%-70% of the prussic acid.

Nutrient Preservation. Silage nutrient losses (i.e., DM) are divided into two categories: (1) unavoidable and (2) avoidable (Table 12.6).

Unavoidable losses include field losses and those from plant respiration and primary fermentation. Avoidable losses include those from effluent, secondary fermentation, and aerobic deterioration. Estimates of unavoidable losses generally range from 5%-15% of the DM, and those of avoidable losses generally range from 2%-25% of the DM.

The importance of quickly achieving and maintaining O_2-free conditions has led to improved equipment and techniques for precision chopping, better compaction, rapid filling, and complete sealing. Slow silo filling and delayed or inadequate sealing subject a silage to respiration losses higher than those expected, excessive surface wastes, and large aerobic deterioration losses during storage and feedout.

A series of experiments with grass or grass-legume silages in several European countries show in-silo losses were decreased as DM content of the ensiled forage increased (Zimmer and Wilkins 1984). When direct-cut silages made with formic acid as an additive were compared with wilted silages made without additives, average in-silo DM losses were 16.3% and 7.9%, respectively. In a summary of 25 silages made without additives and ensiled in 3-m by 15-m concrete stave tower silos, Bolsen et al. (1992) report in-silo DM losses of 9.8% for corn silages and 16.9% for forage sorghum silages. Presumably, the differences were due to the lower DM content of sorghum silages (31.4%) than of the corn silages (37.8%) and a longer fermentation phase for the sorghums. Construction of silos to minimize air entry, in combination with harvesting crops at the correct moisture content, will reduce both in-silo DM losses and variations in silage quality, particularly for hay-crop haylages and silages.

Surface Spoilage. During the past three decades, large horizontal silos (i.e., bunkers, trenches, and stacks) have become the most common means of storing large amounts of ensiled feeds (Pitt 1990; Rotz and Muck 1993). Although these structures are economically attractive, their design allows a high percentage of the silage to be exposed to environmental and seasonal weather effects. Because these structures are relatively shallow, 20%-25% of the original ensiled volume can be within the top 1 m.

Often, the top layer of ensiled material is protected by polyethylene sheeting weighted with tires. However, the protection provided is highly variable and depends on sealing techniques, especially along the edges, and the physical properties of the sheeting. Also, the labor required to apply and remove the sheeting and tires, or other weighting materials (i.e., soil, manure, or other organic waste), has deterred many silage makers from using a seal or cover of any kind. Only limited information is currently available to document the DM or organic matter (OM) losses at well-defined depths in commercial silos under field conditions.

Ashbell et al. (1990) report that the relationship between the ash content in a silage sample and the estimated additional spoilage loss of OM in the surface (in excess of that lost from the presumably well-preserved feedout face sample) can be expressed as

$$\text{Additional spoilage} = 1 - \left[\frac{(\text{AF} \times \text{OMS})}{(\text{AS} \times \text{OMF})}\right] \times 100$$

where AF = % ash in the face sample; OMF = % OM in the face sample; AS = % ash in the surface sample; and OMS = % OM in the surface sample. The relationship is based on the assumption that OM disappears as spoilage occurs, but the absolute amount of ash remains constant. In theory, regardless of the % ash in the face sample of silage at feedout, a small increase in ash content in the deteriorated silage sample would represent a large percentage unit increase in loss of OM.

Dickerson et al. (1992a) conducted two surveys to estimate OM loss from the top layer of silage in farm-scale, horizontal silos and to

TABLE 12.6. Silage DM losses

Process	Classified as	Approximate DM loss, %
Field loss by wilting (or effluent)	Unavoidable (or avoidable)	2 - >6
Respiration	Unavoidable	1 - 2
Fermentation	Unavoidable	2 - 7
Secondary fermentation	Avoidable	0 - >5
Aerobic deterioration during storage	Avoidable	2 - >10
Aerobic deterioration after unloading	Avoidable	0 - >10
Total		7 - >40

Source: Adapted from Zimmer (1980).

compare the losses from unsealed and sealed corn and forage sorghum silages. Only 22% of the silos were sealed with polyethylene sheeting. The top 1 m of silage from each of 30 horizontal silos was sampled at three locations across the width of the silo for 2 consecutive years. Losses of OM were estimated by using ash content as an internal marker as proposed by Ashbell et al. (1990). Sealing reduced spoilage losses of OM at the 0- to 50-cm depth by an average of 27 percentage units compared with losses from silages with no seal. Similarly, losses at the 50- to 100-cm depth were reduced by an average of 9 percentage units when silages were sealed.

In other studies using alfalfa, corn, and sorghum silages made in simulated, farm-scale, bunker silos, Dickerson et al. (1992b) report OM losses of 32%-72% within the 0- to 67-cm depth from the original surface in unsealed silos after 180 d of storage. In silos sealed immediately after filling with polyethylene sheeting, OM losses were reduced to 8%-14% within the 0- to 67-cm depth. When the silos were sealed 7 d after filling, the OM losses were 18%-24%.

Developing alternatives to polyethylene sheeting and more effective methods for protecting the top layer of silage in bunker, trench, and stack silos is essential. The protection mechanism must be environmentally friendly, safe, and easy to apply. It should have some nutritional value so producers can feed it with the silage to minimize labor and waste disposal. It should also shed water, exclude air, be pliable, and settle with the silage mass. And it should be durable enough to be effective for at least 1 yr.

SILAGE ADDITIVES

In contrast to large industrial fermentations, silage is made from heterogeneous raw crop materials with spontaneous fermentation by epiphytic microflora and under a wide range of environmental conditions. Major efforts have been made throughout the twentieth century to control silage fermentation with additives. These were recently categorized and their effects on silage quality summarized (Pitt 1990; McDonald et al. 1991; Muck and Bolsen 1991; Kung 1992; Muck 1993). Numerous silage additives are available in North America, and they can be separated into three broad types: (1) stimulants, (2) inhibitors, and (3) nutrient sources (Table 12.7).

Inoculants and Nutrient Sources. In the US, where corn is the principal silage crop and where alfalfa or grasses are easily wilted, bacterial inoculants have received fairly wide acceptance by farmers (Bolsen and Heidker 1985; Harrison 1989; Muck and Bolsen 1991; Mahanna 1993). Inoculants reduce risks in silage making and appear to have inherent advantages over other available additives, including low cost, safety in handling, low application rate, and no residue problems. Rather than depend entirely on the natural development of an efficient bacterial population to produce lactic acid, a logical approach is to add specific species and desirable strains of LAB that will dominate the fermentation phase of the ensiling process.

In a series of over 50 experiments conducted during a 3-yr period, the effects of commercial LAB inoculants were tested on both

TABLE 12.7. Types of silage additives and additive ingredients

Stimulants			Inhibitors[d]		Nutrient sources
Bacterial inoculants[a]	Enzymes[b]	Substrate sources[c]	Acids	Others	
Lactic acid bacteria	Amylases	Molasses	Formic	Ammonia	Ammonia
	Cellulases	Glucose	Propionic	Urea	Urea
	Hemicellulases	Sucrose	Acetic	Sodium chloride	Limestone
	Pectinases	Dextrose	Lactic	Carbon dioxide	Other minerals
	Proteases	Whey	Caproic	Sodium sulfate	
	Xylanases	Cereal grains	Sorbic	Sodium sulfite	
		Beet pulp	Benzoic	Sodium hydroxide	
		Citrus pulp	Acrylic	Formaldehyde	
			Hydrochloric	Paraformaldehyde	
			Sulfuric		

Sources: Holland and Kezar (1990), Pitt (1990), and Wilkinson (1990).

Note: Not all additives or ingredients used for silage are listed, not all listed are effective, and not all listed are approved for use on ensiled material intended for livestock feed.

[a]Most contain live cultures of LAB from the genus *Lactobacillus, Pediococcus, Enterococcus,* or *Streptococcus.*

[b]Most enzymes are microbial by-products having enzymatic activity.

[c]Most ingredients can also be listed under nutrient sources.

[d]Some inhibitors work aerobically, suppressing the growth of yeasts, molds, and aerobic bacteria; others work anaerobically, restricting undesirable bacteria (i.e., clostridia and enterobacteria), plant enzymes, and possibly LAB.

rate and efficiency of silage fermentation (Bolsen et al. 1988, 1990). Shown in Figure 12.2 are pH decline and lactic acid production for control and inoculated silages made from wheat harvested at the heading stage. Even though the chopped forage had almost 1 million cfu of epiphytic LAB/g, silages with inoculants had lower pH and higher lactic acid values at 1-, 2-, 4-, and 7-d post-filling. A summary of the results shows that inoculated silages had a lower pH, higher lactic acid content, higher ratio of lactic to acetic acid, and lower ethanol and NH_3 content than control silages.

The data also indicated that knowing only the numbers and species of LAB present on silage crops (i.e., corn, sorghum, wheat, sudangrass, and alfalfa) would not predict either the rate or efficiency of fermentation. Further identification of the epiphytic microflora is essential, the ratio of homofermentative to heterofermentative LAB must be determined, and the growth characteristics of the epiphytic LAB strains must be differentiated in a silage environment.

Other factors, including (1) crop characteristics (i.e., DM and WSC content, buffering capacity, physical structure, and cultivar/hybrid), (2) silage management techniques (i.e., length of chopping, filling rate, packing densi-

ty, and degree of sealing), and (3) climate conditions (i.e., growing season, ambient temperature, and humidity), undoubtedly play roles in both silage fermentation efficiency and response to bacterial inoculants.

Urea and anhydrous NH_3 have been added to low-protein silage crops like corn, forage sorghum, and mature winter cereals (Goodrich and Meiske 1985). The main reasons given for using these nonprotein N sources are to increase the crude protein equivalent of the crop, to reduce the cost of supplemental protein, and to improve aerobic stability of the silage.

Bolsen et al. (1992) summarize results from 26 trials at Kansas State University comparing DM recovery and beef cattle performance for inoculated silages and those treated with nonprotein N. In 23 of 26 trials, silages were made by the alternate-load method, and upright concrete stave silos were used in all but 1 trial. Products from 11 companies were used in the corn silage trials and products from eight companies in the sorghum silage trials. Statistical analysis of the data was conducted using a paired t-test. Only overall mean comparisons were made between paired observations for the nine criteria measured. Treatment means for control and treated silages are summarized in Table 12.8.

The 19 inoculated corn silages had a 1.3 percentage unit higher DM recovery compared with untreated silages, and the inoculated silages supported a 1.8% more efficient gain and a 1.8 kg increase gain/mt of crop ensiled. The addition of anhydrous NH_3 to corn silage increased pH and fermentation acids in the silages, and there was a strong trend for both DM recovery and gain/mt of crop ensiled to be lower, 2.1 percentage units and 3.1 kg, respectively.

When untreated and inoculated forage sorghums were compared, inoculants increased DM recovery, improved feed conversion, and produced 2.3 kg more gain/mt of crop ensiled. The forage sorghum silages treated with anhydrous NH_3 or urea were 5.1 percentage units lower in DM recovery, and cattle fed treated silages gained 0.12 kg/d more slowly, required 1.06 kg more DM/kg of gain, and gained 7.0 kg less/mt of crop ensiled compared with cattle fed untreated silage.

A urea-molasses blend had less of a negative influence on both silage preservation and cattle performance than did anhydrous NH_3 or urea (data not shown). The authors conclude that inoculants consistently improved fermentation, DM recovery, and gain/mt of crop ensiled in both crops. Nonprotein N com-

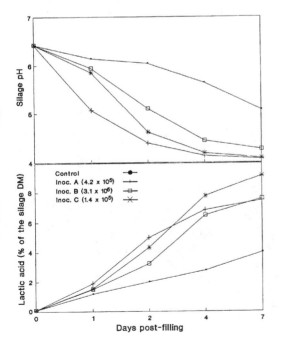

Fig. 12.2. pH decline and lactic acid production in wheat silages. Number of cfu of LAB supplied/g of forage for inoculants A, B, and C is given in parentheses.

TABLE 12.8.   Summary of treatment means for silage fermentation, DM recovery, and cattle performance from inoculant and NPN additions to corn and forage sorghum silages

Crop and silage treatment	Number of silages	DM recovery[a]	Average daily gain	Daily DM intake	DM/kg of gain	Gain/mt of crop ensiled	pH	Lactic acid	Acetic acid	Ethanol[b]
		(%)	(kg)	(kg)	(kg)	(kg)		(% of the silage DM)		
Corn										
Control	15	90.2	1.09	7.73	7.10	49.5	3.82	5.31	2.49	0.77
Inoculant	19	91.5	1.12	7.76	6.97	51.3	3.82	5.45	2.26	0.61
Probability level	...	0.01	NS	NS	0.11	0.01	NS	0.12	0.03	NS
Control	3	91.5	1.04	7.80	7.52	48.1	3.81	4.67	2.01	...
Anhydrous NH_3	3	89.4	1.01	7.96	7.84	45.0	4.19	6.13	2.47	...
Probability level	...	NS	0.16	NS	NS	0.07	0.01	0.01	NS	...
Forage sorghum										
Control	10	83.1	0.75	5.96	8.32	35.3	3.94	5.15	2.58	1.36
Inoculant	10	85.2	0.76	5.85	7.98	37.6	3.93	5.23	2.10	1.20
Probability level	...	0.01	NS	0.20	0.04	0.01	NS	NS	0.02	NS
Control	3	87.7	0.61	5.41	9.52	37.3	3.91	5.14	2.04	...
Anhydrous NH_3 or urea[c]	3	2.6	0.49	5.13	10.58	30.3	4.63	6.07	3.63	...
Probability level	...	0.09	NS	NS	NS	0.24	0.10	NS	0.08	...

Source: Bolsen et al. (1992).
[a]As a percentage of the crop DM ensiled.
[b]Ethanol was not measured in trials conducted prior to 1984.
[c]One trial with anhydrous NH_3 and two trials with urea.

pounds adversely affected fermentation, DM recovery, and gain/mt of crop ensiled, particularly anhydrous NH_3 or urea applied to high-moisture, late-season, forage sorghums.

Research with inoculants and nonprotein N sources between 1985 and 1992 is summarized by Muck (1993). Most studies use perennial grasses, alfalfa, or whole-plant corn or sorghum and were conducted in either North America or Europe. Inoculants significantly improved silage fermentation (i.e., lower pH, higher lactic-to-acetic acid ratio, and lower NH_3) in 67%, increased DM recovery in 60%, but improved aerobic stability in less than 50% of the studies. When inoculants improved DM recovery, increases were generally

Fig. 12.3.   An applicator mounted on the forage harvestor is an ideal way to apply a dry, granular silage inoculant to whole-plant corn. *Pioneer Hi-Bred International, Inc., photo.*

by 2-3 percentage units. Silage DM intake and daily gain were significantly improved in 25% of the studies, while daily milk production and feed efficiency (gain or milk/kg of silage DM) were improved in about 40% of the studies. When significant improvements occurred, DM intake, daily gain, daily milk production, and feed efficiency were increased by 11%, 11%, 5%, and 9%, respectively.

Enzymes. The use of enzymes (Table 12.7) as the only active ingredient in silage additives has received considerable attention in the past 5-10 yr. Most commercial additives contain cell wall-degrading enzymes (i.e., cellulases, hemicellulases, xylanases, and pectinases), and some products also contain starch-degrading enzymes (i.e., amylases). In theory, enzymes would (1) reduce the NDF and ADF content of the silage by hydrolyzing structural carbohydrates, which should increase the rate and/or extent of digestibility, and (2) release additional WSC, which could be used by the LAB to lower the pH of the ensiled material.

In two recent reviews, Kung (1992) and Muck (1993) conclude commercial enzyme additives are more effective in increasing breakdown of fiber and improving fermentation with direct-cut or slightly wilted grasses than with alfalfa or corn. Enzymes have not consistently improved DM digestibility, aerobic stability, or animal performance; however, they have increased effluent production in forages ensiled below 28%-30% DM and decreased DM recovery. Both reviewers conclude that more research is needed in the US on the effects of enzymes on silage preservation and utilization, particularly with alfalfa and corn silages.

Acids. The use of chemicals (Table 12.7), such as formic or sulfuric acids, to prevent clostridial fermentation and to achieve a satisfactory preservation has been common practice in many European countries (Wilkinson 1990). The value of such additives is clearly greatest in climates that make field wilting difficult for grasses, legumes, and cereals and in crops that are low in either DM or WSC content.

CONCLUSIONS

The biological principles of silage can be combined with appropriate management practices to make highly efficient silage systems possible for most livestock enterprises today. Success depends on the suitability of the crop and the technology (management and know-how) used by the silage maker. Quickly achieving and maintaining anaerobic conditions in the silo are essential for maximum nutrient preservation. Using appropriate additives can reduce the risks of poor fermentation and possibly increase the feeding value of the ensiled material.

Silage crops must be selected for both agronomic and nutritional quality characteristics and harvested at the proper stage of growth. Optimum silage harvest and storage practices cannot compensate for the nutritional value lost if the forage was ensiled too mature, and likewise, livestock production potential of a highly nutritious crop can be compromised, or lost entirely, by poor preservation techniques.

QUESTIONS

1. Why has silage become so popular as a method of forage preservation?
2. What are the five activities that occur after the chopped forage is placed in a silo?
3. What is the importance of each of the four phases of silage preservation?
4. Name four genera of LAB, and a species of each, that are important in the ensiling process.
5. Why are the homofermentative pathways preferred over heterofermentative, clostridial, or yeast pathways during silage fermentation?
6. Discuss the factors that affect the aerobic stability of silage during the feedout phase.
7. Why is whole-plant corn the nearly perfect silage crop?
8. Compare the unavoidable and avoidable DM losses in silage preservation.
9. How is ash content used to predict surface spoilage in horizontal silos?
10. What are the three types of silage additives and three examples of each?
11. What effects do LAB inoculants have on rate and efficiency of fermentation?
12. What are the disadvantages of using NPN additives with corn or sorghum silages?
13. In theory, how should enzymes affect the ensiling process and silage quality?

REFERENCES

Ashbell, G, JT Dickerson, KK Bolsen, and C Lin. 1990. Rate and extent of top spoilage losses in horizontal silos. In Proc. 9th Silage Conf., Newcastle-upon-Tyne, UK, 107-8.

Bolsen, KK. 1985. New technology in forage conservation-feeding systems. In Proc. 15th Int. Grassl. Congr., Kyoto, Japan, 82-88.

Bolsen, KK, and JL Heidker. 1985. Silage Additives USA. Church Lane, Canterbury, Kent, UK: Chalcombe Publications.

Bolsen, KK, A Laytimi, R Hart, L Nuzback, F Niroomand, and L Leipold. 1988. Effect of commercial inoculants on fermentation of 1987 silage crops. In Kans. Agric. Exp. Stn. Rep. of Prog. 539, 137-53.

Bolsen, KK, JL Curtis, CJ Lin, and JT Dickerson. 1990. Silage inoculants and indigenous microflora: With emphasis on alfalfa. In TP Lyons (ed.), Biotechnology in the Feed Industry. Nicholasville, Ky.: Alltech Tech Publishing, 257-69.

Bolsen, KK, RN Sonon, B Dalke, R Pope, JG Riley, and A Laytimi. 1992. Evaluation of inoculant and NPN silage additives: A summary of 26 trials and 65 farm-scale silages. In Kans. Agric. Exp. Stn. Rep. of Prog. 651, 101-2.

Dickerson, JT, G Ashbell, KK Bolsen, BE Brent, L Pfaff, and Y Niwa. 1992a. Losses from top spoilage in horizontal silos in western Kansas. In Kans. Agric. Exp. Stn. Rep. of Prog. 651, 127-30.

Dickerson, JT, KK Bolsen, BE Brent, and C Lin. 1992b. Losses from top spoilage in corn and forage sorghum silages in horizontal silos. In Kans. Agric. Exp. Stn. Rep. of Prog. 651, 131-34.

Goodrich, RD, and JC Meiske. 1985. Corn and sorghum silages. In ME Heath, RF Barnes, and DS Metcalfe (eds.), Forages: The Science of Grassland Agriculture, 4th ed. Ames: Iowa State Univ. Press, 527-36.

Harrison, JH. 1989. Use of silage additives and their effect on animal productivity. In Proc. Pac. Northwest Anim. Nutr. Conf., Boise, Idaho, 27-35.

Holland, C, and W Kezar. 1990. Pioneer Forage Manual: A Nutritional Guide. Des Moines, Iowa: Pioneer Hi-Bred International.

Kung, L, Jr. 1992. Use of additives in silage fermentation. In 1993 Direct-fed Microbial, Enzyme, and Forage Additive Compendium. Minnetonka, Minn.: Miller Publishing Company, 31-35.

Lin, C, KK Bolsen, BE Brent, RA Hart, JT Dickerson, AM Feyerherm, and WR Aimutis. 1992. Epiphytic mircroflora on alfalfa and whole-plant corn. J. Dairy Sci. 75:2484-93.

McCullough, ME (ed.). 1978. Fermentation of Silage: A Review. West Des Moines, Iowa: National Feed Ingredients Association.

McCullough, ME, and KK Bolsen (eds.). 1984. Silage Management. West Des Moines: National Feed Ingredients Association.

McDonald, P. 1981. The Biochemistry of Silage. New York: Wiley.

McDonald, P, AR Henderson, and SJE Heron. 1991. The Biochemistry of Silage. 2d ed. Church Lane, Canterbury, Kent, UK: Chalcombe Publications.

Mahanna, WC. 1993. Silage fermentation and additive use in North America. In Proc. Natl. Silage Prod. Conf., NRAES-67. Ithaca, N.Y.: Northeast Regional Agricultural Engineering Service, 85-95.

Muck, RE. 1993. The role of silage additives in making high quality silage. In Proc. Natl. Silage Prod. Conf., NRAES-67. Ithaca, N.Y.: Northeast Regional Agricultural Engineering Service, 106-16.

Muck, RE, and KK Bolsen. 1991. Silage preservation and silage additive products. In KK Bolsen, JE Baylor, and ME McCullough (eds.), Hay and Silage Management in North America. West Des Moines, Iowa: National Feed Ingredients Association, 105-26.

Noller, CH, and JW Thomas. 1985. Hay-crop silage. In ME Heath, RF Barnes, and DS Metcalfe (eds.), Forages: The Science of Grassland Agriculture, 4th ed. Ames: Iowa State Univ. Press, 517-26.

Pahlow, G. 1991. Role of microflora in forage conservation. In G Pahlow and H Honig (eds.), Forage Conservation towards 2000. Braunschweig, Germany: Institute of Grassland Forage Research, 26-36.

Pitt, RE. 1990. Silage and Hay Preservation. Cornell Univ. Coop. Ext. Bull. NRAES-5. Ithaca, N.Y.: Northeast Regional Agricultural Engineering Service.

Pitt, RE, and JY Parlange. 1987. Effluent production from silage with applications to tower silos. Trans. Am. Soc. Agric. Eng. 30:1198-1204.

Rotz, CA, and RE Muck. 1993. Silo selection: Balancing losses and costs. In Proc. Natl. Silage Prod. Conf., NRAES-67, Ithaca, N.Y.: Northeast Regional Agricultural Engineering Service, 134-43.

Sonon, RN, R Suazo, L Pfaff, JT Dickerson, and KK Bolsen. 1991. Effects of maturity at harvest and cultivar on agronomic performance of forage sorghum and the nutritive value of selected sorghum silages. In Kans. Agric. Exp. Stn. Rep. of Prog. 623, 65-69.

Suazo, R, RN Sonon, and KK Bolsen. 1993. Effects of hybrid, growing condition, storage time, and Pioneer 1174 silage inoculant on agronomic performance and nutritive value of whole-plant corn and grain sorghum silages. In Kans. Agric. Exp. Stn. Rep. of Prog. 678, 18-23.

White, J, KK Bolsen, and B Kirch. 1988. Relationship between agronomic and silage quality traits of forage sorghum cultivars. In Kans. Agric. Exp. Stn. Rep. of Prog. 539.

Wilkinson, JM. 1990. Silage UK. 6th ed. Church Lane, Canterbury, Kent, UK: Chalcombe Publications.

Woolford, MK. 1984. The Silage Fermentation. New York: Marcel Dekker.

Zimmer, E. 1980. Efficient silage systems. In Proc. Forage Conserv. in the '80's, Occasional Symp. 11. Brighton, UK: British Grassland Society, 186-97.

Zimmer, E, and RJ Wilkins. 1984. Efficiency of Silage Systems: A Comparison between Unwilted and Wilted Silages. Lanbauforschung Volkenrode, Sonderhelft 69. Braunschweig, Germany.

PART **2**
Forage
Systems

13

Systems of Grazing Management

ARTHUR G. MATCHES
Texas Tech University

JOSEPH C. BURNS
Agricultural Research Service, USDA, and North Carolina State University

THE two most powerful tools producers have for influencing the level of animal output under grazing are (1) the system of grazing management and (2) the concentration of animals per unit area (stocking rate or stocking density). The objectives of this chapter are to describe and compare various systems of grazing management, potential components of forage-livestock systems, and reasons why each is used.

In the past much attention has been given to the total yield potential of a particular species, of a mixture of species, or resulting from management practices. If certain forages have a more favorable distribution of production, so as to provide quality feed during times when good forage normally is not available, it may justify their use over other higher-yielding forages. The producer should consider several alternative forage management schemes and forage-livestock systems so that the one best meeting the producer's needs can be selected.

The selection of grazing systems (manage-

ment and/or forage-livestock systems) may be based solely on personal preference. However, choices based upon knowledge of the forages involved and coupled with the type of livestock and level of animal performance desired are seen as wiser decisions. The choice of a grazing system also hinges upon the managerial skills and ability of producers to evaluate both the forage plants and animal responses over time. Constraints such as competition for managerial time with other farming operations may be a determinant of which system is best for a particular producer. Other factors that must be considered are availability and location of water for cattle; topography of land; available resources, including labor, fencing, and forage-seeding and harvesting equipment; and cost of initiating and maintaining different systems. The system selected should be the most profitable. Each production enterprise is unique, and no one system can meet the needs for all situations.

Among various governmental agencies, scientific societies, and other user groups, many discrepancies have existed in the grassland and grazing terminology used. Therefore, the Forage and Grassland Terminology Committee was formed (31 individuals representing a broad spectrum of grassland activities) in 1989 under the leadership of the American Forage and Grassland Council. This group published, in 1991, "Terminology for Grazing Lands and Grazing Animals" (Forage and Grazing Terminology Committee 1992). Its recommended terminology has been used in this chapter.

GRAZING SYSTEM CONSIDERATIONS

The concept of grazing management im-

ARTHUR G. MATCHES is Thornton Professor in the Department of Plant and Soil Science at Texas Tech University. He received the MS degree from Oregon State University and the PhD from Purdue University. He was Research Agronomist with the ARS, USDA, and Professor of Agronomy at the University of Missouri, Columbia. His research and teaching specialty is forage and pasture management with emphasis on forage-livestock systems.

JOSEPH C. BURNS is Plant Physiologist and Lead Scientist, ARS, USDA, and Professor of Crop Science and Animal Science, North Carolina State University. He holds the MS degree from Iowa State University and the PhD from Purdue University. His major research is devoted to evaluating organic fractions of grazed and stored forages as related to animal intake, digestibility, and performance.

plies decision making. Profitable decision making requires knowledge about pasture species, animal response desired relative to the market, and the pasture-animal interaction. For example, with extensive grazing, ruminant livestock often control their own responses by ranging and selecting their diet (Arnold 1981). Selective grazing is exhibited maximally in the absence of management. Controlled grazing, as practiced in intensive enterprises, does not nullify selective grazing but utilizes such behavior to achieve the animal response desired.

Plants for grazing differ in morphology, inherent quality, rate of quality decline with maturation, and persistence under defoliation. These differences and the season of growth, coupled with animal management factors such as time of breeding, calving interval, age of weaning, and animal condition, provide the management flexibility used to influence subsequent animal response.

Selection of plants for grazing must be based on their inherent quality, as indexed by dry matter (DM) digestibility and relative intake by animals during the season of use, as well as based on their seasonal productivity. The digestibility required can be estimated by first determining the daily energy requirement (NAS 1984) for the animal response expected.

Suggested forage digestibilities needed to meet the requirements of various classes of beef cattle are as follows: dry cows, 46%-56%; cows nursing calves, 56%-66%; heifers nursing first calf or steers weighing at least 270 kg, 62%-68%; and steers weighing 113 kg, greater than 70% (Holt 1977). Forage species that can be managed to maintain high digestibility with time or that offer portions (leaves) of high digestibility are good choices for animal responses requiring high energy intake. Higher-yielding forages that are lower in digestibility may be better choices for animals that require less energy, such as for dry cows and ewes.

The amount of forage available per animal greatly influences animal response. When large amounts are available, animals will select the green herbage that is generally the youngest tissue and of highest quality, whereas with limited amounts available, animals will increase grazing time, exhibit less selectivity among plant species or parts, and may not consume maximum daily energy intake.

METHODS OF GRAZING

Continuous and Rotational Stocking. Grazing management implies a degree of control over both the animals and the sward. Continuous and rotational stocking represent two extremes in grazing management. *Continuous stocking* is the continuous, unrestricted grazing of a specific range or pasture by livestock throughout a year or grazing season (Forage and Grazing Terminology Committee 1992). Continuous stocking generally occurs on range where the expanse between field boundaries is large or essentially nonexistent, but it also may be used in more intensive enterprises where pastures are much smaller. Stocking may be continuous at a fixed or variable number of animals per unit area over the grazing season.

Subdivision of pastures with fences is the major concept that allows grazing management decisions relative to forage utilization. When continuously stocked pasture or range is subdivided into two or more paddocks, the concept of rotational stocking is introduced. (See Figure 13.1*A* and *B* in the next section.) *Rotational stocking* is the grazing of two or more paddocks in sequence followed by a rest period for the recovery and regrowth of the grazed herbage (FGTC 1992). The rotational feature introduces "time" as a management variable for either the stocking duration or pasture regrowth interval of each subdivision.

Duration of stocking and rest in rotational systems generally are governed by the number of paddocks available for stocking and by herbage growth rate, which varies with changing environmental conditions. The number of animals allocated to each paddock may be either fixed or variable. Findings are somewhat inconsistent from experiments comparing continuous and rotational stocking. This is not surprising because experiments have differed markedly in environmental conditions, forage species grazed, number of pasture subdivisions, and rates of stocking. Research facilities seldom are available that permit simultaneous evaluation of all options possible with either rotational or continuous stocking.

In general, rotational stocking in temperate regions is somewhat beneficial in obtaining higher animal response per unit area (8%-10%) when compared with continuous stocking. However, differences of this magnitude in research trials usually are not statistically significant (McMeekan and Walshe 1963; Blaser et al. 1973; Ernst et al. 1980). In the tropics, where animal performance is generally lower, continuous stocking is favored over rotational stocking ('t Mannetje et al. 1976).

Advantages of Rotational Stocking. Three

major advantages of rotational stocking over continuous stocking are (1) improved plant persistence, (2) opportunities to conserve surplus forage, and (3) more timely utilization of forage.

Certain forages will not survive when maintained continually in a defoliated condition. The most-striking example of the benefit of rotational grazing is for improved survival of legumes. For example, continual close defoliation of alfalfa (*Medicago sativa* L.) top growth causes rapid stand depletion (Van Keuren and Matches 1988). Repeated defoliation results in depletion of root reserves and regrowth capacity, eventually leading to the death of plants. With rotational stocking, plants are allowed an adequate rest period between defoliations to replenish root reserves as sufficient leaf area (high-leaf area index) accumulates.

Similarly, certain grasses are favored by controlled stocking. Grazing smooth bromegrass at the time of reproductive tiller elongation removes most of the shoot apices, and recovery growth must come from crown buds. However, at this stage crown buds either are not yet developed to initiate regrowth, or they do so very slowly (Smith 1981). Recovery growth is extremely slow, and stand depletion can occur. Rotational stocking permits scheduling of defoliation when crown buds are ready for growth.

With rotational stocking, excess forage may be harvested as hay or silage to feed later during periods of low forage production. Intensification of forage utilization by rotational stocking may not add more to seasonal productivity, but the total digestible nutrients (TDN) harvested per hectare can range from 11%-22% higher than with continuous stocking (Pigden and Greenshields 1960). Animals provided a daily forage allowance and moved to a different paddock the next day is the extreme of rotational stocking (see strip-grazing). Increasing the proportion of available forage that is consumed in each paddock, i.e., forage utilization, generally favors higher yields of animal product per hectare.

Losses due to trampling and death and decay of ungrazed herbage are reduced by more timely grazing of forage, which occurs when a high degree of forage utilization is practiced. This provides the producer more control of the response per animal, including the planning of production (i.e., calf production, milk production, weight gains, and maintenance of body weight) from different classes of livestock (Burns 1981).

Advantages of Continuous Stocking. Advantages of continuous stocking compared with rotational stocking are (1) lower input costs and (2) fewer management decisions. Less fencing and fewer watering facilities are required with continuous stocking than with rotational stocking because livestock range within one field enclosure. Because livestock are not regularly moved from pasture to pasture, management decisions also are simplified. With set stocking, excess forage frequently may be accumulated and used later. Such understocking and maximum selective grazing frequently result in higher per animal responses than what results from rotational stocking. However, if the pasture varies widely in forage quality, animal response may reflect a decline in diet quality even though selective grazing occurs. Where variable continuous stocking is practiced, i.e., pasture is stocked continuously but stocking rate is varied, some adjustment in grazing pressure (animal/unit of available forage) can be used to reduce the severity of under- and overgrazing.

Limitations of Rotational and Continuous Stocking. Because of additional fencing and watering facilities, rotational stocking requires a high input of both management and capital relative to continuous stocking. Research also shows a continuous daily decline in quality of available forage (Blaser et al. 1986). At first turn-on, animals have access to the leafy, highest-quality forage. Forage remaining on each successive day has progressively lower quality (more stems) as the residue from 1 d becomes the next day's ration. Also at first turn-on, pastures that are late in a rotation sequence can have much lower quality (more stem material) than do pastures grazed early in the rotation.

Forage utilization is a major problem when continuous stocking is used in areas where forage growth is seasonal. Pastures must be stocked for the season at a lower density than desired when forage growth is maximal to avoid overgrazing during periods of minimal forage growth. This frequently results in reduced diet quality as plants mature and a subsequent further reduction in forage quality and increased DM losses due to weathering of the forage and fouling by the animals.

INTENSIVE PASTURE-ANIMAL MANAGEMENT

Several innovative management systems utilize different ways to subdivide and utilize pastures for control of grazing (Fig. 13.1).

Subdivisions are a management tool for controlling the response of different classes of animals (cows, calves, ewes, and lambs). For example, producers can maximize daily energy intake for animal classes requiring high energy and at the same time minimize daily energy intake for animals having low energy requirements. These practices are very useful as a part of yearlong forage-livestock grazing systems.

Strip-grazing. Pastures are subdivided into manageable areas to allow animals access to a restricted land area of fresh herbage (Fig. 13.1C). Once the sequence is initiated a back fence is used to keep animals off the previously grazed strip, allowing it to regrow undisturbed. The strip allocated may be large enough for several days of grazing or small enough for the needs of 0.5-1.0 d. Strip-grazing generally is limited to intensive, land-limiting, high-production enterprises such as dairying to justify the high degree of management and heavy investment required. The practice emphasizes a high degree of forage utilization.

A. Continuous stocking

Animals access total area.

B. Rotational stocking

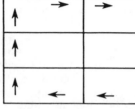

6-day defoliation and 30-day rest.

C. Strip Grazing

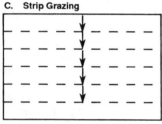

Flexible areas offered each day
(with back fence).

D. First-Last grazing

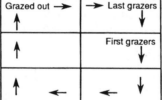

High performers graze first and
low performers last.

E. Forward Creep Grazing

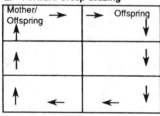

Offspring graze ahead of mothers,
using special creep openings.

F. Creep Grazing

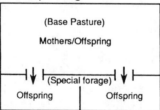

Mothers stay on base pastures,
offspring alternately creep graze
special forage.

G. Early weaning

(Base Pasture)

Dry mothers

Mothers in separate pasture.

(Special forage)

Offspring

Offspring in separate pasture
with high quality forage plus
restricted feeding of concentrate.

Fig. 13.1. Examples of different methods of grazing management.

Comparison among strip-grazing, rigid rotation, and set stocking on perennial ryegrass (*Lolium perrene* L.) shows milk produced per cow for the former two to be similar during all periods (Table 13.1). Set stocking (continuously stocked) produced less milk per cow in mid- to late summer (weeks 11 to 15 and 16 to 20). However, total production per cow for the 20-wk trial was similar in the three grazing methods. The strip and rigid rotation method favored total milk produced per hectare, while milk fat concentration was highest for set stocking. This shows that intensive utilization of forage favors production per land area (quantitative response) while more selective grazing (set stocking) favors milk composition (qualitative response).

Strip-grazing also can be used to limit intake where limited feed supply requires forage conservation (Smetham 1973). The daily animal allowance (kg of DM) is determined based on the response desired per animal and forage available per hectare. The product of daily allowance (kg/animal) and numbers of animals determine the total forage required per day for the herd. The estimated mass of available forage per hectare determines daily strip size. The quality of forage offered may vary, but the factor controlling energy intake is forage availability. See Chapter 7 for a discussion of methods to evaluate forage mass available.

Limitations of strip-grazing lie with the managerial ability of the producer and need for a productive enterprise to justify intensity of management. The manager must have an understanding of forages, be capable of planning animal-grazing schedules weeks in advance, and be able to manage forages properly (fertilizer application, irrigation, etc.) to maintain their productivity and quality.

First-Last Grazing. This method allows the producer to favor a group within the same class of animals or favor higher-producing animals of different classes by providing them the forage of highest quality (Fig. 13.1*D*). An-

imals with highest energy requirements (first grazers) are allowed first access to the pasture. After grazing the higher-quality (leafy) portion of the available forage, the first grazers are moved to the next pasture. A second group of animals (second grazers) with lower energy requirements is given access to residue herbage in the first pasture. The progression continues in that order. To ensure complete forage utilization the system can be extended by grazing with a third group of animals having only a maintenance requirement (e.g., dry females).

Using beef steers and lactating dairy cattle, comparisons of first and second grazers have been made in Virginia (Table 13.2). The advantage in response per animal for first grazers over second grazers is appreciable. Economics of the advantage should dictate its use.

The limitation of this method lies with the managerial aspect of maintaining at least two groups (herds) of the same class or different classes of livestock in a grazing regime. It is complicated by the need for planning appropriately balanced stocking rates and pasture sizes for each growing season segment. This can be partly aided by periodically adjusting the size of each animal group. The payoff for management input is increased productivity.

Forward Creep Grazing. In principle, this technique is similar to that of first-last grazing in that the animal requiring highest energy is given first choice of the available forage. The difference is that two (or more) classes of livestock are grouped together in the same pasture, but a physical barrier is positioned to allow preferential movement of the higher-nutrient-requiring animals into the next pasture in the rotation scheme (Fig. 13.1*E*).

This system is especially well adapted to cow-calf or ewe-lamb enterprises where special creep gates with adjustable openings allow calves or young lambs to slip forward into the adjoining pasture of high-quality forage and graze freely. This allows the preferred an-

TABLE 13.1.   Comparison of milk production and milk fat concentratons by method of grazing from Ayrshire cattle on perennial ryegrass pasture

Grazing method[a]	Milk produced (kg/cow/d)					Milk	
	Weeks 1-5	Weeks 6-10	Weeks 11-15	Weeks 16-20	20-wk mean	kg/ha	Fat (%)
Strip	21.9	18.9	17.4	12.5	17.7	12,400	3.59
Rigid rotation	23.7	20.0	16.3	11.6	17.9	15,500	3.54
Set stocking	23.9	19.5	14.5	8.5	16.6	11,500	4.20

Source: Castle and Watson (1975).

[a]Grazing was initiated in mid-May and terminated September 30. Two fresh strips were offered animals daily in the strip method; rigid rotation offering was one-seventh of a paddock (four-paddock system) daily without back fencing. All methods were stocked at five cows/ha.

TABLE 13.2. Comparison of responses from first and last grazers on bluegrass-white clover, orchard-grass-ladino clover, or orchardgrass-alfalfa pastures

| | Beef steer gains | | Lactating Holsteins Milk/d (kg) |
	Daily (kg)	Seasonal (kg/ha)	
First grazers	0.61	267	13.1
Last grazers	0.37	161	8.5
Difference	0.24	106	4.6
Combined group			
Average	0.49	. . .	10.8
Total	. . .	428	. . .

Source: Blaser et al. (1986).

imal maximum selectivity without competition from the others. The entire group is retained on each pasture in the rotation until all forage is well utilized.

With forward creep grazing, calves and lambs gain at a maximum rate while the dam's or ewe's gain can be controlled at modest levels (Table 13.3). Calves allowed to forward creep graze on Kentucky bluegrass (*Poa pratensis* L.) and white clover (*Trifolium repens* L.) pasture gained more than did control calves that stayed with the dams on similar pasture. Stocking rate on the group pasture had essentially no effect on calf daily gain and subsequent weaning weights. Cow weights varied appreciably, but their weight changes were of little consequence.

Limitations to adapting forward creep grazing are few, once a rotational system is accepted as the grazing method, since only the inexpensive creep opening is required.

Creep Grazing. The principle involved in creep grazing is similar to that of forward creep grazing, but management is quite different. In this case, special creep pastures are established using forage species of inherently high quality. The brood cow and calf (or ewe and lamb) are grazed on a base pasture. Stocking for the season may be continuous or rotational. The brood cow is retained on the base pasture, but the calf has access to the special creep pasture through appropriately placed creep openings (Fig. 13.1*F*). The benefit of the system is realized only when intensive stocking enables complete forage utilization on the base pasture. In continuous stocking the brood cow is held on the base pasture and may even lose weight during summer periods when forage availability and quality are low.

The creep area will require special management to keep forage productive and of high quality (leafy). This is most easily achieved by rotation between two creep pastures. Once calves have selectively grazed one of the creep areas, and allowed access to the second area that is ready for grazing, the one partially consumed is opened to the base pasture, allowing both cows and calves to graze (several days) until the forage is well utilized. This area is then closed to allow it to regrow, and the brood cows are again retained on the base pasture. An alternative is to clip the grazed over creep area to stimulate high-quality regrowth, but this results in lost forage and added cost.

Creep areas may be seeded to special high-quality perennial legumes or summer or winter annuals that produce high-quality forage

TABLE 13.3. Suckling beef calf gains with a forward-creep system and at different stocking rates on base Kentucky bluegrass-white clover pastures.

	Control	Forward creep
Forward creep-grazing comparison[a]		
Calf daily gain (kg)	0.57	0.82
Calf weaning weight (kg)	227	253
Cow daily gain (kg)	−0.19	−0.01
	(Cow-calf units/ha)	
	1.2	1.5
Stocking rate comparison[b]		
Calf daily gain (kg)	0.96	0.94
Calf weaning weight (kg)	235	244
Cow weight (kg) : December	502	524
April	535	437

[a]Blaser et al. (1980).

at the time of need. Alternatively, creep pastures may be identical to the base pasture but with an excess of forage to allow a high degree of selective grazing. This latter alternative would be applicable for extensive enterprises where it is not reasonable to establish special creep pastures. North Carolina results (Table 13.4) show the benefit on Kentucky bluegrass-white clover base pasture from creep grazing after July 10 using red clover (*T. pratense* L.) and millet (*Pennisetum americanum* [L.] Leeke) as the creep areas.

Creep grazing on millet alone at the high stocking density increased calf gains 106 kg/ha over the control. Calf daily gains changed little with forage species and stocking rates used, whereas calf gain per hectare increased appreciably from 152 up to 250 kg/ha when cow-calf units on the base pasture increased by 1.4/ha. Combining high calf performance using creep pasture and high stocking density on the base pasture gave a 73.6% increase in calf gain per hectare.

Note that cows lost weight at the highest stocking rate for millet. Base pastures were being well utilized, however, and some weight loss of dams in late lactation was of little concern during late summer. Weight recovery by cows could be anticipated with resumption of base pasture growth following fall rains. Of major importance is continued high daily gains of the marketable product, the calf.

The limitation of this practice lies with the additional fencing required and the cost and risk of stand failure associated with continual reseeding when different forages are maintained in the creep area, especially if annual or biennial species are used.

Early Weaning. This practice physically separates the offspring from the mother at 4 to 6 mo of age instead of waiting until the lactation period is finished (Fig. 13.1*G*). This allows special consideration to the nutrient requirements for both classes of animals. The calves (or lambs) in a cow-calf (ewe-lamb) enterprise are placed in a separate area (drylot is desirable) for a 7- to 10-d adjustment period during which they are fed a grain concentrate. Thereafter, they are returned to a separate pasture (away from mother) with free access to graze and to a rationed concentrate supplement. The supplement can be hand-fed, or an intake limiter, such as salt or tallow, can be incorporated to restrict intake to about 1.5% of body weight. Ample energy intake allows continued high daily gain. The dry cow (or ewe) can be stocked according to the response desired, ranging from slight weight losses to some modest level of performance.

Early weaning seems best reserved for emergency situations where forage and feed suddenly become limited, as in unexpected droughts, or it can be used more routinely where midsummer production is normally depressed. In Appalachia, where temperate pastures are dormant in midsummer, early weaning and feeding of concentrates on millet allowed high daily calf gains (Table 13.4) while cow weight was maintained.

A major limitation of early weaning is the large calf to calf variation in consuming and utilizing the concentrate supplement (Burns et al. 1983). Additional problems with this system are (1) ways to limit concentrate intake, as some calves will overconsume on a self-feed system, (2) loss of contributing milk from the dam to the daily energy intake of the calf, and (3) actual cost of the supplement. The economics of a system should dictate its attractiveness and its ultimate use in the enterprise.

FORAGE-LIVESTOCK SYSTEMS

Forage-livestock systems may be defined as integrated forage and livestock management options designed to better meet production and economic goals (Matches 1989).

Nutritional requirements of livestock should be given first consideration in plan-

TABLE 13.4. Simmental-Hereford cow and calf response from noncreep, creep-grazed, and early-weaned treatments on Kentucky bluegrass, orchardgrass, and white clover base pastures after July 10.

Item	Noncreep control	Red clover creep	Millet Creep	Millet Creep	Early Weaned[a]
(cow-calf units/ha)	(1.8)	(2.0)	(1.9)	(3.3)	(2.7)
Calf daily gain (kg)	1.10	1.10	1.09	1.05	0.93
Concentrate fed (kg/d)	. . .	. . .	. . .	. . .	1.94
Cow daily gain (kg)	0.51	0.53	0.46	−0.19	0.12
Calf gain (kg/ha)	144	156	152	250	183

Source: Harvey and Burns (1988).

[a]Weaned calves were placed in drylot for 3 to 5 d then allowed to graze millet (70 to 84 d) and provided ground ear corn daily at approximately 1% of their body weight.

ning a forage program. Unless forage can be managed to meet these needs, it is of little use to the producer. Feed requirements change with the animal's phase of production or growth, making it essential to design forage-livestock systems that adapt to the animal's changing requirements. Shown in Figure 13.2 is the daily TDN requirements for a 450-kg beef cow at two levels of milk production (Thompson 1971).

A high calving percentage (i.e., the percentage of cows that birth a calf within 12 mo of the previous calf) is an important objective that requires proper feeding to ensure high conception rates early in the breeding season. The most critical period in feeding the beef cow is from calving, or slightly before calving, until the cow is rebred and has conceived. Note the high nutritional requirements for milk production (Fig. 13.2) that occur at the same time period. Daily TDN requirements for growing animals, such as backgrounding steers, also are dynamic. As animal weight increases, the energy requirement for maintenance also increases.

With knowledge of the livestock feed requirements (including protein, minerals, and vitamins) both livestock and forage management may be manipulated to provide feed of adequate quality for each time interval in the operation. For example, Holt (1977) lists several ways in which cattle production programs may be altered to more nearly match the nutrient production characteristics of the forage. Holt's suggestions include breeding cows at specific times to regulate calving dates to match (1) maximum animal mass with maximum carrying capacity of the pasture, (2) animal nutrient requirements with nutrient production potential of the pasture, and (3) weaning dates to a time of abundant high-quality forage for weaned calves. These

shifts should occur only if they are in concert with the market demands.

BUILDING FORAGE-LIVESTOCK SYSTEMS

In many geographic regions temperate grasses have an uneven seasonal distribution of forage production, as is shown for tall fescue (*Festuca arundinacea* Schreb.) grown in the Midwest (Fig. 13.3). In the Upper South, peak forage production of temperate forages would be shifted more to the autumn, winter, and spring; but subtropical forages such as bermudagrass (*Cynodon dactylon* [L.] Pers.) and bahiagrass (*Paspalum notatum* Flugge) have high summer production. Thus, temperate, subtropical, and tropical forages can be expected to differ in seasonal production patterns within and among regions. Selected management practices also may influence seasonal distribution of forage production (Chaps. 5 and 12, Vol. 1).

A key principle in developing forage-livestock systems is to utilize inherent differences in seasonal growth patterns among forages for specific forage systems. Systems can be developed for different classes of livestock (i.e., beef cows that calve in spring or fall, stocker cattle production, backgrounding yearling beef animals, replacement heifer production, spring lamb production, etc.) to provide their feed requirements during critical periods of animal growth and development. In addition, there is a rapidly growing interest in producing milk from dairy cattle grazing pastures that are components of forage-livestock systems (Chap. 14, Vol. 1).

The following provides examples of forages that may be manipulated to produce herbage at various times of the grazing season. These potential "components" may be combined to form specific forage-livestock systems.

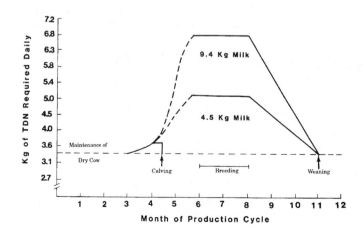

Fig. 13.2. Daily TDN requirements for two levels of milk production for a 450-kg beef cow. (Adapted from Thompson 1971.)

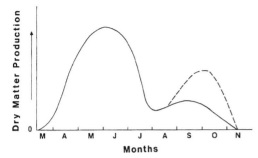

Fig. 13.3. Typical seasonal production of tall fescue in the lower Midwest. The broken line indicates potential production with above normal late summer rains.

Legumes. Growing a legume with a grass, as compared with grass grown with up to 100 kg nitrogen (N)/ha, generally results in higher total forage yields and altered seasonal production. For example, Matches (1979) found tall fescue grown alone and fertilized with 67 kg N/ha yielded 1388 kg/ha less herbage than tall fescue grown with either red clover or birdsfoot trefoil (*Lotus corniculatus* L.). No N fertilizer was used with legume-grass mixtures. More importantly, only 38% of the annual production for tall fescue top-dressed with N occurred after the first spring harvest as compared with 59%-70% for the mixtures. Similar results were obtained in a region where periodic droughts from June to August are common (Matches et al. 1979, 1980). Again, 66%-77% of annual production occurred after the May harvest for tall fescue grown with red clover or birdsfoot trefoil, as compared with only 52% for tall fescue fertilized with 140 kg N/ha.

Growing legumes (red clover or birdsfoot trefoil) in Missouri with switchgrass (*Panicum virgatum* L.) and caucasian bluestem (*Bothriochloa caucasica* [Trin.] C.E. Hubb.), both subtropical C$_4$ grasses, also resulted in forage yields 56%-137% greater than yields from growing these grasses alone without N fertilizer (Matches and Mitchell 1978; Matches and Mowrey 1979). The temperate legumes grown with subtropical grasses resulted in an increase in early growth (prior to June) ranging from 775 to 1715 kg/ha.

Animal performance may be improved by including a legume with a grass. Including ladino white clover with tall fescue in Alabama resulted in higher animal gains, especially in spring and fall (Hoveland et al. 1981). In both Missouri (Hedrick et al. 1982) and North Carolina (Burns et al. 1970, 1973) beef steer or calf gains were increased, especially during

the summer, by adding legumes to grass pastures. In North Carolina, 28% more beef gain per hectare was produced after July where legumes were present (Burns et al. 1970). Since legumes are generally more sensitive to time and frequency of defoliation than are grasses, a greater degree of management is required to sustain legume stands.

Cultivars of Different Maturities. Cultivars within species such as orchardgrass (*Dactylis glomerata* L.), tall fescue, timothy (*Phleum pratense* L.), and smooth bromegrass (*Bromus inermis* Leyss.) may differ markedly in time of maturity (flowering). Matches (1977) observed a wide range in rate of development among 18 cultivars of five temperate forage grasses grown in Missouri. On the same date in late April, phenological development of cultivars ranged from vegetative or nonjointing to heads emerging from the boot stage. By the end of May cultivars ranged from heads emerging from the boot stage to being at full bloom.

On any given date, the in vitro dry matter digestibility (IVDMD) of the later-maturing cultivars within a species was higher than for early-maturing cultivars. Therefore, simple pasture systems may be devised by growing cultivars of different maturities in separate pastures. In this way, both the amount and quality of herbage available for grazing can be maintained in better balance when using rotational stocking. Animals could be allocated to the earliest cultivar in early spring, then rotated to pastures in order of cultivar maturity from early to late. Consequently, a higher-quality feed would be available for a longer time than if the same cultivar was grown in all pastures.

Defoliation and Fertilization Management. Both the timing of initial spring grazing or cutting and the frequency of defoliation can serve as useful tools for influencing seasonal distribution of forage production. By selecting defoliation dates, a producer can partially regulate seasonal distribution of forage production when using rotational stocking. For example, when the first harvest of switchgrass or caucasian bluestem was taken at late heading in Missouri, less than 30% of the total seasonal production was available during late July and August (Anderson and Matches 1983). In contrast, earlier first harvest at the jointing stage resulted in higher summer production (+2000 kg/ha) with over 50% of the total production being available in late summer when herbage is traditionally scarce in this region.

Harvesting the first growth later gave higher total yields for the season than did earlier harvests but resulted in herbage with lower digestibility and crude protein (CP) concentration. Each week's delay in first harvest resulted in declines of 1.5-2.4 IVDMD percentage units and 1.0-1.6 CP percentage units. The lower total yield with earlier harvests was partially offset by higher forage quality of spring growth and greater availability of regrowth in late summer. The principle of regulating seasonal production by defoliation management seems particularly applicable in situations where the producer has a limited number of different forages growing in his or her enterprise.

For grasses, N fertilizer may be applied to enhance plant growth at different times of the growing season. This is a management practice that may be used for increasing the potential carrying capacity of pastures. For example, in North Carolina (Burns et al. 1984), on improved mountain pastures, 67 kg of N/ha increased the stocking rate by 0.37 cow-calf units/ha and calf gain by 32 kg/ha. Murtagh et al. (1980) reports an average increase in pasture carrying capacity of 131% from fertilization of kikuyugrass (*Pennisetum clandestinum* Hochst. ex Chiov.) with over 500 kg N/ha.

A twofold increase in leaf yields of indiangrass (*Sorghastrum nutans* [L.] Nash) and switchgrass were obtained with 45 kg N/ha as compared with yields with no N fertilization (Perry and Baltensperger 1979). These increases were greater in the summer growth of July and August than for the spring growth. Thus, N rates and timing of application are effective management tools for influencing the seasonal distribution of grass growth. Additional information on using fertilization for regulating pasture growth is presented in Chapter 5.

Stockpiled Herbage. *Stockpiled herbage* is defined as "the practice of allowing forage to accumulate in the field until it is needed for grazing" (Mays and Washko 1960; FGTC 1992). Forage generally is stockpiled for fall and winter grazing; however, it may be stockpiled for use in any period when forages have low productivity.

Length of the accumulation period can be used as a management tool for yield and quality of the standing forage. Generally, yields increase with length of accumulation, but forage quality is reduced (Matches et al. 1973;

Ocumpaugh and Matches 1977; Collins 1982). Missouri research (Matches et al. 1973; Matches 1979) shows IVDMD of tall fescue in October averaged 57.9%, 66.4%, and 67.4% when accumulated from May 10, June 21, and August 2, respectively. The respective CP level of the accumulated forages were 8.3%, 8.5%, and 10.1%. Shorter accumulation periods ensure younger tissue and higher forage quality, which is desirable for growing livestock such as beef steers.

Substantial losses in yield and forage quality may occur in tall fescue stockpiled for late fall and winter grazing due to freezing death of upper leaves, decay of lower leaves, and leaching of nutrients by rain and melting snow. Losses of up to 80% of the DM and over 20% of the CP and decreases in IVDMD of 15 percentage units have been reported (Ocumpaugh and Matches 1977; Matches 1979). These losses can be minimized by completely grazing the stockpiled forage before the onset of severe freezing weather.

Where adapted, tall fescue has been shown to be superior to most other grasses for fall stockpiling and winter grazing (Wedin et al. 1966; Bryan et al. 1970; Van Keuren 1972; Reynolds 1975; Taylor and Templeton 1976). Leaves of tall fescue apparently tend to resist frost better and growth continues later in fall, which gives the advantage.

Few legumes can be stockpiled, mainly because most lose their leaves more easily than grasses, either from disease or maturation in summer or from frost in fall. However, birdsfoot trefoil (Mays and Washko 1960; Allison 1971; Cooper 1973; Collins 1982) and sainfoin (*Onobrychis viciifolia* Scop.) (Cooper 1973) appear to have potential as stockpiled herbage for summer grazing. Birdsfoot trefoil can be stockpiled in pure stand until mid-July in the Midwest without excessive losses in quality or palatability (Mays and Washko 1960) or can be cut in May (under Pennsylvania conditions) for silage and then stockpiled for midsummer grazing. In Missouri, it was found that the European-type trefoils (erect growth habit) were better adapted to stockpiling than the 'Empire' types (decumbent growth habit) (Allison 1971). Birdsfoot trefoil stockpiled for 64-80 d in Kentucky provided adequate CP and digestible nutrients during summer for lactating beef cows (Collins 1982).

Temperate and Subtropical Grasses. Temperate or C_3 grasses begin growth much earlier in spring than do subtropical or C_4 grasses, whereas subtropical grasses flourish under high summer temperatures if soil moisture is not limiting. Annual yields of several temperate and subtropical grasses were very similar in Missouri (Table 13.5); however, subtropical grasses yielded 77%-100% of their seasonal production after June 1 as compared with 44%-77% for temperate grasses (Rountree et al. 1974). Similar results have been reported by Overton and Fribourg (1981) in Tennessee.

Likewise, Newell and Moline (1978) in Nebraska compared seasonal trends for yield of digestible DM of smooth bromegrass and intermediate wheatgrass (*Thinopyrum intermedium* [Host] Barkworth), both temperate species, and switchgrass, a subtropical grass. Maximum production of the temperate grasses was 28-42 d earlier than that for switchgrass (Fig. 13.4). This illustrates the principle of using differences in seasonal distribution of forage production in developing forage-livestock systems.

Other Pasture Systems. Other examples of integrating forages into systems include (1) using summer annuals for summer grazing (see Chap. 37, Vol. 1) to supplement temperate pastures and (2) the overseeding of bermudagrass with winter annual legumes, grasses, or cereals to extend the season of production (see Chaps. 14 and 36, Vol. 1). Calf gain per hectare nearly doubled when bermudagrass pastures in Alabama were overseeded with 'Abruzzi' rye (*Secale cereale* L.) and annual clovers (arrowleaf clover [*T. vesiculosum* Savi] and crimson clover [*T. incarnatum* L.]). Furthermore, the

TABLE 13.5. Total yield and distribution of forage production in Missouri

Species	Total DM (mt/ha)	DM production June 1-Aug. 31 (%)
Tall fescue	2.24	42
Reed canarygrass	2.51	74
Bermudagrass	2.44	100
Caucasian bluestem	3.65	93
Indiangrass	2.68	100
Switchgrass[a]	2.90	77

Source: Rountree et al. (1974).
Note: Average for 3 yr and two locations.
[a]Average of four cultivars.

grazing season was extended by 3 mo over 'Coastal' bermudagrass alone (Hoveland et al. 1978).

Goode et al. (1966) compared different forage combinations for yearly beef cattle production in North Carolina. Pastures of orchardgrass and tall fescue grown in a mixture with ladino clover were continuously stocked and were compared with sequence grazing of a tall fescue-Coastal bermudagrass system. In the latter system, cattle grazed tall fescue in the spring, Coastal bermudagrass pastures in the summer, and tall fescue again in late summer. Compared with the continuously stocked treatments, the fescue-bermudagrass system provided a more uniform supply of forage throughout the grazing season. This simplified cattle management and reduced need for accumulated forage, summer annuals, or other supplemental feed.

Advantages are also realized from systems utilizing both native range and seeded pastures. In Wyoming (Hart et al. 1983) a crested wheatgrass (*Agropyron desertorum* [Fisch. ex Link] Schultes) native range system carried 34% more beef cattle and had 90% more calf gain per hectare than native range alone. In North Dakota (Rogler et al. 1962), seeded pastures of a crested wheatgrass, smooth bromegrass, and Russian wildrye (*Psathyrostachys juncea* [Fisch.] Nevski) sequence grazed with native range (1) extended the grazing season, (2) allowed a delay in the turn-out date for cattle going on native range, which permitted a higher stocking density on the native range, and (3) reduced the amount of hay and other feeds necessary.

In nearly all pasture systems that include temperate and subtropical forages, each forage usually is grown in separate pastures (Matches et al. 1975). If temperate and subtropical species are grown together, management becomes very difficult because

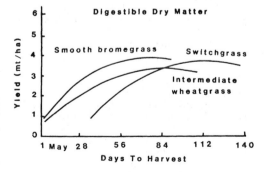

Fig. 13.4. Overlap and extension of season of forage production for grazing by combined use of cool-season (temperate) and warm-season (subtropical) grasses. (Adapted from Newell and Moline 1978.)

(1) species differ in time of growth (Fig. 13.4); (2) early-growing temperate species may deplete available soil moisture before subtropical species begin rapid growth; (3) species differ in optimum time of fertilization, especially with N; and (4) species differ in age and stage of growth and in inherent palatability to livestock, making it difficult to obtain uniform defoliation. Bermudagrass and tall fescue can be successfully grown together and utilized in a grazing system (Mitchell et al. 1986); however, separate pastures provide management and production advantages.

Preserved Forage. Preserved feeds are another means for balancing the seasonal flow of forage available. Hay, grass-legume silage, corn (*Zea mays* L.), and sorghum (*Sorghum bicolor* [L.] Moench) silages are widely used to supply feeds during periods when forages are not producing (see Chaps. 12 and 14, Vol. 1). They are viable components in many forage-livestock systems, especially where producers have excess grass and legume forage at some periods of the year. These excesses may be preserved as hay or silage and fed back during periods of low forage availability.

KEYS TO SUCCESSFUL FORAGE-LIVESTOCK SYSTEMS

Planning of forage-livestock systems should begin well in advance of need. When new forage seedings are necessary, 1 yr or more may be required before all forage components are completely established and ready for grazing. Likewise, shifts in animal management such as from spring to fall calving must be in phase with potential forage and livestock aspects of the system.

In most geographical regions the potential for several forage-livestock systems exists. Knowing (1) characteristics of adapted forages, (2) livestock nutrient requirements, and (3) when forage (standing and/or preserved) is needed are keys to developing successful forage-livestock systems. Each forage system should be specific and may not meet the needs of all producers in a specific area. Producers have different managerial capabilities, monetary and physical resources, and types of farming or ranching operations. Therefore, individual forage-livestock systems should be designed to fit the ability and goals of each producer.

In summary, forage management systems should be aimed at providing the nutritional needs of livestock as well as ensuring the continued productivity of forages used as pasture or preserved feeds. Several management practices are possible, including forms of grazing management and the use of different forage species or manipulation of species to provide herbage at different times of the year.

QUESTIONS

1. What managerial tools do producers use to regulate animal output under grazing?
2. What is the basic difference between rotational and continuous stocking and what are the advantages and disadvantages of each?
3. What is selective grazing? What are the consequences of selective grazing to the animal and to the pasture sward?
4. What is the main difference between forward creep and creep grazing?
5. What principle is useful in designing a pasture system for a specific class of livestock?
6. List and discuss several ways to ensure forage availability for grazing at different times of the year.
7. Why are legumes so useful as components of pasture systems? How may summer and winter annuals be used as components of pasture systems to even out the seasonal availability of grazable herbage?
8. Why are preserved forages valuable in designing forage-livestock systems?
9. Outline what you would consider in planning forage-livestock systems.
10. Based on the concepts presented in this chapter, design a pasture system to extend the grazing season in your area using adapted forage species.

REFERENCES

Allison, LD. 1971. Evaluation of birdsfoot trefoil varieties under various planting and management systems. PhD diss., Univ. of Missouri-Columbia.

Anderson, B, and AG Matches. 1983. Forage yield, quality, and persistence of switchgrass and caucasian bluestem. Agron. J. 75:119-24.

Arnold, GW. 1981. Grazing behavior. In World Animal Science, vol. 1, Grazing Animals. New York: Elsevier, 79-104.

Blaser, RE, DD Wolf, and HT Bryant. 1973. Systems of grazing management. In ME Heath, DS Metcalfe, RF Barnes (eds.), Forages: The Science of Grassland Agriculture, 3d ed. Ames: Iowa State Univ. Press, 581-95.

Blaser, RE, RC Hammes, Jr., JP Fontenot, CE Poland, HT Bryant, and DD Wolf. 1976. Forage-animal production systems on hill land in the eastern United States. In Hill Lands Proc. Int. Symp., 3-9 Oct., Morgan town, W. Va.: West Virginia Univ. Books, 674-85.

———. 1980. Challenges in developing forage-analysis systems. In Proc. Am. Forage and Grassl. Conf., 217-44.

Blaser, RE, RC Hammes, Jr., JP Fontenot, HL Bryant, CE Poland, DD Wolf, FS McClaugherty, RG Kline, and JS Moore. 1986. Forage-Animal Management Systems. Va. Agric. Exp. Stn. Bull. 86-87.

Bryan, WB, WF Wedin, and RL Vetter. 1970. Evaluation of reed canarygrass and tall fescue as spring-summer and fall-saved pasture. Agron. J. 62:75-80.

Burns, JC. 1981. Integration of grazing with other feed resources. In Nutritional Limits to Animal Production from Pastures. Farnham Royal, UK: Commonwealth Agricultural Bureaux, 455-71.

Burns, JC, HD Gross, WB Gilbert, RW Harvey, MB Wise, and DF Tugman. 1970. N.C. Agric. Exp. Stn. Bull. 437.

Burns, JC, L Goode, HD Gross, and AC Linnerud. 1973. Cow and calf gain on ladino clover-tall fescue and tall fescue grazed alone and with Coastal bermudagrass. Agron. J. 65:877-90.

Burns, JC, RW Harvey, FG Giesbrecht, WA Cope, and AC Linnerud. 1983. Central Appalachian hill land pasture evaluation using cows and calves. III. Treatment comparisons of per animal and hectare responses. Agron. J. 75:878-85.

Burns, JC, RW Harvey, DF Tugman, FG Giesbrecht, WA Cope, and AC Linnerud. 1984. Cow and calf performance and productivity of improved mountain pastures in North Carolina. Agric. Res. Serv., N.C. State Univ. Bull. 466.

Castle, ME, and JN Watson. 1975. Further comparison between a rigid rotational "wye college" system and other systems of grazing for milk production. J. Br. Grassl. Soc. 30:1-6.

Collins, M. 1982. Yield and quality of birdsfoot trefoil stockpiled for summer utilization. Agron. J. 74:1036-41.

Cooper, CS. 1973. Sainfoin-birdsfoot trefoil mixtures for pasture, hay-pasture, and hay-stockpile management regimes. Agron. J. 65:752-54.

Ernst, P, YLP Le Du, and L Carlier. 1980. Animal and sward production under rotational and continuous grazing management—a critical appraisal. In Proc. Int. Symp. Role of Nitrogen in Intensive Grassl. Prod., 119-26.

Forage and Grazing Terminology Committee (FGTC). 1992. Terminology for grazing lands and grazing animals. J. Prod. Agric. 5:191-210.

Goode, L, HD Gross, GL Ellis, and WB Gilbert. 1966. Pastures for Yearling Beef Cattle in Piedmont, North Carolina. N.C. Agric. Exp. Stn. Bull. 428.

Hart, RH, JW Waggoner, Jr., JA Hager, and MB Marshall. 1983. Beef cattle production on crested wheatgrass plus native range vs. native range alone. J. Range Manage. 36:38-40.

Harvey, RW, and JC Burns. 1988. Creep grazing and early weaning effects on cow and calf productivity. J. Anim. Sci. 66:1109-14.

Hedrick, HB, JA Paterson, AG Matches, JD Thomas, NG Krause, RE Morrow, and WC Stringer. 1982. The Production, Characteristics, and Utilization of Forage-fed Beef. Univ. Mo. Agric. Exp. Stn. Res. Bull. 1043.

Holt, EC. 1977. Meeting the nutrient requirements of beef cattle with forage. In Forage Fed Beef: Production and Marketing Alternatives in the South. Southern Coop. Ser. Bull. 220:261-85.

Hoveland, CS, WB Anthony, JA McGuire, and JG Starling. 1978. Beef cow-calf performance on Coastal bermudagrass overseeded with winter annual clovers and grasses. Agron. J. 70:418-20.

Hoveland, CS, RR Harris, EE Thomas, EM Clark,

JA McGuire, JT Eason, and ME Ruf. 1981. Tall Fescue with Ladino Clover or Birdsfoot Trefoil as Pasture for Steers in Northern Alabama. Ala. Agric. Exp. Stn. Bull. 530.

McMeekan, CP, and MJ Walshe. 1963. The inter-relationships of grazing method and stocking rate in the efficiency of pasture utilization by dairy cattle. J. Agric. Sci. Camb. 61:147-63.

't Mannetje, L, RJ Jones, and TH Stobbs. 1976. Pasture evaluation by grazing experiments. In Tropical Pasture Research: Principles and Methods, Commonw. Agric. Bur. Bull. 51.

Matches, AG. 1977. Techniques for evaluating cultivars of temperate grasses. Am. Soc. Agron. Abstr. Madison, Wis., 102.

———. 1979. Management. In Tall Fescue, Am. Soc. Agron. Monogr. 20. Madison, Wis., 171-99.

———. 1989. Contribution of the systems approach to improvement of grassland management. In Proc. 16th Int. Grassl. Congr. 4-11 Oct., Nice, France, Vol III:1791-97.

Matches, AG, and ML Mitchell. 1978. Sod-seeding legumes into switchgrass and caucasian bluestem. Univ. Mo. Agric. Coll. Spec. Rep. 215:12-15.

Matches, AG, and D Mowrey. 1979. Legumes frost-seeded into sods of switchgrass and caucasian bluestem. Univ. Mo. Agric. Coll. Spec. Rep. 238:9-14.

Matches, AG, JB Tevis, and FA Martz. 1973. Yield and quality of tall fescue stockpiled for winter grazing. In Research in Agronomy, Univ. Mo. Agron. Misc. Publ. 73-5. 54-57.

Matches, AG, GB Thompson, and FA Martz. 1975. Post-establishment harvesting and management systems of forages. In Proc. No-Tillage Forage Symp., 111-35.

Matches, AG, D Mowrey, and DA Sleper. 1979. Management of tall fescue cultivars grown with legumes. Univ. Mo. Agric. Coll. Spec. Rep. 238:15-34.

———. 1980. Management of tall fescue cultivars grown with legumes. Univ. Mo. Agric. Coll. Spec. Rep. 250:12-22.

Mays, DA, and JB Washko. 1960. The feasibility of stockpiling legume-grass pasturage. Agron. J. 52:190-92.

Mitchell, RL, JB McLaren, and HA Fribourg. 1986. Forage growth, consumption, and performance of steers grazing bermudagrass and fescue mixtures. Agron. J. 78:675-80.

Murtagh, GL, AG Kasser, and DO Huett. 1980. Summer-growing components of a pasture system in a subtropical environment. I. Pasture growth, carrying capacity, and milk production. J. Agric. Sci. 94:645-63.

National Academy of Sciences (NAS). 1984. Requirements of Beef Cattle. Washington, D.C.: National Research Council.

Newell, LC, and WJ Moline. 1978. Forage Quality Evaluations of Twelve Grasses in Relation to Season of Grazing. Nebr. Agric. Exp. Stn. Res. Bull. 283.

Ocumpaugh, WR, and AG Matches. 1977. Autumn-winter yield and quality of tall fescue. Agron. J. 69:639-43.

Overton, JR, and HA Fribourg. 1981. Adaptation and productivity on droughty soils of bahiagrass, bluestem, bermudagrass, and other perennial forages in middle and west Tennessee. Tenn. Farm Home Sci. 100:20-23.

Perry, LJ, and DD Baltensperger. 1979. Leaf and stem yields and forage quality of three N-fertilized warm-season grasses. Agron. J. 71:355-58.

Pigden, WJ, and JER Greenshields. 1960. Interaction of design, sward, and management on yield and utilization of herbage in Canadian grazing trials. In Proc. 8th Int. Grassl. Congr., 594-97.

Reynolds, JH. 1975. Yield and chemical composition of stockpiled tall fescue and orchardgrass. Tenn. Farm Home Sci. 94:127-29.

Rogler, GA, RJ Lorenz, and HM Schaaf. 1962. Progress with Grass. N.Dak. Agric. Exp. Stn. Bull. 439.

Rountree, BH, AG Matches, and FA Martz. 1974. Season too long for your grass pasture? Crops and Soils 26:7-10.

Smetham, ML. 1973. Grazing management. In Pastures and Pasture Plants. Wellington, New Zealand: AH and AW Reed, 179

Smith, D. 1981. Smooth bromegrass. In Forage Management in the North, 3d ed. Dubuque, Iowa: Kendall/Hunt, 167-74.

Taylor, TH, and WC Templeton. 1976. Stockpiled Kentucky bluegrass and tall fescue forage for winter pasturage. Agron. J. 68:235-39.

Thompson, GB. 1971. How can we produce, supplement and utilize forages to make calving season more profitable. In Proc. Univ. Mo. Beef-Cow Calf Clin.

Van Keuren, RW. 1972. All-season forage systems for beef cow herds. In Proc. 27th Annu. Meet. Soil Conserv. Soc. Am., 39-44.

Van Keuren, RW, and AG Matches. 1988. Pasture production and utilization. In AA Hanson, DK Barnes, RR Hill, Jr. (eds.), Alfalfa and Alfalfa Improvement, Am. Soc. Agron. Monogr. 29. Madison, Wis., 515-38.

Wedin, WF, IT Carlson, and RL Vetter. 1966. Studies on nutritive value of fall-saved forage, using rumen fermentation and chemical analysis. In Proc. 10th Int. Grassl. Congr., Helsinki, Finland, 424-28.

14

Cropland Pastures and Crop Residues

WALTER F. WEDIN
University of Minnesota

TERRY J. KLOPFENSTEIN
University of Nebraska

CONTEMPORARY American agriculture has resulted in a reduction in cropland pasture (Wedin et al. 1980) and an increase in crop residues from monoculture aimed at intensive grain production (Vetter and Boehlje 1978). An initial change was the shifting away from crop rotations (and the "ley" influence) in the 1960s and 1970s. There were both changes in pasture needs for ruminant livestock production and perceived economic advantages for row crop production. The latter, however, has run counter to reduction of energy expenditures (Heichel 1979), proper land use (Timmons 1980; Larson et al. 1983), and general well-being of the agricultural economy.

The niche in which cropland pastures fit is discussed in a report of the Council of Agricultural Science and Technology (CAST 1986). In the 1990s and beyond, however, alternative agriculture (NRC 1990) will result in some crop sequencing, i.e., modified crop rotations including forage legumes, and there will be increased availability of crop residues (result-

ing from the productive row crops being grown) for feeding of ruminants.

CONTRIBUTION

Cropland pastures are composed of legume-grass, all-grass, or all-legume swards of improved species grown at varying intervals and durations on land suitable for continuous or intermittent production of agronomic or horticultural crops other than forages. Twenty-seven percent of the 303 million ha of US cropland are used for pasture, with the bulk of pastured cropland in the north central and southern regions (Table 14.1)

Permanent pasture, discussed in Chapter 15, differs from cropland pasture in that the former is maintained indefinitely for grazing.

Crop residues are those plant parts that remain following harvest of the main economic product, as in feed grains and horticultural crops (fruit). The crop residue resource com-

WALTER F. WEDIN is Adjunct Professor, Department of Agronomy and Plant Genetics, University of Minnesota (also Professor Emeritus, Iowa State University). He holds the MS and PhD degrees from the University of Wisconsin. His chief professional interests are pasture and forage production, management, and utilization plus the study and documentation of the role of grasslands and forages as a resource base for ruminant animal production in the US and abroad.

TERRY J. KLOPFENSTEIN is Professor of Animal Science at the University of Nebraska. He received the MS and PhD degrees from Ohio State University. His research activity relates to the growing calf. Major areas of research are improvement of feeding value of crop residues, protein utilization, and increasing feeding value of silages.

TABLE 14.1. US land in cropland pasture as related to harvested cropland, 1987 data

	Harvested	Used only for pasture	% pasture
	(ha × 10^6)		
US[a]	134.0	26.3	2
Northeast	4.8	0.9	19
Lake states	13.0	1.1	8
Corn belt	29.8	4.2	14
Northern Plains	35.3	4.0	11
Appalachian	6.6	3.5	53
Southeast	4.2	1.7	40
Delta states	6.3	1.6	25
Southern Plains	11.8	5.9	50
Mountain	14.3	2.2	15
Pacific	7.7	1.1	14

Source: USDA (1992).
[a]Alaska and Hawaii data included only in total US.

prises almost 65% of all agricultural residues available from crops, manures, and forestry (Table 14.2).

TABLE 14.2. US crop residues available by regions (mt × 10⁶)

New England/Mid-Atlantic	5.6
Southeast	25.8
Corn Belt	98.3
Plains	94.3
Intermountain/West	28.0
Total	252.0

Source: Vetter and Boehlje (1978).

Meeting Animal Needs. Cropland pastures and crop residues offer several alternative uses. Cropland pastures are higher yielding than permanent pastures, and excess forage can be either machine or animal harvested. More than one type of cropland pasture may be needed to provide a balance of forage supply and demand for various classes of livestock. Crop residues can be used to extend the grazing season. Various mixes of these two resources are possible (Fig. 14.1).

A southern Iowa site provides an example wherein the cropland pasture portion is shown to make a substantial contribution to the overall program. Total pasture needed from permanent or cropland pasture to meet annual nutrient needs of 100 beef cows (*Bos taurus*) and calves was 75 ha; 14 ha of this total was cropland pasture (Wedin 1970). Twelve hectares of cornstalk residues were used.

Climate Limitations. Climate imposes a limit on adaptation and persistence of species for use in cropland pastures. The selection of species should be on the basis of their individual and collective utilities in meeting animal needs. Also, the geographical areas where certain species can be effectively utilized may change. It was shown that perennial ryegrass (*Lolium perenne* L.) would persist well in Pennsylvania (Jung and Kocher 1974) and perform as well or better than orchardgrass (*Dactylis glomerata* L.) in combination with alfalfa (*Medicago sativa* L.) in cropland pastures (Jung et al. 1982). Advances in forage breeding have extended the limits of adapted growth for some species. For example, alfalfa

Fig. 14.1. Crop residues, especially from corn, and cropland pastures make an efficient beef production system.

Fig. 14.2.   Factors of wetness and slope may be detrimental to corn but an advantage to cool-season grass cropland pastures with nitrogen.

has shown greater persistence and productivity in the South (Brummer and Bouton 1991), even with continuous stocking (Smith et al. 1992).

Through use of more than one adapted forage species, a hedge against fluctuations in climatic conditions can be made. For example, the distinctive characteristics of four grasses (orchardgrass, tall fescue [*Festuca arundinacea* Schreb.], reed canarygrass [*Phalaris arundinacea* L.], and smooth bromegrass [*Bromus inermis* Leyss.]) allow an appropriately managed combination of them to be used to advantage (Wedin et al. 1971; Marten and Hovin 1980). Crop residues also vary because of climatic conditions.

Land-Use Factors. No two farms or ranches are comparable. Pasture and forage needs must fit the operation in question. Of the adapted forages in any area, the species and cultivars best suited to soil conditions and proper land use on an individual farm or ranch must be selected.

In planning forage production, consideration of the USDA land-use capability classes is helpful (Klingebiel and Montgomery 1961). Cropland pastures fit best in Classes II and III and only sparingly in Class IV. Class I land will be used primarily for intertilled row crops and thus will provide substantial quantities of crop residues. Cropland pastures must be highly productive and well managed if they are to be economic alternatives on Class I land. Factors of wetness and slope may be detrimental to corn (*Zea mays* L.) production (and therefore to accumulation of crop residues) but may be of no consequence or may even be an advantage for cropland pastures composed of tall grasses fertilized with nitrogen (N) (Wedin and Vetter 1970) (Fig. 14.2). Cropland pastures thus offer excellent alternatives on those sites where row crops are commonly grown, but because of various factors corn or other row crop yields are restricted (Fenton et al. 1971).

Cropland pastures also aid in curbing environmental pollution. Research has shown that swards of sown legumes and grasses are effective "living filter" systems for municipal sewage effluent (Kardos et al. 1971; Marten et al. 1979). Such effluent and animal manures also may be used effectively on row crops from which crop residues result. These uses are discussed extensively in a report by CAST (1975).

Cropland Pasture

TYPES

Cropland pastures can be of short-term or long-term duration. Specific species used alone or as mixtures in pasture and hay seedings are discussed in Chapters 16-37, Volume 1, and in systems in Chapters 1-23, this volume. Here, considerations are related to legume-grass, all-grass, or all-legume species or mixtures as they fit for cropland pastures.

Legume-Grass Mixtures. Properly compounded legume-grass mixtures have provided the basis for economical utilization of pastures on cropland. Legume-grass mixtures have been superior to N-fertilized grass in pastures for one or more of the following factors: ground cover potential, forage production, animal performance of cattle grazed thereon, and economic returns (Wagner 1954a, 1954b; Stricker et al. 1979; Petritz et al. 1980; Smith et al. 1992; Van Keuren and Matches 1988; Casler and Walgenbach 1990).

Frequently, adequate legume stands cannot be maintained in pastures in temperate areas of the US. Thus, there is need for N fertilization to optimize production of cropland pastures destined for long-term stands. Research has shown that this is a practical way to extend the life of a stand (Wedin et al. 1965; Wedin 1966). In Indiana research, however, it was not beneficial to apply N when alfalfa comprised more than 50% of the mixture (Rhykerd et al. 1967).

Efforts to maintain the legume were enhanced by liming plus phosphorus (P) and potassium (K) applications in Indiana research (Rhykerd et al. 1967) where alfalfa was managed for hay harvest. Though such data suggest that alfalfa can be maintained in a rotational grazing management with long recovery periods for the alfalfa, both grazing results in Minnesota and a severe monthly cutting regime used in Alabama reduced alfalfa stands in the third year (Hoveland and Evans 1970; Wedin et al. 1971).

In situations typified by these experiments, either N applications to the remaining grass or reestablishment of alfalfa would be necessary. For reestablishment, research in Iowa showed that precision seed placement, good seed to soil contact, and reduced competition from other species were as important when introducing legumes into a grass sward as when seeding into a conventional seedbed (Barnhart and Wedin 1983). Having grazing-type alfalfa cultivars closer to general availability is a definite boon to maintaining legumes (Brummer and Bouton 1991; Bouton et al. 1991; Smith et al. 1992).

All-Grass Swards. Tall cool-season grasses are excellent in cropland pastures because they offer alternative pasture management schemes when adequate, timely N applications are made (Wedin et al. 1971). In the South, the Alabama station has demonstrated that fall and winter applications of N to grass are necessary for winter grazing (Hoveland et al. 1970). Approximately 6 mo of grazing is possible in southern Iowa using a three-season grazing system; 4-yr averages for a grazing study in this region are given in Table 14.3.

The four grasses used in the study (smooth bromegrass, orchardgrass, reed canarygrass, and tall fescue) each fitted differently in the system. Nitrogen applications (kg/ha) were 45 kg in early spring, 90 kg in June, and 134 kg on August 1. Two grazing periods were used: spring-summer (late April to August 1) and fall (mid-October to early December).

All-Legume Swards. Two developments relating to cropland pasture suggest that all-legume pastures are feasible. Bloat prevention is now possible. Work in Utah has shown that irrigated alfalfa pastures, when properly managed with cattle fed a bloat preventative, permit attainment of liveweight gains exceeding 1500 kg/ha for cattle grazing thereon

TABLE 14.3.   Tall cool-season grasses in a three-season pasture system, Iowa, 1967-69

	Grazing period[a]	Tall fescue	Reed canarygrass	Smooth bromegrass	Orchard-grass	Mean of grasses
Average daily gain (kg)	I	.48	.55	.69	.55	.57
	II	.79	.64	.92	.76	.78
Steer-d/ha	I	788	703	666	751	727
	II	417	349	242	299	327
Total		1205	1052	908	1,050	1054
Liveweight gain/ha (kg)	I	380	388	457	413	410
	II	328	223	222	227	250
Total		708	611	679	640	660

Source: Wedin et al. (1971).

[a]*Grazing period and days of grazing*

	Spring-summer (I)	Late fall-early winter (II)
1967	May 4-Aug. 3	Oct. 23-Dec. 18
1968	Apr. 25-Aug. 1	Oct. 21-Dec. 16
1969	May 1-July 31	Oct. 10-Dec. 5

(Acord 1969). Alfalfas with low crowns and spreading characteristics are of increasing value in pastures, as they are more adapted to grazing than are the bulk of the alfalfa cultivars on the market (Brummer and Bouton 1991).

The second development is greater use of nonbloating legumes such as crownvetch (*Coronilla varia* L.) and birdsfoot trefoil (*Lotus corniculatus* L.). While primarily recommended for rolling land in permanent pastures, their use is certainly not restricted to those sites.

PRODUCTION AND UTILIZATION

Cropland pastures occupy land that has other alternatives. Economical use of these pastures requires good stands of adapted forages. Stand failures are largely due to forage species and cultivar selection, seeding method, and climatic conditions during and following establishment. (See Chap. 7, Vol. 1, and Chap. 3, Vol. 2, on establishment.)

Benefits to Other Crops. Systems including cropland pasture usually include row crops grown either at set intervals or intermittently. Benefits to the other crops include increased yields and reduced soil erosion.

Recently there has been an increasing awareness that the beneficial effect of legumes is very important to the row crops that follow (Heichel et al. 1983). Alfalfa managed for hay production fixed 160 kg/ha of N in the seeding year, 161 kg/ha in the third year of the stand, and 224 kg/ha in the fourth year of the stand. For birdsfoot trefoil, N fixation was quite constant for years 1, 3, and 4 at about 105 kg/ha each year.

In cropland pastures, the benefit of the legume to companion grasses has been recognized but is difficult to measure quantitatively. Using the ^{15}N isotope-dilution technique, West and Wedin (1985) report on pasture studies in southern Iowa where 90% of the N in alfalfa forage and 38% of the N in alfalfa-orchardgrass forage were symbiotically derived. Alfalfa comprised only 35% of the pasture forage in this study. The legume proportion in pasture mixtures is generally recommended to be 50%, if possible.

Both the residual N and an effect of a legume other than N on subsequent corn yields have been shown. Voss and Shrader (1982) report yields of first-year corn following a grass-legume meadow to be greater than subsequent years of corn, particularly continuous corn, regardless of the amount of N used on continuous corn. These workers point out that it is possible for the legume to provide all of the N for a following corn crop. At present N rates for corn, the residual value of cropland pasture is of considerable significance.

Losses from soil erosion on sloping croplands have been of concern for 200 yr in the US, but these losses were not seriously addressed in public policy until the 1930s. Thereafter, data as presented in Table 14.4 clearly show that forages alleviate soil erosion (Hays et al. 1949).

TABLE 14.4.   Cropping systems for the control of soil losses.

Cropping system[a]	Average soil loss[b] (mt/ha/yr)
Corn, annually	199
Corn, barley, sweetclover	74
Corn, barley, hay	49
Corn, barley, hay, hay	29
Corn, barley, hay, hay, hay	18
Corn, barley, hay, hay, hay, hay	16

Source: Hays et al. (1949).
[a]Hay = clover-timothy
[b]16% slope, plots 22.1 m long, silty clay loam.

While soil erosion seemingly was of less concern during the period of the 1950s to the 1980s, an increasing concern is evident as described by the timely appraisal of Larson et al. (1983). They state, "In general, erosion is greatest on land with row crops, less with close-seeded crops, and least with grass and legume forage crops."

Overview of Grazing Managements. Cropland pastures produce more dry matter (DM) than do permanent pastures. Grazing managements must therefore be geared to utilize as high a percentage of the forage produced as possible. Thus, as indicated in Chapter 13, the increased efficiency resulting from widening the period of rest to grazing should be observed. For example, in grazing of an alfalfa-grass cropland pasture, the period of grazing should perhaps be 1 wk followed by 3 wk of rest. With this type of management the expected efficiency, i.e., the percentage of forage utilized by the animal of the total forage produced, would be in the order of 65%-70%. If the cropland pasture is used for green chopping or zero grazing, however, these efficiencies could be greater than 90%.

Crop Residues

Crop residues, which comprise almost 65% of all agricultural residues in the US (Vetter and Boehlje 1978), are mostly from grain crops. With the production of essentially all grains, 1 kg of crop residue is produced for each kilogram of grain produced (Vetter and Boehlje 1978). Over 250 million mt are produced annually in the US. Given that about 4.5 mt would maintain a beef cow for 1 yr, over 55 million cows could be supported on the residues produced. This exceeds the number of cows presently in the US. Over 1 billion mt of residues are produced worldwide (Kossila 1984) and provide much of the feed resources for ruminants in developing countries (Owen and Jayasurirya 1989; Wanapat and Denendra 1985; Wedin and Hoveland 1987).

TYPES

Corn production results in the greatest quantity of total residues available. The cornstalk is of good quality at physiological maturity of the corn plant, but quality decreases with time as the grain dries. At high-moisture grain harvest (25%-30% grain moisture), the stalk is still of good quality. At dry grain harvest, quality has decreased considerably. The corn husk, which makes up a much smaller proportion of the corn residue, is highly digestible, often being above 60% in dry matter digestibility.

Other residues are primarily the small-grain residues, wheat (*Triticum aestivum* L. emend. Thell.), barley (*Hordeum vuglare* L.), oat (*Avena sativa* L.), and grain sorghum (*Sorghum bicolor* [L.] Moench). There is considerable residue produced in soybean (*Glycine max* [L.] Merr.) production, but it is of low quality and needs to be left on the soil to aid in controlling soil erosion. Barley and oat straws probably are slightly higher in quality than wheat straw. These straws are generally lower in quality than corn residues.

Grain sorghum is produced primarily in Texas, Kansas, Nebraska, and Oklahoma. The sorghum plant is different in that it does not "die" at physiological maturity of the grain but instead continues to synthesize new vegetative material. Generally, the quality of grain sorghum stubble is not as high as that of cornstalks at physiological maturity; however, the grain sorghum stubble does not decrease in quality as much or as rapidly (Bolsen et al. 1977).

UTILIZATION VIA GRAZING

The primary problems with utilization of crop residues are low quality and harvesting difficulties. Yet an advantage of using crop residues is that the cost of production is usually charged against the production of grain. By utilizing the crop residues via grazing, considerable economic advantage can be realized.

While small-grain residues do not lend themselves well to grazing, corn and grain sorghum residues are well utilized in forage systems, particularly for beef cows (Ward 1978). At present the most common means of feeding beef cows during late fall and early winter in Iowa, Nebraska, and Kansas is by grazing of corn or sorghum residues (Clanton 1989; Russell 1990; Russell et al. 1993).

A beef cow can be maintained for about 80 d on 0.9 ha of corn or sorghum residue (Vetter and Boehlje 1978; Ward 1978). Calves are often weaned in mid-October when corn and sorghum are being harvested. The cows can then graze the residues until spring thaw. The biggest problems are related to the weather. Often snow cover is the limiting factor in determining how long feed will be available. If there is too much rain, the resulting mud and trampling will greatly reduce the feed supply. Greater chances of mud and the need (or desire) for fall tillage reduce the opportunities for stalk grazing in the eastern Corn Belt.

Often 130-380 kg/ha of corn grain remain in the field after harvest. While corn grain is an excellent feed supply for cattle, overconsumption of grain can produce acidosis or founder in cows. This can partially be overcome by limiting daily the time cows are on newly grazed stalk fields. Obviously, this usually is impractical because of labor shortages and because of a lack of alternative fields or pastures for cows.

All of the grain remaining in the field is available to the cows on the first day of grazing, with less available each succeeding day. Also, cows select the most digestible forage portions early, especially the husk. The net result is decreasing feed quality the longer cows are on the stalk fields (Fernandez-Rivera and Klopfenstein 1989a; Gutierrez-Ornelas and Klopfenstein 1991; Russell et al. 1993). Nutritionally, this is contrary to the needs of the cow. In midgestation, the cow's needs are increasing as the fetus grows. Often cows will gain 0.45-0.9 kg daily during the first 30 d on cornstalks. For the next 30-50 d, weight will just be maintained.

Little supplementation is needed during the first 30 d, but some protein supplement should be fed thereafter. Alfalfa hay fed two to three times per week is an excellent supplement to cornstalks or grain sorghum stubble. An excellent practice is grazing of stalk fields and alfalfa fields after the fall-hardening period for alfalfa.

Stocking rate influences the quality of the diet and cow performance. If cows are in good condition when stalk grazing is initiated, stocking rates can be high. Conversely, if cows are in poor condition, stocking rate should be relatively lower (Fernandez-Rivera and Klopfenstein 1989b; Irlbeck et al. 1991; Russell et al. 1993).

There are fewer problems with grain remaining in the field if it is grain sorghum than if it is corn. Excessive grain usually remains only if snow is early or if another factor has caused low grain harvest efficiency. Performance on grain sorghum stubble is comparable to that of cornstalks, and equal weight changes of beef cows can be expected (Irlbeck et al. 1990). Supplementation should be similar.

UTILIZATION VIA MACHINE HARVEST

Beef Cow Needs. During late winter and early spring months it is necessary to feed harvested forage to beef cows in the northern US. Cornstalk or grain sorghum residue grazing is usually not feasible after March 1, and pasture is not available until late April or early May. This is an expensive period for feeding beef cows because feed must be harvested, and the cows' nutrient needs are high during late gestation and early lactation. Harvesting and processing of crop residues provide the main costs because the production cost is accounted for in grain production.

The low quality of harvested crop residues generally will not meet the energy needs of the beef cow. Two practical possibilities for enhancing the quality of residues are (1) management schemes to increase residue quality or (2) chemical treatment, primarily ammonia. Both cornstalks and grain sorghum stubble are of good quality at plant maturity. If the grain is harvested at greater than 20% moisture, the residues will likely contain more than 50% moisture. These residues, if harvested immediately, can be chopped and ensiled (Berger et al. 1979). The resulting silage has good feed value and probably will meet the energy needs of the beef cow. Some protein supplementation is needed and could be in the form of alfalfa hay or silage or a commercial protein supplement.

Ammonia treatment of dry-harvested crop residues holds considerable potential for increasing feeding value (Fig. 14.3). The straw of small grains do not lend themselves to early harvest but instead are easily stacked or baled. Ammonia treatment is accomplished by putting bales or stacks into a larger stack of 14-27 mt, covering with a sheet of 12.6- by 30-m plastic, and sealing the edges with soil (Sundstol et al. 1978; Klopfenstein and Owen 1981; Ward et al. 1981). Anhydrous ammonia is injected at the rate of 30 kg/mt of dry residue. Depending upon temperature, 2-3 wk are needed for the reaction to be completed. Cornstalks stacked after dry grain harvest also can be treated in a similar manner; however, low temperatures may limit the treatment effect (Nelson et al. 1984).

The treated residue, after aeration, contains 10%-12% crude protein (CP) equivalent. The DM digestibility should be increased by 5-15 percentage units. Treated residues are quite palatable, which is a great improvement compared with the relatively unpalatable, untreated straw. Treated residues are probably equivalent to average quality hay, depending upon the quality of straw or stalks treated and the effectiveness of the treatment.

Either early-harvested corn or grain sorghum stalks or ammonia-treated residues have great potential to decrease the cost of beef cow feeding during the harvested feed period. While residue might make up the majority of the feed supply, some supplemental feed in the form of high-quality hay or concentrates may be needed. Cows should gain some

Fig. 14.3. Up to 50 bales of crop residues can be ammoniated under one 12.6 × 30 m plastic sheet.

weight during this period to help in rebreeding.

Crop Residues in Growing-Finishing Systems. Many people, including producers and researchers, assume that it is either nutritionally or economically impossible to use cornstalk grazing in backgrounding programs for calves. It has been clearly demonstrated, however, that these perceptions are incorrect. The economics of using stalk grazing as a part of a backgrounding program are favorable (Stock 1990; Klopfenstein 1991).

Calves are inexperienced at stalk grazing and need a few days to learn to find and eat the grain (Klopfenstein et al. 1990). Maximum grain intake may not occur for 10-20 d. Thereafter, grain intake declines to about zero at 40 d. The diet may be 40%-50% grain at peak intakes.

The protein requirements of the calf are high, but supplemental protein is expensive. Crude protein in the diet of calves declines with time of grazing. The grain serves as both a source of energy and protein. Alfalfa is a good source of protein for calves but may not supply enough rumen escape protein. Generally the more rumen escape protein in a supplement, the higher the calf gains and the higher the cost.

Stalk grazing can be an economical component of a growing program for calves, but fairly intensive management is required. The following points need to be considered. (1) What is the desired level of performance? If the calves are "going to grass," then the rate of gain when grazing stalks is much less important than wintering cost because the calves make compensatory gain on grass. (2) Lower stocking rates will increase calf gains, but more hectares will be required. (3) Up to 15 cm of snow cover may not reduce grazing. Sorghum stubble may be better than corn residue during snow cover. (4) Trampling during muddy conditions is a problem, and management to minimize the problem will increase quality and quantity of residue grazed. (5) Protein supplementation is a major expense. Up to 0.14 kg/d of rumen escape protein is required to maximize gains. More economical wintering may be possible, however, with little or no supplementation of rumen escape protein. (6) A reserve feed supply of alfalfa, ammoniated straw, grass hay, etc., must be available during snow cover and muddy conditions and after spring thaw.

Forage Systems

The quantity and quality of forage adequate to meet the nutritional needs of livestock throughout both the grazing season and the entire year can be achieved in many areas with planned forage systems. A single pasture species or mixture of grasses and legumes under one management system cannot provide adequate forage over the entire grazing season, much less an entire year. In the design of forage systems, careful consideration must be given to livestock feed requirements throughout the grazing season or year; growth characteristics, especially the seasonal distribution of growth, of forage species adapted to the region; and the influence of different management and fertilization practices on forage production. Forage systems should be planned to ensure a supply of forage for both summer and winter and to cover unforeseen difficulties.

Cropland pastures are important in forage systems, either as the basic plan around which the rest of the system is developed or as a stopgap during specific periods when forages are often in short supply from permanent pastures.

In many regions, midsummer and winter are the deficit periods in forage availability for which pasture systems must be programmed. For winter, crop residues fit well in many cases. These shortage periods differ not only among regions but also within a region for different classes of livestock. For example, lactating dairy cows require a continuous supply of high-quality feed throughout the year, but forage needs of a steer backgrounding operation may be limited to a grazing period of 6 mo or less (i.e., April to September). An alternative using crop residues has been discussed earlier in this chapter.

Forage demands for a beef cow-calf enterprise are closely geared to calving date. Beef cows should be in favorable condition during the breeding season (first 3 mo after calving) to ensure conception. Failure or delay in conception is costly to the producer. The Illinois station reports when calving percentage is reduced by 10 percentage points, the producer not only loses 10 calves for 100 cows but also pays the room and board for 10 unproductive cows.

Cropland pastures and crop residues can, in combination, provide all of the energy needs and a major portion of the protein needs for livestock, particularly those being grown for reproductive purposes. The mix can be visualized as varying in at least three major ways, as depicted in Figure 14.4.

PASTURE: ROW-CROP RESIDUE FEED SUPPLY SYSTEMS FOR BEEF COWS

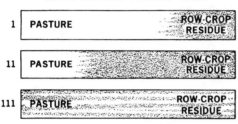

Fig. 14.4. A conceptual look at cropland pasture and crop residue supply systems.

Ideally, the mix of the two resources should be based first on the land resource available. Figure 14.4 indicates the proportion of cropland pasture will be less on the most level land (as in Case III). That land is less subject to erosion and therefore is utilized more intensively for row crops from which a higher proportion of crop residues result. On sites subject to erosion, one excellent alternative is to increase the proportion of land in pasture, as in Case I. These cases may also be visualized as varying for reasons other than the land resources.

Specific examples of systems that include one or more resources—cropland pasture, crop residue, hay or grass silage, and/or nonpasture feeds—are listed in Table 14.5.

TYPES OF FORAGE SYSTEMS

Species Mixed. Species mixed, as described in Table 14.5, is a grass-legume mixture. Legumes not only provide the grass with N but enhance the quality of pasture herbage and provide more forage available for grazing in the midsummer period. Alfalfa, birdsfoot trefoil, red clover (*Trifolium pratense* L.), crownvetch, and lespedeza (*Kummerowia stipulacea* [Maxim.] Makino) are especially useful legumes for bridging the summer slump in pasture productivity. Grasses fertilized with N in combinations such as birdsfoot trefoil with grass or alfalfa with smooth bromegrass or orchardgrass (Table 14.6) result in increased forage availability in July and August from the inclusion of legumes in the pasture mixture (Schaller 1967).

Separate Pastures of Different Forages. Separate pastures of different forages permit the development of a more-detailed system if each pasture is used for grazing during particular periods of the season. Separate pastures are planted to forages of differing growth patterns. Perennial and annual types of cool- and warm-season forages are commonly used. For example, as shown in Table 14.6, a pasture of cool-season grasses fertilized with N can be grazed during May and June, cattle can then be shifted to sudangrass pastures from July to mid-September, and then cattle can be returned to tall grasses until regrowth is utilized to the extent desired.

In the southern Corn Belt, perennial warm-season grasses like switchgrass (*Panicum vir-*

TABLE 14.5.   Cropland pasture and pasture:residue utilization systems

System	Description
Cropland pasture only	
Species mixed	Different forage species grown in mixtures for extending the grazing season
Separate pastures of different forages	Different types of forages grown in separate pastures for grazing at different times of the year
Overseeding annuals	Annual species overseeded into established stands of perennial forages to extend the grazing season
Separate pastures— varying management and fertility	Separate pastures of the same species under different managements for grazing at different times of the year
Cropland pasture and crop residues	
Crop residues	Use of crop residues as extenders of the grazing season
Pasture and conserved forages or nonpasture feeds	
Supplement with conserved forage	Supplemental feeding of preserved forage harvested during period of surplus production
Supplement with nonpasture feeds	Supplemental feeding of nonpasture forage sources or energy concentrates

Source: Adapted from Wedin and Matches (1973).

TABLE 14.6.   Estimated monthly availability of forage for grazing

Type of pasture	% available, by month							
	May	June	July	Aug.	Sept.	Oct.	Nov.	Dec.
Kentucky bluegrass-white clover, unimproved	25	30	10	5	15	10	5	...
Kentucky bluegrass-white clover + N, P	35	35	8	5	10	4	3	...
Renovated (continuous grazing)								
Birdsfoot trefoil-grass	10	25	25	21	10[a]	5[a]	5	...
Birdsfoot trefoil-grass, deferred for midsummer grazing	...	15	35	25	15[a]	5[a]	5	...
Cool-season grasses + N[b]	30	30	10	5	10	10	5	...
Cool-season grasses + N, deferred for fall grazing[b]	30	30	...	...	...	25	15	...
Renovated (rotational grazing)								
Alfalfa with smooth bromegrass or orchardgrass	20	25	25	15	5[a]	5[a]	5	...
Supplemental								
Sudangrass or sorghum-sudan hybrids	...	...	40	40	15	...[c]	5	...
Sudangrass or sorghum-sudan hybrids, deferred to fall and winter grazing	...	...	...	...	...	30	45	25
Winter rye	50	20	...	...	5	15	10	...
Miscellaneous								
Meadow aftermath—following one cutting	...	20	30	25	5[a]	15[a]	5	...
Meadow aftermath—following one cutting, to be plowed	...	20	30	10	20	20	...	...
Meadow aftermath—following two cuttings	...	...	10	35	25[a]	25[a]	5	...
Meadow aftermath—following two cuttings, to be plowed	...	...	10	25	35	30	...	...
Cornstalks	...	...	...	...	...	10	60	...

Source: Schaller (1967). Compiled originally by W.F. Wedin, Department of Agronomy, Iowa State University.

[a]Allowances have been made for winter hardening of legume from about Sept. 15 to Oct. 15.

[b]Smooth bromegrass, orchardgrass, tall fescue, reed canarygrass, or combinations.

[c]Grazing must be avoided between first frost and definite killing frosts because of prussic acid content in regrowth shoots.

gatum L.) and caucasian bluestem (*Bothriochloa caucasia* [Trin.] C.E. Hubb.) provide good midsummer grazing in pasture systems that include cool-season grasses fertilized with N for spring and fall grazing (Matches et al. 1971). Temporary pastures of fall-sown grain and pastures of perennial forages such as tall fescue are commonly grown in the South for fall, winter, and early spring grazing, with a warm-season forage such as bermudagrass (*Cynodon dactylon* [L.] Pers.) for summer grazing.

Small grains (oat, barley, wheat, and rye [*Secale cereale* L.]) are especially good sources of winter pastures (Elder 1967; King et al. 1971). They produce an abundance of high-quality pasture when forage is generally in short supply. With proper management they may be grazed during winter and then harvested later as a cash crop. Average gains above 0.6 kg/d are not uncommon for cattle grazing small grains.

In the western US, cropland pastures often are used in combination with other types, e.g., range. Raguse et al. (1980) report that perennial irrigated pastures could be used effectively with winter annual range.

Overseeding Annuals. Overseeding annuals into stands of perennial forage is common in the South and Southwest. Small grains (oat, barley, wheat, and rye) and legumes such as crimson clover (*T. incarnatum* L.), vetch (*Vicia sativa* L.), and winter peas (*Pisum sativum* L.) are the annuals often used. When seeded into bermudagrass sods, they extend the grazing period into the winter months, thus increasing total production per hectare (Walker et al. 1966; Decker et al. 1969; Holt et al. 1969; Hoveland et al. 1978; Kalmbacher et al. 1980). For example, overseeding Coastal bermudagrass in the fall with crimson clover has increased animal liveweight gain by as much as 300 kg/ha (Walker et al. 1966).

Old World bluestems are warm-season perennial grasses that appear to have much potential for late summer pasture, particularly in Oklahoma and parts of Missouri, Kansas, and Texas. Since the Old World bluestems do not initiate vigorous growth until late spring, overseeding with a small grain in the fall provides additional early spring grazing (Matches et al. 1971).

As may be apparent from the examples cited here, overseeding of annuals works best with forage species that have contrasting periods of production. With warm-season perennials such as bermudagrass and Old World bluestems, the overseeded cool-season annuals are the pasture extenders for fall, winter, and early spring grazing.

Separate Pastures with Varied Management and Fertility. Separate pastures with varied management and fertility can be developed based on Table 14.6. For example, the Ohio station has reported steer gains of 0.49-0.6 kg/d during parts of July, August, and September with deferred grazing of Kentucky bluegrass (*Poa pratensis* L.)-birdsfoot trefoil (Van Keuren 1970).

Summer grazing may be provided by round baling the first growth of the tall grasses and leaving the bales in the pasture and utilizing both regrowth and round bales when summer pasture is needed. This method has been researched by Van Keuren (1970) in Ohio with orchardgrass, tall fescue, and timothy (*Phleum pratense* L.) and in Missouri by Matches et al. (1971) with tall fescue and an orchardgrass-annual lespedeza mixture. In Missouri the average beef and dairy heifer gains ranged from 0.41-0.7 kg/d during July and August.

By using the same species but varying management and fertilization, cool-season grasses can be used for fall and winter grazing (Schaller 1967; Burton 1970; Wedin 1970; Wedin et al. 1971). *Late summer-saved forage, accumulated growth, stockpiled growth, forage on the stump,* and *fall-saved forage* are some of the terms describing the practice of deferring late summer grazing until fall or winter. This management is best adapted to grasses because, unlike legumes, they retain most of their leaves after a killing frost.

Tall fescue is superior for stockpiling because leaves remain green longer and forage declines less in digestibility in the fall than does orchardgrass, smooth bromegrass, timothy, or reed canarygrass. Spring growth may be grazed or cut as hay. Leaving the hay as round bales in the field for grazing along with the aftermath is practiced in many areas of the Midwest, particularly for beef cows. Such systems have yielded 10 mt/ha of DM, which, when used by the grazing animals, provided over 530 cow-d/ha. Only 0.26-0.29 ha/cow was required for fall and winter grazing from November 15 through April 15 in Ohio experiments (Van Keuren 1970).

Harvesting a seed crop from some tall fescue fields and stockpiling the regrowth for winter grazing is practiced in many seed-producing areas. To ensure a good seed crop the following year, the crop residue must be removed; therefore, it is usually clipped, round baled, and left in the field for winter utilization.

Summer annuals such as sudangrass, sorghum-sudangrass hybrids, and pearlmillet (*Pennisetum americanum* [L.] Leeke) also may be deferred for fall and winter grazing (Table 14.6) (Van Keuren 1970; Wedin 1970). In Missouri research, heifer daily gains on a stockpiled pearlmillet and sorghum-sudangrass hybrid combination ranged from 0.45 to over 1.3 kg during a 2- to 6-wk grazing period from mid-November to January (Matches et al. 1971). Highest gains were obtained when pearlmillet had fully developed seed heads. Generally one could not expect stockpiled summer annuals to be more than a maintenance ration for cattle.

Crop Residues. Crop residues are integral components of many forage systems. Their general use is discussed earlier in this chapter.

Supplementing with Conserved Forage and Nonpasture Feeds. Supplementing with conserved forage and supplement with nonpasture feeds offer many possibilities. For conserved forage, preserved forage (hay or silage) is either removed from the pasture for feeding at a later period of need or fed within the pasture itself, as with round bales.

The producer should consider utilizing preserved forage or other feeds in planning pasture systems. (See Chaps. 10-12 for methods of preserving and handling forage.) Cropland pastures have considerable flexibility in duration of use, and they can be visualized in systems utilizing conserved forages.

Output per animal (liveweight increase or milk production), animal output per hectare, and carrying capacity of the pasture are all increased by supplemental feeding on pasture (Bryant et al. 1965; Donker et al. 1968; Hammes et al. 1968; Hart et al. 1971; Mott et al. 1971). Supplementing with conserved forage may be planned only during times when pasture is in short supply, whereas with nonpasture feed, cattle may receive supplemental feed for the entire grazing season or for a only a portion.

The type of supplemental program selected is largely dependent on the class of livestock and the economics of feeding on pasture. Lactating dairy cattle may require feeding of concentrates throughout the entire season if the producer wishes to maintain high milk production (Donker et al. 1968). In backgrounding beef cattle on pasture followed by drylot finishing, however, carryover effects of supplementation must be considered. Beef cattle gain during a drylot-finishing period is negatively correlated with the rate of gain on pasture. In Indiana, researchers report that for each kilogram that cattle gained during a 58-d pasture season, they gained 0.2 kg less than when cattle from various levels of concentrate feeding were all fed the same daily concentrate ration during the drylot-finishing period (Perry et al. 1971).

Good advance planning is the key to successful pasture systems involving cropland pasture and crop residues. Each system must be individualized in accordance with the particular farm enterprise. Agricultural college specialists and Natural Resource Conservation Service personnel have developed guidelines for several planning system methods. By utilizing a planning method, the producer can select from among the several basic systems suggested.

QUESTIONS

1. Define cropland pastures in terms of types, species, and use.
2. From what crops does the major portion of crop residues result?
3. What is the animal production potential of crop residues in the US?
4. On what land classes do cropland pastures fit best? Crop residues?
5. What are important criteria in selecting forage cultivars for cropland pasture? Give examples.
6. Would a grass pasture seeded to tall cool-season grasses alone be justified for a short-term cropland pasture? Why?
7. Contrast the differences in using crop residues extensively in a forage program for overwintering beef cows as compared with growing-finishing cattle.
8. Why are pasture systems necessary at all? What is meant by a "year-round forage system"?
9. Using the percentage distribution of forage availability as given in Table 14.6, outline a tentative 6-mo grazing sequence for 100 beef cows with calves. Be specific as to which system or systems will be used to provide sufficient pasture without producing excess.

REFERENCES

Acord, CR. 1969. Beef production on irrigated pastures. Utah Sci., Mar., 7-9.

Barnhart, SK, and WF Wedin. 1983. Reduced-tillage pasture renovation in the semihumid temperate region of the U.S.A. In JA Smith and VW Hays (eds.), Proc. 14th Int. Grassl. Congr., Lexington, Ky. Boulder, Colo.: Westview, 545-47.

Berger, LL, JA Paterson, TJ Klopfenstein, and RA Britton. 1979. Effect of harvest date and chemical treatment on the feeding value of corn stalklage. J. Anim. Sci. 49:1312-23.

Bolsen, KK, C Grimes, and JG Riley. 1977. Milo stover in rations for growing heifers and lambs. J. Anim. Sci. 45:377-84.

Bouton, JH, SR Smith, Jr., DT Wood, CS Hoveland, and EC Brummer. 1991. Registration of 'Alfagraze' alfalfa. Crop Sci. 31:479.

Brummer, EC, and JH Bouton. 1991. Plant traits associated with grazing-tolerant alfalfa. Agron. J. 83:996-99.

Bryant, HT, RC Hammes, Jr., RE Blaser, and JP Fontenot. 1965. Effects of feeding grain to grazing steers to be fattened in drylot. J. Anim. Sci. 24:676-80.

Burton, GW. 1970. Symposium on pasture methods for maximum production in beef cattle: Breeding and managing new grasses to maximize beef cattle production in the South. J. Anim. Sci. 30:143-47.

Casler, MD, and RP Walgenbach. 1990. Ground cover potential of forage grass cultivars mixed with alfalfa at divergent locations. Crop Sci. 30:825-31.

Clanton, DC. 1989. Grazing cornstalks—a review. Nebr. Beef Cattle Rep. MP 54:11.

Council of Agricultural Science and Technology (CAST). 1975. Utilization of Animal Manures and Sewage Sludges in Food and Fiber Production. Rep. 41. Ames, Iowa.

———. 1986. Forages: Resources for the Future. Rep. 108. Ames, Iowa.

Decker, AM, HJ Retzer, FG Swain, and RF Dudley. 1969. Midland Bermudagrass Forage Production Supplemented by Sod-seeded Cool-Season Annual Forages. Md. Agric. Exp. Stn. Bull. 484.

Donker, JD, GC Marten, and WF Wedin. 1968. Effect of concentrate level on milk production of cattle grazing high-quality pasture. J. Dairy Sci. 51:67-73.

Elder, WC. 1967. Winter Grazing Small Grains in Oklahoma. Okla. Exp. Stn. Bull. B-654.

Fenton, TE, ER Duncan, WD Shrader, and LC Dumenil. 1971. Productivity Levels of Some Iowa Soils. Iowa Agric. Exp. Stn. and Agric. Ext. Serv. Spec. Rep. 66.

Fernandez-Rivera, S, and TJ Klopfenstein. 1989a. Diet composition and daily gain of growing cattle grazing dryland and irrigated cornstalks at several stocking rates. J. Anim. Sci. 67:590.

———. 1989b. Yield and quality components of corn crop residues and utilization of these residues by grazing cattle. J. Anim. Sci. 67:597.

Gutierrez-Ornelas, E, and TJ Klopfenstein. 1991. Changes in availability and nutritive value of different corn residue parts as affected by early and late grazing season. J. Anim. Sci. 69:1741.

Hammes, RC, Jr., RE Blaser, JP Fontenot, HT Bryant, and RW Engel. 1968. Relative value of different forages and supplements for nursing and early weaned beef calves. J. Anim. Sci. 27:509-15.

Hart, RH, J Bond, GE Carlson, and TR Rumsey. 1971. Feeding corn or molasses to cattle on orchardgrass pasture. Agron. J. 63:397-401.

Hays, OE, AG McCall, and FG Bell. 1949. Investigations in Erosion Control and the Reclamation of Eroded Land at the Upper Mississippi Valley Conservation Experiment Station near La Crosse, Wis., 1933-1943. USDA Soil Conserv. Serv. Tech. Bull. 973.

Heichel, GH. 1979. Stabilizing agricultural energy needs: Role of forages, rotations, and nitrogen fixation. J. Soil Water Conserv. 33:279-82.

Heichel, GH, CP Vance, and DK Barnes. 1983. Symbiotic nitrogen fixation of alfalfa, birdsfoot trefoil, and red clover. In JA Smith and VW Hays (eds.), Proc. 14th Int. Grassl. Congr., Lexington, Ky. Boulder, Colo.: Westview, 336-41.

Holt, EC, MJ Norris, and JA Lancaster. 1969. Production and Management of Small Grains for Forage. Tex. Agric. Exp. Stn. Bull. B-1082.

Hoveland, CS, and EM Evans. 1970. Cool-Season Perennial Grass and Grass-Clover Management. Ala. Agric. Exp. Stn. Circ. 175.

Hoveland, CS, EM Evans, and DA Mays. 1970. Cool-Season Perennial Grass Species for Forage in Alabama. Ala. Agric. Exp. Stn. Bull. 397.

Hoveland, CS, WB Anthony, JA McGuire, and JG Starling. 1978. Beef cow-calf performance on Coastal bermudagrass overseeded with winter annual clovers and grasses. Agron. J. 70:418-20.

Irlbeck, N, J Ward, and T Klopfenstein. 1990. Corn, sorghum, or combinations for residue grazing. Nebr. Beef Cattle Rep. MP 55:55.

Irlbeck, N, T Klopfenstein, M Sindt, and R Stock. 1991. Quality and quantity of corn residue on gain of grazing cattle. Nebr. Beef Cattle Rep. MP 56:19.

Jung, GA, and RE Kocher. 1974. Influence of applied nitrogen and clipping treatments on winter survival of perennial cool-season grasses. Agron. J. 66:62-65.

Jung, GA, LL Wilson, PJ LeVan, RE Kocher, and RF Todd. 1982. Herbage and beef production from ryegrass-alfalfa and orchardgrass-alfalfa pastures. Agron. J. 74:937-42.

Kalmbacher, RS, P Mislevy, and FG Martin. 1980. Sod-seeding bahiagrass in winter with three temperate legumes. Agron. J. 72:114-18.

Kardos, LT, WF Sopper, EA Meyers, and JM Bollog. 1971. "Living filter" system curbs water pollution by removing effluent. Pa. Agric. Exp. Stn. Sci. Agric. 18:6-7.

King, CC, Jr., WB Anthony, SC Bell, LA Smith, and H Grimes. 1971. Beef Cow Grazing Systems Compared on Entaw Clay. Ala. Agric. Exp. Stn. Bull. 424.

Klingebiel, AA, and PH Montgomery. 1961. Land Capability Classification. USDA Agric. Handb. 210. Washington, D.C.: US Gov. Print. Off.

Klopfenstein, TJ. 1991. Low-input, high-forage beef production. In Sustainable Agriculture Research and Education in the Field—A Proceedings. Washington, D.C.: National Academy Press.

Klopfenstein, TJ, and FG Owen. 1981. Value and potential use of crop residues and by-products in dairy rations. J. Dairy Sci. 64:1250-68.

Klopfenstein, TJ, B Vieselmeyer, N Irlbeck, E Gutierrez-Ornelas, V Wilkerson, and R Stock. 1990. Stocker calf cornstalk grazing. Low Input Sustainable Agric. Beef and Forage Conf., 13-14 June, Omaha, Nebr.

Kossila, VL. 1984. Location and potential feed use. In F Sundstol and E Owen, (eds.), Straw and Other Fibrous By-production as Feed. Amsterdam: Elsevier, Chap. 2.

Larson, WE, FJ Pierce, and RH Dowdy. 1983. The threat of soil erosion to long-term crop production. Sci. 219:458-65.

Marten, GC, and AW Hovin. 1980. Harvest schedule, persistence, yield, and quality interactions among four perennial grasses. Agron. J. 72:378-87.

Marten, GC, CE Clapp, and WE Larson. 1979. Effects of municipal wastewater effluent and cutting management on persistence and yield of eight perennial forages. Agron. J. 71:650-58.

Matches, AG, FA Martz, M Mitchell, and S Bell.

1971. Research in Agronomy: Southwest Center Grazing Trials. Univ. Mo. Agron. Dept. Misc. Publ. 71-3.

Mott, GO, CJ Kaiser, RC Peterson, R Peterson, Jr., and CL Rhykerd. 1971. Supplemental feeding of steers on *Festuca arundinacea* Schreb. pastures fertilized at three levels of nitrogen. Agron. J. 63:751-54.

National Research Council (NRC). 1990. Alternative Agriculture (Committee on the Role of Alternative Farming Methods in Modern Production Agriculture). Board on Agriculture. Washington, D.C.: National Academy Press.

Nelson, M, R Gates, N Voyles, T Klopfenstein, R Britton, and J Ward. 1984. Methods of ammoniation of roughages. Nebr. Beef Cattle Rep. MP 47:37.

Owen, E, and MCN Jayasurirya. 1989. Use of crop residues as animal feeds in developing countries. Res. and Dev. Agric. 6(3):129.

Perry, TW, DA Huber, GO Mott, CL Rhykerd, and RW Taylor. 1971. Effect of level of pasture supplementation on pasture drylot and total performance of beef cattle. I. Spring pasture. J. Anim. Sci. 32:744-48.

Petritz, DC, VL Lechtenberg, and WH Smith. 1980. Performance and economic returns of beef cows and calves grazing grass-legume herbage. Agron. J. 72:581-84.

Raguse, CA, JL Hull, and RE Delmas. 1980. Perennial irrigated pastures: III. Beef calf production from irrigated pasture and winter annual range. Agron. J. 72:493-99.

Rhykerd, CL, CH Noller, JE Dillon, JB Ragland, BW Crowl, GC Naderman, and DL Hill. 1967. Managing Alfalfa-Grass Mixtures for Yield and Protein. Ind. Agric. Exp. Stn. Res. Bull. 839.

Russell, JR. 1990. Beef cow cornstalk grazing. In Proc. Low Input Sustainable Agric. Beef and Forage Conf., 13-14 June, Omaha, Nebr., 115.

Russell, JR, MR Brasche, and AM Cowen. 1993. Effects of grazing allowance and system on the use of corn-crop residue by gestating beef cows. J. Anim. Sci. 71:1256.

Schaller, FW. 1967. The Beef Cow Herd in Iowa—The Forage Supply. Iowa Agric. Ext. Serv. Pam. 369.

Smith, SR, Jr., JH Bouton, and CS Hoveland. 1992. Persistence of alfalfa under continuous grazing in pure stands and in mixtures with tall fescue. Crop Sci. 32:1259-64.

Stock, RA. 1990. Low input growing-finishing systems. Proc. Low Input Sustainable Agric. Beef and Forage Conf., 13-14 June, Omaha, Nebr.

Stricker, JA, AG Matches, GB Thompson, VE Jacobs, FA Martz, HN Wheaton, HD Currence, and GF Krause. 1979. Cow-calf production on tall fescue-ladino clover pastures with and without nitrogen fertilization or creep feeding: Spring calves. J. Anim. Sci. 48:13-25.

Sundstol, F, E Coxworth, and DN Mowat. 1978. Improving the nutritive value of straw and other low-quality roughages by treatment with ammonia. World Anim. Rev. 26:13-21.

Timmons, JF. 1980. Protecting agriculture's natural resource base. J. Soil Water Conserv. 35:5-11.

USDA. 1992. Agricultural Statistics. Washington, D.C.: US Gov. Print. Off.

Van Keuren, RW. 1970. Symposium on pasture

methods for maximum production in beef cattle: Pasture methods for maximizing beef cattle production in Ohio. J. Anim. Sci. 30:138-42.

Van Keuren, RW, and AG Matches. 1988. Pasture production and utilization. In AA Hanson, DK Barnes, and RR Hill, Jr. (eds.), Alfalfa and Alfalfa Improvement, Am. Soc. Agron. Monogr. 29. Madison, Wis., 239-57.

Vetter, RL, and M Boehlje. 1978. Plant and Animal Products in the U.S. Food System. Washington, D.C.: National Academy of Science.

Voss, RD, and WD Shrader. 1982. Crop Rotations: Effect on Yields and Response to Nitrogen. Iowa Coop. Ext. Serv. Pam. 905.

Wagner, RE. 1954a. Influence of legume and fertilizer nitrogen on forage production and botanical composition. Agron. J. 46:167-71.

———. 1954b. Legume nitrogen versus fertilizer nitrogen in protein production of forage. Agron. J. 46:233-36.

Walker, O, C Maynard, and B Brant. 1966. Economics of Winter Pasture. Okla. Agric. Ext. Facts 105.

Wanapat, M, and C Denendra. 1985. Relevance of Crop Residues as Animal Feeds in Developing Countries. Bangkok: Funny Press.

Ward, JK. 1978. Utilization of corn and grain sorghum residues in beef cow forage systems. J. Anim. Sci. 46:831-40.

Ward, JK, G Llamas-Lamas, DB Faulkner, TJ Klopfenstein, and IG Rush. 1981. Techniques for ammonia treatment of straw. In Midwest Sect., Am. Soc. Anim. Sci. Abstr. 79:100.

Wedin, WF. 1966. Legume and inorganic nitrogen for pasture swards in subhumid, microthermal climates of the United States. In Proc. 9th Int. Grassl. Congr., Sao Paulo, Brazil, 1163-69.

———. 1970. What can Iowa do with a million acres of forage? Iowa Farm Sci. 24(9):3-8.

Wedin, WF, and CS Hoveland. 1987. Cereal vegetation as a source of forage. In Nutritional Quality of Cereal Grains: Genetic and Agronomic Improvement, Am. Soc. Agron. Monogr. 28: Madison, Wis., 83-99.

Wedin, WF, and AG Matches. 1973. Cropland Pastures. In ME Heath, DS Metcalfe, and RF Barnes (eds.), Forages: The Science of Grassland Agriculture, 3d ed. Ames: Iowa State Univ Press, 607-16.

Wedin, WF, and RL Vetter. 1970. Pasture for beef production in the western Corn Belt, USA. In Proc. 11th Int. Grassl. Congr., Surfers Paradise, Queensland, Australia. St. Lucia: Univ. of Queensland Press, 842-45.

Wedin, WF, JD Donker, and GC Marten. 1965. An evaluation of nitrogen fertilization in legume-grass and all-grass pasture. Agron. J. 57:185-88.

Wedin, WF, RL Vetter, and IT Carlson. 1971. The potential of tall grasses as autumn-saved forages under heavy nitrogen fertilization and intensive grazing management. In Proc. 32d Conf. N.Z. Grassl. Assoc. (1970), 160-67 (Lincoln, New Zealand).

Wedin, WF, HJ Hodgson, JE Oldfield, and KJ Frey. 1980. Feed Production. In WG Pond et al. (eds.), Animal Agriculture: Research to Meet Human Needs in the 21st Century. Boulder, Colo.: Westview, 153-92.

West, CP, and WF Wedin. 1985. Dinitrogen fixation in alfalfa-orchardgrass pasture. Agron. J. 77:89-94.

15 Permanent Pasture Ecosystems

DWAYNE A. ROHWEDER
University of Wisconsin

KENNETH A. ALBRECHT
University of Wisconsin

PASTURE ecosystems include plants, animals, insects, organic residues, atmospheric gases, water, minerals, and their interactions that are involved in the flow of energy and the circulation of matter. Pasture ecosystem management manipulates this soil-plant-animal-environment complex on lands devoted to grazing for desired economic and ecological returns.

The average level of output from grassland is normally only about half that achieved by the best grassland farmers, and the best farmers achieve only about half that shown to be theoretically possible (Hodgson 1990). Because land is a major cost in forage-animal production, economic returns from such enterprises should be reported per unit of land area. The management system used should be economically relevant and maintain or improve soil productivity, as well as control or minimize surface runoff, erosion, and pollution. Management practices also should discourage, for example, destroying plant cover, encroachment of weeds and brush, and uneven distribution of animal excreta.

Permanent pastures are composed of perennial grasses and legumes or self-seeding annuals—frequently both—and are maintained

DWAYNE A. ROHWEDER is Emeritus Professor of Agronomy-Extension, University of Wisconsin, Madison. He holds the MS and PhD degrees from Iowa State University. He has researched production, management, and utilization of forages.

KENNETH A. ALBRECHT is Associate Professor of Agronomy, University of Wisconsin, Madison. He holds the MS degree from the University of Minnesota and the PhD from Iowa State University. His research is on forage management and utilization.

through several years for grazing. A permanent pasture may be an indigenous stand, may have been seeded, or may have resulted from occupation of onetime cultivated fields by forage plants that were sufficiently aggressive to spread without man's assistance.

Pastures are classified as cropland used as pasture, woodland pasture, open permanent pasture on land other than woodland, or cropland used for pasture or grazing. Improved pasture is that portion of open permanent pasture that has received one or more of the following practices: liming, fertilizing, seeding, irrigating, draining, or clearing and control of weed and brush growth. Pastures also are classified as to need for erosion control and improvement.

A 1982 inventory of permanent pasture resources shows that about 54% needed some form of improvement (USDA-SCS 1982). Stocking and grazing full, productive swards in a schedule that optimizes the potential of the forage species will assist enterprise managers achieve their "pasture and livestock" potential.

IMPORTANCE OF PASTURE

In the US, pasture forage is produced on 53.4% of the 390.6 million ha of land in farms (Table 15.1). Area in pasture declined nearly 17% from 1949 to 1987. Nearly one-half of the decrease occurred in the 10-yr period 1978-87, due largely to governmental policies and economic incentives to produce row crops. All pasture totaled 208.9 million ha in 1987. Permanent pasture and rangeland on farms totaled 166.2 million ha or 79.5% of all pasture in 1987.

In 1974, pastures contributed 40% of the

TABLE 15.1   Hectares (millions) of US cropland pasture, other pasture, and woodland pasture on farms, 1949-87

Year	Cropland (ha)	Other[a] (ha)	Woodland (ha)
1949	28.1	168.8	54.7
1959	26.5	188.8	37.5
1964	23.3	198.6	33.3
1969	37.5	. . .	. . .
1974	33.5	. . .	. . .
1978	30.8	176.9	19.6
1987	26.3	166.2	16.4

Sources: USDC (1978, 1987).
[a]Includes rangeland.

feed consumed by livestock in the US (Table 15.2), and harvested forages (hay and silage) contributed an additional 20%. The balance, or 40%, was obtained from concentrate feed, including grain and protein supplements. In 1988, concentrates composed 42%, harvested forage 16%, and pasture 42% of the feed units consumed by all livestock. All forage furnished 58% of the feed units (USDA 1992).

Ruminant livestock are efficient consumers of forages. Generally, sheep and goats obtain 90%, beef over 80%, and dairy more than 60% of their feed needs from both stored forages and pasture.

The amount of feed obtained from pasture varies with different classes of livestock as follows: sheep and goats, 80%; beef cattle, 74%; horses, 51%; dairy cattle other than lactating cows, 43%; and milking cows, 18%. The proportion of feed obtained from forage for all livestock increased by 10% over the 5-yr period 1969-74, while the proportion of concentrates decreased.

Forage produced on permanent pasture has little economic value unless consumed by ruminant animals. Ruminants also convert crop and animal residues not suitable for people into nutritious animal products for human consumption. Ruminants need not compete with humans for plant protein and energy since forage grown on land not suited for row crop production can furnish a major portion of the nutritional needs for beef cow and sheep enterprises (Blaser et al. 1986).

Since the number of animal units has decreased by 14% during the past four decades, methods of pasture utilization are changing. For example, since 1965, dairy cow and sheep numbers decreased 34% and 55%, respectively. Conversely, beef cow numbers remained relatively steady with about 40% of the beef cows in the US being located east of the Missouri River. Beef cow-calf units use large amounts of forage from pasture (Table 15.2).

PERMANENT PASTURE REGIONS OF THE US

Climatic conditions divide the US into five major pasture regions (Fig. 15.1). Only the permanent pasture of the humid regions (Regions 1, 2, and 5) are considered in this chapter. Permanent pastures in much of the humid region of the US are often located on lower-productive (lower-quality) sites and receive low management input. These pastures generally are not improved; consequently, yields are low. Thus, it is critical to use ecology-based management to retain desirable species.

The practice of using abandoned or unproductive cropland for pasture without improvement is a waste of a natural resource. It is estimated that 38% of permanent pasture needs treatment through improvement or reestablishment of vegetative cover. Only 46% is adequately treated to control erosion (USDA-SCS 1982). Production from open permanent pastures where Kentucky bluegrass (*Poa pratensis* L.) and white clover (*Trifolium repens* L.) are the dominant species generally ranges from 610 to over 3180 kg/ha of dry matter, or 115-160 kg/ha of beef (Table 15.3).

Quality Influencing Pasture Production

SOIL FACTORS. Soil factors influencing pas-

TABLE 15.2.   Consumption of all feed by kind of livestock, 1969 and 1974.

Kind of livestock	Concentrates[a] 1969	1974	Hay 1969	1974	Other harvested forage 1969	1974	Pasture 1969	1974	All forage[a] 1969	1974
Cattle on feed	74	61	13	18	8	17	5	4	26	39
Other beef cattle	8	7	15	13	4	6	73	74	92	93
Milk cows	37	40	23	21	20	21	20	18	63	60
Sheep and goats	10	11	5	5	4	4	81	80	90	89
Other dairy	17	16	30	26	7	15	46	43	83	84
Horses	21	. . .	18	. . .	10	. . .	51	. . .	79	. . .
All livestock	45	40	12	12	6	8	37	40	55	60

Source: Rohweder and Van Keuren (1985).
[a]Concentrates + all forage = 100%; hay + other harvested forage + pasture = all forage.

Fig. 15.1. Five major pasture regions of the US with type of forage that provides the majority of pasturage. (Semple et al. 1946).

ture yields include slope, moisture status, texture, depth, degree of erosion, soil fertility, and soil acidity. In general, yields increase with increasing soil depth and fineness of texture. In southwestern Wisconsin, yields decreased as slope increased.

Environmental concerns may require areas on many farms be devoted to grasslands for erosion control, stream bank stability, and riparian strips for maintaining water quality. These areas may be excluded from pasture operations, or controlled stocking and grazing practices will be required to minimize pollution.

TREES AND WOODED AREAS. Trees and thin wooded areas are often a part of permanent pastures. They are rarely cut for hay due to machinery limitations. These limitations also affect fertilizing, liming, weed and brush con-

trol, and insect spray operations. Thus, alternatives must be used. Wisconsin studies show that production from pastures on comparable soils and slopes, but with some tree cover, was only 310 kg/ha compared with an average of 1625 kg/ha with no tree cover (Ahlgren et al. 1946).

WEEDS, BRUSH, AND INSECTS. Weeds, brush, and insects such as white grubs often reduce forage production and quality in permanent pastures, yet they are rarely sprayed. Sward improvement as well as improved stocking and grazing practices must be practiced to improve pasture production.

VEGETATIVE COVER. Production also varies with composition of vegetative cover. West Virginia studies have shown the carrying capacity of an individual pasture to be related to

TABLE 15.3. Dry matter and beef production from pastures on selected soils in the humid US

Soil and location	Unimproved (kg/ha)	Fertilized (kg/ha)	Renovated (kg/ha)
Dry matter			
Shallow soils—Ont.[a]	610	. . .	2060
Gale loam—Wis.			
15%-25% slope	1990	. . .	3770
26%-35% slope	1365	. . .	3420
wooded[b]	310	. . .	. . .
Fayette-Dubuque—Wis.	2195	5875	. . .
Miami silt loam—Wis.	3180	. . .	5155
Dewey-Humphries—Ala.	. . .	4075	3360
Beef			
Weller-Lindley—Iowa	120	. . .	375
Nappanee-Brookston—Ind.	160	350	390
Norfolk sandy loam—Ala.	115	330	. . .
Fayette-Dubuque—Wis.[c]			
(1)	(est) 110	348	345
(2)	. . .	538	507
(3)		747	697

Source: Rohweder and Van Keuren (1985).
[a]Twenty-one sites and 78 site yr.
[b]Approximately 50% tree cover.

type of vegetation (Pierre et al. 1937). An Iowa study also indicated a highly significant association between percentage of ground covered by economic species and yield of forage on four different soils (Rohweder 1963).

Beef and Sheep Enterprises. The beef industry constitutes the largest market for forage. Beef cow-calf and sheep-raising enterprises are effective users of permanent pastures. Improving permanent pastures through establishment of productive legumes and grasses, fertilization, and improved stocking in turn improves animal production and performance in beef- and sheep-feeding enterprises. Highly productive permanent pastures containing productive legumes and grasses can be effective components in a complete feeding program for dairy enterprises. Pastures having a tough, resilient grass turf also can be used in horse programs.

BEEF PASTURES. Production from pasture regions 1-b and 2-b (Fig. 15.1) verify the data in Table 15.3. Although orchardgrass (*Dactylis glomerata* L.) and 'Kentucky 31' tall fescue (*Festuca arundinacea* Schreb.) fertilized with 224 kg nitrogen (N)/ha produced 27% more forage resulting in 31% more grazing days by steers than did ladino clover (*T. repens* L.) -grass pastures without N, the legume-grass pastures had 18% higher average daily gain. Therefore, animal gains per hectare averaged only 10% higher for N-fertilized grass (Blaser et al. 1986).

Forage production from sainfoin (*Onobrychis viciifolia* Scop.) -wheatgrass (*Agropyron* and *Thinopyron* spp.) mixtures usually did not exceed that of the grasses fertilized with N, but intake was 13%-29% higher for the mixtures because of higher forage quality, except under drought conditions (Griggs and Matches 1991; Karnezos and Matches 1991; Mowrey and Matches 1991).

Tennessee data show that ladino clover-orchardgrass pastures produced 561 kg beef/ha in spring. From 175 to 200 kg N/ha was needed on bermudagrass (*Cynodon dactylon* [L.] Pers.) pastures to achieve a high summer forage yield potential and to fill the summer gap in forage-beef enterprise systems with yields near 300-350 kg beef/ha. About 400 kg N/ha were needed on bermudagrass pastures to equal the beef yields from ladino clover-orchardgrass pastures in spring (Fribourg et al. 1979).

In Florida, average daily gain for beef calves on aeschynomene (*Aeschynomene*

americana L.) -limpograss 'Floralta' (*Hemarthria altissima* [Poir.] Stapf & C.E. Hubb.) pastures was 80% greater than from limpograss pastures fertilized with N. This resulted in nearly 43% higher steer gains per ha (377 versus 264 kg) (Rusland et al. 1988).

The data from pastures in the humid regions of the US show that, in general, higher production of forage resulting in higher carrying capacity is gained from N fertilization of grasses; however, higher intake and ADG is usually achieved with legume-grass pastures resulting in animal yields equal to N-fertilized grasses.

CREEP FEEDING. Creep feeding or creep grazing will improve calf or lamb gains when available pasture is low or when low forage quality inhibits digestible dry matter intake. Creep grazing permits calves or lambs to graze pastures high in quality adjacent to those grazed by nursing cows or ewes. The management principle states that calves, lambs, or young stock require higher nutrition than cows or ewes especially as their rumen begins to function and milk production of the dam decreases. Creep grazing improves weight gains and weaning weights of calves or lambs (Chap. 13).

Dairy Enterprises. Since World War II, high-producing dairy enterprises have shifted from use of rotational stocking and strip-grazing to intensive feeding of forage that is harvested at optimum quality, stored to preserve that quality, and fed in drylot with total mixed rations balanced for energy, protein, and minerals. Milk production increased concurrently. Year-round drylot feeding may be the best economic system for soils and climates that can produce high crop yields, as well as for maintaining soil productivity and the environment. But most dairy farms also have some land with soil and topography best suited to pasturing perennial forage plants. These pastures are usually used to supplement the stored forage in feeding lactating cows, dry cows, and young stock.

DAIRY PASTURES. Rotationally stocked legume and grass pastures probably should not be the only forage in rations of high-producing dairy herds. Milk production by cows receiving their entire diet from pasture-based operations generally averaged 5000 to 7300 kg/cow as an annual rolling herd average (RHA) compared with highs of 10,454 to 13,846 kg/cow under intensive feeding (Blaser et al. 1986; McVickar 1974; Howard 1993). An Ohio study shows milk production decreased as forage quality of the species in the pasture declined due to advancing maturity, requiring marked increases in grain to maintain optimum production. High-producing cows will not reach their genetic potential for milk production when fed forage alone (Van Keuren et al. 1966).

A more recent Ohio study evaluating seasonal milk production with cows fed supplement along with intensively stocking them in legume-grass pastures showed an annual RHA of only 6129 kg/cow. Fixed costs were 32%-35% below conventional enterprise budgets, but total costs of producing milk on pasture were 40%-50% above the price received for milk. Protein concentration in the grazed forage ranged from 15% to 23%. Obtaining and maintaining body condition on the cows was a continual challenge (Shoemaker et al. 1992). However, milk production and body condition were maintained in cows pastured on alfalfa (*Medicago sativa* L.) and alfalfa-grass pastures in Wisconsin when rations were adequately balanced for energy and protein (Vaughan et al. 1992).

Species in Permanent Pastures. In general, climate in each broad pasture region determines botanical composition of the prevailing permanent pasture. Grasses and legumes persisting in permanent pastures depend upon climatic and soil characteristics, palatability, fertilizer treatments, cultural practices, and grazing management.

Kentucky bluegrass with varying amounts of white clover is widely distributed on the more fertile soils in Region 1-a and the northern parts of Regions 1-b and 5-a, especially on pastures stocked continuously for several months. Smaller percentages of such grasses as orchardgrass, tall fescue, bentgrass (*Agrostis* L. spp.), redtop (*A. gigantea* Roth), timothy (*Phleum pratense* L.), Canada bluegrass (*Poa compressa* L.), and others also are found in permanent pastures. The fine-leaved fescues (*Festuca* L. spp.), Canada bluegrass, redtop, and timothy predominate on drier, less fertile soils, along with broadleaf weeds and annual grass weeds, which often have low quality. Bentgrass, timothy, and redtop predominate on more moist soils.

In the central belt, between the North and South (1-b), white clover and annual lespedezas (*Kummerowia* spp.) also are found in many of the permanent pastures. Bermudagrass, bahiagrass (*Paspalum notatum*

Flugge), annual lespedezas, and crimson clover (*T. incarnatum* L.) are found in upland pastures of the South (Regions 2-a and 2-b). In lowland pastures, carpetgrass (*Axonopus affinis* Chase), dallisgrass (*P. dilatatum* Poir.), and several clover species predominate. Winter annuals predominate in Region 5-b.

IMPROVED LEGUMES AND GRASSES

Improved, high-yielding, more palatable legume and grass species are becoming widely distributed in permanent pastures as a result of continuing research and pasture improvement programs. New cultivars having improved yield, palatability, winterhardiness, disease resistance, and quality are being developed.

Alfalfa, red clover (*T. pratense* L.), ladino clover, smooth bromegrass (*Bromus inermis* Leyss.), timothy, and reed canarygrass (*Phalaris arundinacea* L.) are being used to improve pastures in Regions 1-a and 1-b. Alfalfa is generally the most productive legume available for pasture renovation, but it requires well-drained soils, high pH, and optimum fertility. It frequently lacks persistence, even with rotational grazing, although new grazing tolerant varieties may improve this characteristic (Brummer and Bouton 1991, 1992). And reseeding back into old alfalfa stands often is not satisfactory due to autotoxicity.

Red clover is easier to establish and to reseed in permanent pastures but is shorter lived. Birdsfoot trefoil (*Lotus corniculatus* L.), increasing in prominence, offers advantages over alfalfa in Region 1-a. It persists longer than alfalfa and red clover due to natural reseeding under controlled, continuous stocking or rotational stocking when properly managed and fertilized and does not cause bloat. With special grazing management to permit seed production, birdsfoot trefoil is being used in some areas of the southern part of Region 1-b.

Tall fescue is a basic winter grass in permanent pastures of Regions 2-a and 2-b and for cool-season grazing in 1-b. It is usually grown with ladino clover, red clover, or annual lespedeza. Ladino clover and white clover are grown in combination with tall fescue, bermudagrass, dallisgrass, and bahiagrass to improve forage quality and production in southern pastures.

Sainfoin is suggested for use with wheatgrasses to improve forage quality and intake. 'Coastal' bermudagrass, dallisgrass, annual and perennial ryegrass (*Lolium multiflorum*

Lam. and *L. perenne* L.), crimson clover, arrowleaf clover (*T. vesiculosum* Savi), subterranean clover (*T. subterraneum* L.), and sericea lespedeza (*Lespedeza cuneata* [Dum.-Cours.] G. Don) are increasing in importance to replace lower-producing permanent pasture in the South. Red clover, orchardgrass, and timothy often are used in renovating cool-season pastures in Regions 1-b and 2-a. Red clover has proved satisfactory in winter forage programs with tall fescue for beef cows, utilized as fall-saved regrowth and field-stored round bales.

'Pangola' digitgrass (*Digiteria eriantha* Steud.) has become important in the lower part of Region 2-b. Alfalfa, tall fescue, and ladino clover are found on the heavier, more fertile soils of the region. Warm-season grasses such as switchgrass (*Panicum virgatum* L.) and Caucasian bluestem (*Bothriochloa caucasica* [Trin.] C.E. Hubb.) are being used increasingly in upper southern and lower northern areas. In these regions, they are frequently used for summer stocking to complement the production during spring and fall on cool-season pastures.

Several species are suggested as alternatives or additions to 'Pensacola' bahiagrass pastures in the South. Grass species such as 'Tifton 44' bermudagrass and limpograss and legume species such as aeschynomene, 'Florida' carpon desmodium (*Desmodium heterocarpon* [L.] DC.), and 'Florigraze' rhizoma peanut (*Arachis glabrata* Benth.) offer potential for pasture improvement. Although forage quality and, sometimes, yield are increased, even on wetter soils, establishment and persistence of many of these species are problems (Aiken et al. 1991; Dunavin 1992; Rusland et al. 1988).

In general, legumes are seeded with grasses to improve yield, quality, and seasonal distribution of pasturage; however, N-fertilized grasses are best suited for some situations.

Planting endophyte-free (*Acremonium coenophialum* Morgan-Jones & Gams) seed of tall fescue cultivars has increased beef gains per hectare over 75%; however, beef gains may not always be increased on fungus-free pastures. State recommendations should be followed for changing present stands to those containing fungus-free cultivars. Using low-alkaloid reed canarygrass cultivars also improves the potential for this grass.

POTENTIAL FOR IMPROVEMENT

Climate and soil characteristics make the humid region ideally suited for forage production. Rising land prices, higher meat and milk prices, water quality laws, erosion control, interest in wildlife, and the pressure placed on arable cropland for grain production are causing renewed interest in making permanent pastures more productive and profitable (Fig. 15.2).

More than 76% of permanent pasture in humid regions of the US is located on soils with great potential for improvement (Classes I-IV) (Diderickson 1980). Of this pasture, over 57% is found on soils suitable for regular cultivation (Classes I-III), with hazards to cultivation varying with slope, soil depth, topography, and available moisture. The remaining 19% is located on soils suitable for occasional or limited cultivation (Class IV).

New technology is indicating how the permanent pasture can be improved to better meet the needs of the beef enterprise. Considerable evidence indicates that most permanent pastures can be improved. Legume-grass mixtures produce as much beef per hectare as grasses fertilized with 132-220 kg N/ha (Table 15.3: Fayette-Dubuque [1] and [2]). Improving soil fertility with grazing management and providing additional energy to the grazing animal improved beef yields about 165 and 220 kg/ha, respectively, Fayette-Dubuque (3).

A pasture can be improved by complete seedbed preparation and reseeding, renovation, overseeding or sod-seeding, or fertiliza-

Fig. 15.2. Permanent pasture improved by adding legumes means more profit through more and better cattle. *Univ. of Kentucky photo.*

tion and control of undesirable vegetation. The alternative selected will depend on soil capability, quality of present vegetative cover, livestock enterprise, use for the pasture, and need for forage.

Renovation. Renovation is the improvement of a pasture by partial or complete destruction of a sod plus liming, fertilizing, weed control, and seeding as required to establish desirable forage plants (Leonard et al. 1968). Renovation can restore permanent pasture to greater productivity with little danger of soil loss because the dead or killed vegetation remains on or near the soil surface (Fig. 15.3). Production from renovated permanent pasture may be increased two- to fivefold depending on the soil characteristics and the condition of the sod. Animal production may be further increased by addition of energy from legume forage and/or grain supplements to livestock rations (Table 15.3).

Steps to follow for a successful renovation program include

1. Select the area to be renovated a year before the planned date of seeding. Before seeding, control noxious weeds such as Canada thistle (*Cirsium arvense* [L.] Scop.)

Fig. 15.3. An improved and properly managed permanent pasture can be one of the most profitable parts of the farm enterprise. Such a pasture is this tall fescue field following renovation, seeded to red and ladino clover in second growth. *Univ. of Kentucky photo.*

and ironweed (*Vernonia baldwinii* Torr.) using herbicides.

2. Test the soil to determine soil nutrient deficiencies, and apply nutrients needed. Apply lime at least 6 mo prior to seeding; incorporation is preferred. Apply phosphorus (P) and potassium (K) prior to or at seeding time.

3. Seed early in spring for success. To permit more effective seeding, allow close grazing (to about 5 cm) of existing grass sward before renovating. Greater success in seeding establishment will be obtained if much of the tillage is done in late summer and fall in the North.

   a. Early spring renovation in Region 1-a and either early spring or late summer renovation in Region 1-b are satisfactory. Late summer seeding is risky if precipitation is less than normal.

   b. Under average to above average rainfall conditions, the new seeding will be established by midsummer and can be grazed without injury when properly managed. This results in essentially no loss in production during the year of renovation.

4. Prepare field for seeding. Several options are available.

   a. Soils not low in pH and fertility: on permanent pastures containing largely Kentucky bluegrass and common pasture weeds such as dandelion (*Taraxacum officinale* Weber) and plantain (*Plantago major* L.) and where soil pH and fertility are not low, seedbed preparation can be minimized or eliminated by use of a sod-penetrating no-till drill with chemical herbicides.

   • Short-lived herbicides that control broad-leaved weeds and burn back and retard existing grass vegetation are effective in suppressing the grass sod and permitting legume establishment. Top-dress needed fertility when seeding to ensure success.

   • Where perennial broadleaf weeds are a problem, start the control program 1 or 2 yr before seeding.

   • Herbicides can be used to manage vegetation in thick, vigorous sods when interseeding winter annuals into dormant warm-season grasses such as bermudagrass in Region 2.

   b. Soils low in pH and fertility: combining "minimum" fall tillage (e.g., disking or field cultivation) followed by a herbicide application prior to seeding with a no-till drill in spring has often improved success in the North. This tillage step permits incorporation of lime and fertilizers on soils having low pH and fertility, as surface-applied lime moves into the soil very slowly. Fall tillage helps destroy old sod, which reduces competition in spring and improves legume establishment. "Minimum" tillage leaves a sod surface or mulch, thus reducing soil erosion during winter.

• Where vigorously growing perennial grasses such as quackgrass (*Elytrigia repens* [L.] Nevski) are present, more than one application of a short-lived herbicide may be required to achieve grass suppression.

• Herbicides to eliminate grass also can be used. However, this practice often substitutes a sward of competitive weeds for the grass, resulting in seeding failures.

• A better approach, where practical, would be to plant a no-till row crop for 1 yr to suppress vegetation and then to reseed to productive pasture species.

c. Consult current state recommendations for herbicides to use in pasture renovation.

5. Plant seed of adapted cultivars of high-yielding legumes and grasses at recommended rates in early spring for best results. Place seed at 0.5 to 1.5 cm depth in a firm seedbed for good seed-soil contact and best establishment. Seedbeds in renovated sods often are too loose for successful establishment because of surface trash; thus, they should be firmed with a culti-packer or similar piece of equipment. Inoculate legumes with proper N-fixing bacteria.

6. Graze renovated pastures short until livestock begin to defoliate new legume seedlings. Then remove livestock to permit forage regrowth. Thereafter, control stocking to best suit regrowth of the legume. Proper stocking will permit grazing of some species throughout the summer.

7. Renovated and reseeded pastures will revert to their former botanical composition within 1 to 3 yr unless fertility, weed and brush control, and proper grazing management are continued on a regular basis (Ahlgren et al. 1944).

• Continuous close grazing and failure to maintain fertility will cause tall-growing legumes to die, resulting in decreased pasture production. Select a grazing management program consistent with the growth habit of the legume.

• Annually apply topdressing to replace nutrients removed in meat, milk, and/or forage the previous year.

Complete Seedbed Preparation with Reseed-

ing. Complete seedbed preparation followed by reseeding has been superior to renovation in several areas of the North where perennial grass weeds in the sod are difficult to subdue. Legume establishment is markedly improved by plowing as compared with tilling with a field cultivator (Scholl et al. 1970a). Studies also indicate that 20%-60% less work is required to prepare a seedbed by plowing than may be required with any one of several surface-cultivating tools (Sprague et al. 1947). Steps in reseeding the pasture are identical to those of renovation except for tillage.

Perennial grass weeds should be subdued in late summer by timely and repeated use of a heavy disk or field cultivator, and the seedbed should be prepared in late fall by plowing on the contour to minimize erosion losses. Freezing and thawing over the winter, followed by a light disking the following spring, provide a firm seedbed.

Use of an intervening row crop for 1 yr may or may not be desirable, depending on soil conditions (whether or not sloping soils permit row crop productivity), to reduce weed population and competition from the sod and to help decrease costs.

Sod-seeding or Overseeding. Reseeding of legumes and/or grasses into thin permanent pasture sods by no-till sod-seeding or overseeding without complete seedbed preparation can be successful in the humid region. Success depends on the ability of the introduced seedlings to become established in competition with the old pasture sod. Once established, the persistence and vigor of the introduced species will be determined largely by pasture management.

In the North, new seedings are made in the spring at a time coinciding with the beginning of the maximum growth period of grasses in the old sod. Placing lime and fertilizer—especially N—in a band near the seed stimulates the grass sod and results in increased competition. Consequently, the introduced seedlings are seldom able to become established. However, "minimum" tillage to reduce vigor of the grass sod, and to incorporate broadcast lime and fertilizer on soils having low pH and fertility, improves establishment. No-till seeding is most successful in pastures having optimum fertility.

Several studies indicate that chemical herbicides, tilled strips of varying widths, and tillage methods have reduced competition from the existing vegetation and improved seedling establishment in the southern part of Region 1-a and in Region 1-b.

Sod-seeding of legumes such as white clover, alfalfa, and birdsfoot trefoil offers an opportunity to economically increase yields from low-producing, cool-season, permanent pastureland on neglected hillsides, steep slopes, and eroded or stony soils where plowing may be difficult.

Spring seedings generally are most successful (Decker et al. 1969; Taylor et al. 1969; Martin et al. 1979), and the length of time required for legumes to become well established varies with species (Van Keuren and Triplett 1970). The most important factors for successful establishment are proper depth of seeding, good seed-soil contact, and subduing of the existing vegetation to reduce competition. Seeding followed by trampling from livestock often improves on-the-farm results, especially if legumes are seeded prior to the start of competitive growth from the grass in the sod and if the soil is not too wet.

Chemicals used to kill or suppress existing vegetation in a band along the row have enhanced legume establishment only when grass stands were dense and growth vigorous. The most satisfactory width for the tilled strip varies and appears to be dependent on soil conditions. Sod-seeding has been the most successful in the South and/or subtropical regions where introduced legumes and grasses are seeded at a time that does not coincide with the maximum growth period of the existing sward (Coats 1957). Seedings generally are made in late summer or early fall. Winter annuals or perennial grasses and legumes can be incorporated into the sod to provide winter grazing. Fall plantings can be made directly into thin, dormant, or heavily grazed sod. However, one should allow close grazing of vigorous swards prior to and following seeding to reduce competition.

Overseeding with pelleted seed of legumes such as white clover and alfalfa from an airplane also has been successful in parts of the humid region (Robinson and Cross 1961; White 1970).

Fertilizing and Liming. Large areas of permanent pasture have adequate cover but are low in productivity because they lack sufficient fertility. Fertilization can be an economical method of increasing pasture production. When soil fertility is increased and maintained with regular or annual applications of fertilizer, improved, highly productive species can be seeded and maintained.

Legume-grass mixtures have nutritional,

utilization, and yield advantages; however, it is often easier to manage N-fertilized grasses than legume-grass mixtures. In the absence of legumes, N is the factor most limiting to growth of pasture grasses, especially tall-growing species. However, adequate pH, P, and K soil levels must be maintained to achieve top results from N.

Pasture yields can be increased two- to fivefold by fertilization with N at 134 kg/ha, particularly where the botanical composition is predominantly grass. However, considerably higher yields of forage and livestock products have been reported when higher rates of N have been applied to improved tall-growing grasses.

Soil and climate characteristics, species, time of fertilizer application, economics, and managerial ability will influence the application rate of N. Higher N rates may be used when top management is applied to the total forage-livestock enterprise. However, less N is needed for grazing than for hay production because part of the N is recycled through manure, urine, and unconsumed forage. But these are not uniformly distributed on the pasture (see animal effects).

Nitrogen fertilization not only increases yield per hectare but encourages growth earlier in the spring, sustains production later in the fall, and increases carrying capacity. The effect of N on occurrence of high nitrate levels in the forage also must be considered. Studies indicate high nitrates are not a problem in tall grasses fertilized with N at rates of 269 kg/ha, particularly when N was applied in split applications.

To equal yields by legume-grass pastures, 134-269 kg N/ha are needed on grass pastures (Alexander and McCloud 1962; Scholl et al. 1970b). Pasture swards having more than 50% legumes usually do not respond to N fertilization. However, response can be obtained if the sward is less than 50% legumes. Nitrogen fertilization of short grasses such as Kentucky bluegrass generally will not produce yields equal to those of legume-grass mixtures receiving no N.

Pastures also may respond to lime, P, and K. Lime and P-K fertilizer encourage legume growth and increase N production. In West Virginia, pastures treated with lime and P yielded 75% more than untreated pastures, while similar pastures treated with lime, N, P, and K gave increases of almost 125% (Schaller et al. 1945).

Proper liming and fertilization increases the proportion of desirable pasture plants and sward density. In North Carolina the total vegetative cover increased 33% by use of fertilizer. Annual lespedezas, white clover, and Kentucky bluegrass plant cover have been increased as much as 65%. Fertilization enables earlier spring grazing and extends the pasture season (Woodhouse and Lovvorn 1942). Environmental and water quality concerns encourage the pasture manager to apply rates of lime as well as N, P, and K fertilizers as determined by soil tests and consistent with needs of the pasture species and the production level. Fertilizers should be applied on the contour, where soil and plant cover are adequate, and in a season when plants will make optimum use of the nutrients to minimize runoff and pollution.

PLANT AND ANIMAL FACTORS IN GRAZING MANAGEMENT

Perennial forages are a renewable resource. They require infrequent planting and grow in predictable annual cycles. Pasture researchers and managers have come to rely heavily on legumes for pasture improvement, as other means of increasing livestock production have generally proven more costly (Harper 1978). With a basic understanding of how legumes and grasses grow and regrow after defoliation, plus a knowledge of how the animal can alter these processes, the manager can enhance plant growth and production. Separating the pasture system into the activities of its various components and species will assist in developing this understanding.

Managers frequently pay more attention to the livestock portion of their enterprise than to their pastures and stored forages. This may not be from a lack of interest but rather from a lack of information on the basic concepts of legume and grass growth, their development, and differences. Lack of proper grazing management results in reduced plant and animal productivity.

Following are the seven factors that need to be considered in grazing management (Blaser et al. 1986; Hodgson 1990; Langer 1973; Voisin 1988; Waller et al. 1985; Watkins and Clements 1978).

Carbohydrate and Bud Management. A pasture sward (green leaf area) obtains its energy for growth from sunlight through photosynthesis, which occurs in the green leaves. The pasture manager supplies the land, labor, fertilizer, irrigation, pesticides, etc., as "limiting and expensive" inputs to optimize the production of pasture forage. Grazing management is the means by which the manager controls the "costs" of producing and using the plant production. In summary, the plant produces during one-half of the day while the animal may graze 24 h a day, anytime forage is available.

Carbohydrates are used to maintain the plant and to support growth after defoliation, mowing, or dormancy and are in highest concentration when growth is most active. Perennial forage plants have storage organs—stem bases, roots, rhizomes, stolons, or corms. Buds on these storage organs or at leaf axils initiate growth in spring or after defoliation. Active buds are needed for efficient production.

Growth usually takes priority over storage for carbohydrate use. During the growing season, when leaf area is large and photosynthesis produces more energy than required for growth, part of that carbohydrate is shifted into storage for later use. But, when photosynthesis is not sufficient, such as after grazing, that carbohydrate is brought back out of storage to support top growth until adequate leaf area is again formed. Uncontrolled grazing can deplete a plant's energy reserve. Storage must be permitted for future productivity. Single species pasture swards are easier to manage than multispecies swards. A knowledge of energy storage cycles for each species is necessary to properly manage each sward.

With the onset of the shorter days and cool temperatures of fall, perennial plants must prepare for winter and accumulate carbohydrate to support growth the following spring. Extensive defoliation, lack of a sufficient fall rest period, environmental stress, and pest infestations during the growing season can markedly reduce carbohydrate production and storage. Therefore, proper fall management is critical.

A rest period of 4 to 6 wk before growth ceases in autumn is necessary to permit legume plants to store food reserves in the crown and roots over the winter and for making vigorous growth the following spring (Smith 1964). This relationship is not as well determined for grasses. Loss of legumes from a sward eliminates a source of N for associated grasses. Reduction of ground cover also exposes the crown, tiller buds, and tender growth to frost injury. In the North, stubble and unused vegetation hold snow and provide protection for overwintering plants. In the South, late summer grazing to reduce the competition of warm-season grasses with cool-season legumes such as white, hop (*T.*

agrarium L.), and Persian (*T. resupinatum* L.) clovers is a desirable practice.

Tiller Management. Useful criteria for making grazing decisions are

1. Perennial forage plants have stems (legumes) or tillers (grasses) composed of a growing point, stem, leaves, nodes, and axillary buds. Dormant or inactive buds have the potential of producing a new stem or tiller with a growing point. The number of buds determines the potential number of new shoots when the growing point of the initial stem is removed, and the number is the basis for perennial plant survival.

2. Full season grazing plants are nonexistent in climates with hot summers (Waller et al. 1985). Forage plants are divided into cool-season (C_3) plants that have an optimum growth temperature of 18°-24°C and warm-season (C_4) plants that have an optimum growth temperature of 32°-35°C. A full season grazing program may well use both types or provide for stored forage of the major forage type for use during periods of lower production and dormancy.

3. Early in the season, all legumes and grasses have their growing point close to or at ground level. Since most grazing animals cannot physically graze any closer than about 2.5 cm from the ground, there is little danger of removing the growing point at this time. White clover, Kentucky bluegrass, tall fescue, and several other grasses have no internode elongation until about the time the growing point enters the reproductive phase. Grazing can occur with little damage to the plant. Also, white clover and Kentucky bluegrass maintain their growing point below or at ground level throughout most of the grazing season to tolerate continuous, close grazing.

4. On the other hand, species such as alfalfa, red clover, smooth bromegrass, orchardgrass, and switchgrass elevate their growing point early in their development by elongation of the internodes and require more careful management. Knowledge of the cycle of energy storage and use in the crowns or underground stems is necessary to manage these crops efficiently. Alfalfa, red clover, and smooth bromegrass may be grazed or cut after first flower or early heading when the plant has replenished much of the stored energy used to initiate growth (Smith 1981).

Temperature during grazing or cutting also will influence tiller response. Grazing alfalfa during periods of cool temperatures may not be as detrimental as during periods of warm temperatures (Feltner and Massengale 1965). Virginia research (Allen et al. 1986a, 1986b) has shown that alfalfa can be used flexibly for grazing if adequate recovery of regrowth is permitted. Five years of research with both sheep and cattle indicates that grazing of alfalfa can begin with a plant height of 7.6-13.0 cm in spring and 13.0-18.0 cm in summer. Grazing can be conducted for up to 6 wk in spring with little detrimental effect to the stand as long as available forage has been maintained at a leaf area index (LAI) of 1. Newer grazing cultivars may extend this season. Overgrazing may reduce productivity and increase weed encroachment. Summer grazing for 1 wk at early bloom can be accomplished. Grazing at other stages of maturity and periods of duration may decrease production and increase weed encroachment. In one trial, summer grazing reduced stem density by 34%-50%.

5. Strategies for control of potato leafhopper on alfalfa will have to be changed when alfalfa is grazed.

Remaining Leaf Area. When all environmental factors are favorable, optimum growth occurs when leaf area, expressed as leaf area index, is sufficient to intercept 90% or more of the incident light. Perennial grasses with semierect leaves need larger LAI than legumes with horizontal leaves in order to intercept the light and optimize yields. A very high LAI does not give added increases in forage production because basal leaves are shaded, become mature and inefficient, and die as new leaves form, thus nullifying added production (Blaser et al. 1986). Pasture managers may waste as much as 50% of the spring growth because of understocking. As spring advances, the shift from slow or little to rapid growth must be controlled by early and high density stocking.

Developing reproductive tillers in grasses should be grazed while they are still leafy, palatable, and high in digestibility. Grazing below the flower buds will kill these tillers and permit the shorter vegetative tillers to persist and produce high-quality, leafy forage. Managers should stock pastures so the residue maintains sufficient LAI to generate new growth.

Perennial legumes and grasses tend to adapt to harvest method. Under heavy, continuous stocking, shoots and leaves are smaller and more prostrate than under infrequent harvest for hay and silage harvest. Grasses that elevate their growing point early in the growth cycle must be managed differently than those that elevate the growing point late in the growth cycle. Kentucky bluegrass and some other grasses have no internode elongation until about the time the growing point enters the reproductive phase. Such grasses maintain their growing point below or at ground level throughout most of the growing season and are resistant to continuous close grazing. If Kentucky bluegrass-white clover pastures are grazed to 2.5 cm, growth of grass is depressed more than that of clover, causing clover dominance as small white clover leaves escape grazing. But leaving a 5-cm or greater stubble of leafy grass suppresses the white clover. Close, continuous grazing of tall-growing grasses, i.e., orchardgrass and tall fescue, suppresses new growth because the animals graze most of the leaves and basal tillers that are high in stored carbohydrates.

Species such as smooth bromegrass, switchgrass, alfalfa, and red clover elevate their growing point early in development with stem elongation. Grazing an alfalfa-orchardgrass sward to 4 cm favors the alfalfa because new growth originates from crown buds at soil level; grazing to 8 cm favors the orchardgrass because the remaining leaf area supports continued growth. Smooth bromegrass, reed canarygrass, and western wheatgrass (*Pascopyrum smithii* [Rydb.] Löve) have elongated stem growth without seed head formation during the remainder of the year until spring flowering.

Intensive and rotational stocking involving grazing periods and rest intervals permits managers to optimally use different forage species. The grazing period of a paddock should be sufficiently short such that a plant grazed on the first day or at the beginning of occupation is not grazed again by the animals before they leave the paddock. A sufficient interval must elapse between two grazings to permit plants to accumulate stored energy reserves necessary for vigorous regrowth and high yield. The length of interval will vary with season of the year, climate, and environmental conditions.

Tall-growing species, such as alfalfa and smooth bromegrass, can be grazed close but not often, while birdsfoot trefoil, orchardgrass, and tall fescue can be grazed often but not close. Birdsfoot trefoil dictates a different grazing management. During the growing season trefoil stores little carbohydrate in the crown; thus, regrowth occurs from axillary buds on the stem.

Adequate leaf area remaining after grazing will minimize reserve carbohydrate depletion. Remaining leaf material also enhances the microclimate for growth during the growing season and improves interception of precipitation, insulation, and snow cover.

Defoliation. Optimum grazing management avoids repeated, severe defoliation of a tiller without a recovery period (planned rest). Fresh, immature growth is highly palatable, and livestock will selectively graze it. Therefore, the duration of livestock occupation must be controlled to optimize plant and animal production. Repeated severe defoliation of desirable plants or areas in pasture can be reduced by increasing stocking density and reducing the duration of grazing.

Plant growth rates vary, and the animal stocking rate must be adjusted accordingly. Close defoliation during the growing season is at the expense of maximum yield and stand persistence. Overgrazing during periods of low rainfall reduces animal performance and photosynthesis, surface soil temperatures may become too high, and the forage plants may lack vigor because of weakened root systems.

Grazing of half or more of the foliage of grasses (both cool- and warm-season species) during the growing season has been found to cause root growth to stop for a time after each removal, with the exception of orchardgrass after the first clipping (Crider 1955). Repeated clipping (similar to continuous overgrazing) has also been found to result in complete and permanent cessation of root growth and poor plant development. Annual and perennial weed species become established under these conditions and compete with forage plants for water and nutrients.

Productive pasture is less likely to be damaged by undergrazing than by overgrazing; however, under these conditions tall weeds often increase in number. Excessive wastage of forage occurs because of trampling, fouling, and shading of low-growing species. Insect and rodent infestations also are encouraged by undergrazing, and foliage and root diseases often are more prevalent.

When pastures are grazed too early, productivity of forage plants is lowered by premature leaf removal. Root development is reduced, and dry matter production is hampered. With plants at less than optimum height, few leaves remain, sunlight is used inefficiently, and moisture conditions invite growth of weedy annual grasses. Plants also are destroyed by trampling because the ground is soft from winter freezing and thawing.

Livestock Nutritional Needs. To optimize animal performance and pasture performance (gain/ha), the selection of grazing species and duration of rest period is critical. Nonuse periods should be long enough to permit the plants to recover from defoliation, but short enough to avoid the plants becoming too mature when pastures are grazed more than one time each season. Successful grazing management must also consider the type of livestock and their nutritional needs. Managers must match the nutritive needs of their livestock and management goals for livestock performance with the seasonal availability and quality of forage.

Grazing Program. Appropriate grazing management is dependent on the individual enterprise. Controlled stocking programs permit stocking rates to be sustained at higher densities compared with season-long, continuous stocking because of improved harvest efficiency. Grazing distribution, season of grazing, and degree of use must all receive emphasis in the grazing program. Legume-grass pastures provide high yields of nutritious forage during midseason, or in some regions (always in the South) warm-season forages may be used for summer stocking. Nitrogen-fertilized cool-season grass is ideal for grazing during the early and late season when the legume should not be grazed. If the grass has been properly managed in previous years, or during the summer period, it will recover from this late-season grazing; however, the same pasture should not be the last pasture grazed the following year.

Animal Effects. Plant relations in pastures cannot be understood without reference to animals. Productivity and botanic composition of pastures can be rapidly and substantially altered by grazing animals—the effects may be either harmful or beneficial. During grazing, there are three fundamental processes that affect the plants:

1. The grazing animal will either clip or tear off selected plant parts. Beef and sheep may complement each other in the manner in which each removes forage growth: (a) cattle leave patches of forage almost untouched while sheep graze more evenly, (b) sheep use their narrow muzzle and greater lip mobility to select plant parts while cattle take large bites with little selection, (c) sheep graze many common weeds even if good pasture is available while cattle usually avoid weeds, and (d) there is less parasite infection in each animal species on pasture than on stored feeding in enclosures. The defoliation pattern is probably the most important animal factor to consider.

2. Plants are trampled and can suffer some mechanical damage.

3. Fouling (manure and urine) will occur.

Livestock are selective in their choice of plants and will eat the most palatable plants first. Selective defoliation can be very important in maintaining composition of multiple plant swards through its differential effect on individual species. Sheep and cattle select leaf before stem and young leaves before old leaves, particularly when a pasture has reached advanced maturity. The selected forage is frequently higher in crude protein, minerals, soluble carbohydrates, digestibility, and gross energy as well as lower in lignin and structural carbohydrates than the whole pasture sward.

A seeding mixture should contain plants of similar palatability and growth form. If a less palatable species is included with more palatable species, the less palatable will soon dominate. Likewise a species that grows very rapidly early in the season and gets ahead of livestock will soon dominate the sward. As forage availability decreases so will livestock selection, grazing time, and bite size, whereas number of bites increase. In extreme cases, the animals reduce intake.

Gross effects from treading including soil erosion, altered grazing behavior, and hoof cultivation have been noted for many years (Watkins and Clements 1978). Observations on temperate pastures have shown that treading damage may pass unnoticed or its extent may not be fully appreciated. Treading usually produces a significant and progressive reduction in forage yield of temperate pastures as stocking rates increase. Yield reductions of at least 20% were obtained at commercial stocking rates. The magnitude depends only slightly on soil type and fertility, is moderately influenced by sward height, and depends strongly on plant species and soil moisture.

Numbers of plants may decline immediately after treading, and tiller density is reduced. Even though damaged tillers may recover, the vigor appears to be less over the season than in the old tillers. Perennial ryegrass and Kentucky bluegrass are the species most resistant to treading. Indirect effects from treading exist in the form of soil compaction and puddling, increased mechanical resistance to root penetration, reduced aeration, and increased runoff.

The grazing animal removes only a small quantity of the nutrients from the pasture in the form of mineral product; the remainder are excreted. But livestock do not spread their excretions uniformly over a pasture. They tend to concentrate manure in primary focal points of livestock concentration (e.g., watering sites and shade areas). These areas tend to be enriched while areas most distant from these focal sites decline in fertility. Potassium levels 10 times greater and P levels 4 to 5 times greater in these areas than the pasture as a whole are noted in a Missouri study (Gerrish et al. 1993). These areas should be treated differently from the remainder of the pasture when determining fertilization programs. Conditions favoring significant nutrient gradients are immovable water sources, pastures having high length-to-width ratios, and extreme distances of animal travel to water.

Even though the nutrients returned to the soil are recycled, the urine stimulates grass growth, which in turn depresses legume growth. Dung results in death of covered plants and has limited value to pasture production unless incorporated. While harrowing promotes more even grazing, it slightly depresses pasture growth. Fouled areas also tend to be avoided by grazing animals.

Grazing practices cause dynamic changes in plant composition of a pasture sward, which will affect animal performance and ultimately alter productivity and profitability of the entire enterprise. Economic returns depend on forage quantity, quality, and harvest efficiency.

Rotational versus Continuous Stocking. Rotational stocking is a system of pasture utilization embracing periods of high stocking density followed by periods of rest. This system permits the farmer to match grazing more adequately to the growth habit of the forage species, the condition of the pasture, and animal needs than does continuous stocking. Grazing rotationally provides for a comparatively short period of grazing for each field and a recovery period to allow more effective consumption of forage with less waste by trampling, fouling, or selective grazing.

Rotational stocking also favors legume persistence and improved yields per hectare.

Harvesting mature forage as hay or silage during the maximum growth period when grazing is in surplus supply also aids in feeding animals through crisis periods such as times of drought. The practice also increases carrying capacity and provides time for regrowth and continued production of high-quality pasturage.

Stocking frequency and density depends on the species used and their vigor. Low-growing species such as birdsfoot trefoil, white or ladino clover, and Kentucky bluegrass can be stocked often or continuously. For maximum production, mixtures of Kentucky bluegrass with white or ladino clover should be grazed no closer than 5 cm; with birdsfoot trefoil, no shorter than 7-10 cm. However, tall-growing species such as alfalfa, tall fescue, orchardgrass, smooth bromegrass, and reed canarygrass can be grazed close but not often. Alfalfa and alfalfa-grass mixtures should be allowed to reach 20-45 cm before grazing, and a 5- to 6-wk rest period should be allowed between grazings. Annuals such as sudangrass (*Sorghum bicolor* [L.] Moench) and sorghum-sudangrass hybrids also do best under this type of management.

A combined rotational-seasonal stocking plan may be necessary to achieve maximum production per hectare and most efficient use of management inputs.

Seasonal Stocking. Renovated pastures are more effective when several species are grown as monocultures in separate pastures that are then combined into a pasture system. Thus, the growth habit of each species is exploited to contribute to full season grazing. A farmer can select species adapted for topography and soil conditions on the farm as well as those to hold erosion losses within acceptable limits and to meet the needs of the livestock enterprise in each season of the year. Single species or simple mixtures are easier to manage than complex mixtures.

Grasses start growth earlier in the spring than legumes. Fertilizing grass pastures with a complete fertilizer composed largely of N will encourage earlier grass growth. Split applications of fertilizer provide additional increases in production (Matches 1968).

Legume and legume-grass pastures generally provide the most uniform production for the grazing season. Deeper-rooted legumes provide higher yields during the critical summer period than do the cool-season grasses;

however, production also declines, but at a slower rate than with cool-season grasses. Stockpiling for use in this period has been suggested as one means of obtaining a more stable pasture supply.

Harvesting a portion of the pasture as hay or silage and storing for midsummer use is another form of stockpiling. This also assists the farmer in programming pasture growth so more pasturage is utilized at stages of highest quality. Legume-grass pastures also may be supplemented with annual pastures such as sudangrass or pearlmillet (*Pennisetum americanum* [L.] Leeke) to provide high carrying capacity in midsummer.

Tall cool-season grasses fertilized with N in late summer help extend forage production and grazing into the fall, when legumes should be rested; also, grasses are more digestible when grazed during this period.

In the South, winter pastures of small grains and annual ryegrass, as well as cool-season legumes and grasses, supplement warm-season grasses to provide year-round grazing.

Effect of Mixed Grazing. Grazing animals influence pasture quantity and quality. Because they are by nature selective short grazers, sheep continuously stocked alone have a negative effect on pasture unless grazing is restricted. Cattle readily select the tender green forage that has been well fertilized, and they prefer forage to be at least 10-12 cm high. Tall-growing palatable grasses are more competitive when grazed by cattle than by sheep (Wedin and Vetter 1970).

By selective and close grazing, animals overgraze the palatable species in a pasture and leave unpalatable ones. Continuous overstocking eventually will eliminate more palatable species, resulting in a pasture composed principally of those having low palatability.

Little definitive research on grazing with mixed groups of animals has been conducted in the US. However, Ohio studies show that pastures were grazed more completely and uniformly by cattle and sheep together than if grazed by cattle alone. Cattle tend to graze in patches, and sheep will graze areas avoided by cattle. Sheep also tend to be selective of plant parts, whereas cattle are less selective. Furthermore, sheep and cattle prefer different plant species. With proper stocking rates, animal gains can be increased (Van Keuren and Parker 1967).

Studies in South Africa and Scotland also indicate that sheep grazed in conjunction with cattle increased livestock production and had a beneficial effect on botanical composition of the pasture. In a Canadian study, sheep consumed plants that cattle did not consume (Hidiroglou and Knott 1963). Sheep also reduce weed and brush populations and in that way also improve the pasture sward.

Weeds. Weeds are indicative of poor pasture management. Many low-quality grassy weeds invade pastures largely because of low soil fertility. These weeds can be easily controlled and eliminated from pastures by liming, fertilizing, and seeding improved species of grasses and legumes on infested areas.

Timely mowing, adequate fertilization, and good grazing management are the best methods of controlling weeds and brush. Herbicides may be applied for control of difficult species. Missouri studies show a higher proportion of economic species, fewer weeds, and highest forage consumption from Kentucky bluegrass pastures receiving a combination of high-fertility and herbicide treatments (Peters and Stritzke 1971).

QUESTIONS

1. Which legume-grass mixture prevails in permanent pastures on the more fertile soils of the North? On the less fertile upland soils? On low or bottomland pasture? Why?
2. Which annual legume is recognized for its pasture value in the central humid region? Which perennial legumes are found in permanent pastures? Why are more biennial and perennial legumes being used in pastures today?
3. Relate the species found in permanent pastures in the South to soil condition.
4. What changes in the livestock population have occurred during the past two decades? Why the renewed interest in pastures?
5. Compare the probable value of reseeding legumes with that of N fertilization when renovating permanent pastures.
6. How would you proceed to renovate an unproductive upland permanent pasture in your area?
7. Has rotational stocking of permanent pastures usually resulted in increased animal production? Of renovated pastures? Has alternate stocking of such pastures usually resulted in increased animal production? Explain. What is seasonal grazing?
8. Develop a full season pasture system for beef cows for your area. Include species, system of grazing, and other management practices. What will this system do for seasonal production of forage, forage quality, animal production, and forage yield? How will this vary from a pasture system for dairy cows?
9. How do plant and animal factors interact in management of a pasture ecosystem?

10. Name all the factors in an ecosystem affected by grazing.

REFERENCES

Ahlgren, HL, ML Wall, RJ Muckenhirn, and FV Burcalow. 1944. Effectiveness of renovation in increasing yields of permanent pastures in southern Wisconsin. J. Am. Soc. Agron. 36:121-31.

Ahlgren, HL, ML Wall, RJ Muckenhirn, and JM Sund. 1946. Yields of forage from woodland pasture on sloping land in southern Wisconsin. J. For. 44:709-11.

Aiken, GE, WD Pitman, CG Chambliss, and KM Portier. 1991. Plant responses to stocking rate in a subtropical grass-legume pasture. Agron. J. 83:124-29.

Alexander, CW, and DE McCloud. 1962. Influence of time and rate of nitrogen application on production and botanical composition of forage. Agron. J. 54:521-22.

Allen, VE, DD Wolf, JP Fontenot, J Cardina, and DR Notter. 1986a. Yield and regrowth characteristics of alfalfa grazed with sheep. I. Spring grazing, 1981-82. Agron. J. 78:974-79.

Allen, VE, LA Hamilton, DD Wolf, JP Fontenot, TH Terrill, and DR Notter. 1986b. Yield and regrowth characteristics of alfalfa grazed with sheep. II. Summer grazing, 1981-82. Agron. J. 78:979-85.

Blaser, RE, RC Hammes, Jr., JP Fontenot, HT Bryant, CE Polan, DD Wolf, FS McClaugherty, RG Kline, and JS Moore. 1986. Forage-Animal Management Systems. Va. Agric. Exp. Stn. Bull. 86-87.

Brummer, EC, and JH Bouton. 1991. Plant traits associated with grazing—tolerant alfalfa. Agron. J. 83:996-1000.

———. 1992. Physiological traits associated with grazing-tolerant alfalfa. Agron. J. 84:138-143.

Coats, RF. 1957. Sod-seeding—Brown Loam Tests. Miss. Agric. Exp. Stn. Bull. 554.

Crider, FJ. 1955. Root Growth Stoppage Results from Defoliation of Grass. USDA Tech. Bull. 1102.

Decker, AM, HJ Retzer, ML Sorna, and HD Kerr. 1969. Permanent pastures improved with sod seeding and fertilization. Agron. J. 61:243-47.

Diderickson, RI. 1980. l977 National Resources Inventory (NRI). Washington, D.C.: SCS/USDA.

Dunavin, LS. 1992. Florigraze rhizoma peanut in association with warm-season perennial grasses. Agron. J. 84:148-51.

Feltner, KC, and MA Massengale. 1965. Influence of temperature and harvest management on growth, level of carbohydrate in the roots, and survival of alfalfa (Medicago sativa L.). Crop Sci. 5:585-88.

Fribourg, HA, JB McLaren, KM Barth, JM Bryan, and JT Connell. 1979. Productivity and quality of bermudagrass and orchardgrass—ladino clover pastures for beef steers. Agron. J. 71:315-20.

Gerrish, JR, JR Brown, and PR Peterson. 1993. Impact of grazing cattle on distribution of soil minerals. In Proc. Am. Forage Grassl. Counc., 29-31 March, Des Moines, Iowa, 22.

Griggs, TC, and AG Matches. 1991. Productivity and consumption of wheatgrasses and wheatgrass-sainfoin mixtures grazed by sheep. Crop Sci. 31:1267-73.

Harper, JL. 1978. Plant relations in pastures. In Wilson, JR (ed.), Plant Relations in Pastures. CSIRO. East Melbourne, Australia, 3-14.

Hidiroglou, M, and HJ Knott. 1963. The effects of green, tall buttercup in roughage on the growth and health of beef cattle and sheep. Can. J. Anim. Sci. 1:68-71.

Hodgson, J. 1990. Longman Handbooks in Agric. New York: Wiley.

Howard, WT. 1993. Summary of forage use in top DHI herds. In Proc. 23d Nat. Alfalfa Symp., Appleton, Wis., 44-49.

Karnezos, TP, and AG Matches. 1991. Lamb production on wheatgrasses and wheatgrass-sainfoin mixtures. Agron. J. 83:278-86.

Langer, RHM. 1973. Lucerne. In RHM Langer (ed.), Pastures and Pasture Plants. Wellington, New Zealand: AH and AW Reed, 357-60

Leonard, WH, RM Love, and ME Heath. 1968. Crop Terminology Today. Crop Sci. 8:257-61.

McVicker, MH. 1974. Pastures for dairy cattle. In Approved Practices in Pasture Management. Danville, Ill.: Interstate Printers and Publishers, 181-95.

Martin, NP, DL Rabas, CC Shaeffer, and DL Wyse. 1979. A critical look at minimum-till pasture improvement: What makes it work! In Proc. 3d Wis. Forage Counc. Forage Prod. Use Symp., 1-14.

Matches, AG. 1968. Performance of four pasture mixtures defoliated by mowing or grazing with cattle or sheep. Agron. J. 60:281-85.

Mowrey, DP, and AG Matches. 1991. Persistence of sainfoin under different grazing regimes. Agron. J. 83:714-16.

Peters, EJ, and JF Stritzke. 1971. Effects of Weed Control and Fertilization on Botanical Composition and Forage Yields of Kentucky Bluegrass Pastures. USDA Tech. Bull. 1430.

Pierre, WH, JH Longwell, RR Robinson, GM Browning, I McKeever, and RF Copple. 1937. West Virginia Pastures: Type of Vegetation, Carrying Capacity, and Soil Properties. W.Va. Agric. Exp. Stn. Bull. 280.

Robinson, GS, and MW Cross. 1961. Improvement of some New Zealand grassland by oversowing and overdrilling. In Proc. 8th Int. Grassl. Congr., 402-5.

Rohweder, DA. 1963. The nature and productivity of southern Iowa pastures as affected by soil type, slope, erosion, fertility, and botanical composition. PhD diss. Iowa State Univ. Diss. Abstr. 24:3909.

Rohweder, DA, and RW Van Keuren. 1985. Permanent pastures. In ME Heath, RF Barnes, and DS Metcalfe (eds.), Forages: The Science of Grassland Agriculture, 4th ed. Ames: Iowa State Univ. Press, 487-95.

Rusland, GA, LE Sollenberger, KA Albrecht, CS Jones, Jr., and LV Crowder. 1988. Animal performance on limpograss-aeschynomene and nitrogen-fertilized limpograss pastures. Agron. J. 80:957-62.

Schaller, FW, GG Pohlman, HO Henderson, and RA Ackerman. 1945. Pasture Fertilization Experiments at Reymann Memorial Farm. W.Va. Agric. Exp. Stn. Bull. 324.

Scholl, JM, MF Finner, AE Peterson, and JM Sund. 1970a. Pasture Renovation in Southwestern Wisconsin. Wis. Agric. Life Sci. Res. Rep. 65.

Scholl, JM, WH Paulson, and VH Brungardt. 1970b. A Pasture Program for Beef Cattle in Wisconsin. In Proc. Wis. Lancaster Cow-Calf Day, 2-7.

Semple, AT, HN Vinall, CR Enlow, and TE Woodward. 1946. A Pasture Handbook. USDA Misc. Publ. 194.

Shoemaker, DE, SR Shoemaker, and DL Zartman. 1992. Seasonal Milking with Intensive Grazing: The Ohio Experience. P-38. American Dairy Science Association.

Smith, D. 1964. Winter injury and the survival of forage plants. Herb. Abstr. 34:203-9.

———. 1981. Forage Management in the North. Dubuque, Iowa: Kendall/Hunt.

Sprague, VG, RR Robinson, and AW Clyde. 1947. Pasture renovation. I. Seedbed preparation, seedling establishment, and subsequent yields. J. Am. Soc. Agron. 39:12-25.

Taylor, TH, EM Smith, and WC Templeton, Jr. 1969. Use of minimum tillage and herbicides for establishing legumes in Kentucky bluegrass swards. Agron. J. 61:761-66.

USDA. 1992. Agricultural Statistics.

USDA-SCS. 1982. Basic Stat. Nat. Res. Invent. ISU Stat. Bull. 756. Ames, Iowa.

USDC. 1978. Census of Agriculture. Washington, D.C.: US Gov. Print. Off.

———. 1987. Census of Agriculture. Washington, D.C.: US Gov. Print. Off.

Van Keuren, RW, and CF Parker. 1967. Grazing Sheep and Cattle Together. Ohio Farm and Home Sci. 57(1):12-13.

Van Keuren, RW, and GB Triplett. 1970. Seeding legumes into established grass swards. In Proc.

11th Int. Grassl. Congr., Surfers Paradise, Queensland, Australia. St. Lucia: Univ. Queensland Press, 131-34.

Van Keuren, RW, ET Shaudys, RH Baker, and AD Pratt. 1966. Economy of grazing, soilage, and stored feeding methods for summer grazing of dairy cattle. In Proc. 10th Int. Grassl. Congr., Helsinki, Finland, 2:505.

Vaughan, KK, DK Combes, and KA Albrecht. 1992. Performance of dairy cattle on an intensive rotational grazing system versus a conventional forage system. Am. Soc. Agron. Abstr., Madison, Wis., 185.

Voisin, André. 1988. Grass Productivity. Washington, D.C.: Island Press, 131-34.

Waller, SS, LE Moser, PE Reece, and GA Gates. 1985. Understanding Grass Growth: The Key to Profitable Livestock Production. Kansas City, Mo.: Trabon Print Company.

Watkins, BR, and RJ Clements. 1978. The effects of grazing animals on pastures. In JR Wilson (ed.), Plant Relations in Pastures, CSIRO East Melbourne, Australia, 273-89.

Wedin, WF, and RL Vetter. 1970. Pasture for beef production in the western Corn Belt, USA. In Proc. 11th Int. Grassl. Congr., Surfers Paradise, Queensland, Australia. St. Lucia: Univ. Queensland Press, 842-45.

White, JGH. 1970. Establishment of lucerne (*Medicago sativa* L.) in uncultivated country by sod-seeding and oversowing. In Proc. 11th Int. Grassl. Congr., Surfers Paradise, Queensland, Australia. St. Lucia: Univ. Queensland Press, 134-38.

Woodhouse, WW, Jr., and RL Lovvorn. 1942. Establishing and Improving Permanent Pastures in North Carolina. N.C. Agric. Exp. Stn. Bull. 338.

Rangeland and Agroforestry

16

R. DENNIS CHILD
Agricultural Research Service, USDA

HENRY A. PEARSON
Agricultural Research Service, USDA

THE largest of the terrestrial ecosystems, rangelands comprise nearly 400 million ha in the continental US. About 80% of the lands in the 17 western states and about 55% of all US lands are classified as rangeland. These lands are a primary source of food, fuel, fiber, recreation, and quality water supply. They are the frontline defense against soil erosion, ecosystem instability and degradation, and inadequate supplies of quality water. Periodic drought, atmospheric change, and increased public interest require innovative management strategies that will improve the stewardship of public and privately owned rangelands. Strategies are needed to improve the ecological stability and productivity of rangelands and maintain the diversity of rangeland ecosystems. Three fundamental issues drive rangeland management strategies in the US: sustainability, environmental stewardship, and profitability.

Rangeland is land on which the indigenous vegetation (climax or natural potential) is predominantly grasses, grasslike plants, forbs, or shrubs and is managed as a natural ecosystem; if plants are introduced, they are managed as indigenous species (FGTC 1992). Range ecosystems encompass both rangelands and grazable forestlands. Range management can be described as the science and art of manipulating range ecosystems to optimize the returns in combinations most desirable and suitable to society; it encompasses many aspects of the biological, physical, and social sciences (Stoddart et al. 1975; Heady 1975; Heady and Child 1994; SRM 1964, 1974, 1989).

Agroforestry, on the other hand, is a land-use system in which woody perennials are grown for wood production in combination with agricultural crops, with or without animal production (FGTC 1992). Agroforestry generally occurs on croplands and/or pasturelands where woody perennials are introduced into crops and pasture or vice versa. However, forest grazing (silvo-pastoral), combined wood and animal production with grazing on coexisting indigenous forage, is often classified under agroforestry systems.

R. DENNIS CHILD is a Range Scientist, Agricultural Research Service, USDA, National Program Staff, Beltsville, Maryland, and previously with Winrock International, Morrilton, Arkansas. He conducted research in range management and ecosystem analysis and has worked in research administration and international agricultural development.

HENRY A. PEARSON is Range Scientist, Agricultural Research Service, USDA, South Central Family Farm Research Center, Booneville, Arkansas. Previously with the Forest Service, USDA, Pineville, Louisiana, Flagstaff, Arizona, and Washington, D.C., he conducted research in forest-range management, agroforestry, and relationships among trees, livestock, wildlife, and watersheds.

Rangeland Ecosystems

Many rangeland ecosystems exist in the US. Eight regions were used to address these ecosystems in the fourth edition (Chaps. 37-43) of this book (Heath et al. 1985). Some of the background information, ecological descriptions, and developmental history for this edition are excerpted from these earlier writers and other classical rangelands writings (Heady and Child 1994; Stoddart et al. 1975; Heady 1975; Sampson 1952). The 1994 Society for Range Management publication, *Rangeland Cover Types of the United States,* was also used as a primary resource for a brief description of the rangeland ecosystems and cover types of the US.

Rangeland vegetation can be grouped into associations based on the plants that are currently present (SRM 1994). The 1994 Society for Range Management publication describes 175 unique cover types found in nine regions of the US. Using the Society for Range Management definition, the *cover type* of an area is the existing vegetation. A brief description of these regions and a list of the cover types within each region follows.

PACIFIC NORTHWEST

The Pacific Northwest region constitutes the area north of the Great Basin of Nevada, west of the Rocky Mountains in Idaho, and south of the Okanogan Highlands and Cascade Mountains separating Canada and the US. Topographically the rangelands pertaining to this region are located east of the Cascadian crest and west of the Rockies on the central and eastern portions of Washington and Oregon. Major physiographic provinces included in this region are the Okanogan Highlands and Columbia Basin in the state of Washington and the Great Basin and Basin range, High Lava Plains, and Blue Mountains of central and eastern Oregon.

Annual precipitation ranges from lows of 17.5-20.0 cm (7 to 8 in.) in the driest portions of the Basin and Columbia River basin provinces to highs of up to 163 cm (65 in.) in the subalpine of the Wallowa Mountains in northeastern Oregon. Most of the grasslands and shrublands of the region receive less than 50 cm (20 in.) of precipitation a year. Temperatures are extreme with hot, dry summer months in the Great Basin and Basin range province while cold, dry winter months prevail from the continental climate.

The native flora is well adapted to the warm, dry summer drought period and can tolerate the cold, dry winter extremes as well. A diverse vegetation occupies the grasslands and shrublands of the Pacific Northwest, reflecting the ability of plants to persist in environments with climatic extremes and the principal modifying agents—fire, grazing, and browsing.

Cover types: bluebunch wheatgrass, Idaho fescue, greenleaf fescue, antelope bitterbrush—bluebunch wheatgrass, antelope bitterbrush—Idaho fescue, bluegrass scabland, western juniper—big sagebrush—bluebunch wheatgrass, alpine Idaho fescue, ponderosa pine shrublands and grasslands, ponderosa pine—shrubland, ponderosa pine—grassland.

PACIFIC SOUTHWEST

All but a small part of the Pacific Coast region is arid or semiarid, with dry summers; much of the land is noncultivatable forest, mountains, or desert unsuitable for tillage (Marble et al. 1985). Grazing of rangelands contributes a major feed component for beef cattle and sheep. In California, rangelands support nearly 14 million animal unit mo of feed annually for beef breeding herds, stocker cattle, and sheep (Reed 1974). Temperature and precipitation extremes result from great differences in latitude and elevation (Marble et al. 1985). Growing season varies from 90 to 340 d annually. Rainfall ranges from 50 to 75 mm in low desert valleys to over 2500 mm on the northwest coast.

California has a rich diversity of vegetation types. Some communities are rare and endemic to the state; others are well-known. These diverse vegetation types are a function of biological and physical factors including slope, elevation, aspect, and climatic factors such as precipitation and temperature. The majority of the region is influenced by a Mediterranean climate characterized by long dry summers and mild winters. However, higher elevations, areas east of the Sierra Nevada, and the southern deserts are more affected by continental weather patterns.

Rangeland cover types for California have been divided into four major community types. Conifer and hardwood woodlands occupy approximately 6.3 million ha (15.5 million A), shrublands (including desert shrublands) occupy approximately 16.0 million ha (40.0 million A), and grasslands cover about 4.0 million ha (9.6 million A) of the 40.5 million ha (100.0 million A) of the state.

Cover types: blue oak woodland, coast live oak woodland, riparian woodland, north coastal shrub, coastal sage shrub, chamise chaparral, scrub oak mixed chaparral, ceanothus mixed chaparral, montane shrubland, bitterbrush, creosotebush scrub, blackbush, alpine grassland, coastal prairie, valley grassland, montane meadows, wetlands.

NORTHERN ROCKY MOUNTAINS

The cover types in this section are the nonforested grasslands and shrublands of the northern Rocky Mountains physiographic province (Fenneman 1931). This region includes mountains and valleys of western Montana, central and northern Idaho, and the Okanogan Highlands of northeastern Washington. Much of the region is dominated by forest types, which are described elsewhere (Society of American Foresters 1980).

The grass- and shrub-dominated cover types in this region are most extensive in southwestern Montana, in the wide intermountain valleys and extensive foothills common to the eastern extension of the province. The remainder of the area is characterized by essentially continuous mountain and gorge topography, so the grass and shrub cover types are limited in abundance and size.

The intermountain region lies between the Sierra Nevada and Cascade mountains on the west and the Rocky Mountains on the east and extends from northern New Mexico and Arizona into southern British Columbia (Eckert and Klebesadel 1985). Precipitation averages about 350 mm/yr but varies from less than 120 mm to over 1250 mm. Summers are usually dry although some thunderstorms may occur in the southern and eastern parts. Hay, pasture, and rangeland are essential for efficient, year-round livestock operations. Rangeland forage sources are primarily from low-elevation salt desert shrub during winter, midelevation sagebrush-grass and pinyon-juniper vegetation types for spring-fall grazing, and high-elevation grasslands, meadows, and open timber for summer grazing.

Large numbers of livestock in the mid-1800s and poor judgment in management resulted in overstocking and deterioration of most rangelands. These depleted rangelands are characterized by dense stands of unpalatable plant species or, worse still, denuded sites. Vegetation improvement techniques included weed control, revegetating with desirable range plants, seeding techniques, and grazing management. Of these, probably most range improvements fail for lack of posttreatment grazing management.

Cover types: bluebunch wheatgrass—blue grama, bluebunch wheatgrass—Sandberg bluegrass, bluebunch wheatgrass—western wheatgrass, Idaho fescue—bluebunch wheatgrass, Idaho fescue—Richardson needlegrass, Idaho fescue—slender wheatgrass, Idaho fescue—threadleaf sedge, Idaho fescue—tufted hairgrass, Idaho fescue—western wheatgrass, needle-and-thread—blue grama, rough fescue—bluebunch wheatgrass, rough fescue—Idaho fescue, tufted hairgrass—sedge, big sagebrush—bluebunch wheatgrass, big sagebrush—Idaho fescue, big sagebrush—

rough fescue, bitterbrush—bluebunch wheatgrass, bitterbrush—Idaho fescue, bitterbrush—rough fescue, black sagebrush—bluebunch wheatgrass, black sagebrush—Idaho fescue, curlleaf mountain mahogany—bluebunch wheatgrass, shrubby cinquefoil—rough fescue, threetip sagebrush—Idaho fescue.

GREAT BASIN

The term *Great Basin* was first applied by John C. Frémont following his 1843-44 expedition along the edges of this region. Physiographically the Great Basin lies west of the Colorado Plateau and is bordered on the west by the Sierra Nevada Mountains, on the east by the Wasatch Mountains in Utah, on the north by the lava flows of the Snake River plains, and on the south by the arbitrary designation of 35′30′ latitude (Fenneman 1931).

Topographically, this region consists of numerous parallel north-south-trending isolated mountain ranges separated by nearly level intermountain basins (Cronquist et al. 1972). Elevations range from the 1300- to 2000-m (4000- to 6000-ft) valley floors to the 4000-m (12,000-ft) mountain peaks.

In general, the Great Basin is an area of low rainfall—a desert in the rain shadow of the Sierra Nevada and the Cascade ranges. Precipitation in this region ranges between 15 and 28 cm (6 to 11 in.) annually in the lowlands but may reach up to 63-75 cm (25-30 in.) in the higher mountainous areas. A majority of this moisture comes as winter snow, although summer thunderstorms are common and often violent, as is evidenced by the deeply cut canyons that dissect the alluvial terraces at the mountain-valley interfaces. The natural flora consists of plant species that are well adapted to the cold winters as well as the warm, dry summer seasons.

Cover types: Basin big sagebrush, mountain big sagebrush, Wyoming big sagebrush, threetip sagebrush, black sagebrush, low sagebrush, stiff sagebrush, other sagebrush types, tall forb, alpine rangeland, aspen woodland, juniper—pinyon woodland, gambel oak, salt desert shrub, mountain mahogany types, curlleaf mountain mahogany, true mountain mahogany, littleleaf mountain mahogany, mountain brush types, bigtooth maple, bittercherry, snowbush, chokecherry—serviceberry—rose, riparian.

SOUTHWESTERN US

The southwestern US often is called "Big Country" due to the sparseness of urban centers (Herbel and Baltensperger 1985). Annual average precipitation is less than 250 mm in the lower elevations of Arizona, New Mexico, and western Texas. About 70% of the precipitation occurs during springtime in the Great Plains. Arizona and New Mexico precipitation often occurs during summer, with spring normally dry. Frequent droughts are common. A high percentage of the region is used for ranching, while farming is limited in the region because of low precipitation, shallow and rocky soil, or rolling terrain.

This is a diverse region of plateaus, plains, basins, and many isolated mountain ranges. The land resources in western Texas, southern and western New Mexico, Chihuahua, Arizona, southeastern California, northern Sonora, and Baja California are southern desertic basins, plains, and mountains; southeastern Arizona basin and range; Arizona, Chihuahua, New Mexico, and Sonora mountains and plateaus; central Arizona basin and range; San Juan River Valley plateaus; Colorado River plateau; and Sonoran basin and range (Herbel 1979).

Only a few southwestern drainages presently contain any extensive linear riparian development. These habitats were originally populated with cottonwoods (e.g., Fremont poplar [*Populus fremontii*]) and willows (e.g., Goodding willow [*Salix gooddingii*]).

Cover types: saltbush—greasewood, grama—galleta, Arizona chaparral, juniper—pinyon pine woodland, grama—tobosa shrub, creosotebush—bursage, palo verde—cactus, creosotebush—tarbush, oak—juniper woodland, and mahogany—oak.

NORTHERN GREAT PLAINS

The northern Great Plains region is bounded on the west by the intermountain region and on the east by the northeastern deciduous forest region; the eastern boundary is the transition zone between native midgrass prairie on the west and tall-grass prairie on the east at approximately the 98th meridian (Moore and Lorenz 1985). The midgrass zone gives way to the short-grass prairie and continues to the western boundary. The climate of the region is semiarid to arid with annual precipitation decreasing from east to west and south to north and varying from 250 mm to 750 mm. Elevations above sea level range from 425 m in the east to more than 1500 m in the west and freeze-free days range from 175 d in the southwest to 110 d in the northwest. Introduced species and tall- and midgrass native species are seeded for hay, pasture, and cash crop-livestock operations in

the eastern portion of the region. Grains such as wheat (*Tricicum aestivum* L. emend. Thell.), corn (*Zea mays* L.), sorghum (*Sorghum bicolor* [L.] Moench), sunflower (*Helianthus annuus* L.), oat (*Avena sativa* L.), rye (*Secale cereale* L.), and barley (*Hordeum vulgare* L.) are grown as cash crops. Farther west, more than half the land remains as native midgrass or short-grass prairie with cattle and sheep ranches common.

Cover types: bluestem prairie, bluestem—prairie sandreed, prairie sandreed—needlegrass, bluestem-grama prairie, sandsage prairie, wheatgrass—bluestem—needlegrass, wheatgrass—needlegrass, wheatgrass—grama—needlegrass, wheatgrass—grama, wheatgrass, blue grama—buffalograss, sagebrush—grass, fescue grassland, crested wheatgrass, wheatgrass—saltgrass—grama.

SOUTHERN GREAT PLAINS

The southern Great Plains includes Texas, Oklahoma, and portions of New Mexico. The southern Great Plains is a diverse area due to soils, climate, and geographic location. This region ranges in elevation from sea level along the Gulf of Mexico, to over 2000 m (6000 ft) in New Mexico and to 2917 m (8751 ft) at Guadelupe Peak in far west Texas. The growing season extends from an average of 330 d per year in the Lower Rio Grande Valley of Texas to less than 180 d per year in northern New Mexico and the northwestern Panhandle of Texas. Generally, the rainfall is quite variable, ranging from about 20 cm (8 in.) per yr in far west Texas to more than 100 cm (40 in.) per year along the Gulf Coast.

Geographically, the southern Great Plains includes eastern New Mexico from Raton to Carlsbad; the plains east of the Sangre de Cristo, Capitan, and Sacramento mountains. Also included is Oklahoma with the exception of some of the post oak and blackjack oak forests in the eastern part of the state where very little, if any, grazing lands occur. Texas is included with the exception of the Chihuahuan Desert of the Trans Pecos region and the piney woods of east Texas. Grasslands in the Trans Pecos region, however, are included.

Cover types: alkali sacaton—tobosagrass, black grama—alkali sacaton, black grama—sideoats grama, blue grama—western wheatgrass, blue grama—galleta, blue grama—sideoats grama, blue grama—sideoats grama—black grama, bluestem—dropseed, bluestem—grama, bluestem prairie, bluestem—sacahuista prairie, galleta—alkali sacaton, grama—muhly—three-awn, grama—bluestem, grama—buffalograss, grama—feathergrass, little bluestem—indiangrass—Texas wintergrass, mesquite—grama, mesquite—live oak—seacoast bluestem, sand bluestem—little bluestem dunes, sand bluestem—little bluestem plains, sand sagebrush—mixed prairie, sea-oat, sideoats grama—New Mexico feathergrass—winterfat, vine mesquite—alkali sacaton, cordgrass, mesquite—buffalograss, mesquite-granjeno—acacia, mesquite, sand shinnery oak, cross timbers—Oklahoma, cross timbers—Texas little bluestem—post oak, juniper—oak, mesquite—oak, sideoats grama—sumac—juniper.

SOUTHEAST

This region includes eastern Texas, eastern Oklahoma, and states east, including Missouri, that have significant range cover types. This is a diverse area in that it extends from the Florida flatwoods and freshwater marshes to the Missouri savanna and prairies. The gulf salt marsh that runs along the Gulf of Mexico from Texas to Florida is also included. Most of the true range cover types in the Southeast have been plowed up and are now classified as cropland or introduced forages (pasture).

The humid South has a relatively high rainfall (1250 mm or more), with uniform annual distribution even though drought periods are not uncommon (Chamblee and Spooner 1985). The annual freeze-free period ranges from 200 to 365 d (Holechek et al. 1989). Temperatures are usually high with a yearly average of 21°C. Soils are highly leached and usually acid. Due to the warm temperatures and relatively high precipitation, range forage growth of this region is potentially greater than in any of the other regions.

Cover types: savanna, Missouri prairie, Missouri glades, tall fescue, riparian, Gulf Coast salt marsh, Gulf Coast fresh marsh, sand pine scrub, mixed hardwood and pine, longleaf pine—turkey oak hills, south Florida flatwoods, north Florida flatwoods, cutthroat seeps, cabbage palms flatwoods, upland hardwood hammocks, cabbage palm hammocks, oak hammocks, Florida salt marsh, freshwater marsh and ponds, everglades flatwoods, pitcher plant bogs, slough.

ALASKA

In Alaska, the first cattle were brought to Kodiak Island in 1794 during the Russian period and were kept at several agricultural settlements along the southern coast (Eckert and Klebesadel 1985). Alaska's climate, seasonal and photoperiodic patterns, and soil processes are similar to those of other circumpolar areas, such as Norway, Sweden, Finland, and Iceland but dissimilar from those of the lower 48 states. Alaska has a significant amount of native rangeland existing on a number of islands, from Kodiak to Umnak in the western Gulf of Alaska, which is grazed virtually year-round by sheep and cattle (Eckert and Klebesadel 1985). A considerable number of native grasses, bluejoint (*Calamagrostis canadensis* [Michx.] P. Beauv.), dune wildrye (*Leymus mollis* [Trin.] Pilger subsp. *mollis*), Bering hairgrass (*Deschampsia beringensis* Hult.), alpine timothy (*Phleum alpinum* L.), etc., and numerous forbs and sedges provide forage for range livestock.

Some ranchers bale native grass hay where topography and accessibility are favorable. Along Alaska's western coastal region, the herding of reindeer is common; federal laws restrict ownership of reindeer to Alaskan natives. Reindeer exist on a variety of sedges, forbs, and slow-growing lichens (*Cladonia* spp. and *Cetraria* spp.), especially during winter. Grasses are often less abundant on reindeer ranges than forbs or sedges.

Cover types: alder, alpine herb, beach wildrye—mixed forb, black spruce—lichen, bluejoint reedgrass, broadleaf forest, dryas, fescues, freshwater marsh, hairgrass, lichen tundra, low scrub shrub birch—ericaceous, low scrub swamp, mesic sedge—grass—herb meadow tundra, mixed herb—herbaceous, sedge—shrub tundra, tall shrub swamp, tussock tundra, wet meadow tundra, white spruce—paper birch, willow.

NORTHEAST

The Northeast region is not included in the 1994 Society for Range Management publication *Rangeland Cover Types of the United States*. The northeastern deciduous forest region provides important livestock grazing areas, especially in Missouri, Indiana, Ohio, Kentucky, Virginia, and Wisconsin (Holechek et al. 1989). Typically pasture production and introduced forages predominate for livestock production in the region compared with the native range forages, especially in the eastern part. Dairying accounts for nearly half the cash receipts from agriculture in many of the states (Baylor and Vough 1985).

Average annual precipitation varies from less than 800 mm to nearly 2000 mm in the Northeast region and is uniformly distributed throughout the year. The freeze-free period ranges from 115 to 240 d (Baylor and Vough 1985; Van Keuren and George 1985; Holechek et al. 1989). Much of the region has been modified by farming, logging, and industrialization. Because of the high precipitation, pastures with introduced forages are widely used and nitrogen (N) and phosphorus (P) fertilization is common and usually effective. Soils are commonly acid and relatively unfertile.

Native grasses in the region include the bluestems (*Schizachyrium* spp. and *Andropogon* spp.), eastern gamagrass (*Tripsacum dactyloides* [L.] L.), indiangrass (*Sorghastrum nutans* [L.] Nash), and switchgrass (*Panicum virgatum* L.). Introduced grasses are predominately tall fescue (*Festuca atundinacea* Schreb.), timothy (*P. pratense* L.), smooth bromegrass (*Bromus inermis* Leyss.), reed canarygrass (*Phalaris arundinacea* L.), orchardgrass (*Dactylis glomerata* L.), and bermudagrass (*Cynodon dactylon* [L.] Pers.). Legume-grass mixtures are commonly used in improved pastures.

RANGELAND CONDITION ASSESSMENT AND MANAGEMENT

The debate about the health of the nation's rangelands has become a national issue, and the need to have an acceptable rangeland condition assessment has been called urgent (NRC 1994). Heady and Child (1994) identify two problems of rangeland condition evaluation for each rangeland site: (1) determining potential conditions and (2) determining management objectives. The potential is not accurately known for many rangeland sites. Objectives of management may change, and vegetation to meet the new objectives is often selected based on guesswork.

Reporting range condition on a management unit or nationwide cannot be accurate without stating the management objectives. Rangeland managers must be able to determine the current state of their resource and predict the future state when certain management practices are applied. At this writing, the actual measurement and use of range condition has seen little change.

However, federal agencies are working together to update and standardize assessment techniques, and with the publication of the National Research Council report (1994) the controversy over the concepts and suggested methodologies has been brought to the forefront.

NEW DIRECTIONS IN RANGE CONDITION ASSESSMENT

For the most part, vegetational changes have been described on the basis of time, the scale being hourly to geologic. Each type of change is also associated with a space or *site*. The three terms, *ecological site, range site,* and *habitat,* are of long standing. (The definition and usage of other terms are still being determined.) These terms have essentially the same definition: a unit of land supporting or capable of supporting a distinctive climax vegetation (Shiflet 1973). Ecological sites and range sites may be less extensive and included with others in the habitat type. *Plant community* is a general term commonly referring to a collection of plants with no successional status or size implied, seeded or nonnative vegetation included. More specific usage is not widely accepted.

These concepts have much in common as a conceptual basis for classifying natural ecosystems. The concepts differ more in their application to land classification for multiple-use purposes than to ecological differences (Leonard and Miles 1989). They are based on the proposition that vegetation is an integrated expression of abiotic and biotic characteristics and that stability, as succession to climax, reflects the site and habitat potentials.

Range sites and habitat types emphasize potential vegetation while community (type) and ecological site focus on the present vegetation. The purpose of using these concepts is to provide a basis for inventory and land management, not to prove succession and climax. Site definition in terms of potential can hardly be used in rangeland inventory and management because present vegetation must be defined as well as its relation to the potential. This is range condition: an inseparable complex of site, present vegetation, and potential vegetation.

The habitat-type method of classifying land was first described by Daubenmire (1952). It has become a widely used system for land classification in forests and more slowly for shrublands and grasslands. The concept of potential vegetation has been gradually enlarged to include seral vegetation, soils, landforms, and management (Wellner 1989). Classification of habitat types is ecological and not colored by "for what use."

An effort has been made to separate the range condition concept of Dyksterhuis (1949) as used by the Natural Resource Conservation Service into two concepts. One is *ecological status,* defined as the present state of vegetation and soil protection of an ecological site in relation to the potential natural community for the site (Jacoby 1989). The second redefines range condition by giving the vegetation on an ecological site a resource value rating

(RVR) for a particular use or benefit (Jacoby 1989). Both concepts are applied to the same range site. If the vegetation is nonnative, RVR is used. The revised concepts emphasize and separate plant succession from multiple-use values.

The *potential vegetation* is defined as the stable vegetation community that could eventually occupy a site without human influence. It is currently called *potential natural vegetation* or *potential natural community*. With or without human influence, potential plant community is one of several that may become established on an ecological site. Still another is *desired plant community,* one identified in a management plan as desirable.

The Society for Range Management (1991), by action of its board of directors, has accepted and encouraged new directions in the assessment and reporting of range condition. Rangeland should be classified by *ecological sites,* and the management objective should be defined in terms of *desired plant community* for each site. Protection of the site against erosion should be assessed in terms of a *site conservation rating,* and where accelerated erosion begins is the *site conservation threshold*. Procedures based on these concepts are still not widely accepted. However, terms and concepts are needed that give clear interpretation of present vegetation, likely changes, possible stability, planned use, and potential vegetation.

Two site classifications have made their way into rangeland inventory and management practice. Ecological site (Bureau of Land Management) is similar to the range site (Natural Resource Conservation Service) and to the percentage of similarity in potential natural community or climax (Table 16.1).

Westoby et al. (1989) suggest a state-and-transition model for research and management of rangelands. Data are catalogs of different states of the vegetation and of the possible transitions among the states. The states are illustrated as a series of boxes that describe different combinations of dominant plants. Arrows between the boxes suggest different states when factors such as fire, herbi-

cides, and grazing change the vegetation.

Lower successional steady states, for example, a dense sagebrush (*Artemisia* spp.) stand, can be shown with reference to various treatments. An advantage of the model is that several steady states can be shown for an ecological site—not just one climax. Incorporation of this type of model into range condition assessment is yet to be done (Laycock 1991). Procedures for field sampling and determination of each condition class are defined within the using agency. The terms for range site emphasize use for what purpose and for the ecological condition currently accepted.

MANAGEMENT PRACTICES TO IMPROVE OR MAINTAIN RANGELAND HEALTH

The recent report by the National Research Council on rangeland health recommends that the term *rangeland health* be used to indicate the degree of integrity of the soil and ecological processes that are most important in sustaining the capacity of rangelands to satisfy values and produce commodities. The council recommends further that the minimum standard for rangeland management should be to prevent human-induced loss of rangeland health (NRC 1994). There are a number of management practices that can be used to improve or maintain rangeland health. Three of these management practices, (1) controlling problem plants, (2) seeding, and (3) grazing systems, are briefly discussed in the following. Much of the text for this section is taken from Heady and Child 1994.

Controlling Problem Plants. Undesirable woody shrubs, herbaceous weed species, and poisonous plants for livestock grazing cover much of the earth's land surface. In the US, the estimated area of brush is 130 million ha. *Artemisia* spp. occupy nearly 40 million ha, and *Prosopis* spp. in Texas alone inhabit 22 million ha (Sampson and Schultz 1957). Over 80% of Texas rangeland is brush infested. Half of it has brush cover greater than 20% (Smith and Rechenthin 1964).

Most noxious weeds on US rangelands are aliens. Including *Bromus tectorum,* they probably occur on more than 50 million ha. Welsh et al. (1991) list 310 species (not all weeds) in 47 plant families that have been introduced into Utah, and new ones continue to arrive. For the intermountain states Callihan and Evans (1991) estimate new arrivals at the rate of 6 or more per yr over the last 100 yr.

Annual loss of forage production and direct losses from poisonous plants severely impact the range livestock industry. For example,

TABLE 16.1. Comparative relationships among descriptors for similarity in potential natural community or climax, range site or ecological site

% similarity	Range condition	Ecological condition
76%-100% of climax	Excellent potential	Natural community
51%-75%	Good	Late seral
26%-50%	Fair	Mid-seral
0%-25%	Poor	Early seral

leafy spurge (*Euphorbia esula*) is increasing rapidly. It is especially harmful to forage yield in the northern Great Plains, where losses were estimated in excess of $14 million annually (Lym 1991). The weed is difficult to control, but it is palatable to sheep and goats and may be reduced by them.

Occupation of the land by weeds and brush in the 1960s resulted in plant poisoning, physical injury, and increased costs of management, estimated at $250 million annually on western US rangelands (USDA 1965). Frandsen and Boe (1991) estimate the 1989 loss in the 17 western states at $340 million because of noxious weeds and poisonous plants.

The kinds of undesirable plants and the problems they cause are as diverse as the soils and climates they inhabit. The list of shrub species identified as undesirable over the years has changed very little. However, the 1991 book on noxious range weeds (James et al. 1991) lists the following herbaceous species that were not prominent in lists of noxious weeds developed in the 1950s and earlier:

Centaurea maculosa	Spotted knapweed
Centaurea diffusa	Diffuse knapweed
Centaurea repens	Russian knapweed
Centaurea solstitialis	Yellow starthistle
Euphorbia esula	Leafy spurge
Isatis tinctoria	Dyers woad
Chondrilla juncea	Rush skeletonweed
Cirsium arvense	Canadian thistle
Carduus nutans	Musk thistle
Onopordum acanthium	Scotch thistle

Undesirable plants are often called "weeds" or "noxious weeds." *Weed* and *noxious weed* refer to a plant that detracts from the use objective. *Noxious* also designates those so declared and restricted in some way by law. A plant species may be desirable, a weed, or a noxious weed according to area and type of use. Rangeland management includes techniques that are used to directly modify the species composition and/or structure of the vegetation. Manipulation of grazing can be used along with such tools as machines, chemicals, fire, and organisms to reduce undesirable species and thereby encourage desirable vegetation.

Seeding. Rehabilitation of rangelands by seeding began in the western US in the late 1800s. More literature exists on range seeding than on any other practice in range management. Practical trials and experiments have failed and succeeded in numbers sufficient to serve as a foundation for seeding practices. Specific recommendations can be made with assurance of adequate stands in the seeding of many range sites. However, the environmental movement during the two decades after 1970 demanded less seeding of rangeland, less monoculture of introduced species, and more seeding of native species.

Rehabilitation and prevention of erosion after wildfire and other land disturbances have replaced forage production as the principal objectives of seeding public land. Success with the native species has been difficult for reasons of low seed availability due to high seed shattering, low seed retention and hence low harvesting amounts, high seed dormancy, and lack of knowledge about seedbed requirements.

Before the actual seeding can begin, Heady and Child (1994) suggest that several questions must be answered: Is seeding necessary? Is the climate favorable? Is the habitat favorable? What species should be seeded? Is proper management possible? Is it economical?

Grazing Systems. Grazing by wild and domestic animals exerts an influence upon the productive rangeland systems by the animals' defoliation of plants through eating and physical damage, by their digestive processes, and by their movements. Consideration of animals only as products is not enough; grazing animals are tools to attain vegetational production goals. When a grazing animal eats, it selects certain plants or plant parts and removes them to a definite degree or intensity. Grazing includes four aspects of defoliation: intensity, frequency, seasonality, and selectivity. Each of these factors influences the growth and reproduction of the plants differently and, hence, the vegetation being grazed (Heady and Child 1994).

Each species of range herbivore has its own peculiar behavioral characteristics, some inherited and some learned, that determine part of its total impact on the habitat. Sheep (*Ovis aries*) often graze into the wind, many species prefer specific types of cover, some establish territories, and herding instincts are common. Sheep and cattle (*Bos taurus*) differ in their seasonal preference of riparian and upland sites. Animals exert a physical impact by trampling, which damages plants, compacts soil, makes trails, churns soil surfaces, and covers seeds. Other physical actions by animals include the burrowing activities of rodents and the mixing of organic materials with mineral soil by invertebrates.

The objective of seasonal grazing plans may be any one or combination of the following (Heady and Child 1994):

1. To improve range condition including attainment of less erosion, to increase soil cover, to provide fuel for prescribed fire, to maintain and increase plant vigor, to obtain seed and seedling establishment, to promote vegetational succession, and to achieve other goals associated with proper stewardship of the rangeland resources.

2. To achieve regular distribution of grazing animals through careful attention to pasture size and shape.

3. To promote uniform forage utilization by reducing selectivity of forage.

4. To coordinate domestic animal grazing with habitat needs of wildlife and other uses of the land.

5. To increase animal performance either individually or in terms of production per land unit, thereby increasing ranch income and decreasing costs of pest control, supplemental feeding, and labor.

6. To increase flexibility and decrease risk in the ranch operation.

7. To improve quality and quantity of forage and to provide reserve feed for emergencies.

Grazing systems have had a long history. The Hema System, gradually regaining support in the Middle East, predates the Islamic era and has been in use for centuries. Active ahemia (plural) occur in Saudi Arabia, Syria, Lebanon, Tunisia, and other Middle East countries. They are commonly recognized as grazing cooperatives.

Rotational stocking has been advocated in Europe for over 200 yr and in southern Africa since 1887. Jared Smith (1895) first suggested seasonal stocking plans on rangelands in the US when he advocated rotational stocking as one means of improving range conditions in the southern Great Plains. Sampson (1913, 1914), after considerable ecological research in the Wallowa Mountains of Oregon, recommended rotational deferred stocking as a general practice. Shortly thereafter, Jardine (1915) and Jardine and Anderson (1919) presented that schedule in diagram form and suggested it for use on national forests. Since that early beginning, evolution in design of plans and seasonal stocking effects have been the subject of considerable discussion, research, and argument, with gradually increasing acceptance.

Agroforestry Systems

Agroforestry has the potential of minimizing costs in sustainable agriculture, reducing soil erosion, improving water quality, providing wildlife habitat, allowing economic flexibility, and stabilizing family farm operations. Family farming, ranching, and wood product (pulp, timber, firewood, etc.) enterprises continue to undergird rural community development and food and fiber production for the nation. Important problems facing people now and in the future are shortages of food and uncertainties in energy supplies (Gold and Hanover 1987). Agroforestry provides flexible alternatives for family farms to improve their economies and survive poor markets while increasing the nation's food, forage, and fiber supplies. Land use for two or more commodities complicates management and affects each of the resources producing those commodities, beneficially in some cases, detrimentally in others. The increasing demands from the land for combined supplies of food and fiber place greater demands on the environment.

AGROFORESTRY DEFINED

In preparation for the next Resource Conservation Act appraisal, a committee report, *A Comprehensive Assessment of US Agroforestry* (Garrett et al. 1994), defines *agroforestry* as an intensive land management system that optimizes the benefits from the biological interactions created when trees and/or shrubs are deliberately combined with crops and/or livestock. Four key components of this definition are (1) land use is intensive, (2) benefits are optimized, (3) biological interactions are increased, and (4) trees and/or shrubs are deliberately combined with crops and/or pastures for livestock.

1. Intensive land use—Agroforestry utilizes management practices for producing trees with crops and/or livestock. Management schemes incorporate intensive practices such as tree planting, annual cultivation, fertilization, irrigation, weed control, liming, grazing, or combinations of these and other practices. Therefore, agroforestry manipulates the agro-

nomic ecosystem to achieve optimal benefits. Grazed forests or woodlands that are extensively managed as natural ecosystems also produce wood products and forage for livestock or wildlife but without using intensive approaches such as annual inputs of nutrients, herbicides, or mechanical treatments. These latter systems are often excluded from the agroforestry classification, but because of similar interactions between woody plants, forage supplies, and livestock, they are considered a subset of the agroforestry land-use classification for management purposes.

2. Benefits optimized—As a land-use system, agroforestry provides many benefits for humankind. These benefits may be economic, environmental, biological, or social. Optimization implies that the various factors can be combined, through compromise, to best serve the interests of the user and society. Since each user may have different objectives, such as soil conservation, minimization of inputs, integrated pest management, greater profit, or aesthetics, optimization of system components will vary.

3. Biological interactions—Protective and productive benefits are realized from agroforestry practices. They are the products of biological interactions resulting from the proper mix of woody perennials, herbaceous species, and livestock. These interactions affect soil, water and air quality, biological diversity, wildlife habitat, aesthetics, economics, and, ultimately, rural community development. As an example, riparian filter strips may help reduce soil erosion from cropland and improve habitat for some particularly desired fish or wildlife species while enhancing the visual appearance of landscapes.

4. Trees and/or shrubs deliberately combined with crops and/or livestock—An intentional combining of trees and/or shrubs with crops and/or pasture for livestock is an essential element of agroforestry systems. This aspect mimics the multispecies, multistoried characteristic found in natural ecosystems while providing a variety of marketable commercial products. Crops may include conventional agronomic commodities such as corn, cotton (*Gossypium* spp.), and soybeans (*Glycine max* [L.] Merr.); specialty commodities such as ginseng (*Panax guinquefolium*), golden seal (*Hydastis canadensis*), shiitake mushrooms (*Lentinula edodes* [Berk.] Pegler), and honey; both warm- and cool-season tame pasture; and managed native forages—forestland, woodland, pastureland, and rangeland. Trees and shrubs may consist of high-value timber/veneer, nut-/fruit-producing woody perennials, and others. Livestock are animals kept for use on the farm or raised for sale and profit.

While few data exist that would allow a reliable estimation of the degree of current use of various agroforestry practices, vast acreages of forestland, pastureland, rangeland, cropland, and stream corridor suitable for agroforestry occur throughout the US (see Table 16.2). The adoption and application of practices varies by region and is driven by local tradition, economic factors, and land ownership patterns (government versus private).

INTEGRATED LAND USES

Agroforestry has been practiced for many years throughout third world countries but only in recent times have these practices been investigated by researchers in developed countries (Lassoie and Buck 1991). Agroforestry research in the southern US started during the mid-1940s but has gained considerable interest in recent years (Lewis et al. 1991). The "rediscovery" of agroforestry by the global scientific community dates back to 1977 (Raintree 1991). This growing interest since the 1980s spawned development and conduct of three North American conferences on agroforestry: 1989 (Guelph, Ontario), 1991 (Springfield, Missouri), and 1993 (Ames, Iowa) (Williams 1991; Garrett 1991a). Also a significant regional conference was held in the mid-South during 1990 at West Memphis, Arkansas (Henderson 1991). Proceedings from these conferences and other literature, including several books, provide a foundation for agroforestry system research and management practices in the developed world, including temperate zone climates (Gholz 1987; Linnartz and Johnson 1984; MAART 1986; Reid and Wilson 1985).

Integrated forestry-farming (agroforestry) management typically involves planting trees at wide spacing with intercrops grown in alleys between trees (Garrett et al. 1991b). Intercropping, strip-cropping, planting specialty crops, grazing with sheep, cattle, or goats (*Capra hircus*), and a variety of other agroforestry options appear feasible for the temperate zones of North America (Williams and Gordon 1991; Countryman and Krambeer 1991; Dawson and Paschke 1991; Hill 1991; McDonald and Fiddler 1991; Pearson and Martin 1991). Tree density and pattern are significant factors in successful agroforestry system design; tree-planting configuration and orientation patterns to optimize both trees and ground-level crops appear to favor triple rows compared with uniform grids (Sharrow 1991).

Table 16.2. Total hectares of nonfederal pastureland and rangeland and those with high or medium potential for conversion to cropland or agroforestry

Farming region and state	Total (1000s)		Pastureland (1000s)		Rangeland (1000s)	
	Pastureland	Rangeland	High	Medium	High	Medium
Northeast						
Connecticut	46.1	0.0	10.8	17.0	0.0	0.0
Delaware	14.2	0.0	4.6	5.5	0.0	0.0
Maine	230.3	0.0	24.6	66.5	0.0	0.0
Maryland	216.2	0.0	28.5	52.2	0.0	0.0
Massachusetts	81.6	0.0	8.3	25.9	0.0	0.0
New Hampshire	50.4	0.0	8.9	7.6	0.0	0.0
New Jersey	97.0	0.0	6.4	17.4	0.0	0.0
New York	1566.8	0.0	115.6	558.9	0.0	0.0
Pennsylvania	1049.2	0.0	138.8	321.2	0.0	0.0
Rhode Island	14.5	0.0	0.8	1.8	0.0	0.0
Vermont	202.6	0.0	25.9	48.1	0.0	0.0
Region	3569.0	0.0	373.2	1122.2	0.0	0.0
Appalachia						
Kentucky	2379.5	0.0	392.3	651.0	0.0	0.0
North Carolina	801.4	0.0	111.6	286.3	0.0	0.0
Tennessee	2167.4	0.0	421.1	586.9	0.0	0.0
Virginia	1372.7	0.0	94.3	33.3	0.0	0.0
West Virginia	756.4	0.0	45.4	221.5	0.0	0.0
Region	7477.4	0.0	1064.6	2079.1	0.0	0.0
Southeast						
Alabama	1544.6	0.0	329.6	497.9	0.0	0.0
Florida	1729.3	1539.4	169.8	671.3	35.0	408.5
Georgia	1204.6	0.0	221.2	396.3	0.0	0.0
South Carolina	488.9	0.0	48.6	168.4	0.0	0.1
Region	4967.4	1539.4	769.2	1834.0	35.0	408.5
Lake states						
Michigan	1178.1	0.0	177.6	346.4	0.0	0.0
Minnesota	1452.8	80.3	277.7	452.9	4.4	15.0
Wisconsin	1373.6	0.0	215.0	325.8	0.0	0.0
Region	4004.5	80.3	670.3	1125.1	4.4	15.0
Corn Belt						
Illinois	1277.7	0.0	189.9	348.4	0.0	0.0
Indiana	895.1	0.0	170.1	258.5	0.0	0.0
Iowa	1835.7	0.0	317.2	589.9	0.0	0.0
Missouri	5088.2	67.9	958.0	1898.3	19.3	13.1
Ohio	1098.3	0.0	157.3	290.8	0.0	0.0
Region	10,195.1	67.9	1792.4	3385.8	19.3	13.1
Delta states						
Arkansas	2344.7	66.6	177.5	706.6	1.0	3.0
Louisiana	958.5	97.7	180.5	300.9	0.4	4.4
Mississippi	1608.7	0.0	255.0	457.2	0.0	0.0
Region	4912.0	164.3	613.0	1464.7	1.4	7.4
Northern Plains						
Kansas	906.9	6842.9	168.5	355.4	461.1	1301.9
Nebraska	860.1	9346.7	153.3	314.6	261.3	1507.0
North Dakota	514.7	4430.8	93.3	170.8	189.2	783.3
South Dakota	1093.9	9220.4	235.5	487.3	278.9	1492.8
Region	3375.6	29,840.8	650.5	1328.1	1190.4	5085.1
Southern Plains						
Oklahoma	2888.7	6094.5	296.7	863.7	294.5	1017.8
Texas	6897.1	38,588.9	562.6	1888.3	1007.7	4562.4
Region	9785.8	44,683.4	859.3	2752.0	1302.2	5580.2
Mountain						
Arizona	32.1	12,524.6	17.1	13.0	133.4	689.3
Colorado	509.8	9802.7	34.0	94.1	112.1	718.5
Idaho	515.7	2724.8	47.8	206.9	46.3	156.6
Montana	1228.7	15,312.4	177.3	414.1	304.9	1484.7
Nevada	123.2	3201.1	6.6	31.2	18.8	68.8
New Mexico	66.1	16,585.1	7.7	13.0	72.0	383.6
Utah	198.4	3435.6	11.4	57.5	15.8	77.5
Wyoming	305.5	10,892.4	38.9	127.0	75.2	584.5
Region	2979.4	74,477.8	341.0	918.6	778.6	4163.6
Pacific						
California	563.5	7334.9	85.3	171.5	128.7	454.4
Oregon	795.6	3800.9	98.9	203.8	27.3	247.9
Washington	544.2	2281.3	71.8	207.4	59.7	218.2
Region	1903.3	13,417.1	255.9	382.7	215.7	920.5

Source: Based upon the 1982 NRI.

A key feature in developing agroforestry systems in North America appears to be sustainable agriculture or land-use systems that allow increasing food, forage, and fiber production without causing environmental degradation (Lassoie and Buck 1991). Several agroforestry programs have demonstrated that economies are greater with the combination of crops than when either is singly grown (Garrett 1991b; Kurtz et al. 1984; Pearson 1991a, 1991b).

SILVO-PASTORAL SYSTEMS

Although of minor consequence, types of agroforestry systems are defined differently due to forage crops or animal management. *Silvo-pastoral*, or *forest grazing*, refers to the grazing of natural or indigenous forages that exist under naturally occurring forests; *agro-silvo-pastoral* refers to the combining of trees, forage crops (primarily tame pastures but also row crops), and livestock (Lewis and Pearson 1987). The main distinction relates to the type of understory forage crop, native versus introduced species involving cultural practices. This "silvo-pastoral" section relates to native forage management systems; if forage species are discussed that are introduced, then their management is similar to native vegetation management.

Woody Perennials. Density of overstory woody perennials is the most powerful factor influencing the capacity of forested areas to produce forage for livestock or wildlife. On the other hand, competing ground vegetation negatively influences survival and growth of newly planted trees. Further complicating tree-forage-animal relationships, livestock and/or wildlife continually loom as a potential threat for animal damage to newly planted woody vegetation. Judicious planning for use of forages by animals throughout timber rotations requires a thorough understanding of tree-forage-animal interrelationships and dynamics.

Site preparation, tree regeneration densities, timber harvesting, fertilization, and other forest and animal management practices significantly affect the forest floor and the understory forage supplies. Successful tree regeneration and establishment in tame pastures is highly dependent upon competition control and livestock management practices. New markets and innovative marketing are necessary approaches to achieving economic stability for family farming operations in an ever-expanding, competitive world market environment.

Woody perennials most prevalent in agro-forestry programs in the US are conifers; however, hardwoods, especially nut trees, are also common in some systems. In the southern US, grasses and legumes growing with pines are commonly grazed in the coastal plains and interior highlands. Southern pine species commonly include longleaf *(Pinus palustris* Mill.) and slash (*P. elliottii* Engelm.) pines in the lower coastal plain and loblolly (*P. taoda* L.) and shortleaf (*P. echinata* Mill.) pines in the upper coastal plain and interior highlands. Douglas-fir (*Pseudotsuga menziesii* [Mirb.] Franco) forests are grazed by sheep and cattle in the Pacific Northwest while ponderosa pine (*P. ponderosa* Dougl. ex Laws.) forests have been traditionally grazed in the West and Southwest (Sharrow 1991; Pearson 1964).

Other successful agroforestry systems, including nut producers, have incorporated hardwood trees with forage crops; the native pecan (*Carya illinoensis* [Wangenh.] K. Koch) agroecosystems represent a common form of agroforestry in North America (Reid 1991). Eastern black walnut (*Juglans nigra* L.) alleycropping is a viable land-use strategy, especially on marginal lands where both a wood-based and nut-related industry can provide short- and long-term returns to the landowner (Garrett et al. 1991a). In sustainable agriculture the commercial nut crops, forage crops, livestock grazing, wildlife, row crops, and/or wood harvests are combined to maximize returns on landowner investments.

Other useful trees and specialty crops for agroforestry programs are the English walnut (*J. regia* L.), royal paulownia (*Paulownia tomentosa* [Thunb.] Seib. & Zucc.), hazel or filbert (*Corylus americana* Marsh., *C. cornula* Marsh., and others), black cherry (*Prunus serotina* Ehrh.), honeylocust (*Gleditsia tricanthos* L.), American beech (*Fagus grandifolia* Ehrh.), chestnut (*Castanea dentata* [Marsh.] Borkh., *C. mollissima* B1., *C. crenata* Seib. & Zucc.), persimmon (*Diopyros virginiana* L.), willow (*S. nigra* Marsh.), sycamore (*Plantanus occidentalis* L.), sweetgum (*Liquidambar styraciflua* T.), and cottonwood (*P. deltoides* Bartr. ex Marsh.) (Stanley L. Krugman, personal communication, July 14, 1992; Krugman 1988; Burns and Honkala 1990; USDA Forest Service 1974; Hill 1991). Other specialty crops, such as ginseng, golden seal, and shiitake mushrooms, are used in conjunction with the woody perennials (Hill 1991).

Management. Grazing forage plants in the forest understory is a practical, efficient, and economical method to simultaneously produce beef and trees from the same land. Trees need to be thinned and/or planted in widely spaced rows to maintain forage yields, and animals must be stocked to be in balance with forage supplies. Several kinds of animals are incorporated into agroforestry systems: sheep, cattle, goats, and wildlife such as deer (*Odocoileus virginianus)*, quail (*Colinus virginianus*), turkey (*Meleagris gallopavo*), and/or nongame species.

PASTURES/CROPS UNDER PINES/HARDWOODS

Some of the earliest attempts at agroforestry in the southern US began in the 1940s when several grasses and legumes were tested for their adaptability to southern conditions (Burton and Matthews 1949; Halls et al. 1957). These early studies show the need for good site preparation and fertilization plus open tree canopies if exotic forages were to persist. Best results were achieved with common carpetgrass (*Axonopus affinis* Chase), 'Pensacola' bahiagrass (*Paspalum notatum* Flugge), annual lespedeza (*Kummerowia striata* [Thunb.] Schindler), and white clover (*Trifolium repens* L.).

Grasses. In the South, Pensacola bahiagrass has been found to be the most shade tolerant warm-season perennial grass. With a single fertilization (34-15-28 kg/ha N-P-K) bahiagrass yields increased in agroforestry environments by threefold (3600 versus 1100 kg/ha) during the initial year, while residual yields in the second year were double the yields in unfertilized areas (Lewis and Pearson 1987). 'Kentucky 31' and 'Kenwell' varieties of tall fescue make tall fescue the most shade tolerant cool-season perennial grass. Both perennial ryegrass (*Lolium perenne* L.) and tall fescue exhibited shade tolerance in Mississippi when grown under artificial shades (Watson et al. 1984); total yields for ryegrass with 0%, 50%, and 75% shade were 9122, 7031, and 4206 kg/ha, respectively; tall fescue yields were 5859, 4368, and 4026 kg/ha, respectively.

Legumes. Cool-season legumes for use in agroforestry environments in the South appear to be limited to a few cultivars of subterranean clover (*T. subterraneum* L.), berseem clover (*T. alexandrinum* L.), and crimson clover (*T. incarnatum* L.) (Watson et al. 1984; Sawyer 1978; Haines et al. 1978).

Alleycropping. Several N-fixing tree species have shown some promise for alleycropping; these include Siberian pea tree (*Caragana arborescens*), black locust (*Robinia pseucoac-*

cale), and Japanese pagoda tree (*Saphora japonica*) (Janke 1991). Intercropping with barley and black locust and with corn and black walnut and red oak (*Quercus rubra* L.) has shown potential value in agroforestry (McLean et al. 1991; Ntayombya and Gordon 1991).

Management. Agroforestry pasture management requires judicious planning to realize full benefits from both the pasture and the woody perennial. With protection, some pastures with improved forages and southern pines may be grazed while the trees are small (Pearson et al. 1990a, 1990b).

PINES/HARDWOODS IN PASTURES/CROPLANDS

An agroforestry program for the interior highlands and southeastern US was initiated during 1991 to provide additional income alternative opportunities for family farms (Pearson 1993). Objectives were to develop new information on the establishment, maintenance, and utilization of conifer and hardwood tree stands in open pastures; to develop new knowledge in multiple-use management of land and associated responses in trees, forages, crops, livestock, wildlife, recreation, and watersheds and the environmental impacts of these components on surface and groundwater quality; and to facilitate multidisciplinary networks and partnerships of farmers, technical specialists, scientists, and managers for accomplishing agroforestry research and technology transfer.

Establishment. Control of vegetation competition is needed for effective establishment of woody perennials in improved pastures. Competition control can be accomplished by intensive culture techniques such as clean tilling or cultivating 0.5 to 1.0 m on either side of the young seedlings. Chemical control of the competing pasture vegetation may be a more practical method; heavy stocking prior to planting, close mowing, or mulching young trees with hay or pine needle straw also reduces pasture competition to some extent.

Management. Agroforestry pasture management requires judicious planning to realize full benefits from both the pasture and the woody perennial. With protection, some pastures with improved forages and southern pines may be grazed while the trees are small (Pearson et al. 1990a, 1990b). The success of grazing young pines in pastures depends largely on balancing forage availability and animal stocking rates. In heavily stocked pastures, electric fences or other protective devices are required to prevent livestock grazing or trampling damage to the young tree seedlings.

TREE-PLANTING CONFIGURATIONS/ ORIENTATIONS

Although some information is available regarding plants, animals, and their interactions, information on a wider array of vegetation species, configurations, combinations, and interrelationships is needed to attain full value of mixed tree-crop-animal systems in the US (Long and Nair 1991; McLean et al. 1991). Some success has been experienced using southern pines and warm-season Pensacola bahiagrass for summer grazing and cool-season subterranean clover, tall fescue, and perennial ryegrass under slash pine for supplemental feed during winter (Lewis and Pearson 1987; Lewis et al. 1991). East-west row orientation of trees appears to provide maximum light for meeting photosynthetic needs of intercrop plants (Garrett et al. 1991b).

WINDBREAKS

Living windbreaks provide a conservation measure to protect the soil resources (Baldwin and Nanni 1991). Windbreaks have proven to reduce wind erosion; consequently, windbreaks allow a full range of intensive agricultural practices on prime agricultural land—plowing, discing, tillage, and crop rotation—without subjecting the soil to erosion by wind (Boysen 1991); efforts in Canada are focused on establishing and maintaining single row crop windbreaks.

Soil-Water Protection. Tree windbreaks can be a vital part of sustainable agriculture, especially in dryland regions of the US (Rietveld and Schaefer 1991). These vegetative barriers provide the tallest structure to control wind erosion and allow for greater field widths. Trees are long-lived and deep rooted, which helps hold soil in place. Trees, shrubs, and/or herbaceous cover reduce(s) water pollution by acting as buffer strips for intercepting sediment, chemicals, or animal effluents from rivers or streams (Schultz et al. 1991).

Integrated Systems. Windbreaks add stability to agricultural systems by buffering crops, livestock, soil, water, wildlife, and humans from cumulative effects and extreme weather events (Rietveld and Schaefer 1991; Schultz et al. 1991). Trees and shrubs are effective

windbreaks, enhance wildlife, produce wood products, improve biodiversity of landscapes, and effectively support sustainable agriculture.

Socioeconomics. Agroforestry provides economic opportunities for flexibility in the family farm enterprise, which undergirds the social well-being, wealth, and stability of individual family farms and rural community development. Economic analyses of agroforestry systems in the temperate zone are somewhat lacking, with the only well-studied systems being pine trees in pastures and lines of timber trees in cropland for income or crop protection (Scherr 1991). Generally livestock or crops exceed wood product returns during the early years in the agroforestry rotation; during later years wood products exceed livestock or crop returns (Pearson 1991b).

Alternative economic products include some marginal businesses such as Christmas tree production (Pearson et al. 1990a; Lamont et al. 1993), growing shiitake mushrooms, and beekeeping. The domestic honeybee not only pollinates billions of dollars worth of crops annually but also produces millions of dollars worth of honey (Schmidt et al. 1992). However, this important resource faces several enemies including varroa mites, tracheal mites, and the Africanized bee.

Shiitake mushrooms are native to Japan and are grown on small diameter hardwood logs (Miller 1992; Hale 1987; Leatham 1982). The shiitake mushroom business not only provides direct family farm business sale opportunities but also increases the need to utilize small diameter hardwoods.

Pine straw provides another economic alternative: it is a clean, attractive, bright brown mulch that can be used in most landscaping situations (Pope 1990). Pine straw mulch protects the soil surface, maintaining favorable plant growth conditions, and has become widely used for residential complexes, industrial complexes, highway landscape projects, and single family residential sites. Pine straw harvesting on the East Coast is well developed, with economic returns exceeding several hundred dollars per acre annually (Mills and Robertson 1991). The pine straw enterprise appears to be profitable; however, frequent rakings may reduce tree growth and affect wildlife, forest plants, soil nutrients, and water stresses (Jamison 1943; Koch and McKenzie 1976; Reinke et al. 1981; Van Cleve and Dyrness 1983; Ginter et al. 1979; McLeod et al. 1979).

QUESTIONS

1. What is the definition of *rangeland?*
2. Distinguish the difference between the terms *ecological site, range site,* and *habitat.*
3. What is a noxious weed? Name four common rangeland weeds.
4. What questions must be considered before seeding any rangeland site?
5. What is the Hema System?
6. What are the differences/similarities between agro-silvo-pastoral and silvo-pastoral systems?
7. What are some woody perennials that are N-fixing plants?
8. What are some of the important conifer trees found in agroforestry systems of the South, Northwest, and intermountain regions?
9. List several important cool- and warm-season forages grown in agroforestry systems.

REFERENCES

Baldwin, CS, and C Nanni. 1991. Living windbreaks: the basis of future crop production. In P Williams (ed.), Proc. 1st Conf. Agrofor. North Am., 13-16 Aug. 1989, Guelph, Ontario, Canada. Guelph, Ontario: Univ. of Guelph, 258.

Baylor, JE, and LR Vough. 1985. Hay and pasture seedings for the Northeast. In ME Heath, RF Barnes, and DS Metcalfe (eds.), Forages: The Science of Grassland Agriculture, 4th ed. Ames: Iowa State Univ. Press, 338-47.

Boysen, E. 1991. Windbreaks in eastern Ontario. In P Williams (ed.), Proc. 1st Conf. Agrofor. North Am., 13-16 Aug. 1989, Guelph, Ontario, Canada. Guelph, Ontario: Univ. of Guelph, 259.

Burns, RM, and BH Honkala. 1990. Silvics of North America. Vol. 2, Hardwoods. USDA For. Serv. Agric. Handb. 654.

Burton, GW, and AC Matthews. 1949. A Study of Species, Seeding Methods and Fertilizing Practices for Use on Piney Woods Ranges. Ga. Coastal Plain Exp. Stn. Tech. Mimeo Pap. 1. Tifton, Ga.

Callihan, RH, and JO Evans. 1991. Weed dynamics on rangeland. In LF James, JO Evans, MH Ralphs, and RD Child (eds.), Noxious Range Weeds. Boulder, Colo.: Westview, 55-61.

Chamblee, DS, and AE Spooner. 1985. Hay and pasture seedings for the humid South. In ME Heath, RF Barnes, and DS Metcalfe (eds.), Forages: The Science of Grassland Agriculture, 4th ed. Ames: Iowa State Univ. Press, 359-70.

Countryman, DW, and CL Krambeer. 1991. Stripcropping with trees. In HE Garrett (ed.), Proc. 2d Conf. Agrofor. North Am., 18-21 Aug. 1991, Springfield, Mo. Columbia: Univ. of Missouri, 176-82.

Cronquist, A, AH Holmgren, NH Holgren, and JL Reveal. 1972. Intermountain Flora—Vascular Plants of the Intermountain West, US. Vol. 1. New York: Hafner.

Daubenmire, R. 1952. Forest vegetation of northern Idaho and adjacent Washington, and its bearing on concepts of vegetation classification. Ecol. Monogr. 22:301-30.

Dawson, JO, and MW Paschke. 1991. Current and potential uses of nitrogen-fixing trees and shrubs in temperate agroforestry systems. In HE Garrett (ed.), Proc. 2d Conf. Agrofor. North Am., 18-21 Aug. 1991, Springfield, Mo. Columbia: Univ. of Missouri, 183-209.

Dyksterhuis, EJ. 1949. Condition and management of range land based on quantitative ecology. J. Range Manage. 2:104-15.

Eckert, RE, Jr., and LJ Klebesadel. 1985. Hay, pasture, and rangeland of the intermountain area and Alaska. In ME Heath, RF Barnes, and DS Metcalfe (eds.), Forages: The Science of Grassland Agriculture, 4th ed. Ames: Iowa State Univ. Press, 389-99.

Fenneman, NM. 1931. Physiography of the Western United States. New York: McGraw-Hill.

Forage and Grazing Terminology Committee (FGTC). 1992. Terminology for grazing lands and grazing animals. J. Prod. Agric. 5:191-210.

Frandsen, E, and D Boe. 1991. Economics of noxious weeds and poisonous plants. In LF James, JO Evans, MH Ralphs, and RD Child (eds.), Noxious Range Weeds. Boulder, Colo.: Westview, 442-58.

Garrett, HE (ed.). 1991a. Proc. 2d Conf. Agrofor. North Am., 18-21 Aug. 1991, Springfield, Mo. Columbia: Univ. of Missouri.

Garrett, HE. 1991b. The role of agroforestry in low-input sustainable agriculture. In DR Henderson (ed.), Proc. Mid-South Conf. Agrofor. Pract. and Policies, 28-29 Nov. 1990, West Memphis, Ark. Morrilton, Ark.: Winrock International Institute for Agricultural Development, 1-11.

Garrett HE, JE Jones, J Haines, and JP Slusher. 1991a. Black walnut nut production under alley-cropping management: An old but new cash crop for the farm community. In HE Garrett (ed.), Proc. 2d Conf. Agrofor. North Am., 18-21 Aug. 1991, Springfield, Mo. Columbia: Univ. of Missouri, 159-65.

Garrett HE, JE Jones, WB Kurtz, and JP Slusher. 1991b. Black walnut (*Juglans nigra* L.) agroforestry—its design and potential as a land-use alternative. For. Chron. 67(3):213-18.

Garrett, HE, WB Kurtz, LE Buck, JP Lassoie, MA Gold, HA Pearson, LH Hardesty, and JP Slusher. 1994. Agroforestry: An Integrated Land-Use Management System for Production and Farmland Conservation. A Comprehensive Assessment of US Agroforestry—Long Version. USDA Soil Conservation Service (Account 68-3A75-3-134).

Gholz, HL (ed.). 1987. Agroforestry: Realities, Possibilities, and Potentials. Dordrecht, Netherlands: Martinus Nijhoff Publishers.

Ginter DL, RW McLeon, and C Sherrod, Jr. 1979. Water stress in longleaf pine induced by litter removal. For. Ecol. and Manage. 2:13-20.

Gold, MA, and JW Hanover. 1987. Agroforestry systems for temperate zone. Agrofor. Syst. 5:109-21.

Haines SG, LW Haines, and G White. 1978. Leguminous plants increase sycamore growth in northern Alabama. Soil Sci. Soc. Am. J. 42:130-32.

Hale, W. 1987. Shiitake mushrooms. Adapt. 2, 1987, 100-101.

Halls, LK, GW Burton, and BL Southwell. 1957. Some Results of Seeding and Fertilization to Improve Southern Forest Ranges. USDA For. Serv., Southeast For. Exp. Stn. Res. Pap. 78. Asheville, N.C.

Heady, HF. 1975. Rangeland Management. New York: McGraw-Hill.

Heady, HF, and RD Child. 1994. Rangeland Ecology and Management. Boulder, Colo.: Westview.

Heath, ME, RF Barnes, and DS Metcalfe (eds.). 1985. Forages: The Science of Grassland Agriculture. 4th ed. Ames: Iowa State Univ. Press.

Henderson, DR (ed.). 1991. Proc. Mid-South Conf. Agrofor. Pract. and Policies. 28-29 Nov. 1990, West Memphis, Ark. Morrilton: Winrock International Institute for Agricultural Development.

Herbel, CH. 1979. Utilization of grass and shrublands of the Southwestern United States. In BH Walker (ed.), Management of Semi-arid Ecosystems. Amsterdam: Elsevier, 161-203.

Herbel, CH, and AA Baltensperger. 1985. Ranges and pastures of the southern Great Plains and the Southwest. In ME Heath, RF Barnes, and DS Metcalfe (eds.), Forages: The Science of Grassland Agriculture, 4th ed. Ames: Iowa State Univ. Press, 380-88.

Hill, DB. 1991. Agroforestry specialty crops. In HE Garrett (ed.), Proc. 2d Conf. Agrofor. North Am., 18-21 Aug. 1991, Springfield, Mo. Columbia: Univ. of Missouri, 210-20.

Holechek, JL, RD Pieper, and CH Herbel. 1989. Range Management Principles and Practices. Englewood Cliffs, N.J.: Prentice Hall.

Jacoby, PW. 1989. A Glossary of Terms Used in Range Management. 3d ed. Denver, Colo.: Society for Range Management.

Janke, RR. 1991. Alley cropping with nitrogen-fixing trees and vegetables in northeast USA. In P Williams (ed.), Proc. 1st Conf. Agrofor. North Am., 13-16 Aug., Guelph, Ontario, Canada. Guelph, Ontario: Univ. of Guelph, 254.

James, LF, JO Evans, MH Ralphs, and RD Child (eds.). 1991. Noxious Range Weeds. Boulder, Colo.: Westview.

Jamison, GM. 1943. Effect of litter removal on diameter growth of shortleaf pine. J. For. 41:213-14.

Jardine, JT. 1915. Improvement and management of native pastures in the West. In USDA Yearbook, 299-310.

Jardine, JT, and M Anderson. 1919. Range Management on the National Forests. USDA Bull. 790.

Koch, P, and DW McKenzie. 1976. Machine to harvest slash, brush, and thinnings for fuel and fiber—a concept. J. For. 74:802-12.

Krugman, SL. 1988. Woody fiber crops. In J Janick and JE Simon (eds.), Advances in New Crops. Portland, Oreg.: Timber Press, 275-77.

Kurtz, WB, HE Garrett, and WH Kincaid, Jr. 1984. Investment alternatives for black walnut plantation management. J. For. 82:604-8.

Lamont, WJ, DL Hensley, S Wiest, and RE Gaussoin. 1993. Relay-intercropping muskmelons with Scotch pine Christmas trees using plastic mulch and drip irrigation. HortSci. 28(3):177-78.

Lassoie, JP, and LE Buck. 1991. Agroforestry in North America: New challenges and opportunities for integrated resource management. In HE Garrett (ed.), Proc. 2d Conf. Agrofor. North Am., 18-21 Aug. 1991, Springfield, Mo. Columbia:

Univ. of Missouri, 1-19.

Laycock, WA. 1991. Stable states and thresholds of range condition on North American rangelands: A viewpoint. J. Range Manage. 44:427-33.

Leatham, GF. 1982. Cultivation of shiitake, the Japanese forest mushroom, on logs: A potential industry for the United States. For. Prod. J. 32(8):29-35.

Leonard, SG, and RL Miles. 1989. Range sites/ecological sites—a perspective in classification and use. In Proceedings—Land Classifications Based on Vegetation: Applications for Resource Management, USDA For. Serv. Gen. Tech. Rep. INT-257, 150-53.

Lewis, CE, and HA Pearson. 1987. Agroforestry using tame pastures under planted pines in the southeastern United States. In HL Gholz (ed.), Agroforestry: Realities, Possibilities, and Potentials. Dordrecht, Netherlands: Martinus Nijhoff Publishers, 195-212.

Lewis, CE, GW Tanner, and HA Pearson. 1991. Agroforestry in the southern US. In P Williams (ed.), Proc. 1st Conf. Agrofor. North Am., 13-16 Aug., Guelph, Ontario, Canada. Guelph, Ontario: Univ. of Guelph, 258.

Linnartz, NE, and MK Johnson (eds.). 1984. Agroforestry in the Southern United States. Baton Rouge: Louisiana State Univ. Agricultural Center.

Long, AJ, and PK Nair. 1991. Agroforestry system design for the temperate zone: Lessons from the tropics. In HE Garrett (ed.), Proc. 2d Conf. Agrofor. North Am., 18-21 Aug. 1991, Springfield, Mo. Columbia: Univ. of Missouri, 133-39.

Lym, RG. 1991. Economic impact, classification, distribution, and ecology of leafy spurge. In LF James, JO Evans, MH Ralphs, and RD Child (eds.), Noxious Range Weeds. Boulder, Colo.: Westview, 169-81.

Monsoon Asia Agroforestry Joint Research Team (MAART). 1986. Comparative Studies on the Utilization and Conservation of Natural Environment by Agroforestry Systems. Kyoto, Japan: Kyoto Univ.

McDonald, PM, and GO Fiddler. 1991. Grazing with sheep: Effect on pine seedlings, shrubs, forbs, and grasses. In HE Garrett (ed.), Proc. 2d Conf. Agrofor. North Am., 18-21 Aug. 1991, Springfield, Mo. Columbia: Univ. of Missouri, 221-31.

McLean, HDJ, KM King, AM Gordon, and TJ Gillespie. 1991. Width and orientation of corn rows affect growth of interplanted hardwood seedlings. In P Williams (ed.), Proc. 1st Conf. Agrofor. North Am., 13-16 Aug., Guelph, Ontario, Canada. Guelph, Ontario: Univ. of Guelph, 255.

McLeod, KW, C Sherrod, Jr., and TE Porch. 1979. Response of longleaf pine plantations to litter removal. For. Ecol. and Manage. 2:1-12.

Marble, VL, WS McGuire, CA Raguse, and DB Hannaway. 1985. Hay and pasture seedings for the Pacific Coast States. In ME Heath, RF Barnes, and DS Metcalfe (eds.), Forages: The Science of Grassland Agriculture, 4th ed. Ames: Iowa State Univ. Press, 400-410.

Mills, R, and DR Robertson. 1991. Production and Marketing of Louisiana Pine Straw. La. Coop. Ext. Serv. Publ. 2430 (10M) 2/91.

Miller, J. 1992. Shiitake Mushroom Cultivation. Meadowcreek Tech. Brief 3 (1992).

Moore, RA, and RJ Lorenz. 1985. Hay and pasture seedings for the central and northern Great Plains. In ME Heath, RF Barnes, and DS Metcalfe (eds.), Forages: The Science of Grassland Agriculture, 4th ed. Ames: Iowa State Univ. Press, 371-79.

National Research Council (NRC). 1994. Rangeland Health—New Methods to Classify, Inventory, and Monitor Rangelands. Washington, D.C.: National Academy Press.

Ntayombya, P, and AM Gordon. 1991. In P Williams (ed.), Proc. 1st Conf. Agrofor. North Am., 13-16 Aug., Guelph, Ontario, Canada. Guelph, Ontario: Univ. of Guelph, 254.

Pearson, HA. 1964. Studies of forage digestibility under ponderosa pine stands. In Proc. Soc. Am. For., Denver, Colo., 71-73.

———. 1991a. Silvopastoral management potential in the mid-South pine belt. In HE Garrett (ed.), Proc. 2d Conf. Agrofor. North Am., 18-21 Aug. 1991, Springfield, Mo. Columbia: Univ. of Missouri, 232-41.

———. 1991b. Silvopasture: Forest grazing and agroforestry in the southern coastal plain. In DR Henderson (ed.), Proc. Mid-South Conf. Agrofor. Pract. and Policies, 28-29 Nov. 1990, West Memphis, Ark. Morrilton, Ark.: Winrock International Institute for Agricultural Development, 25-42.

———. 1993. Agroforestry: A family farm alternative. Temperate Agrofor. 1(2):2.

Pearson, HA, and A Martin, Jr. 1991. Goats for vegetation management on the Ouachita National Forest. In SG Solaiman and WA Hill (eds.), Using Goats to Manage Forest Vegetation. Tuskegee, Ala.: Tuskegee Univ. Press, 59-73.

Pearson, HA, TE Prince, Jr., and CM Todd, Jr. 1990a. Virginia pines and cattle grazing—an agroforestry opportunity. South. J. Appl. For. 14(2):55-59.

Pearson, HA, VC Baldwin, and JP Barnett. 1990b. Cattle grazing and pine survival and growth in subterranean clover pasture. Agrofor. Syst. 10:161-68.

Pope, TE. 1990. Pine straw mulch in the landscape. La. Coop. Ext. Serv. Publ. 2425 (5M) 11/90.

Raintree, JB. 1991. Agroforestry in North America: The next generation. In P Williams (ed.), Proc. 1st Conf. Agrofor. North Am., 13-16 Aug., Guelph, Ontario, Canada. Guelph, Ontario: Univ. of Guelph, 1-14.

Reed, AD. 1974. The Contribution of Rangelands to the Economy of California. Univ. of Calif. Coop. Ext. Serv. MA-82.

Reid, R, and G Wilson. 1985. Agroforestry in Australia and New Zealand. Box Hill, Victoria, Australia: Goddard and Dobson Publishers.

Reid, WR. 1991. Low-input management systems for native pecans. In HE Garrett (ed.), Proc. 2d Conf. Agrofor. North Am., 18-21 Aug. 1991, Springfield, Mo. Columbia: Univ. of Missouri, 140-58.

Reinke, JJ, DC Adriano, and KW McLeod. 1981. Effects of litter alteration on carbon dioxide evolution from a South Carolina pine forest floor. Soil Sci. Soc. Am. J. 45(3):620-23.

Rietveld, WJ, and PR Schaefer. 1991. Trees and shrubs: An integral component of sustainable agricultural land-use systems. In HE Garrett (ed.), Proc. 2d Conf. Agrofor. North Am., 18-21

Aug. 1991, Springfield, Mo. Columbia: Univ. of Missouri, 20-26.

Sampson, AW. 1913. Range Improvement by Deferred and Rotation Grazing. USDA Bull. 34.

———. 1914. Natural revegetation of range lands based upon growth requirements and life history of vegetation. J. Agric. Res. 3:93-147.

———. 1952. Range Management Principles and Practices. New York: Wiley.

Sampson, AW, and AM Schultz. 1957. Control of Brush and Undesirable Trees. Rome: FAO Forestry Division.

Sawyer, GEJ, Jr. 1978. Evaluation of four clovers on a forest soil. In SG Haines (ed.), Nitrogen Fixation in Southern Forestry. Bainbridge, Ga.: International Paper Company, 108-11.

Scherr, SJ. 1991. Economic analysis of agroforestry in temperate zones: A review of recent studies. In HE Garrett (ed.), Proc. 2d Conf. Agrofor. North Am., 18-21 Aug. 1991, Springfield, Mo. Columbia: Univ. of Missouri, 364-91.

Schmidt, JO, TE Rinderer, and WT Wilson. 1992. Honey bees—the good, the bad, and the ugly. Res. Prog., 1990-91, USDA Agric. Res. Serv., May, 19.

Schultz, R, JP Colletti, CW Mize, A Skadberg, MF Christian, W Simpkins, M Thompson, and B Menzel. 1991. In HE Garrett (ed.), Proc. 2d Conf. Agrofor. North Am., 18-21 Aug. 1991, Springfield, Mo. Columbia: Univ. of Missouri, 312-26.

Sharrow, SH. 1991. Tree density and pattern as factors in agrosilvopastoral system design. In HE Garrett (ed.), Proc. 2d Conf. Agrofor. North Am., 18-21 Aug., 1991, Springfield, Mo. Columbia: Univ. of Missouri, 242-47.

Shelton, MG. 1984. Effects of the initial spacing of loblolly pine plantations on production and nutrition during the first 25 years. HSU/PhD diss., Mississippi State Univ.

Shiflet, TN. 1973. Range sites and soils in the United States. In Proc. 3d Workshop US/Australia Rangelands Panel, Tucson, Ariz., 26-33.

Smith, HN, and CA Rechenthin. 1964. Grassland Restoration. The Problem. USDA-Soil Conservation Society.

Smith, JG. 1895. Forage conditions of the prairie regions. In USDA Yearbook, 309-24.

Society for Range Management. 1964. A Glossary of Terms Used in Range Management. Denver, Colo.: Range Glossary Committee, Society for Range Management.

———. 1974. A Glossary of Terms Used in Range Management. 2d ed. Denver, Colo.: Range Glossary Committee, Society for Range Management.

———. 1989. A Glossary of Terms Used in Range Management. 3d ed., Range Glossary Committee, Society for Range Management. Denver, Colo.: Edison Press.

———. 1991. SRM recommends new range terminology. Trail Boss, Sept., 1-3.

———. 1994. Rangeland Cover Types of the United States. Ed. TN Shiflet. Denver, Colo.: Society for Range Management.

Society of American Foresters. 1980. Forest Cover Types of the United States and Canada. Ed. FH Eyre. Washington, D.C.: Society of American Foresters.

Stoddart LA, AD Smith, and TW Box. 1975. Range Management. New York: McGraw-Hill.

USDA. 1965. Losses in Agriculture. USDA-ARS

Handb. 291. Washington, D.C.: US Gov. Print. Off.

USDA Forest Service. 1974. Seeds of Woody Plants in the United States. USDA Agric. Handb. 450. Washington, D.C.: US Gov. Print. Off.

Van Cleve, K, and CT Dyrness. 1983. Effects of forest-floor disturbance on soil-solution nutrient composition in a black spruce ecosystem. Can. J. For. Res. 13:894-902.

Van Keuren, RW, and JR George. 1985. Hay and pasture seedings for the central and lake states. In ME Heath, RF Barnes, and DS Metcalfe (eds.), Forages: The Science of Grassland Agriculture, 4th ed. Ames: Iowa State Univ. Press, 348-58.

Watson, VH, C Hagedorn, WE Knight, and HA Pearson. 1984. Shade tolerance of grass and legume germplasm for use in the southern forest range. J. Range Manage. 37:229-32.

Wellner, CA. 1989. Classification of habitat types in the western United States. In Proceedings—

Land Classifications Based on Vegetation: Applications for Resource Management. USDA For. Serv. Gen. Tech. Rep. INT-257, 7-21.

Welsh, SL, CR Nelson, and KH Thorne. 1991. Naturalization of plant species in Utah—1842 to present. In LF James, JO Evans, MH Ralphs, and RD Child (eds.), Noxious Range Weeds. Boulder, Colo.: Westview, 17-29.

Westoby, M, B Walker, and I Noy-Meir. 1989. Opportunistic management for rangelands not at equilibrium. J. Range Manage. 42:266-74.

Williams, P (ed.). 1991. Agroforestry in North America. Proc. 1st Conf. Agrofor. North Am., 13-16 Aug., Guelph, Ontario, Canada. Guelph, Ontario: Univ. of Guelph.

Williams, PA, and AM Gordon. 1991. The potential of intercropping as an alternative land use system. In HE Garrett (ed.), Proc. 2d Conf. Agrofor. North Am., 18-21 Aug. 1991, Springfield, Mo. Columbia: Univ. of Missouri, 166-75.

Forages for Conservation and Soil Stabilization

17

W. CURTIS SHARP
Natural Resource Conservation Service, USDA

DAVID L. SCHERTZ
Natural Resource Conservation Service, USDA

JACK R. CARLSON
Natural Resource Conservation Service, USDA

NO single tool is better or more cost-effective than plants for conserving and stabilizing soil. Seventy-three percent (Bureau of the Census 1987) of the landmass of the US is covered by permanent vegetation consisting of forest-, range-, and pastureland. An additional 2.7%, or 25.6 million ha, is used for rotational hay and pasture (SCS 1980) and is covered with forages year-round. Perennial forages, whether managed for pastureland, hayland, rangeland, or grazed forestland, play a dominate conservation role on about 445.8 million ha, or 48.5% of the landmass of the US (SCS 1987). Silage corn (*Zea mays* L.) adds another 1.4 million ha of forage-producing land. The quality of vegetative cover on these lands, which depends on the species and how they are managed, has a greater impact on the conservation and stabilization of agricultural land than any other factor.

In addition to forage production, many exotic and native forage plants are used specifically for conservation and soil stabilization

purposes. For example, across much of the eastern US, tall fescue (*Festuca arundinacea* Schreb.) is an excellent forage, but it is also used extensively as a component of conservation plantings (SCS 1992a). Bahiagrass (*Paspalum notatum* Flugge) and bermudagrass (*Cynodon dactylon* [L.] Pers.) play similar roles in the Southeast. The Conservation Reserve Program, enacted as a part of the 1985 Farm Bill (Food Security Act 1985), shifted 13.2 million ha of highly erodible cropland to permanent vegetative cover between 1986 and 1991. Forage grasses and herbs were planted on 89% of the land (ASCS 1991).

HISTORY OF FORAGES FOR CONSERVATION AND SOIL STABILIZATION IN THE US

As early European settlers in the eastern and southern US cleared forests to plant their crops and feed their livestock, they found few, if any, suitable forage plants among the native flora. Thus, most of the forage plants grown in those areas have been introduced (exotics). Foremost among these, in addition to tall fescue, bahiagrass, and bermudagrass, are orchardgrass (*Dactylis glomerata* L.), timothy (*Phleum pratense* L.), smooth bromegrass (*Bromus inermis* Leyss.), Kentucky bluegrass (*Poa pratensis* L.) (considered by some to be indigenous), red and white clover (*Trifolium pratense* L. and *T. repens* L.), and alfalfa (*Medicago sativa* L.). Many of these plants, particularly the grasses, had been exposed to hundreds of years of intensive grazing in Europe, Asia, or Africa and immediately found a niche in the similar climates of the US.

Plant successions growing today in the

W. CURTIS SHARP is National Plant Materials Specialist, NRCS, USDA, Washington, D.C. He received his MS degree from Pennsylvania State University. He has worked at the local, state, regional, and national level in developing plants for conservation since 1957.

DAVID L. SCHERTZ is National Agronomist, NRCS, USDA, Washington, D.C. He received his PhD degree from Purdue University. He has worked in conservation agronomy at the local, state, regional, and national level since 1969.

JACK R. CARLSON is Decision Support Team Leader, NRCS, USDA, Fort Collins. He received his MS degree from Oregon State University. Since 1970 he has worked at the local and regional level in developing plants for conservation and at the national level developing software applications for using plants in conservation programs.

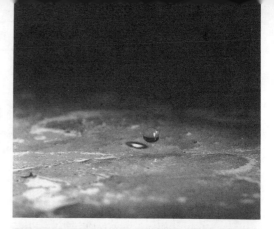

semiarid and arid parts of the US have a similar history. The plants comprising these native grasslands, which had evolved under extensive grazing pressure from native herbivores, were displaced by unpalatable plants that provide poor soil erosion protection when the area was grazed intensively by domestic livestock. For example, annual grasses have replaced bluebunch wheatgrass (*Pseudoroegneria spicata* [Pursh] A. Love subsp. *spicata*) on many western North American ranges as a result of overgrazing (Harris 1967). Annual species that replaced the perennial grasses are lower in biomass and ground cover; this increases the potential for runoff and erosion (Ndawula-Senyimba et al. 1971).

Runoff and erosion are lessened by changing to less intensive management or by re-establishing plants that improve soil protection and have higher tolerance to grazing, or both. While the availability of native plants with desirable traits for forage production was much greater in the range area than in the humid regions of the US, the grasses used initially to protect degraded range and pastureland were exotics from Asia, primarily crested (*Agropyron cristatum* [L.] Gaertn.), intermediate (*Thinopyrum intermedium* [Host] Barkworth), and tall wheatgrasses (*T. ponticum* [Podp.] Barkworth & D.R. Dewey) (Carlson 1988).

VEGETATION AND SOIL CONSERVATION

A 40 mm rainstorm falling at the rate of 40 mm h^{-1} exerts almost 700,000 kg m of energy on the surface of 1 ha (Wischmeier and Smith 1958). See Figure 17.1. Where the soil is bare, the falling rain puddles it, increasing both runoff and soil loss. In Iowa, annual soil loss due to sheet erosion was 0.67 mt ha^{-1} from a field-sized watershed of well-managed smooth bromegrass pasture, compared with 67.3 mt ha^{-1} for corn under conventional tillage (Saxton et al. 1971).

Vegetation cover enhances water movement into the soil by protecting the surface from the damaging effects of pounding raindrops: vegetation retards runoff and keeps the surface open and receptive to infiltration. For exam-

Fig. 17.1. The camera catches the raindrop (*top*) as it hits the soil surface. As the drop strikes, it disperses surface materials (*bottom*) much as a small explosion does. *Dept. of Navy photo.*

ple, the rate of water infiltration in an undisturbed sod of mixed grasses in an orchard was 12.7 cm h^{-1} (Browning and Suds 1942), about five times that of adjacent areas that were cultivated and subjected to disturbance and compaction by farm equipment.

Roots play an important role in water infiltration because they spread through the soil profile to varying degrees depending on the plant species. Root systems of most perennial forages, particularly of grasses and deep-rooted legumes, are often longer and more extensive than the roots of annuals. Living root tips force their way through minute cracks in the soil, enlarging these cracks and breaking the soil aggregates into smaller pieces. Roots of perennial grasses are replaced each year or two, and those that die and decay leave channels through which water can penetrate the soil.

A cropping system of continuous grass cover resulted in the least soil erosion, followed by rotational hay, in Wisconsin, Oklahoma, and Iowa studies summarized by Wadleigh et al. (1974). Soil losses were greater from con-

tour row crops than from rotational grain and rotational hay (Table 17.1). A summary of several studies by Browning (1962) compares the difference in erosion and runoff on the same soil from a variety of geographical locations. These data, which include several kinds of rainfall patterns and intensities, show that close-growing perennial vegetation is most effective in reducing erosion. In all cases, the differences between row and sod crops were found to be dramatic (Table 17.2).

Intensity of grazing on permanent grasslands affects soil loss and runoff. Average annual runoff during the studies by Browning, which covered a 10-yr period, was 4.3 cm, and sediment yield was 0.3 mt ha^{-1} on a rangeland watershed in Oklahoma that was stocked moderately and grazed rotationally, compared with 9.8 cm and 8.1 mt ha^{-1} on a watershed area that was overgrazed. Hanson et al. (1970) found the seasonal (May 14 to October 31) runoff was 198, 140, and 105 mm from heavily, moderately, and lightly stocked rangeland, respectively (Fig. 17.2). An overgrazed pasture in southern Illinois lost 3973

TABLE 17.1.   Effects of cropping systems on soil erosion

Location	Treatment	Annual soil loss (mt ha^{-1})
La Crosse, Wis.	Contour row crop	561.30
	Rotation, row crop	118.60
	Rotation, grain	66.80
	Rotation, hay	1.50
	Continuous grass	0.20
Guthrie, Okla.	Contour row crop	74.00
	Rotation, row crop	7.20
	Rotation, grain	15.50
	Rotation, hay	5.60
	Continuous grass	0.04
Clarinda, Iowa	Contour row crop	82.70
	Rotation, row crop	41.20
	Rotation, grain	22.60
	Rotation, hay	12.10
	Continuous grass	0.07

Source: Adapted from Wadleigh et al. (1974).

TABLE 17.2.   Effects of row and sod crops on erosion and runoff (% of rainfall)

Location	Soil texture	Slope (%)	Soil loss (mt ha^{-1}) Row crop[a]	Sod	Runoff (%) Row crop[a]	Sod
Iowa	Silt loam	9.0	85.6	0.07	18.7	1.30
Missouri	Loam	8.0	114.1	0.36	27.1	8.10
Ohio	Silt loam	12.0	222.6	0.05	40.3	4.80
Oklahoma	Fine sandy loam	7.7	42.4	0.05	12.5	1.00
North Carolina	Clay loam	10.0	70.0	0.70	12.4	1.90
Texas	Fine sandy loam	16.5	137.0	0.01	14.4	0.30
	Sandy loam	10.0	14.6	0.01	13.9	0.30
	Clay	4.0	46.2	0.05	13.6	0.05
	Black clay	2.0	17.5	0.18	10.5	1.20
Wisconsin	Silt loam	16.0	250.4	0.22	29.2	0.55

Source: Browning (1973).

[a]Row crop was continuous corn on the Iowa, Missouri, Ohio, and Wisconsin soils and the Texas clay soils. The sod crop was either Kentucky bluegrass or bermudagrass.

kg ha^{-1} of soil and 17.3% runoff compared with only 390 kg ha^{-1} soil loss and 3.4% runoff from a well-fertilized and moderately grazed pasture (Gard 1943).

Approximately 25% of pollutants in rivers and 15% in lakes are sediments from agricultural land. Factors that improve infiltration or reduce soil erosion and runoff contribute to protecting surface water from nonpoint pollution (Baker and Laflan 1982; Carey 1991).

ROLE OF FORAGES IN SUSTAINABLE AGRICULTURE SYSTEMS

Sustainable agriculture, as defined by the American Society of Agronomy, is "one that, over the long-term, enhances environmental quality and the resource base on which agriculture depends, provides for basic human food and fiber needs, is economically viable, and enhances the quality of life for farmers and society as a whole" (Schaller 1990). Developing sustainable agriculture systems de-

pends on in-depth understanding of management practices involved in agriculture systems. An agriculture system can be viewed as an ecosystem, a unit composed of associated organisms and their physical and chemical environment. The use of ecological principles, directly or indirectly, in the formulation of these management practices is essential to meeting the goal of sustainable agriculture.

The components of sustainable agriculture (environmental quality, plant and animal productivity, and socioeconomic viability) (Francis and Youngblood 1990) depend on several important ecological processes that are influenced by the use of forages in an agriculture system. Three of these are nutrient cycling (primarily nitrogen [N]), hydrology, and soil microbial population dynamics (Neher 1992).

Nutrient Cycling. Measurable N inputs to the agriculture system include rainfall, chemical fertilizers, and animal manure or sewage

Fig. 17.2. Overgrazing, as shown on the left, results in excessive runoff and soil loss. When soil is protected with vegetation through proper management, runoff is minimized, as demonstrated on the right side of the fence. *Nebr. NRCS photo.*

sludge. In agriculture systems, nutrients are removed from the cycle by harvestable products or are exported from the field through leaching, denitrification, volatilization, and erosion (King 1990). To improve the sustainability of agriculture, the goal of nutrient management should be to achieve maximum nutrient uptake and recycling efficiency, to increase nutrient storage in active available soil pools, and to minimize nutrient loss.

Biological, chemical, and physical processes in soil affect nutrient cycling through influences on the physical properties of the soil. One mechanism of particular importance in agricultural soils is the formation of water-stable aggregates. Soil aggregates are basic units of soil structure, composed of mineral and organic particles held together by a variety of forces. The stability of these aggregates depends on physical and chemical characteristics of the soil, but their formation appears to depend largely on biological activity in soils (Boyle et al. 1989).

Formation of water-stable aggregates represents a key ecosystem process through which organic matter accumulates because it is protected from rapid decomposition. This has important implications for nutrient management in agriculture. The natural cycle of aggregate formation and disruption via freezing/thawing or wetting/drying cycles and root proliferation may indirectly regulate the movement of nutrients between organic and inorganic pools. Management practices that protect soil aggregates or improve soil biological activity tend to increase the organic content of soil over the long term.

Research in temperate agriculture systems has shown how cropping systems influence soil organic matter levels and soil fertility. Carbon (C) accumulated rapidly from 0.6% to 2.3% organic matter over 4 yr in a sod of alfalfa and tall fescue. Conversely, plowing and winter bare-fallow culture in organic-rich forest soils resulted in a rapid loss of C from 2.3% to an equilibrium of 0.7% over 3 yr. The data suggest the lower and upper equilibrium levels for C in the soils (Doran and Werner 1990), based on fertilizer-subsidized, in situ production of organic outputs.

Crop rotation is both an ecological and economic way to control soil erosion (Table 17.2). Corn following wheat (*Triticum aestivum* L.) produced greater yields than continuous corn when the same amount of fertilizer was applied (Power 1987). The increase of organic matter in sod-based rotations, and the N produced by leguminous crops, both contribute to crop yield increases from rotations. First-year yields of corn, after 3 yr of alfalfa, were as great as that of corn receiving 100 kg ha^{-1} of N (Classen 1980). This is similar to results reported by Lamond et al. (1988) of a 112 kg ha^{-1} N credit following a legume in the rotation. The increase in crop yields following a leguminous crop is usually greater than expected from the estimated quantity of N supplied. In fact, yields following legumes are as much as 10%-20% greater than continuous grain regardless of the amount of fertilizer applied (Cook 1984; Goldstein and Young 1987; Heichel 1987).

Hydrologic Cycle. Vegetation is an important factor in the hydrologic cycle and the variable most affected by human activity. Dense canopies of vegetation protect the soil from erosion and increase water infiltration. Soil moisture, which influences infiltration and percolation, is a function of infiltration and evapotranspiration. Both are dependent on vegetation.

A bare soil surface dries rapidly, reducing the amount of water returning to the atmosphere. Plant root systems occupy a large volume of soil and extract moisture from deeper locations over time, providing that the atmosphere demands it and that root zone water is available. Grasses and other perennial plants use water during a higher proportion of the year than most annual field crops such as corn.

Stream flow, erosion, and sedimentation are affected by many factors: rainfall, soil erodibility, slope gradient, slope length, farming practices, and conservation treatment. In Iowa runoff and erosion from a field-sized watershed in perennial grass was found to be much less than from a comparable watershed cropped to continuous corn (Saxton et al. 1971). During the period 1964-69, soil loss from sheet erosion averaged 67.3 mt ha^{-1} for corn and only 0.67 mt ha^{-1} for well-managed smooth bromegrass pasture. Water yields likewise were affected. The 1964-69 water yield from base flow, i.e., that portion of the stream discharge that is derived from groundwater, was 17% higher from the smooth bromegrass pasture than from the contoured corn. In 1967, base flow from pasture was 45% greater than from corn and reflects the higher infiltration rate under smooth bromegrass.

Peak rates of stream flow depend largely on the amount of surface runoff, the time distribution, and the channel complex through which it flows. Rainfall rate less infiltration rate determines the amount and time distribution of runoff. Dense vegetation cover, such as perennial forages and grass, and conservation practices will reduce runoff volumes and peak rates (Browning 1962). Based on 82 storms, the peak runoff rates from a grassed watershed in Iowa were less than 10% of those from a similar area planted to contoured corn (Saxton and Spomer 1968).

Population Dynamics. Population dynamics of organisms in all ecosystems are important in maintaining a sustainable balance between populations, their respective food sources, and space requirements. The population dynamics of crops are regulated by natural and improved factors such as hydrology, soil structure and fertility, climate, fertilizers, crop diversity, management practices, and other organisms. Population dynamics of insects, pathogens, and weeds are regulated by predator-prey interaction, competition, and cultural, chemical, and biological control practices (Neher 1992).

Biological complexity and diversity, which are essential components of sustainable agriculture, require the maintenance of a wide range of plant types and habitats on the farm (Hauptli et al. 1990). Increased diversity of species within an agriculture system (e.g., polycultures or hedgerows) may decrease risks or production failure by providing alternate crops and by promoting natural predators of pests (Hendrix et al. 1990). For example, in California, the harvesting schedule of alfalfa growing adjacent to cotton (*Gossypium hirsutum* L.) influences the damage from the lygus bug (*Lygus lineolaris* P. deB) on cotton. The lygus bug favors the lush, nutritious growth of alfalfa over cotton. By staggering the harvest schedule of alfalfa so that there is a continuous presence of unharvested alfalfa, lygus bug populations on cotton can be kept in check (Flint and Roberts 1989).

The establishment and maintenance of a complex and diverse biological population requires a sophisticated understanding and management of relationships among hosts, pests, and predators. This, in turn, serves to minimize major disruptions that require other kinds of intervention, such as pesticide applications (Hauptli et al. 1990). Indicators of agriculture system biodiversity include management practices such as strip-cropping, crop rotation, trap crops, and intercropping; fragmented agricultural landscapes; and the quality of wildlife habitats.

USING FORAGES FOR SOIL CONSERVATION ON CROPLAND

Forages can be used independently or in a variety of ways to complement other conservation practices. For example, cover crops, such as cereal rye (*Secale cereal* L.), annual ryegrass (*Lolium multiflorum* [Lam.]), hairy vetch (*Vicia villosa* Roth), or crimson clover (*T. incarnatum* L.), are usually planted in the fall following the harvest of such cash crops as corn or vegetables. Cover crops are especially useful following low-residue crops such as cotton, soybeans (*Glycine max* [L.] Merr.), vegetables, and silage corn. When used in high-rainfall areas or on sandy soils, the cover crops utilize plant nutrients that otherwise might be leached out of the root zone. April-harvested cereal rye, annual ryegrass, hairy vetch, and crimson clover captured 64, 43, 10, and 1 kg ha^{-1} of N respectively, when 336 kg ha^{-1} N was applied to corn the previous year (Shipley et al. 1992).

Cover crops also use excess profile moisture during production of dry matter (Meisinger et al. 1991), thus reducing the quantity of soil water available for leaching. In subhumid and semiarid farming zones, cover crops may deplete the profile of soil moisture and reduce crop performance. In these zones, high crop yields depend on the amount of rainfall received during the growing season and that stored in soil moisture during the previous fallow season.

Legume cover crops can contribute N to the rotation; the amount being determined by the species used, the success of seed inoculation with the N-fixing bacteria at planting time, the temperature and length of the growth period, and total biomass produced. Estimates of the N produced vary from 220 to 330 kg ha^{-1} from hairy vetch (Ebelhar et al. 1984; Hargrove 1986; Funderburg 1987; Touchton et al. 1984; Shurley 1987) and 200 kg ha^{-1} from crimson clover (Funderburg 1987). Forage from these and other cover crops may be harvested or grazed in the late spring. Management of the crops is covered in detail in other chapters.

A cover crop can become part of an integrated pest management system. For instance, it was noted orchards and vineyards with cover crops in northern California often had less colonization and infestation of pests, more soil-dwelling predators, and higher removal rates of artificially placed prey than they did without cover crops. Legume cover crops with extended bloom supported the largest reservoir of natural enemies of deleterious pests (Altieri and Schmidt 1985).

Cover Crop Selection. The best cover crop for a given situation depends on the environment and the goals of using the cover crop. Five of the most widely used cover crops are described in the following, along with the crops associated with them, the environmental conditions where they are most effective, and their principal function as a cover crop. Information about their use and management for forage production is covered in other chapters.

RYE AND OTHER CEREAL CROPS. Cereal rye is one of the most widely used winter cover crops within the cool temperature region of the US. It is used primarily as a cover crop following late-harvested crops such as potatoes (*Solanum tuberosum* L.), silage corn, and sugarbeets (*Beta vulgaris* L. subsp. *vulgaris*), but it is also widely used following grain corn. Rye will germinate and produce fall growth at lower temperatures than other commonly used cover crops. It can be planted as late as September 15 in northern climates and into late October in its southern range, or it can be planted into a standing crop before harvest; after harvest, the best method is to broadcast seed on a prepared seedbed or lightly disk or drill the seed into the soil. Nitrogen uptake and retention are good. It provides rapid spring growth and excellent spring grazing, and it can be used as a green manure crop. Rank spring growth must be controlled if the following crop is to be no-tilled. The cultivar 'Aroostook' was developed for use as a cover crop in northern climates. The seeding rate is 50-100 kg ha^{-1}.

Barley (*Hordeum vulgare* L.) is also used as a winter cover crop and is adapted climatically to most of the US except the most northern parts. It requires higher soil and air temperature for fall growth than rye, but it is tolerant to saline soils and generally grows better than other cereals on light, sandy soils. The cultivar 'Seco' was developed for use in the Southwest to provide short-term soil protection for land going out of irrigated crop production. Management of barley is similar to that of rye. The seeding rate range is 60-110 kg ha^{-1}.

Winter wheat is a good substitute for rye or barley in areas with cold winters, hot summers, and moderate precipitation (35-75 cm). Fall growth and early spring growth are less than for rye.

HAIRY VETCH. This is the most widely used legume cover crop in the southeastern US where it is a winter cover following cotton, soybeans, tobacco (*Nicotina tabacum* L.), rice (*Oryza sativa* L.), and vegetable crops. Vetches are more suited to wetter soils and colder winter temperatures than other winter active legumes, and they are more productive on sandy soils than crimson clover. Vetch is planted in early fall, as soon as possible after the previous crop is removed, by being broadcast on a prepared seedbed, by being lightly disked into the soil, or by drilling. Soil pH 6.0 or above is desirable. Fall growth is generally slow; spring growth is rapid. It is less effective than grasses in capturing soil N when killed for no-till planting of the crop or when turned under for green manure. Its decomposition is rapid, making it a desirable cover crop for use in no-till systems. The seeding rate is 15-20 kg ha^{-1}.

CRIMSON CLOVER. A winter annual legume with leaves and stems similar to those of red clover, crimson clover is adapted to the Southeast, north into Virginia and west into the humid parts of Texas. Fall-sown seedlings grow rapidly, with leaves developing from the crown and forming a rosette that enlarges as the weather warms in spring. Crimson clover grows on both sandy and clayey soils, provided they are well drained. It is tolerant of a pH range of 5.5-7.0 but performs poorly on soils low in phosphorus (P) and potassium (K). It should be planted in August and September. It can be broadcast between crop rows of standing soybeans prior to leaf drop. After harvest, seed should be broadcast on a prepared seedbed or lightly disked into the soil. A seeding rate of 22-28 kg ha^{-1} is appropriate.

ANNUAL RYEGRASS. A vigorous annual that can be fall or spring seeded, ryegrass is well adapted to the Southeast, north to Maryland and west to Texas as a winter cover crop. It can be planted as a summer annual somewhat farther north. Annual ryegrass is used as a cover in orchards and vineyards and as a winter cover crop following corn, soybeans, peanuts, and vegetables. On erosion-prone orchard and vineyard land, ryegrass makes rapid growth with the first fall rains and subsequently provides excellent protection against winter erosion. It also inhibits nitrate leaching. Annual ryegrass has a wide range of adaptability to soils, but it does not withstand hot, dry weather or severe winters. When planted as a winter cover crop, the soil surface should be loosened and the seed broadcast and mixed into the soil surface. It establishes rapidly and is very competitive; it may retard the establishment of other species if seeded too heavily in a mixture. The seeding rate is 20-30 kg ha^{-1} when seeded alone.

SOFT CHESS. Soft chess (*Bromus mollis* L.), a winter-growing, self-seeding annual grass, grows to generally 40-60 cm. Soft chess is well suited as a reseeding annual for orchards and vineyards, primarily in California. It withstands excessive mowing and matures seed under minimum moisture conditions. It is also suited for an erosion control plant on bare slopes. 'Blando' is the principal cultivar available; it is adapted to areas with annual precipitation of more than 25 cm and a range of soils from coarse to fine, having a near-neutral pH. Areas to be seeded should have a firm seedbed that has been previously roughened by disking, harrowing, raking, or otherwise working to a depth of approximately 5 cm. Seeding should be made in the fall. The preferred seeding method is with a drill. Broadcast seeding followed by a culti-packer, spike-tooth harrow, or similar tool to cover the seed to a depth of 1.2 cm is satisfactory. The seeding rate is 5-7 kg ha^{-1}.

CROP RESIDUE MANAGEMENT

Crop residue management is an economical and practical practice to help minimize the erosion processes (detachment, transport, and deposition) associated with crop production compared with conventional tillage without surface residue cover (Baker and Laflan 1982). See Figure 17.3. Some form of residue management is currently used on approximately one-third of annually planted cropland in the US and has doubled in the last 4 yr (CTIC 1992). Crop residue from the previous crop left on the soil surface absorbs the impact of falling raindrops, slows surface runoff, provides a greater opportunity for surface water to infiltrate, and provides protection against the forces of wind (Fig. 17.3). A residue covering 10% of the soil surface can result in 30% less soil loss than erosion from bare soil, with greater reductions occurring as the amount of residue increases. For instance, increasing surface residue cover from 20% to 30% reduces sheet and rill erosion by 50% and 65%, respectively (Renard et al. 1993). Due to the rapid decomposition of residue, these estimated reductions are based on the Revised Universal Soil Loss Equation and are accurate only for the time period that these amounts of residue are present. Similar reductions in wind erosion are also possible with surface residue.

Other benefits of maintaining a portion of crop residue on the surface include mulch effects such as postponing the detrimental effects of a short-term drought and increasing crop yield by conserving soil moisture and reducing surface evaporation. Even with as little as 20% of the soil surface covered by residue, relative potential evaporation can be reduced by 22% (Linden et al. 1987) (Fig. 17.4).

In the northern Great Plains, good crop residue management increased effectiveness equivalent to adding 6-10 cm of soil moisture, compared with conventional tillage where there is little or no surface residue cover. This is accomplished by an increase in snow catch, an increase in infiltration, and a reduction in evaporation. Just 2.5 cm of added moisture can result in about 340 kg ha^{-1} more wheat or 470 kg ha^{-1} more barley in the northern Great Plains (A. L. Black 1992, personal communication).

Contour Strip-cropping. Contour strip-cropping is the growing of crops in strips or bands on the contour or across the general slope of the land. Crops are arranged so that a strip of perennial forage crop, usually hay, is alternated with a strip of clean-tilled crop or fallow. The velocity of any runoff water from the cultivated part of the field is reduced, and the otherwise concentrated flow is dispersed as the water enters the perennial forage crop. This improves infiltration and prevents concentrated water flow in low areas, where gullies can develop (Fig. 17.5). Studies have shown that the average annual soil loss for six locations, involving 34 cropping seasons, was 5.5 mt ha^{-1} for strip-cropped fields compared with 17.0 mt ha^{-1} for up-and-down-the-hill farming. The water loss in percentage of the rainfall for the two methods was 5.5% and 12.0%, respectively (Tower 1946).

Grass Barriers. In the semi-arid northern (Black and Siddoway 1971; Aase et al. 1985) and southern Great Plains (Sajjadi and Zartman 1990), perennial forage grasses are planted perpendicular to the prevailing wind to prevent erosion from wind and to conserve moisture (Fig. 17.6). Two rows of tall wheatgrass were planted 0.9 m apart. The wind speed in open field conditions was reduced by 17% to 70% from the leeward side of one barrier to the windward side of the next. Soil and water storage in strip-cropped fields doubled

Fig. 17.3. Two examples of conservation tillage: *top,* soybeans in corn residue—Iowa; bottom, corn in wheat stubble—Illinois. *NRCS photos.*

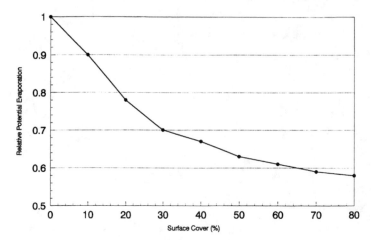

Fig. 17.4. Effectiveness of crop residue in reducing surface evaporation. (Adapted from Linden et al. 1987.)

in the first 9 mo compared with storage in unprotected fields. This conserved moisture made intensive cropping feasible in each year of the 8-yr study and made alternate year fallowing unnecessary (Black and Siddoway 1971).

Tall wheatgrass is typically planted in rows that are 15-100 cm apart or planted on a clean, prepared seedbed at the rate of 40-50 pure live seed per meter of row. Closely spaced rows (15-30 cm) inhibit weedy grasses and can be managed as a single row.

In the southern Great Plains, adapted varieties of annual grasses, such as small grains,

Fig. 17.5. These contoured strips of corn and hay also utilize crop rotations in an excellent resource-conserving system.

Fig. 17.6. This grass barrier in Montana prevents wind erosion and traps and holds snow on the land. *Mont. NRCS photo.*

sorghums, or sudangrass (*Sorghum bicolor* [L.] Moench) are planted 1.0 m apart, or two rows of perennial grasses such as weeping lovegrass (*Eragrostis curvula* [Schrad.] Nees), 'Blackwell' switchgrass (*Panicum virgatum* L.), or kleingrass (*P. coloratum* L.) are planted 0.5 m apart. The characteristics of these plants are discussed in other chapters. The distance between grass barrier strips is farther for heavy-textured than for light-textured soils, for crops that provide quicker and earlier ground cover in the spring, and when long-term average wind speeds are lower.

An emerging use of grass barriers, frequently involving forage species, involves planting a narrow band of vegetation at appropriate intervals across the slope of a clean-tilled crop field (Kemper et al. 1992). These barriers diffuse and spread any concentrated flows of runoff water, thus allowing the sediment to be deposited and the water to infiltrate. The sediment, over time, will build bioterraces, which reduce slope length and prevent sediment from leaving the field. Single- or double-row barriers are planted on the contour or across the slope, with each plant spaced 30-45 cm apart. Stems or 6- to 8-wk-old seedlings are ideal for planting. This developing technology will have widespread application in much of the US.

The four plants being evaluated most wide-ly for this use are vetiver grass (*Vetiveria zizanoides* [Linn.] Nash), switchgrass, eastern gamagrass (*Tripsacum dactyloides* [L.] L.), and silver plumegrass (*Miscanthus sinensis* Anderss.). Vetiver grass, an exotic, has been used widely in Southeast Asia and parts of Africa as a vegetative barrier (Pohl 1980). Its adaptation to the US is limited to areas where minimum temperatures do not go below 5°C. It is a densely tufted, perennial grass, with coarse and relatively stiff leaves. The foliage is mostly basal, creating a physical barrier at the ground surface. It is a deep-rooted, very fast-growing tropical grass, reaching 1.5 m in 1 yr in Louisiana.

Switchgrass, discussed more in Chapter 32, Volume 1, has characteristics that make it desirable for use as a grass barrier, including the production of dense, stiff stems (which remain erect during the winter better than those of most grasses), long life, deep roots, and tolerance to some herbicides used in crop production. Some ecotypes and cultivars for regions of the US have unique characteristics for grass hedge use, including 'Shelter' (Northeast), 'Alamo' (South), and 'Cave-In-Rock' (Midwest).

The description and area of adaptation of eastern gamagrass is covered in Chapter 31, Volume 1. Those characteristics that make it desirable for use as a barrier plant are similar to those of switchgrass. Silver plumegrass is an exotic, tall, robust perennial grass, used mainly as an ornamental. It is a C_4 grass, rapid growing and long-lived, with an adaptation similar to that of switchgrass. As with the other plants used for grass barriers, silver plumegrass produces a tight clump that is resistant to breakage by flowing water. A wide range of growth forms are available.

CROPLAND CONVERSION

The Conservation Reserve Program, already referred to, is an example of a government program that supports the conversion of highly erodible cropland to a permanent cover, usually with forage plants. Even without government-supported programs, such land-use conversion may be both environmentally and economically desirable. For example, silage corn, the principal forage used by dairy farmers in the northeastern US, is often not economically feasible to produce. The breakeven yield in New York is 8960-9600 kg ha^{-1}, yet the average production in 1979-88 was 8874 kg ha^{-1} (New York Agriculture Statistics 1992). The estimated erosion rate from silage corn averages about 9 mt ha^{-1} annually in New York (Dickerson et al. 1991).

The use of perennial forage grasses in place of corn may provide the farmer with a cost-effective soil-conserving system. One example is eastern gamagrass. A dry matter yield of 11,000-15,900 kg ha^{-1} of eastern gamagrass has been measured in New York, which is equivalent to the dry matter yield contained in 11,500-16,000 kg ha^{-1} of silage corn (J. A. Dickerson 1992, personal communication). Other aspects of eastern gamagrass as a forage are discussed in Chapter 31, Volume 1.

USING FORAGES FOR SOIL CONSERVATION ON DISTURBED LAND

In addition to protecting against soil erosion on pastures and rangelands, forages are also used extensively to stabilize areas against erosion resulting from some man-induced disturbances. Those protected areas not used for forage production generally contain the same plant species recommended for pasture, hayland, or rangeland plantings (SCS 1992b).

Following mineral mining, road building, and other development that destroys the surface vegetation, concentrated efforts are required to restore that protective cover. Forage plants are widely used for stabilizing such disturbed areas because they are easily established and are adapted to varying soil and en-

vironmental conditions. More than 120 improved cultivars of grasses, legumes, herbs, and shrubs have been developed for use in soil stabilization and forage production in the semiarid and arid parts of the US (Alderson and Sharp 1994; Sharp et al. 1992).

Although disturbed lands can seldom be returned to their original condition, with planning most can be reclaimed to the level of previous production. Before mining or construction begins, it is essential to document fully the soils, topography, overburden materials, groundwater aquifers, and existing vegetation of the present land use and the desired postconstruction land use. Because disturbed sites are reclaimed through the manipulation of the original soil materials and overburden, i.e., material moved from the top—the goal is to ensure that the best conditions for supporting plant growth are present in the root zone and that the slopes are stable and compatible with the planned use.

Generally, restoring surface material back to the surface retains the productivity capacity; in some locations, however, the subsurface materials are superior to the surface materials. Due to climate, soil characteristics, or inadequate planning, it may not be possible to return some disturbed sites to their original productivity or species diversity for several years following reclamation. In other areas, the land can be returned to greater productivity than existed prior to the disturbance.

Postconstruction land use of disturbed sites may not always involve the production of forages. Nevertheless, forage plants will be used on most sites where the primary objective is stabilization. Revegetation techniques are very similar to those used for pasture or range reseeding; the seeding rate is usually double the rate used for forage plantings. Excellent guidelines for re-establishing vegetation on disturbed sites are available (Vogel 1981; Cook et al. 1974; Thornburg 1982; Sedgley 1974; SCS 1992a).

In the humid regions of the US, grass-legume mixtures are usually seeded for stabilization purposes. A quick-starting, generally short-lived grass, such as perennial ryegrass, weeping lovegrass, tall fescue, or redtop (*Agrostis gigantea* Roth), is included to provide immediate cover along with a longer-lived grass and legume to provide a semipermanent cover and a source of N. Unless managed for agricultural production or other uses, sites in the humid parts of the country will gradually convert back to forest.

In the semiarid and arid regions of the US,

the mixture used for stabilization may contain several forage species, but for different reasons than in the humid areas. A quick-starting but short-lived species may use available moisture to limit or even to prevent the establishment of those species in the mixture that are expected to provide long-term cover. However, the mixture may contain species that fit various niches in the local environment (such as an understory grass with a tall-growing one), forbs (which may or may not be legumes) to improve the quality of the stand for grazing purposes, or species for wildlife browse. Reclamation plantings containing only one species are common, however.

Because of increasing societal concern about the use of exotics, native plants should be given priority for soil stabilization work. Some introduced plants may be overly aggressive and may replace stands of native plants and/or become unmanageable weeds. Because most sites requiring stabilization,

such as mine spoils and road banks, provide a hostile environment for plants, plant aggressiveness is an asset. And in most rainfall areas, introduced plants are generally more aggressive (Baldridge and Lohmiller 1990).

There are virtually no native cool-season grasses or legumes for use in the humid and subhumid parts of the US, and therefore, most stabilization work is done with introduced plants (Ruffner 1978) (Table 17.3). However, the number of available warm-season native grasses is increasing as more knowledge of their culture and value is acquired (Sharp and Shiflet 1986; Reid et al. 1988; Everett 1991).

In contrast, commercial seed production techniques have been developed for many native species, and their use is increasing in the semiarid and arid regions (Thornburg 1982; Lohmiller et al. 1990; Baldridge and Lohmiller 1990). Even then, the most widely used plant for range reseeding and other soil

TABLE 17.3.   Commonly used plants for revegetating disturbed sites (e.g., strip mine land and road banks)

Species	Exotic or native	Seeding rate[b] (kg ha⁻¹)	Outstanding attributes and limitations
Humid and subhumid regions[a]			
1. Crownvetch (*Coronilla varia* L.)	Exotic	15	Long-lived, attractive, fair forage quality, excellent stabilizing plant in cooler parts of humid region, cultivars available. Seeded with tall fescue, perennial ryegrass, or redtop.
2. Flatpea (*Lathysus sylvestris* L.)	Exotic	40	Very long lasting, dominates site, tolerates shade, low pH, and drought, may be toxic to livestock if fed in pure diets, one cultivar available. Seeded with tall fescue, perennial ryegrass, or redtop.
3. Sericea lespedeza (*Lespedeza cuneata* [Dum.-Cours.] G. Don)	Exotic	20	Warm-season, deep rooted, tolerates acid and droughty soils. Unattractive after frost, coarse texture, various growth from cultivars available. Seeded with tall fescue or weeping lovegrass for quick cover or with switchgrass for long-term cover.
4. Birdsfoot trefoil (*Lotus corniculatus* L.)	Exotic	15	Excellent forage quality, persistent, less dominating and more acid tolerant than other legumes. Better in mixtures than other legumes, forage cultivars available. Seeded with tall fescue or redtop.
5. Alfalfa (*Medicago sativa* L.)	Exotic	12	Excellent forage value, less tolerant to low pH and other harsh site conditions than other legumes, forage cultivars available. Seeded alone or with tall fescue.
6. Sweetclover (*Melilotus officinalis* Lam.)	Exotic	20	Good nurse crop, usually seeded alone, can become overly aggressive and a weed on the site, cultivars available.
7. Red clover (*Trifolium pratense* L.)	Exotic	15	Biannual, aggressive, lacks acid tolerance, limited use on disturbed sites, forage cultivars available. Used as nurse crop or with tall fescue.
8. Redtop (*Agrostis gigantea* Rotah [syn. *A. alba* L.])	Exotic	1	Fast starting, relatively short-lived, easy to establish, tolerates wet soils, good nurse crop in mixture or as short-term cover, cultivars available.
9. Tall fescue (*Festuca arundinacea* Schreb.)	Exotic	20	Easy to start, relatively long-lived, wide tolerance to site conditions, commonly available, good forage value, serves as nonaggressive nurse crop at low seeding rates, cultivars available. Used as combination nurse crop and long-term with a legume.

TABLE 17.3.   (Continued)

Species	Exotic or native	Seeding rate[b] (kg ha^{-1})	Outstanding attributes and limitations
10. Switchgrass (*Panicum virgatum* L.)	Native	2	Slow to start, tolerates soil acidity, excessive moisture and drought, long-lived, needs to be seeded with nonaggressive nurse crop or with weeping lovegrass at very low seeding rates, several cultivars are available.
11. Big and little bluestem (*Andropogon gerardii* Vitman) and (*Schizachyrium scoparium* [Michx.] Nash)	Native	20	Similar characteristics to switchgrass, more difficult to establish, little bluestem very tolerant to phosphorus deficiency, cultivars available of big bluestem for humid regions, none for little bluestem. Seeded with nurse crop such as low rates of weeping lovegrass and with switchgrass.
12. Bermudagrass (*Cynodon dactylon* [L.] Pers.)	Exotic	12	Limited to Southeast, requires high fertility, gives excellent ground cover, aggressive with adequate fertilizer, limited acid tolerance, excellent forage quality, cultivars available, usually planted alone.
13. Deertongue (*Panicum clandestinum* L.)	Native	15	Slow starting, long-lived, very acid tolerant, excellent in mixtures, one cultivar available. Seeded with tall fescue, weeping lovegrass, or redtop for quick growth and seeded with a long-lived legume.
14. Bahiagrass (*Paspalum notatum* Flugge)	Exotic	70	Limited to Southeast, long-lived, tolerates lower fertility and abusive use better than bermudagrass, cultivars available. Planted alone or with quick cover grass.
Semiarid Regions[a]			
1. Switchgrass (*Panicum virgatum*) L.)	Native	7	See preceding under humid and subhumid heading, usually seeded alone, widely used, cultivars available.
2. Old world bluestems (*Bothriochloa* spp.)	Exotic	2-3	Very aggressive, may tend to be weedy, fair forage value, best adapted to fine-textured soils, several species and cultivars available.
3. Tall wheatgrass (*Thinopyrum ponticum* [Podp.] Barkworth & D.R. Dewey)	Exotic	22	Excellent for wind barrier use, salt tolerant, strong seedlings, widely adapted, cultivars available for use throughout semiarid region.
4. Western wheatgrass (*Pascopyrum smithii* [Rydb.] A. Love)	Native	23	Aggressive sod former, excellent forage, excellent for soil stabilization use, likes heavier soils, very important grass in northern Great Plains, many cultivars available.
5. Big and little bluestem (*Andropogon gerardii* Vitman) and (*Schizachyrium scoparium* [Michx.] Nash)	Native	8	See preceding under humid and sub-humid heading, frequently seeded alone, widely used, cultivars available.
6. Sideoats grama (*Bouteloua curtipendula* [Michx.] Torr.)	Native	10	Broad natural area of adaptation, excellent for grazing, used in mixtures, cultivars available.
7. Indiangrass (*Sorghastrum nutans* [L.] Nash)	Native	11	Common Great Plains grass, excellent forage, less widely used for revegetation, cultivars available.
8. Bluebunch wheatgrass (*Pseudoroegneria spicata* [Pursh] A. Love)	Native	16	Widely distributed native, limited reseeding use to date, excellent forage. Cultivars available.
9. Crested wheatgrass (*Agropyron cristatum* [L.] Gaertn.)	Exotic	11	Quick starting, easy to establish, good spring and fall forage value. Widely used in northern parts of semiarid region.
Arid Regions[a]			
1. Crested wheatgrass (*Agropyron cristatum* [L.] Gaertn.)	Exotic	11	Quick starting, easy to establish, good spring and fall forage value, used extensively.
2. Intermediate wheatgrass (*Thinopyrum intermedium* [Host] Barkworth and D. R. Dewey)	Exotic	16	Widely diverse and adapted species, used extensively for soil conservation and forage, pubescent types more drought tolerant, cultivars available.

TABLE 17.3.   (Continued)

Species	Exotic or native	Seeding rate[b] (kg ha⁻¹)	Outstanding attributes and limitations
3. Thickspike wheatgrass (*Elymus lanceolatus* [Scribner & Smith] Gould subsp. *lanceolatus*)	Native	10	Strongly rhizomatous, widely distributed, more drought tolerant than western wheatgrass, relatively easy to establish, cultivars available.
4. Bluebunch wheatgrass (*Pseudoroegneria spicata* [Pursh] A. Love)	Native	16	Important bunchgrass in western intermountain regions, abundant in some areas, only recently being recognized and domesticated, cultivars available.
5. Black grama (*Bouteloua eriopoda* [Torr.] Torr.)	Native	2	A major species in Great Plains, drought tolerant, produces tight sod, excellent forage, cultivars available.
6. Lovegrasses (*Eragrostis* Wolf)	Exotic	2	Several lovegrasses are used in the Southwest ('Lehmann' being the most drought tolerant), also used in southern Great Plains, excellent on sandy soils, forage value only fair, cultivars available.
7. Sheep fescue (*Festuca ovina* L.)	Native	3	Short bunchgrass, occurs as understory in native grasslands of northern intermountain regions, limited forage value, cultivars available.
8. Indian ricegrass (*Oryzopsis hymenoides* [Roem. & Schult.] Ricker)	Native	10	Densely tufted, easily established, one of most drought tolerant grasses, good winter-standing forage, cultivars available.

Sources: Ruffner (1978); Thornburg (1982); SCS (1992a,b).

[a] Annual rainfall: humid >760mm, subhumid 500-760mm, semiarid 250-500mm, arid < 250 mm.

[b] The seeding rates for the plants listed under humid and subhumid regions are for their inclusion in mixtures. For plants listed for semiarid and arid regions, the rates are for their being seeded alone. The listed seeding rates will result in approximately 535 pure live seeds per square meter.

stabilization purposes in western areas has been an exotic crested wheatgrass. Its ease of establishment, early spring green-up, and tolerance to various soil conditions make it desirable on reseeded range or mine land for early grazing.

Examples of species commonly used for soil stabilization, but which can also be used as forages, are shown in Table 17.3 (Ruffner 1978; Thornburg 1982; SCS 1992a, 1992b). Mixtures containing these and other plants for a specific site, as well as establishment methods, should be formulated only after determining more site characteristics than are listed in Table 17.3. These could include timing and amount of precipitation, soil characteristics (texture, pH, and nutrient content), and intended use. For instance, road banks and much of the mine land in the Northeast are rarely grazed and so are frequently seeded with a long-lived perennial legume such as crownvetch (*Coronilla varia* L.) or sericea lespedeza (*Lespedeza cuneata* [Dum.-Cours.] G. Don) plus a rapid-growing grass such as tall fescue or perennial ryegrass. Mine land in the West, however, will most likely be returned to grazing land, and the site may be replanted with a single forage plant or a mixture.

USING FORAGES TO MAINTAIN WATER QUALITY

Agricultural production is associated with the quality of surface and groundwater as it accounts for 64% of pollution in rivers, of which 47% is sediment, 13% nutrients, and 3% pesticides. Approximately 57% of lake pollution is from agriculture, of which 59% is nutrients and 22% sediment (Carey 1991). When forages are used as a part of crop rotations or combined with strip-cropping, they trap runoff and utilize nutrients that may be in the runoff water. Because forages require only limited applications of fertilizer and pesticides, there is also a reduced likelihood of contaminating surface or groundwater from grazing land.

Using Forages for Water Quality in Agricultural Production. Grant et al. (1982) measured denitrification rates below smooth bromegrass sod and found the rate to be 10 times greater than below corn (11.7 versus 1.1

µg N g soil^{-1} h). Water in tile lines 105 cm below the soil surface had a nitrate concentration from the smooth bromegrass that was only one-third that produced by the corn (2.1 versus 7.7 mg L^{-1}). Nitrate concentration was consistently above 10 mg L^{-1} with the corn both early and late during the growing season. Even though corn removed more N from the soil, the smooth bromegrass transformed more N through denitrification and immobilization. When Barraclouth et al. (1983) applied 250, 500, or 900 kg N ha^{-1} annually as ammonium nitrate to a perennial ryegrass sod, the total mass flow of nitrate from tile lines placed 80 cm deep averaged less than 5 kg ha^{-1}, 27 kg ha^{-1}, and 150 kg ha^{-1}, respectively. This represented 1.5%, 5.4% and 16.7%, respectively, of the total N applied as fertilizer. The optimum N rate for ryegrass forage was deemed to be 350-450 kg ha^{-1}. They conclude nitrate leaching would be an ecological problem only if the amount of fertilizer applied exceeded that required for optimum plant production. Alton and Garcia (1973) found that converting rangeland from dense, continuous stocking to deferred reduced average annual rates of sediment production by 71% on a New Mexico watershed during a 10-yr recovery period.

Using Forages for Water Quality: Managing Animal and Human Wastes. Disposal of manure from livestock feeding operations, litter from poultry production, and effluents from municipal or industrial wastewater treatment facilities used in agriculture can be a major water quality problem. Forage crops are increasingly being used as vegetative covers and filters for such waste disposal areas (Ball et al. 1991). In a 4-yr study, smooth bromegrass was irrigated with 5.25 cm per irrigation of runoff effluent from a beef cattle feedlot (Olson et al. 1982). The soil contained 945 kg N ha^{-1}, 28% of the applied N, labeled as NH$_4^+$ in the soil profile; 41% was taken up and harvested with the forage; 31% was lost (calculated as gas volatilization); and only 0.7% was leached to 105-cm depth. At 10.5 cm per irrigation, 1890 kg ha^{-1}, or 20% of the N, was in the soil, 38% was used by the plant, 40% was lost as gas, and 1.3% was leached.

Brown and Thomas (1978) observe that the percentage uptake of N by bermudagrass is inversely related to the amount applied. Effluent rates were applied in relation to the infiltration and percolation rates of the soils. Up to 25.4 cm h^{-1} on the sandy loam soil of septic tank effluent had only 9% removed by plant

uptake. At 0.3 cm h^{-1} on a clay soil, 46% of the N was harvested in the grass. In a 2-yr New Hampshire study, the average total annual uptake of N by orchardgrass was 290 kg ha^{-1} for three cuttings. The orchardgrass accumulated N for the first 35-40 d after initial cutting in the spring, absorbing 50% and 85% of the applied N (458 and 380 kg ha^{-1}) during the 2-yr period (Palazzo 1981).

Burns et al. (1985) applied swine manure effluent at N rates of 335, 670, and 1340 kg ha^{-1} to 'Coastal' bermudagrass. Harvested forage contained 73% of the applied N at the low application rate, 57% at the medium, and 34% at the high application rate. The plant stand was reduced by the addition of the high rate of N, and foliage concentration of nitrate N approached 0.3%.

Hook and Burton (1979) applied secondary waste water sewage effluent at 7.5 cm per week during the growing season to Kentucky bluegrass. Harvest treatments included no cutting, biweekly mowing, and triennial hay harvest. Over 200 kg N ha^{-1} were applied annually in the irrigation water. Soil water samples were taken 48 h immediately after weekly irrigation from porous-cup suction lysimeters placed 15 and 150 cm below the soil surface in a loam soil. Both ammonium and nitrate plus nitrite content were consistently below 10 mg L^{-1}, regardless of the cutting management. The no-cut and mowed plots were as effective as the harvested plots in not allowing N to leach past the 150-cm depth in the 2-yr study.

Using Forages as Filters for Water Quality. Forage plants may also be used as filter strips along bodies of water down slope from crop fields and other sources of contaminants such as construction sites or mine land. Neibling and Alberts (1979) applied simulated rainfall to an eroded silt loam soil on a 7% slope that had been moldboard plowed and left fallow. The sediment produced was carried across varying widths of Kentucky bluegrass sod strips. For a 127-mm simulated rainfall event applied over a 2-d period with an intensity of 63 mm h^{-1}, the narrowest sod strip (0.61 m wide) drastically reduced the amount of soil passing through the sod filter. The percentage reduction for clay-sized particles (<0.002 mm) was 37% for the 0.61-m strip, 78% for the 1.22-m, 82% for the 2.44-m, and 83% for the 4.88-m sod strip.

Asmussen et al. (1977) found infiltration to be a mechanism for reducing water runoff velocity in a grassed waterway planted to bermudagrass and bahiagrass. In the study of 2,4-D (2,4 Dichlorophenoxyacetic acid) herbicide retention, the infiltration into the waterway reduced runoff by 50%. Sediment was reduced by 98% down gradient in the 24.4-m-wide waterway, and the 2,4-D in the runoff, moving mostly in the soluble form, was either infiltrated into the grass waterway or became absorbed by the plant tissue.

In a study conducted by Doyle et al. (1977), forest buffer strips made up of mixed deciduous trees and strips of tall fescue were found to be effective in reducing nutrient levels and fecal bacteria concentration in surface runoff from manure-treated plots. The most significant reductions in run-on concentration occurred within the first 4 m of the strip. Increasing the flow length down the vegetative filter strip significantly reduced the amounts of surface pollutants. Barker and Young (1984) established a vegetative filter strip of tall fescue, orchardgrass, and ryegrass mixture on a Tate variant rocky loam soil in order to treating wastewater from a dairy farm in North Carolina. The highly permeable soil infiltrated over 94% of the applied milkhouse wastewater, which contained the equivalent of 467 kg N ha^{-1}. When the paved feedlot effluent was added to the runoff, raising the total N applied to 1500 kg ha^{-1}, the filter removed 99% of the N when 95% of the water was infiltrated, and 98% of the total N was absorbed by the filter. Soil nitrate N levels increased over the 2-yr period by sixfold (5 mg L^{-1} to 32 mg L^{-1}), but only in the top of the terraced 79-m grass filter. In the area of nitrate accumulation, the grass was lush and contained 1638 µg g^{-1} of nitrate N, well above the level of 678 µg g^{-1} considered toxic to ruminant animals. Magette et al. (1987) showed that a 4.6-m-wide vegetated filter strip reduced total suspended solids from cropland runoff by 72%, compared with a nonbuffered area on the same slope. Increasing the filter width to 9.2 m reduced total suspended solids by 86%. Total N was reduced by 17% in the 4.6-m strip and 51% in the 0.2-m-wide strip. Subsurface N losses were greater than surface runoff of N. These losses included all N that was infiltrated into the soil.

Using Forages for Water Quality in a Wetland System. There are other methods involving the use of forages for managing the potential of agricultural degradation of water quality. Wengrzynek (1992) developed a Nutrient-Sediment Control System (NSCS) to intercept and treat runoff from cropland in Maine through constructed wetland technology. The NSCS reduced the total P that entered the system by 88%-100% and suspended solids by 95%. Adapted forages were used as the final filter in the system. The construction cost of the system was below that of many less effective land treatment practices for treating runoff, and the system was much more acceptable than taking prime farmland out of production in order to install terraces or to institute other practices.

QUESTIONS

1. Discuss the reasons for preferential use of introduced grasses and legumes for forage and soil conservation in the eastern US.
2. What functions do roots perform in addition to water and nutrient uptake?
3. Discuss the two principal roles forages play in soil conservation work.
4. What is the relationship between residue management and conservation of soil and water?
5. How can forages play a role in sustainable agriculture systems?
6. What are the differences between perennial forages and annual crops for water infiltration and runoff?
7. Discuss the use of grass barriers for wind and water erosion control and for moisture conservation.
8. How does the management of grazing land affect soil conservation?
9. Define cover crops, and discuss how they influence soil and water conservation. How do they affect the plant nutrient supply?
10. Discuss the steps essential for the establishment of soil-conserving vegetation on disturbed sites.
11. How can strip-cropping be combined with crops in rotation to effectively conserve soil and water?
12. How do forage crops influence the quality of surface and groundwater?

REFERENCES

Aase, JK, FH Siddoway, and AL Black. 1985. Effectiveness of grass barriers for reducing wind erosiveness. J. Soil Water Conserv. 40:354-57.

Agriculture Stabilization and Conservation Service (ASCS). 1991. Conservation practice acres by state through signup 11. In LOGO Packet—11th Signup Results, Conservation Reserve Program. Washington, D.C., 1-16

Alderson, JE, and WC Sharp. 1994. USDA Agric. Handb. 170. Washington, D.C.: US Gov. Print. Off.

Altieri, MA, and LL Schmidt. 1985. Cover crop manipulation in northern California orchards and vineyards: Effects on arthropod communities. Biol. Agric. Hortic. 3:1-24.

Alton, EF, and G Garcia. 1973. Seventeen-Year Sediment Production from a Semi-arid Watershed in the Southwest. US For. Serv. Res. Note RM-248.

Asmussen, LE, AW White, Jr., EW Hauser, JM Sheridan. 1977. Reduction of 2,4-D load in surface runoff down a grassed waterway. J. Environ. Qual. 6:159-62.

Baker, EL, and JM Laflan. 1982. Effect of crop residue and fertilizer management on soluble nutrient runoff losses. Trans. Am. Soc. Civ. Eng. 25:344-48.

Baldridge, DE, and RG Lohmiller. 1990. Plant species. In Interagency Plant Materials Handbook, Mont. State Univ. Ext. Serv. OEB69, 8-169.

Ball, DM, CS Hoveland, and GD Lacefield. 1991. Forages and the environment. In Southern Forages. Atlanta, Ga.: Potash and Phosphate Institute, 229-34.

Barker, JC, and BA Young. 1984. Evaluation of a Vegetative Filter for Dairy Wastewater in Southern Appalachia. Raleigh: Water Research Institute, North Carolina State Univ.

Barraclouth, D, MJ Hyden, and GP Davies. 1983. Fate of fertilizer nitrogen applied to grasslands. I. Field leaching results. J. Soil Sci. 34:483-97.

Black, AL, and FH Siddoway. 1971. Tall wheatgrass barriers for soil erosion control and water conservation. J. Soil Water Conserv. 31:101-5.

Boyle, M, WT Frankenberger, Jr., and LH Stolzy. 1989. The influence of organic matter on soil aggregation and water infiltration. J. Prod. Agric. 2:290-99.

Brown, KW, and JC Thomas. 1978. Uptake of N by grass from septic fields in three soils. Agron. J. 70:1037-40.

Browning, GM. 1962. Forages and soil conservation. In HD Hughes, ME Heath, and DS Metcalfe (eds.), Forages: The Science of Grassland Agriculture, 2d ed. Ames: Iowa State Univ. Press, 31-41.

———. 1973. Forages and soil conservation. In ME Heath, DS Metcalfe, and RF Barnes (eds.), Forages: The Science of Grassland Agriculture, 3d ed. Ames: Iowa State Univ. Press.

Browning, GM, and RH Sudds. 1942. Some Physical and Chemical Properties of the Principal Orchard Soils in the Eastern Panhandle of West Virginia. West Va. Agric. and For. Exp. Stn. Bull. 303.

Bureau of the Census. 1987. Census of Agriculture. Washington, D.C.: US Gov. Print. Off.

Burns, JC, PL Westerman, LD King, GA Cummings, MR Overcash, and L Goode. 1985. Swine lagoon effluent applied to Coastal bermudagrass. I. Forage yield, quality, and elemental removal. J. Environ. Qual. 14:9-14.

Carey, AE. 1991. Agriculture, agriculture chemicals, and water quality. In Agriculture and the Environment, USDA Yearbook of Agriculture. Washington, D.C.: US Gov. Print. Off., 78-85.

Carlson, JR. 1988. Theory and Methodology for Grassland Germplasm Enhancement, Cultivar Development, Seed Production, and Use in Revegetation Programs. Grassl. Germplasm Semin., Huhehot, Inner Mongolia.

Classen, MM. 1980. Effect of nitrogen rate on dryland no-till corn following alfalfa. In Kans. Fert. Res. Rep. Prog. 389. Manhattan: Kansas State Univ., 139-40.

Conservation Technology Information Center (CTIC). 1992. National Survey of Crop Residue Management Practices. West Lafayette, Ind.: CTIC.

Cook, CW, RM Hyde, and PL Sims. 1974. Guidelines for Revegetation and Stabilization of Surface Mined Areas in the Western States. Colo. State Univ., Range Sci. Dept. Ser. 16.

Cook, RJ. 1984. Root health: Importance and relationship to farming practices. In Organic Farming: Current Technology and Its Role in a Sustainable Agriculture, Spec. Publ. 46. Madison, Wis.: American Society of Agronomy, 111-27.

Dickerson, JA, DL Emmick, FL Swartz, PR Salon, and FL Gilbert. 1991. Developing a grassland farming system for New York. Am. Soc. Agron. Abstr. Madison, Wis., 2:6 (Append. 2).

Doran, JW, and M Werner. 1990. Management and soil biology. In CA Francis, CB Flora, and LD King (eds.), Sustainable Agriculture in Temperate Zones. New York: Wiley, 205-30.

Doyle, RC, GC Stanton, and DC Wolf. 1977. Effectiveness of Forest and Grass Buffer Strips in Improving the Water Quality of Manures Polluted Runoff. Am. Soc. Agric. Eng. Pap. 77-2501. St. Joseph, Mich.

Ebelhar, SA, WW Frye, and RL Bevins. 1984. Nitrogen from legume cover crops for no-tillage corn. Agron. J. 76:51-55.

Everett, HW. 1991. Native Perennial Warm Season Grasses for Forage in Southeastern United States. Fort Worth, Tex.: Soil Conservation Society.

Flint, ML, and PR Roberts. 1989. Using crop diversity to manage pest problems: Some California examples. Am. J. Alternative Agric. 3:163-67.

Food Security Act. 1985. Food security act of 1985. Public Law, 99-198.

Francis, CA, and G Youngblood. 1990. Sustainable agriculture: An overview. In CA Francis, CB Flora, and LD King (eds.), Sustainable Agriculture in Temperate Zones. New York: Wiley, 1-23.

Funderburg, ER. 1987. Cover Crops. Miss. Coop. Ext. Serv. Inf. Sheet 1552.

Gard, LE. 1943. Runoff from pasture land as affected by soil treatment and grazing management. Agron. J. 35:332-47.

Goldstein, WA, and DL Young. 1987. An agronomic and economic comparison of a conventional and a low-input cropping system in the Palouse. Am. J. Alternative Agric. 2(2):51-56.

Grant, SA, RJ Kunse, and G Asrar. 1982. Irrigation and simulated secondary waste water on tilled soil cropped to bromegrass and corn. J. Environ. Qual. 11:442-46.

Hanson, CL, AR Kuhlman, CJ Erickson, and JK Lewis. 1970. Grazing effects on runoff and vegetation on western South Dakota rangeland. J. Range Manage. 23:418-20.

Hargrove, WL. 1986. Winter legumes as a nitrogen source for no-till grain sorghum. Agron. J. 78:70-74.

Harris, GA. 1967. Some comparative relationships between *Agropyron spicatum* and *Bromus techto-*

rum. Ecol. Monogr. 37:89-111.

Hauptli, H, D Katz, BR Thomas, and RM Goodman. 1990. Biotechnology and crop breeding for sustainable agriculture. In CA Edwards, L Rattan, P Madden, RH Miller, and G House (eds.), Sustainable Agriculture Systems. Ankeny, Iowa: SWCS, 141-56.

Heichel, GH. 1987. Legumes as a source of nitrogen in conservation tillage systems. In JF Power (ed.), Role of Legumes in Conservation Tillage Systems. Ankeny, Iowa: SWCS, 29-35.

Hendrix, PF, DA Crossley, Jr., JM Blair, and DC Coleman. 1990. Soil biota as components of sustainable agroecosystems. In CA Edwards, L Rattan, P Madden, RH Miller, and G House (eds.), Sustainable Agriculture Systems. Ankeny, Iowa: SWCS, 637-54.

Hook, JE, and TM Burton. 1979. Nitrate leaching from sewage-irrigated perennials as affected by cutting management. J. Environ. Qual. 8:496-502.

Kemper, D, S Dabney, L Kramer, D Dominick, and T Keep. 1992. Hedging against erosion. Soil and Water Conserv. 47:284-88.

King, LD. 1990. Soil nutrient management in the US. In CA Edwards, L Rattan, P Madden, RH Miller, and G House (eds.), Sustainable Agriculture Systems. Ankeny, Iowa: SWCS, 89-122.

Lamond, RE, DA Whitnet, LC Bonczkowski, and JS Hickman. 1988. Using Legumes in Crop Rotation. Kans. State Univ. Coop. Ext. Serv. Publ. L-778. Manhattan.

Linden, DR, DM Van Doran, Jr., JK Radke, and DC Reicosky. 1987. Tillage and surface residue sensitivity potential evaporation submodel. In NTRM, A Soil-Crop Simulation Model for Nitrogen, Tillage, and Crop Residue Management, Conserv. Res. Rep. 34-1, 73-85.

Lohmiller, RG, DI Dollhopf, and AH Martinson. 1990. Disturbed Land Rehabilitation. Interagency Plant Materials Handbook. Mont. State Univ. Ext. Serv. OEB69.

Magette, WL, RB Brimfield, RE Palmer, JD Wood, TA Dillaha, and RB Reneau. 1987. Project Summary of Vegetated Filter Strips for Agricultural Runoff Treatment. EPA Rep. X-003314-01. Philadelphia, Pa.

Meisinger, JJ, WL Hargrove, RB Mikkelsen, JR Williams, and VW Bensen. 1991. Effects of Cover Crops on Ground Water Quality. Proc. Cover Crops for Clean Water, Soil and Water Conserv. Soc., Ankeny, Iowa.

Ndawula-Senyimba, MS, VC Brink, and A McLean. 1971. Moisture interception as a factor in the competitive ability of bluebunch wheatgrass. J. Range Mange. 24:190-200.

Neher, D. 1992. Ecological sustainability in agriculture systems: Definition and measurement. In RK Olson (ed.), Integrating Sustainable Agriculture, Ecology, and Environmental Policy. New York: Haworth Press, 51-61.

Neibling, WH, and EE Albert. 1979. Composition and Yield of Soil Particles Transported through Sod Strips. Am. Soc. Agric. Eng. Pap. 79-2065. St Joseph, Mich.

New York Agriculture Statistics. 1992. New York Agriculture Statistics, 1991-1992. Albany: New York Agriculture Statistics Service.

Olson, RV, RV Terry, WL Powers, CW Swallow, and TE Kanemaus. 1982. Disposal of feedlot-lagoon water by irrigating bromegrass: II. Soil accumulation and leaching of nitrogen. J. Environ. Qual. 11:400-405.

Palazzo, AJ. 1981. Seasonal growth and accumulation of nitrogen, phosphorus, and potassium by orchardgrass irrigated with municipal waste water. J. Environ. Qual. 10:64-68.

Pohl, RW. 1980. Family 5, *Gramineae*. New Series No. 4, Burger: Flora Costaricensis. Fieldiana Botany.

Power, JF. 1987. Legumes: Their potential role in agriculture production. Am. J. Alternative Agric. 2(2):69-73.

Reid, RL, GA Jung, and DW Allison. 1988. Nutritive Quality of Warm Season Grasses in the Northeast. West Va. Agric. and For. Exp. Stn. Bull. 699.

Renard, KG, GR Foster, GA Weesies, DK McCool, and DC Yoder. 1993. Predicting Soil Erosion by Water: A Guide to Conservation Planning with Revised Universal Soil Loss Equation. USDA Agric. Handb. 703. Washington, D.C.: US Gov. Print. Off.

Ruffner, JD. 1978. Plant Performance on Surface Coal Mine Spoil in Eastern US. SCS-TP-155.

Sajjadi, A, and RE Zartman. 1990. Wind stripcropping using weeping lovegrass in the southern high plains. J. Soil Water Conserv. 45:397-99.

Saxton, KE, and RG Spomer. 1968. Effects of conservation on the hydrology of loessal watersheds. Trans. Am. Soc. Agric. Eng. 11:848-49, 853.

Saxton, KE, RG Spomer, and CA Kramer. 1971. Hydrology and erosion of loessial watersheds. In Proc. Am. Soc. Civ. Eng. (HY11):1835-51.

Schaller, N. 1990. Mainstreaming low-input agriculture. J. Soil Water Conserv. 45:9-12.

Sedgley, EF. 1974. Revegetation and land rehabilitation of oil shale mining sites. In 2d Res. and Appl. Tech. Symp. Mine-Land Reclamation, Natl. Coal Assoc., Washington, D.C.

Sharp, WC, and TN Shiflet. 1986. Warm-Season Grasses: Balancing Forage Programs in the Northeast and Southern Corn Belt. Ankeny, Iowa: Soil and Water Conservation Society.

Sharp, WC, JR Carlson, and JS Peterson. 1992. Plant Materials Center: Finding Vegetative Solutions to Conservation Problems. Washington, D.C.: USDA-Soil Conservation Society.

Shipley, PR, JJ Meisinger, and AM Decker. 1992. Conserving residual corn fertilizer nitrogen with winter cover crops. Agron. J. 84:869-76.

Shurley, WD. 1987. Economics of Legume Cover Crops in Corn Production, the Role of Legumes in Conservation Tillage Systems. Proc. Soil and Water Conserv. Soc., Ankeny, Iowa.

Soil Conservation Service (SCS). 1980. Basic Statistics: 1977 National Resources Inventory. Washington, D.C.: SCS.

———. 1987. Summary Report, 1987 National Resources Inventory. Iowa State Univ. Stat. Lab. Stat. Bull. 790.

———. 1992a. SCS National Catalog of Conservation Practices and Field Office Technical Guide Critical Area Planting Specification 347. Washington, D.C.: SCS.

———. 1992b. SCS National Catalog of Conservation Practices and Field Office Technical Guide Pasture and Hayland Planting 512 and Range Seeding 512. Washington, D.C.: SCS.

Thornburg, AA. 1982. Plant Materials for Use on Surface-mined Lands in the Arid and Semi-arid Regions. SCS-TP-157.

Touchton, JT, DH Richeri, RH Walker, and CE Snipes. 1984. Reseeding crimson clover as a N source for no-tillage grain sorghum production. Soil Tillage Res. 4:391-401.

Tower, HE. 1946. Strip Cropping for Conservation and Production. USDA Farmers Bull. 1981.

Vogel, WG. 1981. A Guide to Revegetating Minespoils in the Eastern US. USDA For. Serv., Northeast. Exp. Stn. Tech. Rep. 68.

Wadleigh, CM, LM Glymph, and HN Holtan. 1974. Grasslands in relation to water resources. In Grasslands of the United States. Ames: Iowa State Univ. Press, 15-42.

Wengrzynek, RJ. 1992. Constructed Wetlands to Control Nonpoint Source Pollution. US Patent 5,174,897. Washington, D.C.: US Patent Office.

Wischmeier, WH, and DD Smith. 1958. Rainfall energy and its relationship to soil loss. Trans. Am. Geophys. Union 39:285-91.

18

Managing Forages to Benefit Wildlife

STEVE E. CLUBINE
Missouri Department of Conservation

HISTORICAL IMPACT OF GRAZING ON WILDLIFE

THE effect of grazing by large herbivores, native and domestic, is a significant factor affecting wildlife populations in North America. Historically, the distributions of prairie chickens (*Tympanuchus* spp.), sharptail grouse (*Pedioecetes phasianellus* L.), waterfowl, and other prairie wildlife were limited by the effects huge herds of buffalo (*Bison bison* L.) had on nesting cover and native foods (Kirsch and Kruse 1973; Kirsch 1983). "Hundreds of thousands of acres literally eaten to the turf by the immense herds of buffalo. . . . Scarcely a blade of grass standing after the herd passed" (Marlin 1967). "Buffalo had eaten the grass until it was very short, making food scarce. . . . The US Army found it difficult to find sufficient grass for their horses and mules" (Roe 1951). When there was sufficient grass for wildlife to nest and feed, it was often burned by lightning or Native Americans before the nesting season (Townsend 1839; Sauer 1920).

Peak wildlife populations in the US coincided with European settlement, but they were not due to the breaking of the prairie sod and greater food availability from agricultural

STEVE E. CLUBINE is Grassland Biologist, Missouri Department of Conservation. He received a BS degree from Kansas State University and has worked on grassland and wildlife management programs in Kansas and Missouri since 1971.

Appreciation is expressed to Kendall Johnson, Professor and Head, Department of Range Resources, University of Idaho, Moscow, for contributing the section on public rangelands.

grains (Leopold 1931). Peak populations of prairie chicken in Illinois and Indiana occurred from 1860 to 1870 (Sparling 1979), before the prairie was drained and tilled (Weaver 1954). Large numbers of prairie wildlife were killed and shipped out of Nebraska to eastern restaurants in the 1870s (Kobriger 1965) before settlement had a foothold in the plains of the Dakotas and Nebraska. Prairie wildlife was common when the first homesteaders arrived but decreased in the wake of the plow and the cow (Bergerud and Gratson 1988).

Wildlife populations followed the grass. As the buffalo hunter exterminated the buffalo, prairie grouse, waterfowl, upland sandpiper (*Bartramia longicauda* Bechstein), and other prairie wildlife increased in the prairies and expanded ranges in the plains (Kirsch and Kruse 1973; Kirsch 1983; Bergerud and Gratson 1988). With the loss of the buffalo, annual burning of the prairie by Native Americans to attract the grazers to new growth also declined. These idle native grasslands provided increased food and cover for smaller wildlife species, and for a time they flourished.

At least 260 bird species regularly breed in North American grasslands (Johnsgard 1979), 19 have strong associations with grassland ecosystems but also occur in adjacent vegetation types, and 9 species are entirely dependent on short-grass and tall-grass (true prairie) ecosystems (Knopf 1988). A large number of mammals, reptiles, and amphibians are also grassland dependent.

Open vistas and unobstructed horizons are critical to grassland wildlife: they provide security from predators by allowing prey species time to freeze or take flight before the predators approach too closely. Shelterbelts, windbreaks, and woody encroachment have been detrimental to grassland wildlife while favoring woodland edge and eastern forest wildlife species and several species that parasitize and prey on grassland wildlife (Knopf and Olson 1984; Knopf 1988; Johnson and Temple 1990).

CHARACTERISTICS OF WILDLIFE COVER

Nesting Cover. Most ground-nesting wildlife prefer upright cover, 15 to 43 cm tall, and dead material from the previous year. Warm-season bunchgrasses are used by wildlife in preference to cool-season grasses or legumes in the central plains, the prairie regions, and the Southeast. Native cool-season bunchgrasses are used in the absence of warm-season bunchgrasses in the Northern Plains. Bunchgrasses resist compaction over winter and allow wildlife to move between the bunches while providing overhead protection from predators. The densest bunchgrasses, e.g., little bluestem (*Schizachyrium scoparium* [Michx.] Nash), discourage close grazing and provide more residual cover. Most nests of grassland wildlife will be located between two or more clumps of grass with canopies meeting overhead to form a natural tent (Fig. 18.1). Important species are little bluestem, indiangrass (*Sorghastrum nutans* [L.] Nash), big bluestem (*Andropogon gerardii* Vitman), tall dropseed (*Sporobolis asper* [Michx.] Kunth), broomsedge (*A. virginicus* L.), wiregrass (*Aristida* spp.), tobosagrass (*Hilaria mutica* [Buckl.] Benth.), switchgrass (*Panicum virgatum* L.), needlegrasses (*Stipa* spp.), and some wheatgrasses (*Agropyron* spp.) (Rosene 1969; Wiseman 1977; George et al. 1978).

Wildlife species have adapted to using introduced grasses where adequate structure is maintained. Most introduced grasses are sod forming, however, creating dense mats and poor structural cover except where the canopy is supported by forbs, shrubs, or bunchgrasses. Examples of forage grasses that form poor nesting habitat without supporting vegetation are Kentucky bluegrass (*Poa pratensis* L.), cheat (*Bromus* spp.), smooth bromegrass (*B. inermis* Leyss.), bermudagrass (*Cynodon dactylon* [L.] Pers.), bahiagrass (*Paspalum notatum* Flugge), and buffelgrass (*Cenchrus ciliaris* L.). Shrubs and forbs growing with sod-forming grasses improve nesting cover quality in the snow and ice belt (Greenwood et al. 1987) by providing structural support for these grasses. Important species include al-

Fig. 18.1. Nesting cover should be 15-43 cm tall, consisting mostly of dead material from the previous growing season and structured to form a natural tent.

falfa (*Medicago sativa* L.), coralberry (*Symphoricarpus* spp.), and wild rose (*Rosa* spp.).

The optimum quantity of grassland nesting habitat is 5%-20% of the land area for ring-necked pheasant (*Phasianus colchicus* L.), 10%-20% for gray partridge (*Perdix perdix* L.), 30%-40% for bobwhite quail (*Colinus virginianus* L.), and 50%-75% for prairie grouse and waterfowl. Prairie grouse has been extirpated from the eastern true prairie because grasslands suitable for nesting are well below 50% of the land area. Winter wheat (*Triticum aestivum* L. emend. Thell.) fills some of the void left by the loss of quality grasslands for nesting habitat (Christisen 1985). In many regions, the ring-necked pheasant has replaced the prairie grouse. Where both still exist, they may compete for the same limited nesting cover (Sharp 1957; Vance and Westemeier 1979).

Nesting cover should not be grazed or cut for hay to obtain optimum wildlife production (Kirsch et al. 1978). However, nesting cover will not remain in optimum condition without periodic disturbance. Prescribed (i.e., precisely timed) burning or grazing is preferred to haying (Table 18.1) and should be done every 2-4 yr. Higher-rainfall and more fertile sites require more frequent disturbance than drier and less fertile sites. Xeric grasslands may never require treatment. Disturbance is most beneficial and least disruptive if it occurs after the nesting season or in rotation so that a high percentage of the acreage remains undisturbed each year during the nesting period.

TABLE 18.1. The number of bird species and sightings on hayed prairie and grazed prairie in May-June 1977 and 1978

Species	Plot 1 Hayed M-J 1977	Plot 2 Grazed M-J 1977	Plot 3 Grazed M-J 1977	Plot 1 Hayed M-J 1978	Plot 2 Grazed M-J 1978	Plot 3 Grazed M-J 1978
Northern bobwhite			9		1	
Greater prairiechicken	12	32	2			5
Ruby-throated hummingbird				1		
Eastern kingbird		8	1		3	14
Horned lark					3	
Brown-headed cowbird		4	5			
Red-winged blackbird			3			3
Eastern meadowlark	12	21	31	16	24	31
Orchard oriole						1
Savannah sparrow		1			3	
Grasshopper sparrow	7	55	32	11	69	45
Henslow's sparrow	20	1	19	12		6
Dickcissel	2	12	20	—	1	14
Total species	5	8	9	4	7	8
Total sightings	53	134	122	40	104	119

Source: Adapted from Skinner et al. (1984).

Broodrearing Cover. Broodrearing cover should be located near or within the same field as nesting cover. Bobwhite quail, prairie grouse, upland sandpiper, and pheasant chicks leave the nest soon after hatching and forage with their mothers for insects, seeds, and greens. Mother and brood usually move from grass-dominated nesting cover to forb-dominated broodrearing cover where insects, weed seeds, and fruits are more abundant and movement is less restricted by matted vegetation. Insect diversity is significantly higher in grasses that are dense at taller heights compared with grasses that are dense near ground level (Toll 1977).

Bare ground for dusting (cleaning and controlling parasites) and feeding is important to many wildlife species, especially bobwhite quail (Scott and Klimstra 1954). Patches of blackberry (*Rubus* spp.), coralberry, dogwood (*Cornus* spp.), and other young or low-growing brush are used for dusting and loafing on hot days and for escape cover. Alfalfa hayfields and moderate to heavily stocked pastures are good broodrearing areas. Insects feed on the juices and leaves of alfalfa and broadleaf forbs and use grasses for perches and roosts. Livestock grazing and trampling increase the amount of bare ground, aiding wildlife movement, dusting, and the pursuit of insects and seeds.

Grazing is an important disturbance for wildlife habitats in the true prairie and may be more effective in true prairie than in mixed-grass or short-grass prairie (Risser et al. 1981; Skinner et al. 1984). Grazing tall grasses during summer months improves fall and winter habitat by increasing insects, seeds, and green browse (Jackson 1969; Lehmann 1984; Guthery 1986b).

Roosting Cover. Roosting cover varies from dense, erect forbs and low-growing shrubs to dense, upright grass (Klimstra and Ziccardi 1963; Yoho and Dimmick 1972; Skinner et al. 1984; Burger et al. 1990). During mild fall and winter weather, the favored roosting cover for bobwhite quail and prairie grouse is fairly low and open. Unmowed timothy (*Phleum pratense* L.) and orchardgrass (*Dactylis glomerata* L.) or regrowth of native warm-season grass after haying or grazing provides excellent early-season roosting cover. As windchills increase, denser cover is sought, and wildlife moves to unmowed and ungrazed or lightly stocked grasses and forbs and dogwood and plum (*Prunus* spp.) thickets. For bobwhite quail, protection from heavy snow and ice requires dense shrubs, osage orange (*Maclura pomifera* [Raf.] Schneid.) hedgerows, and windbreaks (Roseberry 1964; Wiseman and Lewis 1981) while prairie grouse resort to burrowing into snowdrifts in the grasslands (Bergerud and Gratson 1988).

Most grassland wildlife will use woody cover for escape and protection from severe cold. They also eat fruit, seeds, buds, and bark of woody plants. Thus, woody cover should be present, but it need only compose 2.5%-15.0% of an area for bobwhite quail (Wiseman 1977; Guthery 1980; Wiseman and Lewis 1981) and should be composed mostly of young brush in patches 200 m apart (Guthery 1987). Sharptail grouse is more tolerant of brush in grassland than prairie chicken, but it should still be at minimal levels for both species.

MANAGING FORAGES FOR WILDLIFE

Native versus Introduced Forages. Wildlife evolved with native plants and used them for cover and food. Problems wildlife have experienced with native forages are associated with cultural change. Changes from natural grazing and burning regimes and introduction of nonnative plants have reduced diversity in grasslands and allowed invasion of grasslands by woody species. This has contributed to the degradation of native habitats and decreases in the wildlife species they support.

Obtaining native species for planting has been a problem. Public and private facilities that select, evaluate, and develop species for reseeding have emphasized introduced species over native species. Introduced species have cheaper seed and are quicker to establish, more tolerant of poor grazing management, more responsive to high rates of nitrogen (N) fertilizer, and easier to propagate than native species. These practices should gradually change as a more environmentally aware public places emphasis on sustainable systems and questions the use of nonnative materials. Landowners are becoming more desirous of having wildlife on their land, recreating historic plant communities, and reducing dependence on fossil fuel products.

Plant materials centers and agriculture research stations need to thoroughly evaluate native forage species. Many forage producers will continue to require nonnative species for forage production. Introduced species or ecotypes that are compatible with wildlife and native plants should still be acceptable, but more study is needed to identify problems with introduced species and their control. Landowners need to manage native forages differently than most introduced forages and would benefit from educational programs. Native forages need periods of rest that were naturally provided by free-roaming native grazers. They also benefit from occasional burning since they evolved with fire. The absence of fire puts native species at a competitive disadvantage.

Introduced forage species range from beneficial to harmful, some having no apparent effect on wildlife. Wildlife became adapted to several introduced species when native habitats were reduced and may rely on the new species for food or cover more than on the restricted native species. Examples of introduced species that wildlife use for nesting cover, brood cover, or sources of insects, seeds, or green browse are Kentucky bluegrass, Canada bluegrass (*P. compressa* L.), redtop (*Agrostis gigantea* Roth), timothy, common lespedeza (*Kummerowia striata* [Thunb.] Schindler), korean lespedeza (*K. stipulacea* [Maxim.] Makino), kleingrass (*P. coloratum* L.), blue panicgrass (*P. antidotale* Retz), and

alfalfa. Introduced red clover (*Trifolium pratense* L.) and white clover (*T. repens* L.) are browsed by rabbits, deer, quail, and grouse and can produce many insects during the broodrearing period. However, these species often become too dense at ground level for easy movement and do not produce good structure for nesting. Birdsfoot trefoil (*Lotus corniculatus* L.) has replaced much of the red and white clovers for forage in the Northeast, but the value of birdsfoot for browse or nesting has not been studied. It can produce good nesting structure.

Introduced species that are not a problem for wildlife habitat in one region may be a serious threat in another. Smooth bromegrass provides mediocre nesting habitat and competes with more desirable species in the Northern Plains, but it is readily used for nesting by the prairie chicken in Illinois when ice and snow do not compact it to the ground (Buhnerkempe et al. 1984). Some introduced species may appear to be beneficial or not to be a problem initially, only to become a problem later.

Several introduced forages provide poor habitat, degrade habitat, or are not compatible with native species. 'King Ranch' bluestem (*Bothriochloa ischaemum* var. *songarica* [L.] Keng.), bermudagrass, and buffelgrass are serious problems in Texas (Guthery 1986a). All compete aggressively with native plants that are valuable to wildlife. King Ranch bluestem and bermudagrass form dense mats of vegetation that cover seeds and restrict wildlife movement. Buffelgrass develops heavy leaf litter and may produce allelochemicals that keep other seeds from germinating.

Since its introduction in the early 1900s, tall fescue (*Festuca arundinacea* Schreb.) has become one of the most common forage species in the Midwest and Southeast. Between 1950 and 1978, percentage of total forage seed harvested by Missouri farmers increased from 0% to 92% for tall fescue, while it decreased to less than 3% for lespedeza, red clover, timothy, and orchardgrass. Wildlife population declines have been attributed to land-use changes as to small diverse farms being shifted to vast acreages of corn, soybeans, and tall fescue pastures. Tall fescue pastures support substantially fewer wildlife species than true prairie pastures (Table 18.2) (Skinner et al. 1984). Exact reasons for wildlife declines in the presence of tall fescue have not been well documented, but the competitive grass limits the frequency and production of other plants in its presence. Rabbits eating tall fescue have developed stomach problems and died of stress and diarrhea (Andrews 1974; Anon. 1975; Crowe 1974). The problem may also be partly attributed to the fungal endophyte (*Acremonium coenophialum* Morgan-Jones & Gams) (Morgan-Jones and Gams 1982) in tall fescue.

Wildlife may be similarly adversely affected by eating endophyte-infected tall fescue leaves or seeds as is the case with livestock that exhibit elevated body temperatures, rapid breathing, rough hair coats, excessive salivation, lower average daily gain, and lower conception rates (Blaser et al. 1956; Jacobson et al. 1970). Rodent populations have been higher in endophyte-free than high-endophyte tall fescue (Pelton et al. 1991). Rats fed extracts from endophyte-infected seed gained less than those fed extracts of endophyte-free seed (Jackson et al. 1987).

Other problem forages include reed canarygrass (*Phalaris arundinacea* L.), Lehmann lovegrass (*Eragrostis lehmanniana* Nees), sericea lespedeza (*Lespedeza cuneata* [Dum.-Cours.] G. Don), kudzu (*Pueraria lobata* [Willd.] Ohwi), crownvetch (*Coronilla varia* L.), and Old World bluestems (*Bothriochloa* spp.). Like tall fescue, these highly aggressive introduced plants form monotypic stands that reduce or eliminate plant species diversity (Cable 1971; Cox and Ruyle 1986).

Common forage management of these species does not normally create good habitat. They are neither tolerant of close grazing nor do they have self-protective mechanisms such as tannins or alkaloids, which make livestock prefer not to eat them.

Heavy grazing pressure during the early part of the growing season, then resting, will encourage broadleaf plants, reduce ground cover, expose bare ground, and provide some food and overhead protection (Kirsch et al. 1978; Guthery 1986b, 1987). Native plants will often return following a change in grazing pressure, especially allowing a late-season rest. Prescribed burning in late winter or early spring will encourage native species, especially legumes like tickclovers (*Desmodium* spp.) and lespedezas. Sericea lespedeza vigorously responds to burning, so it will also require chemical control.

Heavily fertilized introduced forages make thick sods, allow less diversity in the plant mixtures, and make poor wildlife habitat. Not fertilizing forage growing near woody cover or retaining patches of native grasses, legumes, and low-growing shrubs, like coralberry and blackberry, will increase the wildlife potential of introduced forage pastures.

Another alternative is to kill 0.5-1.0 ha areas of the undesirable species near woody cover. Protected from grazing, these areas will produce wildlife foods and annual grasses for nesting for about 3 yr without further disturbance.

The best solution for managing forages for wildlife is to avoid planting problem species. Land areas that are already dominated by these species can be converted to more desirable forage species.

Mixtures versus Monocultures. Diversity is an important criterion for wildlife habitat. It provides greater variation in structure, a variety of seeds, browse and insects for food, and greater stability to the plant community. Though studies have shown a difference in nest densities among grass monocultures and a preference by certain wildlife species for specific monocultures (George et al. 1978),

TABLE 18.2.   The number of bird species and sightings on grazed prairie and grazed tall fescue plots in May-June 1977 and 1978 (plots were adjoining)

Bird Species	Grazed prairie M-J 1977	Grazed fescue M-J 1977	Grazed prairie M-J 1978	Grazed fescue M-J 1978
Upland sandpiper	29	1	17	0
Killdeer	4	0	1	0
Mourning dove	4	1	4	0
Northern flicker	2	0	0	0
Eastern kingbird	2	1	2	1
Horned lark	2	4	1	0
Browned-headed cowbird	16	2	0	0
Red-winged blackbird	0	6	0	0
Eastern meadowlark	37	18	50	34
Common grackle	5	2	0	0
Savannah sparrow	0	0	1	4
Grasshopper sparrow	18	21	49	9
Lark sparrow	3	0	0	0
Dickcissel	1	33	10	46
Total species	12	10	9	5
Total sightings	123	89	135	94

Source: Adapted from Skinner et al. (1984).

fields planted to a single species as part of the Conservation Reserve Program (CRP) have lower cover and nesting values for bird species than fields planted to mixtures (King 1991). Native warm-season grass mixtures provide greater long-term benefits to habitat quality than cool-season grasses and provide higher-quality habitat than introduced grasses (Farris and Cole 1981).

Forbs are important for seeds, insects (Panzer 1988), and songbird perches (Risser et al. 1981). Perennial forbs should be included in seeding mixtures. For example, ashy sunflower (*Helianthus mollis* Lam.) is well suited for wildlife and livestock forage in the true prairie region (Skinner et al. 1984).

Native warm-season grass mixtures are commonly planted for livestock forage and wildlife habitat in the central plains states. In the Northern Plains, mixtures of native grasses and introduced cool-season grasses and legumes are planted, while in the Southern Plains both introduced and native warm-season grasses are planted for livestock and wildlife. Monocultures of tall fescue, bermudagrass, buffelgrass, and other introduced grasses are routinely planted in the Northeast and Southeast. A cool-season grass may be combined with a legume, but rarely do eastern plantings contain more than two species.

Few forage agronomists question planting grass monocultures or binary mixtures of a single grass and a single legume instead of a broader mixture of grasses and legumes. The reasoning may be that livestock select the least mature and most palatable species, leaving the others to become overmature. Perhaps this is true with season-long and yearlong stocking using low-palatability species such as tall fescue or bermudagrass, but rotational stocking schemes could change this rationale.

Another reason for planting monocultures may be because forage scientists often study livestock performance on only one species at a time and are hesitant to speculate on animal performance or gain per hectare when species are combined. Scientists at western institutions, however, conduct research on rangeland containing hundreds of species and routinely recommend planting mixtures.

Considerable research has been conducted comparing animal performance and gain per hectare from adding legumes to cool-season grass monocultures. Some studies on compatability of legumes with native warm-season grasses have been published, but to date there is little information available on interrelationships between livestock production and wildlife response.

Grazing Management. Livestock grazing has been implicated in the devastation of thousands of acres of native grasslands and in the decline of prairie grouse, Montezuma quail (*Cyrtonyx montezumae* Vigors), masked bobwhite quail (*Colinius virginianus ridgwayi* Brewster), and the desert tortoise (*Gopherus agassizi* Cooper). Grazing is perceived as detrimental to riparian areas, watersheds, and wildlife habitat (Anderson et al. 1990). Supporting these perceptions are studies comparing no livestock grazing with overgrazing (Kirsch et al. 1978). However, grazing can also be a powerful force in managing wildlife habitat (Guthery 1987). Aldo Leopold (1933) recognized that "game [wildlife] can be restored by the creative use of the same tools which have heretofore destroyed it—axe, plow, cow, fire, and gun."

Livestock grazing is important for managing bobwhite quail in Texas, Oklahoma, and Kansas grasslands. Stocking for moderate to heavy grazing pressure removes grass cover that restricts brood movement and foraging for seeds and insects. It decreases competitiveness of the grass, increases broadleaf forb and legume seeds, green browse, and insects, and concentrates fall and winter bobwhite populations (Hammerquist-Wilson and Crawford 1981; Risser et al. 1981; Wiseman and Lewis 1981; Lehman 1984; Bareiss et al. 1986; Guthery 1986b; Lewis and Harshbarger 1986).

Stocking at light to moderate grazing pressure produces the best nesting structure and roosting cover by creating a mosaic of closely grazed and lightly grazed or ungrazed vegetation (Wiseman and Lewis 1981). However, range or pasture condition may deteriorate under this type of grazing as perennial forage species in closely grazed areas become weak and unproductive. Periodic burning and rest from grazing help equalize grazing pressure on key forage plants and also benefit wildlife (Hurst 1972; Wilson and Crawford 1979; Lewis and Harshbarger 1986). Stocking season-long and yearlong with heavy grazing pressure is detrimental to bobwhite quail, prairie grouse, waterfowl, and most other upland nesting birds (Kirsch et al. 1978; Risser et al. 1981; Barker et al. 1990).

Since the 1940s, a number of grazing systems have been developed to increase livestock production per unit area and restore forage species by providing rest from grazing during the growing season. Some systems, such as alternate stocking and intensive grazing with long rest periods, have improved plant vigor and diversity but have been stressful on livestock. Others, such as high-density stocking for short periods (1 to 2 d) followed by short rest periods (14 to 28 d), have produced good animal production but only moderate improvement in plant species diversity and vigor or have led to declining plant vigor.

Most studies of wildlife response to grazing systems have evaluated wildlife food and cover plants and fall wildlife populations but have not looked at the effects on nesting (Hammerquist-Wilson and Crawford 1981; Bareiss et al. 1986; Schulz and Guthery 1988; Wilkins and Swank 1992). The potential of nest loss due to trampling with short grazing periods at high stocking density (short duration grazing) has been studied using simulated ground nests (Bryant et al. 1982; Koerth et al. 1983; Bareiss et al. 1986; Jensen et al. 1990). Nest loss due to trampling was less under short duration grazing (9%) than under continuous stocking (low stock density) (15%) and is not a significant factor at stock densities up to 1.2 head/ha (Koerth et al. 1983). Nest losses are a significant concern at stock densities over 2.5 animal units/ha (head/ha not specified [Bareiss et al. 1986]), and nests are especially susceptible at stock densities greater than 10 head/ha (Jensen et al. 1990). The first 3 d of grazing are the most critical for nest survival. Stocking densities in drier environments are generally much lower than those that threaten nests but damage could occur frequently in the true prairie or tame pasture regions of the Midwest and East.

Several grazing systems produced more waterfowl than season-long stocking and hatched more ducklings per unit area than idle cover or season-long stocking in south central North Dakota (Barker et al. 1990). Calf gains among systems were not significantly different, but all systems were significantly more productive than season-long stocking. Grazing systems compared were short duration (5 d stocking/35 d rest), twice over (20 d stocking/60 d rest), stocking complementary forages (variable day stocking/variable day rest), and alternate stocking (20 d stocking/20 d rest).

A major benefit to wildlife of using grazing systems with rest periods over season-long stocking is the elimination of stocking on portions of a system until mid-June to mid-July. This period provides undisturbed nesting cover for wildlife in the Northern Plains but would be later than desirable farther south where incubation and peak hatching periods

for grassland wildlife occur from mid-May through mid-July (Rosene 1969; Horak 1985). Still, the advantages of a grazing system are the same as long as undisturbed nesting cover is provided during the nesting period.

Grazing systems studies do not address residual cover needed for nest initiation or whether certain grazing systems leave more residual cover than others. This may be because many grazing system studies have been in regions of the country where the primary cover is warm-season bunchgrasses rather than introduced cool-season grasses and legumes or sod-forming warm-season grasses. The latter are typically grazed much shorter than bunchgrasses. More information is needed on wildlife nesting in these introduced forages under rotation stocking.

Hay Management. Fields used for hay production can provide excellent nesting and broodrearing habitat. Alfalfa and alfalfa-grass mixtures are highly attractive nesting cover to ring-necked pheasant, eastern and western meadowlarks (*Sturnella* spp.), dickcissels (*Spiza americana* Gmelin), grasshopper sparrows (*Ammodramus savannarum* Gmelin), bobolinks (*Dolichonyx oryzivorus* L.), wild turkeys (*Meleagris gallopavo* L.), and waterfowl (Robbins et al. 1986; Warner and Etter 1989; Bollinger et al. 1990; Frawley and Best 1991). Unfortunately, for optimum production and quality these forages are harvested during peak periods of incubation for many species, which destroys nests, chicks, and adult birds (George et al. 1978; Bollinger et al. 1990; Frawley and Best 1991).

Nest mortality due to hay cropping has increased in the last 50 yr due to the development of earlier-maturing forages (Warner and Etter 1989; Bollinger et al. 1990) and a shift to alfalfa from timothy-clover and timothy-lespedeza hay, which mature later. The problem will probably get worse as agricultural research develops new cultivars of alfalfa and other species, which allow earlier mowing, potentially greater production, and additional cuttings.

Solutions to the problem are difficult. For most forage managers wanting to maximize forage yield and quality, there may not be a solution. For managers concerned about the impact of their agricultural operation on wildlife or wishing to share their land with wildlife, there are alternatives.

Delaying hay harvest until after July 1 allows many clutches to hatch, fledglings to fly, and broods to avoid hay equipment or to move out of the hayfield. Current cultivars of alfalfa and cool-season grass will be overmature by this time, but subsequent cuttings will be better quality. First cuttings of alfalfa are difficult to harvest in optimum condition due to wet weather in May and early June. Development of alfalfa cultivars that produce quality forage later in the season could benefit both forage producers and wildlife. Alfalfa hayfields are important enough for waterfowl production that researchers at the Northern Prairie Wildlife Research Station, Jamestown, North Dakota, are searching original alfalfa strains for later maturity (Arnold Kruse, personal communication).

Another alternative is using late-maturing cool-season grasses such as timothy, annual lespedezas, or native warm-season grasses for hay production. Forage quality is not as high as that of alfalfa but is usually adequate for wintering livestock and can be supplemented with small amounts of alfalfa or grain.

Alfalfa can also be planted with a compatible cool-season grass such as timothy, orchardgrass, smooth bromegrass, or intermediate wheatgrass. This will slightly delay alfalfa harvest and extend the productive life of the stand. Several wildlife species prefer grass-dominated alfalfa stands to a legume alone or legume-dominated stands (Bollinger et al. 1990).

Equally as important as time of cutting to avoid peak nesting is cutting early enough so that there is sufficient time for late summer and fall regrowth to provide nesting cover for the next year. Most wildlife build nests from growth remaining from the previous season. If residual cover is not present, nesting may be delayed until there is enough new growth to conceal the nest and incubating female, which increases their vulnerability to haying equipment.

Flushing bars may be mounted on the front ends of haying equipment to reduce incubating hen and brood mortality. Unfortunately nests will still be destroyed.

Escape cover in the form of low, dense shrubs, tall native grasses, or an unmowed strip left around the border of the hayed area will reduce losses to predators as a result of haying. Large hay bales should not be left sitting in hayfields to serve as predator perches or concealment.

NATIVE AND INTRODUCED WILD HERBIVORES

Mule deer (*Odocoileus hemionus* Rafinesque), white-tailed deer (*O. virginianus* Boddaert), and pronghorn antelope (*Antilocarpa americana* Ord) are primarily browsers

having diet preferences that are similar to some domestic livestock but different from cattle. Elk (*Cervus elaphus* L.) are primarily grazers and prefer many of the same grasses as cattle. All wild ungulates seem to have a taste for alfalfa and clover. Reducing damage or excessive loss of forage to wildlife when it is reserved for livestock may be a concern for some livestock producers. Reducing losses can only be effectively accomplished by removal of animals through legal hunting and/or damage control permits or through exclusion by protective fencing.

Managing forages for wildlife may be a more desirable and lucrative pursuit. Managing forage quality for wildlife is not much different than for livestock, i.e., keep it immature and abundant, but the forage species may be different (Bryant et al. 1979; Sowell et al. 1985). Cool-season agronomic grasses are eaten by many native and introduced herbivores to supplement their winter browse diet (Wiggers et al. 1984). If high-quality summer forages are not available, supplementation may be necessary to ensure peak lactation and breeding condition and to promote antler growth. Specific forage mixtures may need to be developed. For example, Schweitzer et al. (1993) recommends seeding a summer-fall forage mixture for the southern mixed prairie of awnless bush sunflower (*Simsia calva* Engeln. & Gray), Illinois bundleflower (*Desmanthus illinoensis* [Michx.] MacM.), and littleleaf lead-tree (*Leucaena retusa* Gray). A winter-spring mixture of subterranean clover (*T. subterraneum* L.), alfalfa, and sainfoin (*Onobrychis viciifolia* Scop.) is recommended. Four-wing saltbush (*Atriplex canescens* [Pursh] Nutt.) provides year-round browse.

Hunting leases and sale of breeding stock have provided significant income to farmers and ranchers (Smith 1986; Steinbach and Ramsey 1988; Nelle 1992). The number of game farms and ranches rearing native and exotic wildlife has increased dramatically in recent years. Competition for available forage among native and exotic wildlife and domestic livestock has sparked controversy and, in some cases, caused serious habitat destruction (Demarias et al. 1990). Exotic wildlife such as axis deer (*Cervus axis* Erxleben) and fallow deer (*Dama dama* [L.] Groves & Grub) may be able to more readily shift from browse to grass once browse has been depleted than can native species such as black-tailed deer (*Odocoileus hemionus hemionus* Rafinesque) and white-tailed deer (Elliott and Barrett 1985; Demarias et al. 1990). Studies of axis and fallow deer diets have shown variation from primarily grass (Smith 1977; Elliott and Barrett 1985) to primarily browse (Butts et al. 1982). Further research is needed on the extent of conflict and how to manage forage and cover to reduce the competition for resources (Demarias et al. 1990).

FEDERAL PROGRAMS AND WILDLIFE

Public Rangelands. The Great Plains were largely converted to cropland upon European settlement, resulting in a general reduction of native wildlife and an introduction of domestic livestock. A general transfer of the public domain to private ownership facilitated agricultural operations in the Great Plains, which continue to center on cropland where rainfall is sufficient, and livestock husbandry elsewhere.

In the intermountain basins and southwestern deserts making up most of the 11 western states, a different pattern emerged. Substantial portions of the land were too arid to support cultivated agriculture, and large blocks of rangeland have remained in public ownership. Historically western rangelands have served primarily to produce food and fiber from domestic livestock. In the late years of the nineteenth century and early years of the twentieth, a pattern of generally unrestrained exploitation developed in which natural resources, including wildlife, were severely affected (Little 1992). Beginning with passage of the Taylor Grazing Act in 1934, public rangelands came under general management of a permit and fee system. Ranchers holding patented private lands acquired the right to graze livestock on adjoining or nearby public lands. The kind of livestock, number of animals, and season of grazing are specified in the permit. The rancher pays a fee for animal unit months (AUMs) of actual use conducted under terms of the permit.

Today, because of wider knowledge of the characteristics of the range resource combined with improved management practices, livestock use generally takes place under a grazing system suited to the site, with shorter grazing seasons and periodic rest (Anderson et al. 1990). In addition, ranchers and public land managers often work together to further improve both public and private rangelands through interventions such as brush control, prescribed fire, reseeding, water development, and land treatments to conserve water and soil. For example, prescribed fire may be used to remove an overabundance of brush, followed by reseeding of adapted grasses and forbs. Thus greater site diversity may be achieved, forage production may be improved, and water and soil may be conserved. Such programs frequently benefit both wild and domestic herbivores (Frisina 1992). Greater opportunities for management of livestock are created, while wildlife may benefit from improved forage production (in both quantity and quality), expanded access to water, and additional "edge effect" for cover.

Better management has brought about improved conditions across most of the western rangelands. Although signs of early damage persist in many areas, and there are still instances of improper grazing management, most range managers agree that rangelands are in the best condition since the turn of the century (Box 1979).

That improvement is making rangelands today many times more productive than at any time since settlement, not only for livestock use but for wildlife habitat as well. The huge increases in populations of native wildlife that have occurred all across the West within the last few decades are in part a reflection of substantial improvement in rangeland condition and productivity.

Public rangelands, in addition to serving as habitat for wild and domestic herbivores, must also serve the uses of water production, mineral development, outdoor recreation, and natural beauty for an expanding human population. The complexity of uses are sometimes complementary, sometimes competitive, and sometimes conflicting, often accompanied by a great deal of controversy. The necessity that public rangelands must provide multiple outputs desired by society will demand a more cooperative and coordinated effort among the several interests involved (Cleary 1988; Box 1993). Through such efforts rangeland productivity will continue to improve for both native wildlife and domestic livestock.

Conservation Reserve Program. One of several goals of the Conservation Reserve Program (CRP), implemented in 1985, was to establish wildlife habitat on 17.4 million ha of cropland converted to permanent cover. This pleased urban taxpayers who funded most of the bill. Memories of wildlife abundance in the Soil Bank lands of the 1950s and 1960s (Schenck and Williamson 1991), an earlier federal land-use program, inspired dreams of repeating an era in which previously barren and eroded cropfields were restored to grasses, legumes, and trees for wildlife.

Options for planting native grasses and establishing wildlife habitat were included in the menu of acceptable permanent covers.

These practices appealed to landowners who wanted to "go back to nature" or who had wildlife interests. The choices of tame and native grasses also gave the farmer-livestock producer the opportunity of tailoring a seeding to fit gaps in a forage program after the 10-yr program expired, e.g., cool-season grasses in the warm-season grass-dominated plains and native warm-season grasses in the tame cool-season grass-dominated East. However, not allowing program participants to hay or graze these lands as a normal agronomic practice during the 10-yr period may have discouraged some participants from using the program to fill their forage gaps.

Hasty implementation of the program did not allow producers time to build up a seed supply, and simultaneous seeding over large areas quickly depleted existing supplies of grass and legume seed. Seed prices increased markedly. High price and scarcity of seed forced many participants to abandon plans to plant what they wanted in favor of species that were cheaper, more available, more open to cost sharing, and easier to plant. In the end, the Southern Plains, already dominated by warm-season grasses were planted to the more expensive native warm-season grasses.

The Northern Plains, already dominated by native cool-season grasses received more crested wheatgrass, and the East and Southeast, already dominated by introduced cool-season species, were planted to more tall fescue (Schenck and Williamson 1991). Further, because seed supplies were limited, most tall fescue was established with endophyte-infected seed.

In a Virginia survey, 72% of CRP contract holders indicated that they wanted to improve wildlife habitat on their retired lands, but 62% indicated that they had not been informed about ways to do so (Miller 1989).

Most early plantings were temporary boons to wildlife because of annual weeds and the lack of disturbance. Once established, native warm-season grasses supported more wildlife species and higher numbers than introduced cool-season grasses (King 1991). A few exceptional cool-season grass seedings resulted from broadcasting seed in strips that did not meet, thus creating a pattern of alternating strips of nesting cover (grass) and broodrearing and roosting cover (forbs or lespedeza) (Fig. 18.2) (Burger et al. 1990).

Plantings that do not meet the economic needs of the landowner will be returned to

Fig. 18.2.   Seeding patterns involving wide (15-m) strips of grass and narrow (5-m) strips of volunteer forb legumes such as lespedeza provided both nesting cover and brood-rearing cover in Conservation Reserve Program seedings.

cropland or other uses when the 10-yr program ends (Bjerke 1991; Laycock 1991). Native warm-season grasses that take longer to establish, but are less demanding of fertilizer, soil moisture, and mowing, generally have more wildlife and may be more likely to be continued by environmentally motivated landowners. US Fish and Wildlife officials in the Northern Plains are hoping to retain CRP land in grassland after contracts expire by demonstrating the income potential of these new grasslands using livestock grazing systems that also favor waterfowl and other wildlife production (Barker et al. 1990).

In future programs, USDA officials need to allow landowners time to make long-term management plans. Governmental programs cost shared with the landowner for conservation practices should result in long-term benefits for wildlife, not just a quick-fix vegetative cover. Programs need sufficient lead time to provide farmers the time and incentive to learn about, evaluate, and implement the more environmentally desirable choices. The public is sensitive to the impact of agriculture on soils, water, and wildlife and is not supportive of farming practices that adversely affect natural resources.

QUESTIONS

1. What types of grasses are most important for wildlife habitat and why? List several examples.
2. How may grazing be managed to benefit nesting wildlife?
3. Discuss why trees or increasing wood cover may be detrimental to grassland wildlife.
4. What impact can forage maturity have on wildlife, and how can it be avoided?
5. How have range management practices changed to improve public rangelands?
6. Describe several important characteristics of broodrearing cover.
7. List several introduced forage species that should not be planted near wildlife habitat, and discuss why not.
8. Describe management practices that may improve habitat qualities of these forages.
9. Describe why using a mixture of many forage species is better for wildlife than using monocultures.
10. How may grazing improve wildlife habitat?
11. How could the Conservation Reserve Program have been more effectively used to solve forage needs, and what factors may have kept it from being more effective?

REFERENCES

Anderson, WE, DL Franzen, and JE Melland. 1990. Rx grazing to benefit watershed-wildlife-livestock. Rangelands 12:105-11.

Andrews, ON. 1974. More on fescue. Letter to the editor. Hounds Hunt Mag. 71(4):8.

Anon. 1975. Fescue means fewer rabbits? Hounds Hunt Mag. 72(1):16.

Bareiss, LJ, P Schulz, and FS Guthery. 1986. Effects of short-duration and continuous grazing on bobwhite and wild turkey nesting. J. Range Manage. 39:259-60.

Barker, TB, KK Sedivec, TA Messmer, KF Higgins, and DR Hertel. 1990. Effects of specialized grazing systems on waterfowl production in southcentral North Dakota. Trans. North Am. Wildl. and Nat. Resour. Conf. 55:462-74.

Bergerud, AT, and MW Gratson. 1988. Adaptive Strategies and Population Ecology of Northern Grouse. Minneapolis: Univ. Minnesota Press.

Bjerke, K. 1991. An overview of the Agricultural Resources Conservation Program. In LA Joyce, JE Mitchell, and MD Skold (eds.), The Conservation Reserve—Yesterday, Today and Tomorrow, Symp. Proc. Gen. Tech. Rep. RM-203. Ft. Collins, Colo.: USDA Forest Service, Rocky Mountain Forest and Range Experiment Station, 7-10.

Blaser, RD, RC Hammes, Jr., HT Bryant, CM Kincaid, WH Skrdla, TH Taylor, and WL Griffith. 1956. The value of forage species and mixtures for fattening steers. Agron. J. 48:508-13.

Bollinger, EK, PB Bollinger, and TA Gavin. 1990. Effects of hay-cropping on eastern populations of the bobolink. Wildl. Soc. Bull. 18:142-50.

Box, TW. 1979. The American rangelands: Their condition and policy implications for management. In Rangeland Policies for the Future: Proceedings of a Symposium, USDA For. Serv. Gen. Tech. Rep. WO-17. Washington, D.C., 16-22.

———. 1993. On rewarding good stewards: A viewpoint. Rangelands 15:181-83.

Bryant, FC, MM Kothmann, and LB Merrill. 1979. Diets of sheep, Angora goats and white-tailed deer under excellent range conditions. J. Range Manage. 32:412-17.

Bryant, FC, FS Guthery, and UM Webb. 1982. Grazing management in Texas and its impact on selected wildlife. In JM Peek and PD Dalke (eds.), Symp. Wildl.-Livest. Relat., Univ. of Idaho, Moscow, 94-112.

Buhnerkempe, JE, WR Edwards, DR Vance, and RL Westemeier. 1984. Effects of residual vegetation on prairie-chicken nest placement and success. Wildl. Soc. Bull. 12:382-86.

Burger, LW, Jr., EW Kurzejeski, TV Dailey, and MR Ryan. 1990. Structural characteristics of vegetation in CRP fields in northern Missouri and their suitability as bobwhite habitat. Trans. North Am. Wildl. and Nat. Resour. Conf. 55:74-83.

Butts, GL, MJ Anderegg, WE Armstrong, ED Harmel, CW Ramsey, and SH Sorola. 1982. Food habits of five exotic ungulates on Kerr Wildlife Management Area. Tex. Tech. Ser. 30. Austin: Texas Parks and Wildlife Department.

Cable, DR. 1971. Lehmann lovegrass on the Santa Rita Experimental Range, 1937-1969. J. Range Manage. 24:17-21.

Christisen, DM. 1985. The Greater Prairie Chicken and Missouri's Land-Use Patterns. Terr. Ser. 15. Jefferson City: Missouri Department of Conservaton.

Cleary, CR. 1988. Coordinated resource management: A planning process that works. J. Soil Water Conserv. 43:138-39.

Cox, JR, and GB Ruyle. 1986. Influence of climate and edaphic factors on the distribution of *Eragrostis lehmanniana* (Nees) in Arizona, USA. J. Grassl. Soc. South Africa 1:25-29.

Crowe, A. 1974. Paradise lost. Hounds Hunt Mag. 71(7):31.

Demarias, S, DA Osborn, and JJ Jackley. 1990. Exotic big game: A controversial resource. Rangelands 12:121-25.

Elliott, HW, and RH Barrett. 1985 Dietary overlap among axis, fallow, and blacktailed deer and cattle. J. Range Manage. 38:435-39.

Farris, AL, and SH Cole. 1981. Strategies and goals for wildlife habitat restoration on agricultural lands. Trans. North Am. Wildl. and Nat. Resour. Conf. 46:130-36.

Frawley, BJ, and LB Best. 1991. Effects of mowing on breeding bird abundance and species composition in alfalfa fields. Wildl. Soc. Bull. 19:135-42.

Frisina, MR. 1992. Elk habitat use within a rest-rotation grazing system. Rangelands 14:93-96.

George, RR, AL Farris, and CC Schwartz. 1978. Native Prairie Grass Pastures as Nesting Habitat for Bobwhite Quail and Ring-necked Pheasant. Iowa Wildl. Res. Bull. 21. Des Moines: Iowa Conservation Commission.

Greenwood, RJ, AB Sargeant, DH Johnson, LM Cowardin, and TL Shaffer. 1987. Mallard nest success and recruitment in prairie Canada. Trans. North Am. Wildl. and Nat. Resour. Conf. 52:298-309.

Guthery, FS. 1980. Bobwhites and brush control. Rangelands 2:202-4.

———. 1986a. Bad Grass. The Kleberg Report. Quail Unlimited Mag. 5(3):12-13.

———. 1986b. Beef, Brush and Bobwhites: Quail Management in Cattle Country. Texas A&I Univ., Kingsville. Ceasar Kleberg Wildlife Research Institute Press.

———. 1987. Perfect grazing for bobwhite. Quail Unlimited Mag. 6(5):16-50.

Hammerquist-Wilson, MM, and JA Crawford. 1981. Response of bobwhites to cover changes within three grazing systems. J. Range Manage. 34:213-15.

Horak, GJ. 1985. Kansas Prairie Chickens. Wildl. Bull. 3. Pratt: Kansas Fish and Game Commission.

Hurst, GA. 1972. Insects and bobwhite quail brood habitat management. In JA Morrison and JC Lewis (eds.), Proc. 1st Natl. Bobwhite Quail Symp., Oklahoma State Univ., Stillwater, 65-82.

Jackson, AJ, RW Hemken, LP Bush, JA Boling, MR Siegal, and PM Zavos. 1987. Physiological responses in rats fed extracts of endophyte infected tall fescue seed. Drug Chem. Toxicol. 10(3 and 4):369-79.

Jackson, AS. 1969. A Handbook for Bobwhite Quail Management in the West Texas Rolling Plains. Tex. Parks and Wildl. Bull. 48.

Jacobson, DR, SB Carr, RH Hatton, RC Buckner, AP Graden, DR Dowden, and WM Miller. 1970. Growth, physiological responses and evidence of toxicity in yearling dairy cattle grazing different grasses. J. Dairy Sci. 53:575.

Jensen, HP, D Rollins, and RL Gillen. 1990. Effects of cattle stock density on trampling loss of simu-

lated ground nests. Wildl. Soc. Bull. 18:71-74.

Johnsgard, PA. 1979. Birds of the Great Plains. Lincoln: Univ. of Nebraska Press.

Johnson, RG, and SA Temple. 1990. Nest predation and brood parasitism of tallgrass prairie birds. J. Wildl. Manage. 54:106-11.

King, JW. 1991. Effects of the Conservation Reserve Program on selected wildlife populations in southeast Nebraska. MS thesis, Univ. of Nebraska, Lincoln.

Kirsch, LM. 1983. Historical ecological records of great plains grassland workshop: Management of the public lands in the northern great plains. Bismark, N.Dak.

Kirsch, LM, and AD Kruse. 1973. Prairie fires and wildlife. In Proc. Annu. Tall Timbers Fire Ecol. Conf. 12:289-303.

Kirsch, LM, HF Duebbert, and AD Kruse. 1978. Grazing and haying effects on habitats of upland nesting birds. Trans. North Am. Wildl. and Nat. Resour. Conf. 43:486-97.

Klimstra, WC, and VC Ziccardi. 1963. Night-roosting habitat of bobwhites. J. Wildl. Manage. 27:202-14.

Knopf, FL. 1988. Conservation of steppe birds in North America. ICBP Tech. Publ. 7:27-41.

Knopf, FL, and TE Olson. 1984. Changing landscapes and the cosmopolitism of the eastern Colorado avifauna. Wildl. Soc. Bull. 14:132-42.

Kobriger, GB. 1965. Status, movements, habitats and foods of prairie grouse on a sandhills refuge. J. Wildl. Manage. 29:788-800.

Koerth, BH, WM Webb, FC Bryant, and FS Guthery. 1983. Cattle trampling of simulated ground nests under short duration and continuous grazing. J. Range Manage. 36:385-86.

Laycock, WA. 1991. The Conservation Reserve Program—How did we get where we are and where do we go from here? In LA Joyce, JE Mitchell, and MD Skold (eds.), The Conservation Reserve—Yesterday, Today and Tomorrow, Symp. Proc. Gen. Tech. Rep. RM-203. Ft. Collins, Colo.: USDA and Forest Service, Rocky Mountain Forest Range Experiment Station, 1-6.

Lehmann, VW. 1984. Bobwhites in the Rio Grande Plain of Texas. College Station: Texas A&M Univ. Press.

Leopold, A. 1931. Report on a Game Survey of the North Central States. Madison, Wis.: Sporting Arms Ammunition Institute.

———. 1933. Game Management. New York: Scribners.

Lewis, CE, and TJ Harshbarger. 1986. Burning and grazing effects on bobwhite foods in the southeastern coastal plain. Wildl. Soc. Bull. 14:455-59.

Little, JA. 1992. Historical livestock grazing perspective. Rangelands 14:88-90.

Marlin, JC. 1967. The Grasslands of North America: Prolegomena to Its History with Addenda and Post Scripts. Gloucester, Mass.: Peter Smith.

Miller, EJ. 1989. Wildlife management on Virginia Conservation Reserve Program land: The farmers' view. MS thesis, Virginia Polytechnic Institute and State Univ., Blacksburg.

Morgan-Jones, G, and W Gams. 1982. Notes on Hyphomycetes. XLI. Mycotaxon. 15:311.

Nelle, S. 1992. Exotics—at home on the range in

Texas. Rangelands 14:77-80.

Panzer, R. 1988. Managing prairie remnants for insect conservation. Nat. Areas 8(2):83-90.

Pelton, MR, HA Fribourg, JW Laundre, and Reynolds. 1991. Preliminary assessment of small wild mammal populations in tall fescue habitats. Tenn. Farm and Home Sci. 160:68-71.

Risser, PG, EC Birney, HD Blocker, SW May, WJ Parton, and JA Wiens. 1981. The True Prairie Ecosystem. Stroudsburg, Pa.: Hutchinson Ross Publishing Company.

Robbins, CS, D Bystark, and PH Geissler. 1986. The Breeding Bird Survey: Its First Fifteen Years, 1965-1979. US Fish and Wildl. Serv. Res. Publ. 157.

Roe, FG. 1951. North American Buffalo. Toronto, Canada: Univ. of Toronto Press.

Roseberry, JL. 1964. Some responses of bobwhites to snow cover in southern Illinois. J. Wildl. Manage. 28:244-49.

Rosene, W. 1969. The Bobwhite Quail, Its Life and Management. Hartwell, Ga.: Sun Press.

Sauer, CO. 1920. The Geography of the Ozark Highland of Missouri. Geogr. Soc. Chicago Bull. 7. Chicago, Ill.: Univ. of Chicago Press.

Schenck, EW, and LL Williamson. 1991. Conservation Reserve effects on wildlife and recreation. In LA Joyce, JE Mitchell, and MD Skold (eds.), The Conservation Reserve—Yesterday, Today and Tomorrow, Symp. Proc. Gen. Tech. Rep. RM-203. Ft. Collins, Colo.: USDA and Forest Service, Rocky Mountain Forest and Range Experiment Station, 37-42.

Schulz, PA, and FS Guthery. 1988. Effects of short duration grazing on northern bobwhites: A pilot study. Wildl. Soc. Bull. 16:18-24.

Schweitzer, SH, FC Bryant, and DB Webster. 1993. Potential forage for deer in the southern mixed prairie. J. Range Manage. 46:70-75.

Scott, TT, and WD Klimstra. 1954. Report on a visit to quail management areas in the southeastern United States. Ill. Wildl. 9:5-9.

Sharp, W. 1957. Social and range dominance in gallinaceous birds—pheasants and prairie grouse. J. Wildl. Manage. 21:242-44.

Skinner, RM, TS Baskett, and MD Blendon. 1984. Bird Habitat on Missouri Prairies. Terr. Ser. 14. Jefferson City: Missouri Department of Conservation.

Smith, JC. 1977. Food habits. In ED Ables (ed.), The Axis Deer in Texas. College Station: Texas A&M Univ. Press, 62-74.

Smith, M. 1986. Wildlife in Texas: Probable and Profitable. Cattleman 72:88-102.

Sowell, BF, BH Koerth, and FC Bryant. 1985. Seasonal nutrient estimates of mule deer diets in the Texas Panhandle. J. Range Manage. 38:163-67.

Sparling, DW, Jr. 1979. Reproductive isolating mechanism and communication in greater prairie chickens (*Tympanuchus cupido*) and sharptail grouse (*Pedioecetes phasianellus*). PhD thesis, Univ. of North Dakota, Grand Forks.

Steinbach, DW, and CW Ramsey. 1988. The Texas "Lease System." History and future. In D Rollins (ed.), Recreation on Rangelands: Promise, Problems, Projections, Symp. Proc. Soc. Range Manage., Corpus Christi, Tex., 54-68.

Toll, PA. 1977. Results of a controlled burn study on Kingman Game Management Area. Unpublished study for Kansas Fish and Game Commission, Pratt.

Townsend, JK. 1839. In Early Western Travels. Cleveland, Ohio: Arthur H. Clark.

Vance, DR, and RL Westemeier. 1979. Interactions of pheasants and prairie chickens in Illinois. Wildl. Soc. Bull. 7:221-25.

Warner, RE, and SL Etter. 1989. Hay cutting and the survival of pheasants: A long-term perspective. J. Wildl. Manage. 53:455-61.

Weaver, JE. 1954. North American Prairie. Lincoln, Nebr.: Johnsen Publishing Company.

Wiggers, EP, DD Wilcox, and FC Bryant. 1984. Cultivated cereal grains as supplemental forages for mule deer in the Texas Panhandle. Wildl. Soc. Bull. 12:240-45.

Wilkins, RN, and WG Swank. 1992. Bobwhite habitat use under short-duration and deferred-rotation grazing. J. Range Manage. 45:549-53.

Wilson, MM, and JA Crawford. 1979. Response of bobwhites to controlled burning in south Texas. Wildl. Soc. Bull. 7:53-56.

Wiseman, DS. 1977. Habitat use, food habits and response to birddog field trials of bobwhite on northeastern Oklahoma tallgrass prairie rangeland. MS thesis, Oklahoma State Univ., Stillwater.

Wiseman, DS, and JC Lewis. 1981. Bobwhite use of habitat in tallgrass rangeland. Wildl. Soc. Bull. 9:248-55.

Yoho, NS, and RW Dimmick. 1972. Habitat utilization by bobwhite quail during winter. In JA Morrison and JC Lewis (eds.), Proc. 1st Natl. Bobwhite Quail Symp., Oklahoma State Univ., Stillwater, 90-99.

PART **3**

Forages
for
Livestock

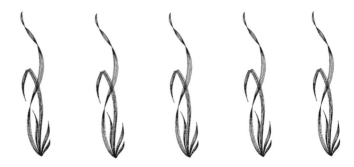

19

Forages for Beef Cattle

J. P. FONTENOT
*Virginia Polytechnic Institute and
State University*

LOWELL L. WILSON
Pennsylvania State University

VIVIEN G. ALLEN
*Virginia Polytechnic Institute and
State University*

BEEF CATTLE INDUSTRY
Demand for Beef

PER capita beef consumption in the US has decreased from 43 kg per person in 1976 to an average consumption of 31 kg per person in 1991 (USDA 1992). However, despite decreased consumption of beef as a food product, the beef industry is extremely important as per capita consumption of beef in the US ranks fourth among all countries in the world. In 1991 consumers spent an average of about $195/person on beef, more than the amount spent for pork and chicken combined (AMI 1991; NCA 1992). The total retail value of fresh beef sales in 1991 was $49.3 billion. Beef is the largest volume item sold in grocery stores. Beef accounted for 53% of the 1991 "center-of-the-plate" restaurant menu sales. Relative to average disposable income, food expenditures continue a declining trend

J. P. FONTENOT is John W. Hancock, Jr., Professor of Animal Sciences, Virginia Polytechnic Institute and State University. He received the MS and PhD degrees from Oklahoma State University. Dr. Fontenot has conducted research on forage utilization by ruminants, underutilized by-products as feedstuffs, mineral metabolism and utilization, and sustainable forage.

LOWELL L. WILSON is Professor of Animal Science, Pennsylvania State University. He holds the MS and PhD degrees from South Dakota State University. Dr. Wilson has conducted beef and sheep forage utilization research.

VIVIEN G. ALLEN is Professor of Agronomy, Virginia Polytechnic Institute and State University. She received the MS and PhD degrees from Louisiana State University. Her research involves development and testing of forage-livestock systems and integrating forage-livestock systems into sustainable, whole-farm systems.

that has been occurring for 30 yr. In 1961 and 1981 Americans spent 17.1% and 13.1%, respectively, of their income on food compared with 11.5%, or $486 billion, in 1991.

It has been demonstrated that every dollar of farm-level cattle sales generates an additional $5 to $6 of business activity in the farm supply, food, and other related businesses. Therefore, cattle sales generate $220 to $250 billion in business activity annually in the US (NCA 1992; USDA 1992). Exports of beef and beef products are also important to the US economy. Exports of all cattle products, including meat and nonmeat commodities such as hides and other coproducts, reached a record level of $3.83 billion in 1991 (USMEF 1991). Imports of meat and nonmeat commodities amounted to $3.07 billion in 1991, leaving a positive balance of trade worth $760 million. In addition to the actual value of cattle products exported, there were $218 million in value of corn (*Zea mays* L.) and $28 million in value of soybeans (*Glycine max* [L.] Merr.) exported through beef and veal. There is potential for further increases in exports of beef and beef products. Positive export balances of commodities such as beef benefit practically all segments of US society.

As indicated by the preceding figures, cattle production and processing provide many jobs for US citizens. Approximately one-half of the people working in production agriculture are directly involved in the beef industry (NCA 1992). The meat-packing and -processing industries employ nearly 250,000 workers with an annual payroll of more than $4 billion.

The preceding economic characteristics of the beef industry are impressive but even more impressive when one considers the nature of the feed resources utilized by beef cattle and the alternatives for use of these resources. Since beef cattle are ruminants, they have the ability not only to utilize forages produced from high-quality land resources but also to directly harvest forages of lower quality from land with no or few alternatives for other crops. These characteristics of feed supply and dedicated land resources suggest that the beef industry will continue to be an important part of US and world agriculture.

Changes in Cattle Numbers. As anticipated from the decreased per capita consumption of beef, there has been a decrease in numbers of cattle. Since beef is a perishable commodity, and only small quantities are exported, essentially all beef produced is consumed within a few months of slaughter. As with any commodity having a cyclic supply, there have been similar lengths of time between price peaks. The supply of beef peaks approximately every 8-12 yr. As indicated in Table 19.1 (USDA 1992), numbers of cows and heifers 2 yr of age and over that have calved decreased 14.3%, and cattle slaughtered decreased 9.1% since 1982. An important criterion reflecting enhanced relative productivity since 1982 is the fact that the nation's cow herd decreased more than did the number of cattle slaughtered. Some of this difference was due to the increased number of male calves of dairy origin fed to conventional beef slaughter weights, but it is also a reflection of increased beef production per cow. Another factor has been the increased average carcass weight from 295 kg in 1986 to 312 kg in 1991. Expressed on a per cow basis, 1991 production was 244 kg of carcass weight compared with 220 kg in 1985 and 204 kg in 1980. According to the National Cattleman's Association (NCA) (1992) beef production from 100 million cows in 1991 was equivalent to the production of 120 million cows during the 1970s.

Nature and Location of Industry Segments. There are three segments of the beef cattle industry that are fairly well-defined: cow-calf, stocker/backgrounding, and finishing. In contrast to several years ago, rarely are all three segments located on the same farm. Instead, the industry has become specialized, depending on the land type, rainfall, topography, source of forage, established markets, and personal interest of the farmer.

COW-CALF. The cow-calf segment is responsible for the cow herd and production of calves through about 7 mo of age, when the calves are weaned. The resulting calves enter either of the next two segments. The average size of the cow herds in the US is only 37 head (NCA 1992). According to the 1987 *Census of Agriculture* (USDC 1987) about 64% of all cow herds in the US had from 1 to 49 head and 28% had 55 to 199 head. Only 1.6% of all herds had 500 head or more; therefore, 98% of all cow-calf producers are small-to moderate-sized single family units.

Beef cow herds are distributed throughout the US with over 31 states having 10,000 or more beef cattle producers. Some of the more important cow-calf states are Alabama, California, Florida, Georgia, Iowa, Kansas, Kentucky, Missouri, Montana, Nebraska, Oklahoma, South Dakota, Tennessee, and Texas. In many of these states, the main reason for the cow herd is to utilize crop aftermath and low-quality forages that have little utility for other purposes. The primary objective is to produce a calf each year that can be marketed. Typically, calves weigh 225 to 300 kg at weaning, but the actual weight can be higher with better management.

BACKGROUNDING/STOCKER. Typical stocker programs involve growing cattle from 250 kg to about 400 kg on a moderate energy density diet for a rate of gain less than that obtained in a feedlot situation. The objective is for the calf to grow without a great emphasis on finishing. Stocker systems are based heavily on forages with minimal or no grain feeding. In the Western wheat (*Triticum aestivum* L. emend. Thell.) belt, stocker cattle graze developing wheat crops in the fall and/or spring. Other typical growing regimens for stocker cattle may be hay-crop silages, hay, or grazing of the corn remaining in fields after grain harvest; the latter is usually supplemented with

TABLE 19.1.   Numbers of cattle of different classes (1000 head)

Class	1982	1986	1991
Cows and heifers calved	39,230	33,633	33,620
Cattle slaughtered	36,158	37,568	32,885
Cattle, 227 kg and over	15,497	15,967	16,548
Calves, 227 kg and less	28,777	24,431	18,720

Source: USDA (1992).

high-quality forages or limited concentrate.

Over the past few years there has been a lower percentage of cattle from the nation's cow herds going through a stocker phase, with more cattle going directly into the feedlot as calves (NCA 1992). However, the stocker phase is still economically significant to producers in areas where forage and land resources may be most efficiently used. Approximately 90% of the feed used for the typical stocker system comes from forages with less than 10% from feed grains.

Backgrounding is a specialized management step between the cow-calf operation and the stocker or finisher operations. A backgrounder typically receives the lighter weight calves (about 150 kg) produced within the cow herd segment. These calves may be lighter weight because of late birth dates, inadequate milk production of individual cows, internal parasite infestation, or mismanagement of the calf crop. These calves typically are placed on a moderate forage, moderate grain diet for periods of 60 to 120 d or until the average weight of the group is between 250 and 300 kg. At that point the calves enter either the stocker or feedlot segment. Backgrounding programs are very specialized and generally involve more intense husbandry and nutritional practices than either of the other postweaning systems.

FINISHING. The finishing systems are the last phase prior to slaughter and processing of the meat product. Interestingly, average slaughter age of cattle has decreased during the past 20 yr, but carcass weight has increased. Slaughter at a younger age has been due to improved genetics and nutrition and enhanced use of businesslike approaches to the cattle industry. The type of finishing ration and the length of the finishing period vary between geographical areas and between feedlots in the same area. For example, there has been a tendency for decreased energy density, reduced proportion of grain, and less time on the finishing ration. Corn silage is the primary forage used in beef finishing programs. Approximately one-half of corn silage is grain and the other one-half is composed of stalks, cobs, and leaves on a dry matter (DM) basis. The latter portion is an important forage component of the finishing ration.

In some areas there has been a renewed interest and practice of finishing cattle on pastures with minimal or no grain. This approach decreases carcass fat deposition, which is of interest to the entire beef industry

since fat represents a by-product that is decreasing rapidly in relative market value. Fat has also been cited as one of the reasons for a diminished perception of beef as a component of a healthy diet for humans. Therefore, the public is demanding more red meat but less fat. However, surveys of beef consumers repeatedly rank eating pleasure as one of the primary reasons for eating beef, and beef of acceptable quality must be produced.

The cattle-feeding segment has shifted from farmer-feeders in the Corn Belt to large western and southwestern feedlots. The slaughter and meat-packing industry has tended to follow the relocation of the feeding segment.

Importance/Dependence on Forages. A primary concern of society is the increasing price and diminishing supply of fossil fuels and the potential environmental degradation from intensive farming systems. As stated earlier, beef cattle are able to utilize forages or other relatively high-fiber feed products or coproducts. Because of the nature of forage production, the cost of forage is less closely related to fossil fuel price than is grain production cost. When grain prices are low because of high grain production, grain is competitive with forages for supplementation of beef cattle rations. However, the grain use pattern for cow herd and stocker systems is erratic, and the beef industry will continue to rely upon forages as the basic feed resource.

Considering the life cycle of beef production, approximately 85% of all feed units are from forages (Hodgson 1978). Forage and beef production will continue to be important in all geographical areas of the US.

ANIMAL CHARACTERISTICS

Life Cycle Production: Measures of Performance. The diversity of US climate and land types dictate that a variety of pasture and range types exist. However, in every segment of the beef industry there are definite biological and economical characteristics that must be measured, recorded, and evaluated. Profit margins in any segment of the cattle industry are narrow and erratic. The basic purpose of each segment must be profit, and the characteristics measured must be reflective of potential profit. Discussed in the following are some of the more important measures of performance.

REPRODUCTION. One of the primary determinants of profit in a cow herd is reproductive efficiency. The most direct measurement is the percentage of cows exposed at breeding time that produce calves at weaning time (7 to 8 mo old). This may vary from 70% for herds receiving a low level of management (which returns a low level of profit) to 95% or more. In turn, one of the primary determinants of weaned calf percentage and calf growth rate is adequate energy intake by the cow from a month before calving to 90 d postcalving. The effects of energy intake during the rebreeding period, about 30 to 90 d after calving, are critical. Rebreeding should be timed to coincide with periods when forage supplies are highest with regard to both quantity and quality. Naturally, the desired calving/rebreeding season depends on the geographical area but is usually in early spring in northern regions and early fall in southern regions of the US.

WEIGHT. Weaning weight of calves is moderately to highly correlated with milk production of the cow, which in turn is determined by genetics of the herd and forage quality and quantity. Milk production is particularly important during the first 90 d of calf age, after which time the calf becomes more dependent on direct consumption of forages. Profitable weaning weights are determined mainly by (in order of priority): quality and quantity of forages for direct consumption by the calf, milk production of the dam through 3 mo of lactation, inherent growth rate of the calf, genetics of the cow, and herd health management practices.

Milk production level may be antagonistic with reproductive efficiency, i.e., high-milk-producing cows may have difficulty conceiving. However, this antagonism can be overcome by providing sufficient quantities of high-quality forage.

GROWTH. Growth of calves is important since cattle are sold primarily by weight. However, there is some differentiation based on feeder calf and slaughter/carcass quality grades. Growth requires acceptable genetic potential although either nutrition or genetics can place a ceiling on realized growth. There is a desirable correlation between rate and efficiency of gain; that is, animals that gain more rapidly because of genetics and/or nutritional regimen will require less feed per unit of weight gain. Therefore, genetic selection for growth also improves feed efficiency and the economics of raising cattle for each segment.

There tends to be genetic antagonism between growth rate and fat deposition: faster-gaining cattle tend to deposit body fat less rapidly. Fat deposition is a major determinant of carcass quality grades, which in turn determine price for the slaughter animal. However, as consumer demand for leaner beef continues to influence the direction of the beef industry, this antagonism is becoming less of a concern. Most beef breeds have shifted to types that gain more rapidly and efficiently, deposit moderate amounts of body fat, and have moderate muscle thickness and mature weights but retain desirable maternal characteristics.

FINISHING. Performance measures of primary importance in the finishing segment are rate of gain and feed efficiency. If feed efficiency is high, then cost of gain tends to be lower. Naturally, the cost of the finishing diet is an important consideration and is determined by feed availability and management proficiency. Although grain is usually competitively priced with forages, some forage in the diet usually enhances profitability and efficiency. Although there is a tendency toward leaner beef production, there continues to be a market for a variety of different carcass qualities. The beef industry is fortunate to have a diversity of desired market endpoints and consumer tastes; hence, it tends to allow flexibility in both cattle and diet types.

Nutrient Requirements. The nutrients required by all classes of beef cattle are protein, minerals and vitamins, and those that supply energy, namely carbohydrates, fats, and protein. The required quantities of the different nutrients vary depending on the function and size of the animal. For example, nutrient requirements will be much higher for a lactating beef cow than a dry pregnant cow. Likewise, level of performance will affect nutrient requirements in stocker or finishing cattle. The quantitative nutrient requirements for different classes and weights of beef cattle are given by the National Research Council (NRC) (1984).

ENERGY. Energy is required for all animal metabolic processes and deposition of internal organ and muscle protein and deposition of internal and external fat. Measurements of gross energy in feedstuffs are not very meaningful since none of the losses is considered. Energy is lost in the feces, urine, combustible gases, and heat production. The energy requirement may be expressed as digestible, metabolizable, or net energy. Energy requirement and energy value of feeds are sometimes expressed as total digestibile nutrients

(TDN), which closely approximate digestible energy.

In many instances the energy required can be supplied by forage alone, depending on the class of animal and quality of the forage. For example, most kinds of hay would supply sufficient energy for dry pregnant beef cows, and high-quality hay would provide sufficient energy for lactating beef cows. However, hay would not supply sufficient energy for finishing cattle. On the other hand, corn silage would supply more energy than pregnant or lactating cows would require if the silage were fed as the major part of the ration.

PROTEIN. Although cattle have a physiological requirement for amino acids, they do not have a dietary requirement for any amino acid because the rumen microorganisms digest the forage protein and synthesize large amounts of all essential amino acids. However, protein solubility or degradability in the rumen is important (NRC 1985), especially for cattle in high production. Because microbial protein is of lower quality than some feed protein, rumen bypass of some high-quality protein will improve efficiency of nitrogen (N) utilization. Heating of forages during drying may reduce protein degradation in the rumen (Britton et al. 1983). This may be beneficial, but overheating may render the protein too undegradable after it leaves the rumen and may be of limited value to cattle.

Cattle can utilize nonprotein N (NPN) such as urea, ammonium salts, and uric acid to meet part of the protein requirement. Rumen microorganisms use NPN to synthesize microbial protein, which is then digested by cattle. At early stages of forage maturity, NPN may make up to 35% of the total N. Some readily digested carbohydrate is essential for optimum N utilization.

For many classes of cattle some high-quality forages may supply sufficient or excess protein to meet the requirement. However, many forages, including crop residues, corn silage, and some grasses, are too low in protein to meet the protein requirement of most classes of beef cattle. Protein supplements, which are usually relatively costly, must be fed for optimal animal performance. Using higher-quality forage is an economic alternative.

MINERALS. Certain inorganic elements are required by cattle. The required minerals are classified as either macro- or micro- (or trace) minerals. Macrominerals that are required are calcium (Ca), phosphorus (P), magnesium (Mg), potassium (K), chlorine (Cl), sodium (Na), and sulfur (S). Required microminerals include iron (Fe), copper (Cu), cobalt (Co), iodine (I), manganese (Mn), zinc (Zn), selenium (Se), and molybdenum (Mo). (See Chap. 4, Vol. 1.) The macrominerals are required and are present in the animal body in larger amounts than the microminerals. General functions of minerals in animals are structural, acid-base balance, and participation in enzymatic reactions.

VITAMINS. Vitamins are organic nutrients that are required by all livestock in very small amounts. They serve a variety of functions associated with enzymatic reactions. Vitamin A is not found in plants, but its precursor, carotene, found in green plant materials, can be converted to vitamin A by the animal. Carotene can be destroyed in cut plants due to solar radiation or heating during hay and silage making. Losses may be severe if the cut forage stays on the field too long, such as during rainy periods. Carotene also is lost during storage of hay.

Although vitamin D is required by ruminants, it is seldom necessary to provide vitamin D as a supplement. It is abundant in sun-cured hays. Animal requirements also can usually be met by nominal exposure to direct sunlight. Vitamin E is a dietary essential, but usually natural feedstuffs contain sufficient amounts.

In ruminants with fully functional digestive tracts, i.e., at approximately 6 wk of age, B-complex vitamins and vitamin K are synthesized in the rumen in amounts sufficient to meet the physiological requirements of the animal. However, the disease, polioencephalomalacia, which occurs in cattle fed high-concentrate rations, has responded to large intravenous doses of vitamin B_1 (thiamine) (Brent 1976). Polioencephalomalacia appears to be due to thiamine destruction by thiaminase in cattle fed these rations. No such condition has been reported in grazing animals or those fed high-forage rations.

Nutrient Utilization. In order for the nutrient requirements of cattle to be met, the nutrients in the forages must be utilized. Fecal excretion of undigested material represents the major loss of most nutrients. For example, the fecal loss in energy may be 30% for very high-quality corn silage and 70% for rye (*Secale cereale* L.) straw. Thus, digestibility or absorption of nutrients is of utmost importance. Digestibility refers to intake minus fecal loss. A number of factors affect digestibility of energy. Deficiencies of certain nutrients such as protein and S may lower digestibility of energy, and high fiber level, especially lignification, will lower digestibility. Feeding liberal amounts of readily digested carbohydrate sources such as grain or molasses may lower fiber digestibility of the forage.

Minimizing fecal losses is also very important for protein and minerals. In some forages protein is not highly digestible, perhaps due to association with indigestible fiber. Low absorption of minerals is a frequent cause of mineral deficiencies. A major problem with hypomagnesemia, for example, is the presence of antiquality factors such as high concentrations of K and fat, which interfere with Mg absorption from the rumen. Absorption of some of the trace minerals is adversely affected by the presence of certain other minerals. For example, high quantities of Ca reduce uptake of P and some microminerals.

Utilization of nutrients after absorption is important, also. Some forms of N may be absorbed efficiently but not utilized efficiently. The S in S-fertilized corn silage seems to be more efficiently utilized than inorganic S (Buttrey et al. 1986). Some antiquality factors in forages may interfere with utilization of other nutrients. For example, high aluminium (Al) does not affect absorption of Ca and Mg, but interferes with utilization of these nutrients after absorption (Allen et al. 1991).

FORAGES TO MEET NUTRITIONAL NEEDS

Forage is broadly defined as edible parts of plants, other than separated grain, that can provide feed for grazing animals or that can be harvested for feeding (FGTC 1992). Forage plants may be annuals, which complete their life cycle in 1 yr, biennials, which require 2 yr, or perennials, which persist for longer periods of time.

Pasture. Forages include a broad range of grasses and legumes and root crops such as turnip (*Brassica rapa* L.), kale (*B. oleracea* L.), and rape (*B. napus* L.). Many crop residues are used as forage. Browse includes leaf and twig growth of shrubs, woody vines, trees, cacti, and other nonherbaceous vegetation available for animal consumption. Herbaceous broadleaf weeds and forbs can also contribute to the diet of grazing animals.

Kinds of Forage. By far, the most important forages are the grasses and legumes. Grasses are broadly classified as cool-season or warm-season species. Cool-season grasses produce most of their growth early in the growing sea-

son, are generally higher in quality, but have a lower potential for growth compared with warm-season species. Warm-season grasses are generally most productive during the mid-part of the growing season and are more tolerant to drought and high temperatures than cool-season grasses.

In the lower South, forage systems are usually based on the perennial warm-season grasses, bermudagrass (*Cynodon dactylon* [L.] Pers.), bahiagrass (*Paspalum notatum* Flugge), and dallisgrass (*P. dilatatum* Poir.). Use of bermudagrass is important across the lower South of the US, extending into California. In the upper South and at higher elevations, the cool-season perennials, tall fescue (*Festuca arundinacea* Schreb.), orchardgrass (*Dactylis glomerata* L.), and Kentucky bluegrass (*Poa pratensis* L.), form the basis of forage systems.

Extending up through the Northeast, bluegrass remains important while timothy (*Phleum pratense* L.) becomes a primary forage for hay production, especially as tall fescue and orchardgrass become less well adapted. Smooth bromegrass (*Bromus inermis* Leyss.) and reed canarygrass (*Phalaris arundinacea* L.) are widely grown across the north central section of the US, with the wheatgrasses (*Agropyron* spp.) increasing in importance as precipitation declines in the more arid West.

Throughout the tall-grass region of the Great Plains, native warm-season grasses occur including switchgrass (*Panicum virgatum* L.), big bluestem (*Andropogon gerardii* Vitman), buffalograss (*Buchloe dactyloides* [Nutt.] Engelm.), indiangrass (*Sorgastrum nutans* [L.] Nash), the gramagrasses (*Bouteloua* spp.), and others. The introduction of Old World bluestems (*Bothriochloa* spp.) has proven useful in the more southern part of this region and have been used successfully in systems in Oklahoma. These species become undesirable as they invade the Flint Hills region in Kansas and compete with native grass species. From East to West, this region generally experiences decreasing precipitation, and grass species shift from the tall grasses of the eastern prairie to the shortgrass, bunch-type species of the West.

Forage systems in the Pacific Coast area may include a variety of the cool-season species found in the East. Under irrigation, many of these species are used even in the more arid inland areas. In most areas, opportunities to use annual grass species exist. These can include sorghums, millets, small grains, or species such as annual ryegrass

(*Lolium multiflorum* Lam.).

Differences in growth potential and quality of the various forage types offer opportunities to combine these into systems to meet needs of animals at different seasons of the year. The species used and the combinations that can help to provide uniform seasonal forage distribution will vary, however, with geographic location and climatic effects.

Legumes, when present, generally increase quality of the forage. Compared with grasses, legumes are generally higher in concentrations of protein, Ca, and Mg as well as certain microminerals. Legumes are also generally higher in rate of passage and digestibility even though they are higher in concentrations of lignin compared with grasses. Animal performance can usually be expected to be higher when the diet includes a legume rather than being all grass. Legumes are also important due to their capacity for N fixation in association with rhizobia bacteria. Legume-fixed N reduces the need for purchased N. There is also generally less opportunity for legume-fixed N to become an environmental pollutant compared with fertilizer-applied N.

Legumes are generally included in forage-beef systems in combination with grasses, but exceptions occur. Alfalfa (*Medicago sativa* L.) grown in monoculture as part of a system offers flexible use for grazing, stored feed, value for N fixation in rotations with crops, and high cash value if excess forage is produced by the system.

Stored Forages. Land area used for producing stored forages is low compared with that used for grazing (Hodgson 1978). However, stored forages are important in beef cattle production systems. Stored forages include hay, silage, and crop residues.

HAY. Hay is forage harvested and preserved by sun drying or mechanical dehydration. In mechanical dehydration more of the nutrients in the crop are retained. In the case of legume forages, there is less shedding of leaves than if processed by sun drying. Dehydration is usually used to process hay from high-value forages such as alfalfa, especially in irrigated areas. Sun drying costs much less, and the hay is higher in vitamin D value than dehydrated hay. However, the risk of losing a hay crop or nutrients due to weather conditions such as unexpected rainfall is much higher for sun drying.

SILAGE. Silage is forage preserved by controlled fermentation of a high-moisture crop under anaerobic conditions (McDonald 1981). The forage is preserved by production of acids, especially lactic acid, which lowers the pH, usually to below 5.0. Well-fermented silage is stable indefinitely if maintained under anaerobic conditions. Optimum moisture level is important in silage making. In order for the forage to ferment satisfactorily to produce good quality silage, a sufficient concentration of readily fermentable carbohydrate is essential.

If the forage is too high in moisture, wilting to reduce the moisture level will improve the quality of the silage. If the forage crop to be ensiled is too low in water-soluble carbohydrates, adding a readily fermentable carbohydrate such as molasses is helpful.

Forage quality varies for both hay and silage, due to the crop and stage of maturity at which the forage is harvested (Table 19.2). Generally, legumes are higher in crude protein and Ca than grasses. For most forages crude protein and energy value decrease as stage of maturity increases. On the other hand, DM yield usually increases as stage of maturity increases. Thus, in deciding on the optimum stage of maturity, the producer must compromise yield and quality to produce forage that will optimize economic efficiency and meet the nutrient needs of the animals.

TABLE 19.2. Nutritional value of some forages for beef cattle

Forage	Dry matter	Crude protein	TDN	Calcium	Phosphorus
			(%)		
Alfalfa hay, sun-cured, early bloom	90	18.0	60.0	1.41	0.22
Red clover hay, sun-cured	89	16.0	55.0	1.64	0.36
Bermudagrass hay, Coastal, sun-cured	90	6.0	49.0	0.43	0.20
Orchardgrass hay, sun-cured, late bloom	91	8.4	54.0	0.26	0.30
Corn silage	33	7.0	8.1	0.23	0.22
Corn stover	85	6.6	50.0	0.57	0.10
Wheat straw	89	3.6	41.0	0.18	0.05

Source: NRC (1984).

CROP RESIDUES. When grain is harvested, 50% or more of the DM remains as the fibrous residue. The crop residues consist of straw for oat (*Avena sativa* L.), wheat, barley (*Hordeum vulgare* L.), and rice (*Oryza sativa* L.) and stover for corn and sorghum (*Sorghum bicolor* [L.] Moench). These residues are low in protein and energy value. Some variation exists in nutritional value among the crop residues (Table 19.2). These are valuable resources that are used extensively in feeding beef cattle. The value of the crop residues will decrease if they are not harvested soon after the grain is harvested.

The value of the crop residues can be enhanced by alkali treatment (Klopfenstein 1981). Treating with sodium hydroxide or a combination of sodium and calcium hydroxide increases the energy value by 10% or more. Similar responses have been seen from treatment with ammonia. In addition, ammonia treatment results in substantial increases in crude protein in the feed. Ammonia treatment can be achieved by treating crop residues with urea, also. The enzyme urease, usually found in crop residues, hydrolyzes the urea, releasing ammonia for treatment of the crop residues (Chap. 14).

Supplementing Forages. The need for supplementing forages will depend on the nutrient requirements of the class of beef cattle and the nutrient composition of the forages. Supplementation may be needed to supply additional protein, energy, minerals, and/or vitamins.

PROTEIN. Some forages are too low in crude protein to meet the requirements of any class of cattle. For example, forages such as corn silage and crop residues such as cereal straw or stover will need to be supplemented with protein. Some high-protein forages, such as legume hay or silage are high in protein and will not need to be supplemented unless the quantities of these feeds in the ration are low. For all forages, the decision concerning supplementation should be based on the requirement of the class of cattle, the protein content of the forage, and the amount of protein supplied by other feeds in the diet.

Oilseed meals from soybeans, cotton (*Gossipium hirsutum* L.), and other crops have been the traditional feeds used to supplement protein to cattle. High-protein cereal-milling by-products such as gluten feeds and by-products from the brewing and distilling industries are good protein sources also. Selection of the protein supplement should be based on price per unit of protein.

Urea, a nonprotein N source, may be used to replace part of the protein in the ration. When fed to livestock with other feeds that supply readily digested carbohydrates, such as grain or molasses, and sufficient minerals, urea is utilized efficiently by cattle. For optimum utilization of the N, no more than one-third of the total N in the ration should be supplied by urea.

Poultry litter, a by-product of the poultry industry, may be used as protein supplement. The crude protein content will usually be 25% to 30%, DM basis (Fontenot 1991). About 50% of the N in broiler litter is in a form other than protein, i.e., uric acid, ammonia, urea, or other (Bhattacharya and Fontenot 1966), but the nonprotein forms of N are efficiently utilized. The litter must be processed to destroy potential pathogens, to stabilize the product, and to enhance palatability. Satisfactory methods of processing are ensiling, deep stacking, or dehydration. As a precaution to avoid medicinal drug residues from the poultry litter being present in the processed beef, the litter must be withdrawn 15 d prior to slaughter. If the price of litter is competitive, relative to other supplemental protein sources, it can be fed to beef cows and stocker and finishing cattle.

ENERGY. Although forages comprise the main feed supply for beef cattle, a more concentrated source of energy may need to be supplemented to achieve certain levels of production. With a high-energy forage such as corn silage, energy supplementation will not be necessary except perhaps for finishing cattle. Yearling cattle initially weighing 300 kg were finished to low choice on a ration of high-quality corn silage and protein supplement (Hammes et al. 1968). For lighter, younger cattle some grain should also be fed, probably at 1% of body weight.

Energy supplementation will be necessary when low-energy hay crop forages are fed to cattle in moderate to high production. Crop residues will need to be supplemented with energy for all classes of cattle except dry beef cows, especially during the early stages of gestation. The level supplemented will be determined by the requirement of the class of cattle and the available energy concentration of the forage. Low-energy forages may be supplemented with higher-energy forages such as corn silage. Usually, concentrates such as cereal grains, by-product feedstuffs, and molasses are used to supplement energy. The choice of energy supplements should be based on the cost per unit of available energy.

MINERALS. Salt (NaCl) should be supplemented to all classes of beef cattle fed any type of forages. Some other minerals may need to be supplemented, depending on the concentrations in the forage and requirement of the cattle. Calcium may need to be supplemented, especially if the stored feed does not contain legumes. Phosphorus will need to be supplemented to forages grown in areas where forages are deficient in P. Generally, legumes are higher in P than grasses. Usually, Mg will need to be supplemented to beef cows during early lactation to prevent grass tetany (Fontenot 1979).

Trace minerals may need to be supplemented in areas in which forages are usually deficient in a given trace mineral. Selenium needs to be supplemented in many areas due to deficiency in forages. Caution should be exercised in supplementing with minerals, since toxicity may result if levels are excessive (NRC 1980).

VITAMINS. If cattle are fed liberal amounts of green, well-preserved forage, usually there is no need for vitamin supplementation. For cattle fed limited forage or hay or silage that was not well preserved, vitamin A should be supplemented. Green forages usually contain large quantities of carotene, a precursor of vitamin A. Conversely, crop residues should be supplemented with vitamin A. The source of vitamin A supplementation are high-potency vitamin A supplements, which are available at modest cost. Vitamin A can be administered by injection if oral supplementation is not practical. Animals in total confinement without access to sunlight that are fed dehydrated forage or silage should be supplemented with vitamin D.

FORAGE/LIVESTOCK SYSTEMS

Systems must allocate quality of forage appropriate to meet various animal requirements. The nutritional needs of the animals must be matched as closely as possible to the potential for forage production. Systems must also have sufficient elasticity to accommodate varying environmental, market, and production conditions. Conservation of excess forage should be planned to provide for needs during forage deficits but should be matched closely to expected needs unless this excess has a viable cash crop potential and is a planned part of the total system. Successful grazing systems must be sustainable environmentally, economically, and in terms of labor demands. These systems integrate soil, plant, and animal characteristics with local environment and management opportunities to result in a profitable, sustainable means of producing the desired plant and animal products (Chap. 14, Vol. 1). Altering stocking rates, fertilizer rates, choice of plants and animals, and grazing methods can have great effects on productivity and profitability of beef-forage systems.

Cow-Calf Systems. Profitable cow-calf systems depend on maximizing use of grazed forages to reduce or eliminate needs for feeding either supplements or conserved forages. The systems should be planned to cover a 12-mo period. Cows need high nutrition during early lactation and rebreeding. The breeding season should be adjusted so that high-quality forages are available to meet the cows' needs during this period. Thus, time of calving is one of the most important decisions in designing cow-calf/forage systems. Choice of calving time is dependent on potential forage growth, marketing options, climate, soil types, and individual management needs. Where there is little potential for forage growth during winter in most regions in the US, spring calving is generally preferred in order to reduce feed needs for the herd during the wintering period. Since most of these calves are weaned and sold in the autumn, oversupplies often exist, resulting in lower calf prices. In the humid South, autumn calving offers opportunities to use dormant perennial warm-season grass pastures that have been overseeded with annuals. Autumn calving in this area can result in higher calf survival rates, heavier weaning weights, and higher prices received for calves weaned in midsummer (Bagley 1988).

The relationship of milk production to calf gain is highest during the first 60 d of the calf's life (Neville 1962). Once the calf reaches 3-4 mo of age, forage quality and quantity for the calf must be emphasized while the cow can profitably use lower-quality forages (Blaser et al. 1986). Grazing methods such as creep grazing or forward creep grazing (a specific type of creep grazing) can help provide high quality and quantity of forage to calves, but response of calves to creep grazing occurs largely after the calves are at least 150 d old (Chap. 13). Furthermore, stocking rates must be sufficiently high to enforce high utilization by cows of the base forage to achieve the appropriate allocation of forage quality and quantity between cows and calves. Calves can creep graze into enclosed areas of the base pasture, adjoining hayfields, conservation strips, firebreaks, or cropping areas during times that these areas are suitable for grazing.

Forages for cow-calf systems should usually be based on perennials that are persistent, are easily managed, and allow flexibility if preserved forages are to be harvested. Use of legumes can extend grazing seasons, aid in increasing calf gains, improve quality of hay or silage, and can help to prevent certain nutritional disorders. The choice of grazing methods should be compatible with the management requirements of the forage base.

In the Lower South, cow-calf systems are usually based on warm-season perennials. Systems based on either 'Alicia' bermudagrass or 'Pensacola' bahiagrass were compared in south Mississippi (St. Louis et al. 1990). Calving season was during February and March with weaning in October. The stocking rate was about 2.5 cow-calf pairs/ha. Each system was equally divided for three-paddock rotational stocking. Creep gates were installed so that calves had access to all three paddocks. Each October, two-thirds of each system was overseeded with 'Marshall' ryegrass. Cows were fed hay, previously harvested when excess forage growth occurred, and protein supplement from 60-d prepartum. Postpartum cows ration-grazed ryegrass when available. As growth increased, cows and calves were continuously stocked on the ryegrass.

The system using bermudagrass produced more hay than the bahiagrass system, but hay feeding requirements were similar between these two systems, largely due to their similar winter management. Quantity of hay harvested from the bahiagrass system was consistantly less than the amount of hay needed for feeding, thus resulting in a need to buy hay each year. Calves from the bahiagrass system had a 9.5 kg heavier weaning weight than those from the bermudagrass system, but this advantage was offset by the additional cost of purchasing hay using the prevailing prices. With the stocking rates, fertilizer rates, and management used, returns per cow/ha were slightly higher for the bermudagrass system.

Allotting 17% of a bermudagrass-based system to double cropping of wheat/ryegrass with millet increased net returns ($) per cow and weaning weights of calves, compared with either an all-bermudagrass system or

systems with a greater percentage of area allotted to the double-cropped annuals (34% or 50%) according to Bagley et al. (1987). Stocking rate was constant at about 2.5 mature animals/ha regardless of area allocated to the double-cropped annuals. Creep-grazing calves also increased weaning weights by 8% over noncreep-grazed calves.

The Middleburg Three-Paddock Grazing System is based on cool-season forages (Fig. 19.1; Blaser et al. 1977; Allen et al. 1992b). Beef cows (1.4/ha) grazed in paddock A during spring, summer, and early autumn. Calves creep grazed into paddocks B and C, which were harvested for hay or grazed by cows whenever there was excess forage. Calves always creep grazed into areas not accessible to cows. In early August, areas B and C were either mowed for hay or were grazed and N was applied. Forage was accumulated ("stockpiled") during late summer and early autumn for late autumn and winter grazing by cows. Calves were weaned in mid-October, and grazing of stockpiled forage began about November 1. Cows were supplemented with hay previously cut from these paddocks, if forage became limited.

Forage species used within this system influenced carrying capacity, amount of hay fed, profitability, and other measures of success (Table 19.3). Weaning weights of calves were similar and averaged 250 kg regardless of forage species since creep grazing allowed free selection of high quality and quantity of forage (Allen et al. 1992b). Harvested tall fescue-red clover (*Trifolium pratense* L.) closely matched hay feeding needs while orchardgrass-red clover did not provide sufficient forage and, hence, required use of purchased

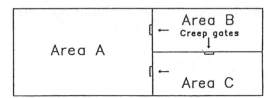

Fig. 19.1. The Middleburg Three-Paddock System for year-round grass-legume and cow-calf management (Blaser et al. 1977). Area *A* composed 55% of the total area and was used for grazing by cows during spring, summer, and early autumn. Areas *B* and *C* composed 45% of the total area, were divided equally, were used for hay and creep grazing by calves, and were stockpiled during autumn for grazing by cows during late autumn and winter (Allen et al. 1992b).

TABLE 19.3.   Influence of three grass-legume mixtures for creep grazing, hay, and stockpiling for winter grazing by cows based on calf performance and requirements for hay harvesting and feeding

Item	Tall fescue-red clover	Orchardgrass	
		Red clover	Alfalfa
Weaning weight, kg	250	251	249
Hay harvested, kg/system	8840	6327	11,338
Hay fed, kg/system	5168	8488	9168
Days hay was fed	73	129	134

Source: Adapted from Allen et al. 1992b.
Note: Cattle grazed either bluegrass-white clover or tall fescue-ladino clover during spring, summer, and early autumn.

hay. Red clover required reseeding almost annually in both systems. Orchardgrass stands declined and required chemical weed control while tall fescue persisted and remained relatively weed free. Combining orchardgrass with alfalfa increased hay yields, compared with red clover, and resulted in a surplus of valuable hay, but it required more mechanical harvesting and more fertilizer compared with the other two systems. Thus, forage species influenced inputs of machinery, energy, seed, chemicals, and labor.

Tall fescue is a major perennial cool-season grass over much of the eastern US. However, its value has been often limited by presence of the endophyte fungus *Acremonium coenophialum* Morgan-Jones & Gams. Creep grazing may be a method to partially offset the loss of gains often associated with infected fescue. Researchers in Missouri found that calf gains were lower when calves were reared on highly (77%) infected tall fescue, compared with gains from calves reared on fescue with a low (21%) level of infection (Tucker et al. 1989). Calves on infected tall fescue that creep grazed into a mixture of birdsfoot trefoil (*Lotus corniculatus* L.), Kentucky bluegrass, and timothy had gains similar to non-creep grazed calves from the low-infected fescue.

Stocker Systems. As with the cow-calf phase, the primary feed should be forage, with the objective to maximize grazing and minimize harvested forage feeding. In the upper South and parts of the Northeast, calves weaned in autumn can utilize stockpiled tall fescue for late autumn and winter grazing, and in some years, little or no hay feeding is required. Combining alfalfa with tall fescue for stockpiling eliminates the need for N fertilizer, improves forage quality, but reduces the length of the grazing season due to more rapid deterioration of alfalfa in winter, compared with tall fescue (Allen et al. 1992a).

Systems can be based on all cool-season forages with excess forage harvested during the early part of the growing season (Fontenot et al. 1993). Systems combining a paddock of a cool-season perennial with a separate paddock of a warm-season perennial can provide winter grazing plus additional forage for grazing during midsummer. Steers stocked at three steers/ha produced 554 kg of gain/ha in a system designed in Virginia (Fig. 19.2). Forty percent of the system area was tall fescue grazed during autumn, winter, and early spring. Steers grazed caucasian bluestem (*B. caucasica* [Trin.] C.E. Hubb.) from late June through late August. Bluestem comprised 30% of the total area. The remaining 30% of

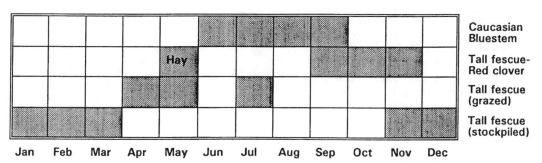

Fig. 19.2.   Growth distribution of tall fescue, fescue plus red clover, caucasian bluestem, and stockpiled tall fescue. Conceived to achieve desirable seasonal distribution of forages for sequence grazing by stocker steers (Fontenot et al. 1993).

the area was tall fescue-red clover, harvested in May for hay and then grazed after growth of bluestem declined in late summer, until cattle were sold in October. The tall fescue-red clover provided hay for winter supplement to stockpiled tall fescue and provided grazing during the time tall fescue in the larger paddock was being stockpiled for the next group of cattle.

In the lower South, autumn-weaned calves make excellent gains during winter on cool-season annuals such as ryegrass or small grains. These annuals, in combination with bermudagrass to provide grazing during the summer, form the basis of many successful systems. Adding rye and ladino clover (*T. repens* L.) to ryegrass increased total forage production and forage quality (Bagley et al. 1988). Steers gained 1.0 kg/d when clover was included, compared with 0.93 kg/d on N-fertilized ryegrass alone. The inclusion of rye increased steer grazing days, particularly during midwinter.

As the stocker period progresses, cattle are growing, and forage consumption increases. Thus, the system must provide for an increasing forage need either by increased forage production, increased area, or decreased animal numbers. Systems using combinations of light and heavy stockers can be designed so that all cattle are present during the peak forage growth period. Heavier animals are marketed as forage growth declines, increasing forage available to lighter animals that are kept longer.

Increased cattle numbers during the early part of the growing season forms the basis for the Intensive Early Stocking system developed for the native bluestem range in the Flint Hills of Kansas (Smith and Owensby 1978). Doubling and even tripling the stocking rate over the traditionally recommended season-long stocking rate of 0.7 steer (200 to 225 kg)/ha, and discontinuing grazing after the first half of the growing season (July), increased efficiency of converting the native range forages into animal gains (Owensby et al. 1988). The increased stocking rates did not reduce individual steer gain; thus, gain/ha was increased linearly. Cattle removed in July can be placed directly into the feedlot. Herbage production appeared to be sustained with all stocking rates tested, but at the highest stocking rates there was a shift in botanical composition indicated by a decrease in percentage indiangrass and an increase in Kentucky bluegrass, an undesirable species in this ecosystem.

Systems can be designed also by altering the size of the grazed area. This can be accomplished by means of a temporary fence that restricts the grazing area during periods of rapid forage growth and expands the grazing area when growth rate of forage declines and/or forage consumption increases. Size of the grazing area can be varied also through use of rotational stocking where paddocks not needed for grazing can be harvested for hay.

Finishing Systems. Several methods for finishing cattle to acceptable weights and grades have been devised to maximize use of forages. These include (1) all-forage grazing systems, (2) forage grazing supplemented by grain supplementation or followed by a short feeding period, and (3) high-forage rations.

The main deterrent to the use of forages for finishing cattle is the low energy content of the ration (Fig. 19.3). For example, metabolizable energy required for 1.1 kg/d gain for finishing a medium frame, 454 kg steer is 2.66 Mcal/kg DM (NRC 1984). The metabolizable energy contained per kg DM in fresh, late vegetative alfalfa or early bloom orchardgrass hay is only 2.28 and 2.35 Mcal, respectively; hence, neither one would meet the energy requirement. Corn, wheat, and sorghum grains contain 3.25, 3.18, and 3.04 Mcal/kg DM, respectively.

In actively growing, well-fertilized forages, ration components other than energy general-

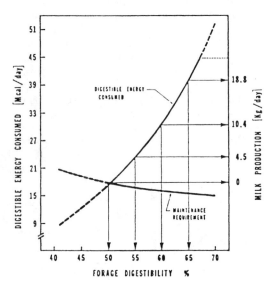

Fig. 19.3. Effect of forage digestibility on digestible energy consumption and weight gain of a 300-kg growing animal (from Bula et al. 1981).

ly are not limiting for animal production. Therefore, increasing the use of forages in the ration usually results in lengthening the time required to reach a given weight and grade and results in an older animal at slaughter. Increasing the energy intake by using forages higher in energy value or energy supplements results in faster animal gains and younger slaughter age.

ALL-FORAGE GRAZING SYSTEMS. In much of the southern US it is possible to finish beef to acceptable grades on forage alone (Stuedemann et al. 1977). Forage finishing systems are more uncertain than feedlot systems. Users of all-forage systems should consider the kinds of forage available and the types of animals to be finished. Most successful systems include overseeding with annuals for high-quality winter or summer grazing, proper sequencing of paddocks based on different forage species, the use of legumes, grazing management to maximize intake of highly digestible forage, and selection of early-maturing animals noted for their marbling characteristics.

The type of forage best suited to forage finishing varies with the environment. Throughout the lower South slaughter beef forage-feeding systems can be based on use of annual species of grasses and legumes. Winter annuals such as annual ryegrass with crimson (*T. incarnatum* L.) or arrowleaf clover (*T. vesiculosum* Savi) provide high-quality pastures well suited to forage finishing. Addition of a cereal forage(s) can help to extend grazing seasons or be targeted to provide high-quality forage at specific times. No one forage combination is consistently superior because of differences in response to varying environmental and management factors. Combinations of winter and summer annuals have supported grass fattening systems in the lower South.

When perennial grasses are incorporated into the system, a short grain feeding period may be required after gains are achieved of about 400 kg on pasture (Blaser et al. 1956; Hoveland and Anthony 1977). In the Northeast, grazing of Kentucky bluegrass-white clover (*T. repens* L.) combined with tall fescue-red clover or orchardgrass-alfalfa worked well (Blaser et al. 1976). Early-season growth of tall fescue-clover pastures is grazed first when quality is highest, but cattle are soon rotated to Kentucky bluegrass-white clover pastures to avoid overmaturity and loss of quality of the bluegrass. If there is excess growth of forage, it should be allowed to occur

in the tall fescue paddocks that are cut for hay. As growth is progressively depressed during hot summer weather, cattle alternately graze both bluegrass and tall fescue paddocks. If orchardgrass-alfalfa is substituted for tall fescue-red clover, quality and quantity of forage can usually be improved, but orchardgrass is of limited value for winter grazing, compared with tall fescue.

Providing high-quality forages for grazing by finishing animals can be enhanced by several different systems of management. First-last grazing versus whole-plant grazing allows animals being fattened (first grazers) to selectively graze the most nutritious, highly digestible plant parts (Blaser et al. 1976). Animals with lower nutrient requirements (last grazers) are used to graze the residues. Either rotational stocking or continuous stocking may provide higher-quality feed, depending upon forage species, season of the year, and forage growth (Chap. 13). Sequence grazing uses two or more land units in succession that differ in forage species composition. This takes advantage of differences among forages in their season of growth and/or quality potential to better match forage quality and quantity with the animals' requirements over the season.

In all-forage finishing programs, there are difficult problems yet to be understood and solved. Uniformity and dependability of product are lacking, and marketing is therefore difficult. All-grass-fed beef also may be characterized by a darker red muscle color, a yellowing of the fat, and a higher percentage of volatile components associated with lower meat palatability and flavor (Hendrick et al. 1982).

LIMITED GRAIN. Feeding forage plus supplemental grain is a common, dependable, and frequently profitable alternative to a grain-based finishing ration. Since all-forage finishing systems require extended periods of high-quality forages available for grazing, opportunities are restricted to regions where these forages, often annuals, can be successfully grown. A much larger part of the US is adapted to perennial forages, where finishing can be accomplished by supplementing a limited amount of grain. It also should be noted that many forage finishing systems become "grain-on-grass" systems under adverse climatic conditions.

Addition of a highly available energy source to forage rations generally improves the rate of animal gain. If grain is fed to grazing animals, the utilization of the forage declines, a phenomenon referred to as "associative effects." However, feeding grain at 1% of body weight to grazing animals results in an increased rate of gain while continuing to place major emphasis on pasture utilization.

Cool-season perennial grasses usually form the basis for "limit grain feeding" on pastures of the upper South, with grain serving to increase the intake of available energy as grass quality decreases late in the growing season (Spooner and Ray 1977). In other parts of the South, the basic forage is more likely to be a warm-season perennial grass such as bermudagrass, bahiagrass, or elephantgrass (*Pennisetum purpureum* Schumach.) (Horn and Taliaferro 1977; Hoveland and Anthony 1977; Monson and Utley 1977). Finishing systems designed in Colorado demonstrated that highly acceptable beef could be produced with minimum supplementation (Cook et al. 1984). Yearling cattle grazed native range plus crested wheatgrass (*A. cristatum* [L.] Gaertn.) during spring. Forage sorghums were grazed during summer and autumn with no grain supplement. Calves were fed in drylot during a final 66-d finishing period. Over 80% of the carcasses graded Select or better.

HIGH-FORAGE RATIONS. Success of high-forage rations depends largely on the energy and protein values of forages. Corn silage is recognized widely as a high-energy forage and is generally the standard to which other silages are compared. Cattle can be finished on high-energy corn silage diets supplemented only with protein (Hammes et al. 1964, 1968). Prior to 1959 corn silage was regarded as a forage to be liberally supplemented with grain for finishing cattle. The introduction of hybrids with a high grain percentage has resulted in production of corn silage that is similar in energy value to some traditional grain-based finishing rations.

If the energy value of the silage is low, a longer feeding period becomes necessary to attain the same weight and carcass grade. Small-grain, sorghum, or hay-crop silages have been tested as the primary forages for finishing cattle. When such silages were fed in combination with corn silage (50:50, dry basis) or were supplemented with ground ear corn at 1% body weight, daily gains of over 1 kg were obtained (Hammes et al. 1968). Daily gains for cattle fed annual ryegrass silage were maximized by including 40% barley grain (dry basis) in the ration (Petchey and Broadbent 1980).

QUESTIONS

1. Describe the three main segments of the beef cattle industry.
2. Explain the importance of forage quality for weaned calf percentage and weaning weight.
3. Explain why cattle do not have a dietary requirement for specific amino acids.
4. Discuss the vitamin requirements of beef cattle.
5. List at least three forage species that are important on pastures in the lower South, Northeast, and Great Plains.
6. What are the primary considerations in designing a 12-mo forage system?
7. Give an example of a successful cow-calf forage system and a stocker system.

REFERENCES

Allen, VG, JP Fontenot, and SH Rahnema. 1991. Influence of aluminum-citrate and citric acid on tissue mineral composition in wether sheep. J. Anim. Sci. 69:792-800.

Allen, VG, JP Fontenot, and A Brock. 1992a. Forage systems for stocker cattle. Va. Agric. Exp. Stn. Anim. Sci. Res. Rep. 10:101-2.

Allen, VG, JP Fontenot, DR Notter, and RC Hammes, Jr. 1992b. Forage systems for beef production from conception to slaughter. I. Cow-calf production. J. Anim. Sci. 70:576-87.

American Meat Institute (AMI). 1991. Meat Facts. Washington, D.C.: AMI.

Bagley, CP. 1988. Twelve-month grazing systems for the humid southeast. In Proc. Am. Forage and Grassl. Conf., Am. Forage and Grassl. Counc., Belleville, Pa., 375-83.

Bagley, CP, JC Carpenter, Jr., JI Feazel, FG Hembry, DC Huffman, and KL Koonce. 1987. Effects of forage system on beef cow-calf productivity. J. Anim. Sci. 64:678-86.

Bagley, CP, JI Feazel, and KL Koonce. 1988. Cool-season annual forage mixtures for grazing beef steers. J. Prod. Agric. 1:149-52.

Bhattacharya, AN, and JP Fontenot. 1966. Protein and energy value of peanut hull and wood shaving poultry litters. J. Anim. Sci. 25:367-71.

Blaser, RE, RC Hammes, Jr., HT Bryant, CM Kincaid, WH Skrdla, TH Taylor, and WH Griffith. 1956. The value of forage species and mixtures for fattening steers. Agron. J. 48:508-13.

Blaser, RE, RC Hammes, Jr., JP Fontenot, CE Polan, HT Bryant, and DD Wolf. 1976. Forage-animal production systems on hill land in the eastern United States. In J Luchok (ed.), Proc. Int. Hill Lands Symp. West Virginia Univ. Books.

Blaser, RE, RC Hammes, Jr., JP Fontenot, and HT Bryant. 1977. Forage-animal systems for economic calf production. Proc. 13th Int. Grassl. Congr. 2:1541-44.

Blaser, RE, and colleagues. 1986. Forage-Animal Management Systems. Va. Agric. Exp. Stn. Bull. 86-7.

Brent, BE. 1976. Relationship of acidosis to other feedlot oilmeats. J. Anim. Sci. 43:930-35.

Britton, R, D Rock, T Klopfenstein, J Ward, and J Merrill. 1983. Harvesting and processing effects on alfalfa protein utilization by ruminants. In JA

Smith and VW Hays (eds.), Proc. 14th Int. Grassl. Congr., Lexington, Ky. Boulder, Colo.: Westview, 653-65.

Bula, RS, VL Lechtenberg, and DA Holt. 1981. Temperate zone cultivated forages. In Winrock Report: Potential of the World's Forages for Ruminant Animal Production, 2d ed. Morrilton, Ark.: Winrock International Livestock Research and Training Center, 7-28.

Buttrey, SA, VG Allen, JP Fontenot, and RB Reneau, Jr. 1986. Effect of sulfur fertilization on chemical composition, ensiling characteristics and utilization of corn silage by lambs. J. Anim. Sci. 63:1236-45.

Cook, CW, DA Cramer, and L Rittenhouse. 1984. Acceptable block beef from steers grazing range and crop forages. J. Range Manage. 37:122-26.

Fontenot, JP. 1979. Animal nutrition aspects of grass tetany. In Proc. Grass Tetany Symp., Spec. Publ. 35. Madison, Wis.: American Society of Agronomy, 51-62.

———. 1991. Recycling animal wastes by feeding to enhance environmental quality. Prof. Anim. Sci. 7(4):1-8.

Fontenot, JP, VG Allen, and A Brock. 1993. Forage systems for production of stocker cattle. In Proc. 7th World Conf. Anim. Prod. 2:398-99.

Forage and Grazing Terminology Committee. 1992. Terminology for grazing lands and grazing animals. J. Prod. Agric. 5:191-210.

Hammes, RC, Jr., JP Fontenot, HT Bryant, RE Blaser, and RW Engel. 1964. Value of high-silage rations for fattening beef cattle. J. Anim. Sci. 23:795-801.

Hammes, RC, Jr., JP Fontenot, RE Blaser, HT Bryant, and RW Engel. 1968. Supplements to corn and hay-crop silages for fattening beef cattle. J. Anim. Sci. 27:1690-94.

Hendrick, HB, AG Matches, JD Thomas, NG Krouse, RE Morrow, and WC Stringer. 1982. The Production, Characteristics and Utilization of Forage-fed Beef. Univ. Mo. Agric. Exp. Stn. Res. Bull. 1043.

Hodgson, HJ. 1978. Feed from plant products—forage. In Plant and Animal Products in the U.S. Food System. Washington, D.C.: National Academy of Sciences, 56-74.

Horn, FP, and CM Taliaferro. 1977. Existing and potential systems of finishing cattle on forages or limited grain rations in the semi-arid southwest. In JA Stuedemann, DL Huffman, JC Purcell, and OL Walker (eds.), Forage-fed Beef: Production and Marketing Alternatives in the South. South. Coop. Ser. Bull. 220:401-18.

Hoveland, CS, and WB Anthony. 1977. Existing and potential systems of finishing cattle on forages or limited grain rations in the Piedmont region of the South. In JA Stuedemann, DL Huffman, JC Purcell, and OL Walker (eds.), Forage-fed Beef: Production and Marketing Alternatives in the South. South. Coop. Ser. Bull. 220:377-418.

Klopfenstein, TJ. 1981. Increasing the nutritive value of crop residues by chemical treatment. In JT Huber (ed.), Upgrading Residues and By-prod-ucts for Animals. Boca Raton, Fla.: CRC Press, 39-60.

McDonald, P. 1981. The Biochemistry of Silage. New York: Wiley.

Monson, WG, and PR Utley. 1977. Existing and potential systems of finishing cattle on forages or limited grain rations in the lower south. In JA Stuedemann, DL Huffman, JC Purcell, and OL Walker (eds.), Forage-fed Beef: Production and Marketing Alternatives in the South. South. Coop. Ser. Bull. 220:391-400.

National Cattlemen's Association (NCA). 1992. Facts, Figures, and Information. Cattle and Beef Handbook. Englewood, Colo.: NCA.

National Research Council (NRC). 1980. Mineral Tolerance of Domestic Animals. Washington, D.C.: National Academy of Sciences.

———. 1984. Nutrient Requirements of Beef Cattle. 6th ed. Washington, D.C.: National Academy of Sciences.

———. 1985. Ruminant Nitrogen Usage. Washington, D.C.: National Academy Press.

Neville, WE, Jr. 1962. Influence of dam's milk production and other factors on 120- and 240-day weight of Hereford calves. J. Anim. Sci. 21:315-20.

Owensby, CE, R Cochran, and EF Smith. 1988. Stocking rate effects on intensive-early stocked flint hills bluestem range. J. Range Manage. 41:483-97.

Petchey, AM, and PJ Broadbent. 1980. The performance of fattening cattle offered barley and grass silage in various proportions either as discrete feed or as a complete diet. Anim. Prod. 31:251-57.

St. Louis, DG, CH Hovermale, JD Davis, and FH Tyner. 1990. Economic Comparison of Intensive Cow-Calf Forage Systems for South Mississippi. Miss. Agric. and Forage Exp. Stn. Bull. 970.

Smith, EF, and CE Owensby. 1978. Intensive-early stocking and season-long stocking of flint hills bluestem range. J. Range Manage. 31:14-18.

Spooner, AE, and ML Ray. 1977. Existing and potential systems of finishing cattle on forages or limited grain rations in the upper south. In JA Stuedemann, DL Huffman, JC Purcell, and OL Walker (eds.), Forage-fed Beef: Production and Marketing Alternatives in the South, South. Coop. Ser. Bull. 220, 363-75.

Stuedemann, JA, DL Huffman, JC Purcell, and OL Walker (eds.). 1977. Forage-fed Beef: Production and Marketing Alternatives in the South. South. Coop. Ser. Bull. 220.

Tucker, CA, RE Morrow, JR Gerrish, CJ Nelson, GB Garner, VE Jacobs, WG Hires, JJ Shinkle, and JR Forwood. 1989. Forage systems for beef cattle: Calf and backgrounded steer performance. J. Prod. Agric. 2:208-13.

USDA. 1992. Agriculture Statistics. Washington, D.C.: USDA.

USDC. 1987. Census of Agriculture. Washington, D.C.: USDC.

US Meat Export Federation (USMEF). 1991. Beef Export Facts and Trivia. Denver, Colo.: USMEF.

Forages for Dairy Cattle

20

LEONARD S. BULL
North Carolina State University

DAIRY cattle, *Bos taurus,* as members of the ruminant family, are physiologically adapted to utilize forages to provide significant portions of the nutrients needed to grow, reproduce, and produce milk. That fact needs no debate or discussion. What is critical, however, is the present and future role of forages in meeting the nutrient requirements of dairy cows. Those include complex questions of economics, forage quality as viewed from nutrient density perspectives, and the absolute requirement of the animal for nutrients.

Questions of types of forages, harvest, storage, and economics are dealt with in other sections of this book, and there is not a duplication of those efforts here. Rather, the unique aspects of the dairy cow, her level of nutrient metabolism, the interactions between forages and concentrates in the digestive system, the role that forages can play in meeting nutrient demands, and the limitations that need critical study are stressed. Those factors ultimately determine the place and extent of use of forages in the diet.

THE US DAIRY INDUSTRY IN BRIEF

There are approximately 9.8 million dairy cows in the US according to the USDA (1993),

LEONARD S. BULL is Professor and Head of Animal Science at North Carolina State University, Raleigh. His MS degree is from Oklahoma State University and PhD from Cornell University. He has postdoctoral experience in medical physiology at the University of Virginia. His research has been in dairy cattle nutrition, with emphasis on forages and protein and energy utilization. He has taught energy, mineral and general nutrition, and animal production.

down about 100,000 from a year earlier, and continuing the decline of several years. Average number of cows per farm continues to increase slightly, and total production is about 68 billion kg. Average production per cow continues to increase and is now at about 700 kg/yr, up slightly from 1992. Prices received by farmers for milk have been lower in recent years than they were in the 1980s, and that trend is expected to continue. As a result, returns to management and profitability have been low.

As an industry, dairying is mature, traditional, and hampered by a product-marketing system that is based on water and fat. Average age of farmers is in the upper 50s, and increasing, indicating that there will be a significant exit of operators from the industry in the near future.

Increasing use of low-fat milk up until the past year has helped producers avoid surplus levels that must be dealt with in the future. Lack of unity and industrywide leadership is a problem, with a price support system that has discouraged needed reform in production and marketing systems.

Concerns about product safety (antibiotic contamination, use of production stimulants such as bovine growth hormone, etc.) have placed stress on growth of sales. If that trend continues, significant stress on prices due to overproduction may occur.

While the preceding are not directly related to the use of forages in dairy rations, they do underscore the fact that a major problem facing the industry is the need to significantly reduce the cost of production. This is important despite the fact that current price support provisions may not encourage that practice now. Forages can play a significant role in reducing cost of production.

Yet all is not gloom and doom. California has developed a dairy industry that threatens to pass that of Wisconsin in quantity. Much of the success of the California industry is due to very high-quality forage (primarily alfalfa [*Medicago sativa* L.]). Herds with production averages of over 14,000 kg milk per cow per year are common. In those herds forages are significant portions of the rations, and forage quality is excellent.

NUTRIENT REQUIREMENTS FOR DAIRY CATTLE, WITH EMPHASIS ON FORAGES

Rations for dairy cows are balanced on the basis of nutrients for which there are known quantitative requirements and those for which there are recommended concentrations in the diet (NRC 1989). Forages contain variable quantities of all of the nutrients and serve as the major source of the nutritive and nonnutritive fiber constituents (including crude fiber, acid detergent fiber, neutral detergent fiber, cellulose, hemicellulose, and lignin). Variations in content of the various nutrients in forages are common. The importance of the extent of that variation is a factor when it becomes impossible to combine other sources of nutrients in such a way that a balanced ration is produced.

The nutrients focused on here are energy and protein, nutrients closely related in discussions of forage. Fiber is dealt with as part of the discussion of energy (density, physical properties, etc.). Questions of what minerals and vitamins to use and how much are easily dealt with by analysis of forage and addition of specific and concentrated sources to ensure that recommended concentrations in the final ration are met.

ENERGY AND FORAGES FOR DAIRY CATTLE

Energy is defined in net energy for lactation terms (Mcal/kg, dry matter [DM] basis) where needed. Other units of energy are used, including digestible energy, metabolizable energy, net energy for maintenance, and total digestible nutrients. The interested reader is referred to *Nutrient Requirements of Dairy Cattle* (NRC 1989).

The critical components of forages relative to energy are energy density (energy per unit forage volume), rate of digestion of energy relative to retention time of forage in the digestive tract, and factors in the diet that influence the energy value of forages (usually called *associative effects of feeds*). These are covered in the order mentioned here.

Energy Density. Traditionally energy density is expressed in terms of calories per unit of weight (i.e., Mcal/kg) or volume (i.e., Mcal/L). These methods are adequate for cereals and other concentrated sources of energy where the density of energy is well above the level where there is a possibility that space in the digestive tract will limit the ability of the animal to consume adequate amounts to meet nutrient needs from those feeds. However, forages present another dimension of concern for dairy cattle. For many combinations of forages and nutrient demands in dairy cattle, forages alone cannot be consumed in adequate quantities to provide for those needs. The reasons for this inability are (1) the relative density of nutrients per unit of volume of feed, and thus space occupied in the digestive tract; (2) the relatively slow rate of digestion of forage fiber in the rumen, and thus reduced rate of creation of space for more intake; and (3) the relatively slow rate of passage of forage material through the digestive tract of the ruminant, creating further physical limitation to intake.

Bull et al. (1976) clearly demonstrate that physical limitation to energy intake in dairy cows is highly correlated to caloric density (Mcal digestible energy/L DM) of forage-based rations (Table 20.1). In those studies energy intake increased with increased caloric density until the production demand of the animal was met. At higher densities calorie intake was constant regardless of density, and dry matter intake declined to maintain constant intake. Similar conclusions are presented by the National Research Council (1989).

Intake of diets by dairy cows has been shown to be closely related to content of neutral detergent fiber (NDF) (Mertens 1985). Since that fiber fraction is also a major component of the variation in bulk or volume of a forage, the similarities between caloric density and NDF relative to intake are to be expected.

Caloric density of forages is related to plant species, but that relationship is determined by composition of the forage. The influence of species will not be dealt with in this chapter. Critical factors in variation in caloric density are concentration of fiber fractions in total dry matter, created by advanced maturity of the plant. While total fiber (NDF) content usually increases with maturity, relative amounts of cellulose and lignin increase also, and per-

TABLE 20.1.   Influence of bulk density of rations on energy intake by lactating dairy cows

	Diet				
Parameter	A	B	C	D	E
Digestible energy:					
Mcal/kg	2.5	2.7	2.8	2.9	3.0
Mcal/L	1.105	1.115	1.249	1.513	1.772
Intake (units/kg$^{.75}$/d):					
Dry matter (g)	129	124	139	146	140
Energy (kcal)	292	324	370	371	369

Source: Bull et al. (1976).
Notes: Diets were based on alfalfa hay of varying maturities and mixed concentrates. Energy need was met by ad libitum intake of diet *C*. Higher caloric density did not increase intake in these cows, and regulation was shown. At lower caloric densities intake was limited by capacity of digestive tract (diets *A* and *B*).

centage of total plant energy that is potentially available to the animal declines.

Mertens (1985) reports that digestibility (extent) is negatively correlated with acid detergent fiber (ADF) content of forages, a fraction made up primarily of cellulose and lignin. In addition, rate of digestion of forage dry matter declines with advanced maturity (Cleale and Bull 1986) (Table 20.2). These events explain the fact that caloric density declines with changes in fiber content and fractions, resulting in reduced intake of forages. For that reason, lower-quality forages are not valuable ingredients of diets for dairy cows.

The influence of moisture content of feeds on intake is important, as forages are usually the primary sources of significant water content of total rations. There is general agreement that moisture content of rations exceeding 50% results in some reduction in intake, but the exact impact is not well-defined. Furthermore, high-moisture forages that are fermented (silages) result in greater declines in intake than those fed without fermentation (NRC 1989). Bull (1974) reports that fresh corn (*Zea mays* L.) plant material fed without

TABLE 20.2.   Composition of mixed forages harvested from same plots at intervals of 21 d

Component	Early-cut	Late-cut
	(% of dry matter)	
Crude protein	13.5	10.1
Acid detergent insoluble nitrogen (% of total N)	12.7	18.1
Neutral detergent fiber	57.7	67.1
Acid detergent fiber	37.5	42.1
Lignin	4.1	6.8
Hemicellulose	20.2	25.0
Cellulose	33.4	35.3
Cell solubles	42.3	32.9
Ruminal digestion rate (%/h)	5.0	3.1

Source: Cleale and Bull (1986).
Note: Ruminal digestion rate is fraction of percentage of potentially digestible DM disappearing per hour.

drying or with variable amounts of heat drying (no fermentation) resulted in equal intakes when fed as the sole source of nutrients. Clancy (1974) identifies nitrogenous fractions resulting from the fermentation of forages that were potent intake depressants.

Clearly, moisture content of forages is important in optimum utilization for dairy cattle, but moisture per se is apparently less critical than other constituents resulting from fermentation of wet forages.

Rate of Digestion. Rate of digestion of forages in the digestive tract of the dairy cow is an important determinant of the intake of forages and thus the potential role of forages in the diet. The unique component of forages is the relatively large concentrations of fiber fractions, compared with those of high-starch cereals and some nonforage by-products fed to cattle. Since the site of digestion of virtually all of the cellulose and much of the hemicellulose in forages is in the rumen, by microbial action, knowledge of the dynamics of rumen function is critical to attempts to optimize forage use in dairy cow rations.

Bull (1980), Cleale and Bull (1986), and Erdman et al. (1987) report on the rate of digestion of fiber fractions in numerous feeds. Rate of digestion of NDF and ADF fractions declined with increased lignin, increased cellulose, and decreased hemicellulose content of total fiber fractions. Cleale and Bull (1986) report that rate of ruminal digestion declined over 40% as maturity of mixed forages increased by 21 d. The data of these authors further emphasize the importance of management in providing forages that will serve as valuable components of dairy cow diets rather than relatively inert diluents for which there is little value.

Factors That Influence Forage Energy Value in the Animal. The discussion to this point is based on the assumption that the "predicted" energy value of a specific feed, and specifically a forage, is constant and can be used casually in balancing rations. Nothing could be further from the truth. In addition to the factors that influence energy value of forages prior to feeding, there are several critical in vivo factors. In some cases these are created as a result of efforts to compensate for factors that occur outside the animal.

The most widely recognized in vivo factor affecting forage energy values is the level of feeding of the diet, usually expressed in multiples of the maintenance requirement of the animal. Wagner and Loosli (1967), Tyrrell and Moe (1975), and Van Soest (1973) present significant integrative discussions on this subject, including effects of physical form, particle size, and chemical or heat treatment. As a result of those observations, the National Research Council (1989) applies a reduction of 4% in the energy value of all feeds for each increase in intake equal to one times the maintenance requirement of the animal. Thus a forage that has a net energy for lactation value of 1.65 Mcal/kg DM when fed at the maintenance level of intake will have a value of 1.58 Mcal/kg DM at twice the maintenance level of intake, and the decline will increase equally for each multiple of maintenance intake increase.

Wagner and Loosli (1967) present data that show that the rate of decline in digestibility of diets increases as percentage of forage in the diet declines. An important aspect of that finding, coupled with the importance of bulk density and physical limitation on intake of diets high in forages, is the fact that at productive levels of intake increasing concentrate or grain content of diets does not increase the percentage of the diet that is digested but, rather, reduces the bulk volume to allow more dry matter to be consumed. That concept is not widely appreciated by nutritionists, but it is important for maximizing forage use in dairy cow diets.

While the assumption of linearity of decline in energy value with increasing intake of a given diet, created almost entirely by decline in digestibility, is probably not correct, there are insufficient data to more clearly define this event at this time. The reader is cautioned to be sure that the basis upon which energy values for feeds are reported is clear so that the appropriate starting point is taken for predicting actual values.

Cleale and Bull (1986) report that for each day of advancement in maturity of mixed perennial forages 1% additional concentrates were needed in the dairy cow diet to meet the nutrient requirements. The reasons are the reduced energy density and reduced rate of digestion of the forages and the fact that concentrates are needed to reduce the bulkiness of the diet and thus enable adequate consumption. This is the practical endpoint of reduced forage quality and is referred to again later in this discussion.

Another factor that influences the energy value of forages is the protein status in the diet, with particular emphasis on conditions in the rumen. In order for fiber fermentation to be optimized in the rumen, adequate nitrogen (N), soluble and available for microbial assimilation during growth, is required. Poos et al. (1979) show that fiber digestion in dairy cattle was significantly influenced by ruminally available N and that dry matter intake was significantly reduced when fiber digestion was reduced by feeding inadequate amounts of ruminally available N.

The National Research Council (1985, 1989) has recommended that between 50% and 60% of the protein fed to dairy cows (depending on level of production) that is not contained in the fraction called *acid detergent insoluble N* (ADIN) should be in forms that are degraded in the rumen to ammonia or other simple compounds that are readily available for use by bacteria. Failure to provide this amount will result in reduced fiber digestion, reduced energy density of the diet, and possibly reduced intake (Poos et al. 1979) (Table 20.3). Even if total protein requirement is balanced in the diet, excess degradation in the rumen leads to inefficient protein use and may induce protein deficiency in the animal at the tissue level.

The last in vivo factor to be mentioned is called the *associative effects of feeds*. There is less systematic data on this concept than on almost any of the critical variables in diet utilization, with the exception that much of the benefit gained from feeding of mineral buffers can be related to this factor. Associative effects (usually considered as negative) involving digestion of diets are situations where the actual digestion of a mixture of feeds is significantly less than the arithmetic mean of the digestibilities of the components would predict, when level of intake is taken into account and the ration is balanced (for protein, etc., as mentioned already) (Table 20.4). For example, if a balanced ration contains 50% concentrate with digestibility of 82% and 50% forage with digestibility of 60%, the predicted digestibility of the ration is 71%. Data of Vandemark (1982) and Stokes and Bull (1986)

TABLE 20.3. Influence of protein content of ration on dry matter and acid detergent fiber (ADF) digestion by lactating dairy cows

Component	Ration protein content (%, DM basis)				
	8.6	11.6	15.2	16.5	17.1
	(digestibility, %)				
Dry matter	59	60	65	68	67
ADF	30	34	39	46	49

Source: Poos et al. (1979).

Note: Rations were based on corn silage, alfalfa hay, and corn grain, with soybean meal or urea added to increase protein content.

have shown that this phenomenon is significant and that the greatest deviation from expected energy value in mixed diets is in the range of about 40% to 60% forages. The importance of this observation is that this is also the most common range used in feeding dairy cows.

While passage of material through the digestive tract may be a factor in the observation of associative effects, with certain mixtures resulting in increased passage rate of fiber, it is probable that the greatest cause is reduced fiber digestion caused by the fermentation of starch. The impact of starch on fiber digestion is well documented (Mertens 1979), but unfortunately it is not considered in ration formulation in a systematic way other than relative to the use of buffers. Data of Erdman et al. (1982) show the opportunity for significantly improving fiber digestion through use of buffers in rations with levels of concentrates in the range noted earlier. It is probable that the associative effects due to adding concentrates to forages in dairy cow diets are related to the buffering capacity of the total diet in the rumen. This is a further

TABLE 20.4. Digestion of rations by lactating dairy cows showing negative associative effects of concentrates on digestion of fiber fractions

Concentrate level	DM	ADF	NDF
(%)	(digestibility, %)		
0	69	71	74
60	74	60	62
0	54	53	56
78	74	43	53
0	62	66	67
70	74	56	61
30	66	54	61
50	68	48	55
70	73	49	58

Sources: Stokes and Bull (1986), Cleale and Bull (1986), and Vandemark 1982.

Notes: Diets were based on ensiled mixed perennial forages and concentrates made up of several grains and protein supplements.

reason for the use of high-quality forages, especially legumes, in feeding programs for dairy cows.

PROTEIN AND FORAGES FOR DAIRY CATTLE

Forage protein may change significantly with stage of maturity as well as method of storage. This subject is covered in depth by the National Research Council (1985), and therefore only the general concepts will be covered here.

It is common for a small portion, 5% or less, of the total N in forages to be associated with structural components of the plant in such a way that it is nearly unavailable to the animal through normal digestive functions. Analytically, this fraction is identified by analysis of the residue remaining after treatment of the feed with acid detergent solution (Goering and Van Soest 1970). As forages become more mature, the percentage of the total N that is found in the unavailable fraction increases (Cleale and Bull 1986). While there are no clear data on the cause of this increase, structural changes and dilution factors are probably involved. In addition, harvest and storage of forages by almost any method increases the unavailable percentage slightly more. While efficient drying as hay results in only small increases, storage in fermented forms where temperatures are significantly elevated may increase the ADIN fraction substantially. Factors influencing this process are dealt with in Chapter 11.

While the ADIN fraction, once thought to be completely unavailable to the animal, may have some modest availability, the nature of that availability is such that it is best to deduct it from the total N (and thus protein) in the forage prior to balancing a ration. In that way, any errors made are in the favor of the animal.

It is absolutely essential that all forages fed to dairy cows be carefully analyzed for protein, including ADIN, prior to any attempt to incorporate those forages into rations.

The other extreme in protein value of forages is that portion that is soluble in the rumen and rapidly converted to ammonia by resident bacteria (N that is already in the form of ammonia is also quickly soluble in rumen fluid and behaves as a part of this fraction [NRC 1985]). Immature forages and those that have undergone fermentation in conditions of high moisture frequently have well in excess of 50% of the total N in a form that is rapidly soluble in the rumen.

In order for ammonia from any source to be used in the rumen, it must be assimilated by bacteria. This is an energy-dependent function. If ammonia availability exceeds the ability to assimilate it, that excess is essentially lost to the animal as far as potential protein value is concerned (NRC 1985). It is therefore important that the total amount of soluble N as well as the portion of the protein that is degraded in the rumen be known in order that rations may be formulated that optimize utilization of protein (NRC 1985, 1989). It is probable that excess degradable protein from high-quality forages preserved as silage is a major opportunity for improvement in the efficiency of nutrient utilization by dairy cattle.

PASTURES FOR DAIRY CATTLE

There is no question that pastures, properly managed, represent an economical way to provide high-quality forage to dairy cattle. Interests in reduced cost of milk production, sustainable agriculture, reduced water and air pollution from high densities of dairy cows in confinement, and other concerns have renewed interest in pastures for lactating cows. Since there is little that is unique to pastures from a nutritional standpoint, the only comments appropriate here are those that relate to management of pastures and animals to enable this practice to be used effectively.

Pasture research with dairy cattle is not new, and there is a significant amount of it. Previous editions of this book have addressed those matters, and there is significant mention in this edition (Chaps. 11 and 14, Vol. 1, and 3, 13, 14, 15, 19, 21, and 22, Vol. 2). Protein nutrition in animals fed pastures high in legumes may be improved by reduction of ruminal degradation of protein (NRC 1985). This is probably due to the increased rate of turnover of material in the rumen with reduced retention time.

Of concern today are the logistics and mechanics of utilization of pastures for dairy cows in large herds and meeting the steady demand for nutrients in diets built around pastures. The ability to predict variation in pasture quantity and quality is also a major concern. If these can be improved upon, pasture use may increase. There are numerous advantages to the use of high-quality pastures in dairy cow rations, and there is need to continue to solve the current limitations.

In some parts of the US, especially the Southeast, moisture variability causing quantity and quality problems in pastures and the possible need for irrigation detract from the economics to be gained by intensive grazing and significant dependence on pastures for forage throughout the year. The fact that there are very few entirely new dairy facilities being built, and that those in place have already made significant investments in equipment for harvest, storage, and feeding of forages in confinement systems, must be taken into account in any attempt to shift to greater use of pastures on existing units.

EXAMPLE RATIONS FOR DAIRY COWS USING FORAGES OF DIFFERING QUALITIES

Table 20.5 contains examples of rations balanced according to the latest documented information on nutrient requirements for dairy cattle (NRC 1989). The only variables are in forage quality. While there are virtually endless combinations of feeds that can be combined to produce balanced rations that can be consumed, it is important to note the impact of slight changes in forage quality on diet composition.

TABLE 20.5. Composition of total rations for dairy cows balanced for equal production using forages harvested at 21-d intervals

Component	Early-cut	Late-cut
	(% of dry matter)	
Concentrate	60.30	77.90
Forage	39.70	22.10
NE (lactation), Mcal/kg	1.64	1.64
Crude protein	15.10	14.40
Neutral detergent fiber	34.90	30.60
Acid detergent fiber	18.60	14.50
Lignin	3.50	3.20
Hemicellulose	9.70	8.20
Cellulose	15.10	11.30

Source: Cleale and Bull (1986).
Note: Concentrate requirement was increased 0.8% for each day of harvest delay.

QUESTIONS

1. What are the real limitations to significantly increased use of forages of high quality in dairy rations?
2. What economic conditions are required to convince dairy producers that forage quality should be a high priority?
3. What key factors must be overcome to eliminate the depression in digestion that takes place with an increased level of feeding of dairy cows?
4. What feeding strategies could minimize the negative impact of cereal grain starch on fiber digestion and thus forage incorporation in the rations of dairy cows?
5. How can the dairy industry nationally alter its marketing structure to provide a basis that encourages reduced cost of production and increased use of forages in all forms?

REFERENCES

Bull, LS. 1974. Unpublished data. Univ. of Maryland, College Park.

———. 1980. Unpublished data. Univ. of Maine, Orono.

Bull, LS, BR Baumgardt, and M Clancy. 1976. Influence of caloric density on energy intake by dairy cows. J. Dairy Sci. 59:1078.

Clancy, M. 1974. Prediction of digestible energy intake of complete diets and conserved alfalfa by sheep. PhD diss., Pennsylvania State Univ., University Park.

Cleale, RM, IV, and LS Bull. 1986. Effect of forage maturity on ration digestibility and production by dairy cows. J. Dairy Sci. 69:1587.

Erdman, RA, RW Hemken, and LS Bull. 1982. Dietary sodium bicarbonate and magnesium oxide for early postpartum lactating dairy cows: Effects on production, acid-base metabolism and digestion. J. Dairy Sci. 65:712.

Erdman, RA, JH Vandersall, E Russek-Cohen, and G Switalski. 1987. Simultaneous measures of rate of ruminal digestion and passage of feeds for prediction of ruminal nitrogen and dry matter digestion in lactating dairy cows. J. Anim. Sci. 64:656.

Goering, HK, and PJ Van Soest. 1970. Forage Fiber Analysis: Apparatus, Reagents, Procedures and Some Applications. USDA Agric. Handb. 379. Washington, D.C.: US Gov. Print. Off.

Mertens, DR. 1979. Effect of buffers upon fiber digestion. In WH Hale and P Meinhardt (eds.), Regulation of Acid-Base Balance. Nutley, N.J.: Church and Dwight, 65-80.

———. 1985. Factors influencing feed intake in lactating dairy cows: From theory to application using neutral detergent fiber. In Proc. 46th Ga. Nutr. Conf. Athens: Univ. of Georgia, 1-18

National Research Council (NRC). 1985. Ruminant Nitrogen Usage. Washington, D.C.: National Academy Press.

———. 1989. Nutrient Requirements of Dairy Cattle. 6th rev. ed. Washington, D.C.: National Academy Press.

Poos, MI, LS Bull, and RW Hemken. 1979. Supplementation of diets with positive and negative urea fermentation potential using urea or soybean meal. J. Anim. Sci. 49:1417.

Stokes, MR, and LS Bull. 1986. Effects of sodium bicarbonate with three ratios of hay crop silage to concentrate for dairy cows. J. Dairy Sci. 69:2671-80.

Tyrrell, HF, and PW Moe. 1975. Symposium—production efficiency in the high producing cow. Effect of intake on digestive efficiency. J. Dairy Sci. 58:1151.

USDA. 1993. Dairy Situation and Outlook Report. June. Washington, D.C.

Vandemark, LL. 1982. Physiology of digestion in the rumen of dairy cows. MS thesis, Univ. of Maine, Orono.

Van Soest, PJ. 1973. Revised estimates of the net energy values of feeds. In Proc. Cornell Nutr. Conf. Ithaca, N.Y.: Cornell Univ. Press, 11-23.

Wagner, DG, and JK Loosli. 1967. Studies on the Energy Requirements of High-producing Dairy Cows. Cornell Agric. Exp. Stn. Memo 400. Ithaca, N.Y.: Cornell Univ. Press.

21

Forages for Horses

JOE L. EVANS
Rutgers University

HORSES (*Equus caballus*) number more than 9 million in the US in the early 1990s. The states of California and Texas are approaching 1 million horses each. Fourteen states—Arizona, Colorado, Florida, Illinois, Kentucky, Michigan, Minnesota, Missouri, Montana, Ohio, Oklahoma, Pennsylvania, New York, and Tennessee—have populations that exceed 200,000 each. Five states—Arkansas, Idaho, Kansas, Louisiana, and Washington—have populations that are approaching 200,000 each. Draft horses, mules/hinnies, and donkeys (*Equus asinus*) make up only a few percentage units of the US horse population, which is mostly light horses (Evans et al. 1990). Many of these light horses serve as family pets and for family recreational activity, while others are workhorses on cattle ranches and in sports racing, shows, and games. Many books on horses have been written, and several are referenced (Crowell 1951; Simpson 1961; Hughes and Heath 1962). (See Fig. 21.1.)

Over the years these herbivorous mammals have consumed large quantities of lower-quality roughages and, when available, higher-quality forages. The larger-type mature horse may require up to 5 mt (dry weight) of nutri-

tionally balanced feedstuffs annually, and more than half of this total is forage. About 60% of the space in the digestive tract of the horse is in the cecum and large intestine; this area is smaller in total capacity but equal in relative size to the ruminant's stomach.

Unlike the rumen in ruminants, the cecum and large intestine in horses are located at the posterior end of the digestive tract, and many of the products of fermentation may be voided in the feces and not absorbed to benefit the animal. Many questions are unanswered concerning the contribution of the lower tract to the nutrition of the horse. With these and other limitations in mind, this chapter deals with nutritional needs such as voluntary intake, digestibility and nutrient requirements as well as maximum forage diets in the life cycle in horses.

VOLUNTARY INTAKE

Limited data from controlled experiments with horses are available concerning the voluntary intake of water and of dry matter in feedstuffs. Morrison (1959) suggested that horses and mules would consume 60-108 $g/W^{0.75}$ (W = body weight) of dry diet per day composed of varying proportions of forage and grain. Standardbred geldings consumed 70-110 $g/W^{0.75}$ of forage and forage plus grain diets (Fonnesbeck 1968a). Voluntary intake ranged from 65 to 95 $g/W^{0.75}$ in smaller ponies that consumed forage (Darlington and Hershberger 1968). Other ponies consumed 75 $g/W^{0.75}$ of 17% protein alfalfa (*Medicago sativa* L.) in long form, 17% more in wafers, and 24% more in pellets (Haenlein et al. 1966). Low forage intake can be maintained by hand-

JOE L. EVANS is Emeritus Professor in Nutrition, Rutgers University. He holds the MS degree from the University of Kentucky and the PhD from the University of Florida. His research was with nutritive evaluation of forages, mineral nutrition and metabolism, digestibility and utilization of different N sources of protein, and the influence of diet composition on intestinal environment and nutritional state. Currently his work involves consulting in nutrition and the ownership-management of a polled Hereford farming operation.

Fig. 21.1.   Donkeys make excellent use of pastures.

feeding and by diluting the diet with fiber from straw or strawlike advanced-maturity forage.

Water intake ranged from 13 to 34 kg/d for 427-kg standardbred geldings at rest and maintenance (Fonnesbeck 1968a). High humidity increased water intake, but temperature was more influential (Yoshida et al. 1967). Both sweating and lactation increased the water requirement. In addition to environmental conditions, physical activity, and the physiological state of the horse, diet composition contributed to differences in water intake. Water intake for forage diets averaged 3.55 kg/kg dry hay. A forage diluted with corn (*Zea mays* L.), barley (*Hordeum vulgare* L.), or oat *Avena sativa* L.) (54% forage to 46% grain) reduced the water intake to 2.90 kg/kg of diet. An increase of fiber or cell wall content in the diet resulted in a reduced digestibility or increased fecal residue and required more water for excretion.

CHEMICAL COMPOSITION AND DIGESTIBILITY OF FEEDSTUFFS

Limitations of the proximate system and the nonuniformity of the crude fiber fraction among forage species are discussed in Chapter 6, and the composition of detergent fiber fractions (Chap. 6) has been analyzed (Colburn and Evans 1967). Lignin is not digested by the horse, whereas the remaining cell wall fiber, composed of cellulose and hemicellulose, is partially digestible (Fonnesbeck 1969). Hemicellulose is slightly, but not significantly, more digestible by the horse than is cellulose,

indicating that the total fiber fraction dilutes the amount of digestible energy (DE) provided in the diet (Fonnesbeck 1969). Crude fiber as determined by the proximate system is both nonuniform in composition and noninclusive of the total diet fiber. Procedures used in the determination of digestibility and specific examples using these procedures in the determination of the digestibility of nutrients of forages in horses have been published (Fonnesbeck et al. 1967; Fonnesbeck 1968b; Maynard et al. 1979).

DIGESTIBLE DRY MATTER AND ENERGY

Table 21.1 gives the daily diet requirements for protein, energy, calcium (Ca), phosphorus, (P), and vitamin A for growing horses and for mature horses at maintenance, work, pregnancy, and lactation (NRC, 1973, 1978). The amounts of daily diet given in Table 21.1 contain 62.5% total digestible nutrient (TDN) and 2.75 Mcal/kg on a dry basis. Most forages and other feedstuffs on an as-fed basis average about 90% dry matter.

The digestibility of most forages declines with advancing maturity. An exception is the silage made from cereal-grain crops. As only limited data are available on the influence of advancing forage maturity on its digestibility in the horse (Darlington and Hershberger 1968), changes in digestible dry matter (DDM) for alfalfa, orchardgrass (*Dactylis glomerata* L.), and corn silage with advancing maturities were estimated (Vander Noot and Gilbreath 1970) (Table 21.2). Equations in which crude fiber was used to represent the total cell wall or fiber in the diet were not considered (Blaxter 1961; Vander Noot and Gilbreath 1970; Vander Noot and Trout 1971).

Many horses performing different types of activity require a diet that provides less than 2.75 Mcal DE/kg. This level of energy may be provided by all-forage diets if an early harvesting schedule is followed for both grasses and legumes. Data in Table 21.2 suggest that cutting orchardgrass at early-head emergence or after 42-d regrowth and cutting alfalfa at 11% bloom will allow an adequate intake of DE for growth and lactation in the horse. Data for DDM in corn silage are lower than expected, but DDM of low-quality reed canarygrass (*Phalaris arundinacea* L.) diluted with 46% corn, barley, or oats averages only 56% in horses (Fonnesbeck 1968a). The influence of total diet fiber is apparent as DDM increases to 66% when alfalfa replaces reed canarygrass.

The ratio of hemicellulose to cellulose

TABLE 21.1.   Nutrient requirements of horses

Body weight (kg)	Age (mo)	Daily gain (kg)	Daily diet (kg)	Daily diet (g/W^0.75)	DP[a] (g)	DE[b] (Mcal)	Calcium (g)	Phosphorus (g)	Carotene[c] (mg)
\multicolumn{10}{c}{*Growing horses (200 kg mature weight)*}									
60	3	0.70	4.0	160	580	11.1	28	18	6
95	6	0.50	3.2	105	470	8.8	19	14	10
145	12	0.20	3.0	73	350	8.1	12	9	14
170	18	0.10	3.0	63	320	8.1	11	7	15
200	42	0.0	3.0	56	320	8.2	9	6	13
\multicolumn{10}{c}{*Growing horses (400 kg mature weight)*}									
125	3	1.00	6.2	166	1050	17.1	42	29	13
185	6	0.65	4.7	94	660	13.0	27	20	19
265	12	0.40	5.0	76	600	13.8	24	17	25
330	18	0.25	5.2	68	590	14.4	22	15	29
400	42	0.0	5.1	56	540	13.9	18	11	25
\multicolumn{10}{c}{*Growing horses (600 kg mature weight)*}									
170	3	1.40	8.0	170	1350	22.0	54	38	17
265	6	0.85	6.1	94	860	16.9	37	27	27
385	12	0.60	6.9	79	900	18.9	35	25	35
475	18	0.35	7.0	68	750	19.1	32	22	34
600	42	0.0	6.8	56	730	18.8	27	17	38
\multicolumn{10}{c}{*Mature horses at light work*}									
200	...	...	4.0	75[d]	340	10.4	9	6	13
400	...	...	6.7	75	570	18.4	18	11	25
500	...	...	8.0	75	680	21.9	23	14	31
600	...	...	9.1	75	773	25.4	27	17	38
\multicolumn{10}{c}{*Mature horses at medium work*}									
200	...	...	5.3	100[d]	450	13.3	10	7	13
400	...	...	8.9	100	756	23.8	19	12	25
500	...	...	10.6	100	900	28.7	23	15	31
600	...	...	12.1	100	1028	33.6	28	18	38
\multicolumn{10}{c}{*Mares (last 90 d of pregnancy)*}									
200	...	...	3.4	63	390	9.2	14	9	25
400	...	...	5.6	63	640	15.5	27	19	50
500	...	...	6.7	63	750	18.4	34	23	63
600	...	...	7.7	63	870	21.0	40	27	75

Body weight (kg)	Age (mo)	Milk (kg)[e]	Daily diet (kg)	Daily diet (g/W^0.75)	DP[a] (g)	DE[b] (Mcal)	Calcium (g)	Phosphorus (g)	Carotene[c] (mg)
\multicolumn{10}{c}{*Mares (first 90 d of lactation)*}									
200	...	8	5.3	100	710	14.6	24	16	33
400	...	12	8.9	100	1120	24.6	40	27	55
500	...	15	10.6	100	1360	29.1	50	34	69
600	...	18	12.1	100	1600	33.3	60	40	83

Sources: NRC 1973, 1978.

[a]Total daily digestible protein (DP) is calculated from data in Table 21.3 and Fig. 21.2.
[b]Based upon 2.75 Mcal DE and 62.5% TDN/kg 100% dry feed.
[c]One mg B-carotene equals 400 IU of vitamin A.
[d]Adjusted to 75 and 100 g/W^0.75 for horses at light and medium work, respectively.
[e]Average milk production for first 90 d of lactation.

varies in grasses and legumes (Colburn et al. 1968). Hemicellulose values, as a percentage of cellulose, for alfalfa, orchardgrass, and corn silage were 43%, 76%, and more than 80%, respectively. These composition data indicate cellulose does not constitute a uniform percentage of all forage fibers and total diet digestibility should not be estimated from cellulose only, but from the total diet fiber.

Coefficients of digestibility for both hemicellulose and cellulose are about 25% lower in mature horses than in growing cattle (Colburn et al. 1968; Fonnesbeck 1968a). True coefficients of digestibility in horses for hemicellulose and cellulose are 49% and 41%, respectively, when the diet contains 25% of each. Coefficients of true digestibility for cel-

lular content and nonfibrous carbohydrate are 90% or greater. In the same digestion trials, the coefficient of true digestibility for forage ether extract (fat) is 75%. Of the total fat in forage, true fat constitutes about 42% and is highly digestible (Morrison 1959; Fonnesbeck 1968a).

Limited data are available on the influence of diet intake on digestibility in the horse. A summary statement of research completed in 1929 states that diet intake has no definite effect on digestibility until it exceeds 2% of body weight (Olson 1969). This does not seem to be an area of great importance in the horse. An exception might be in younger growing horses and lactating mares. These animals have greater diet ingestion capacity and higher nu-

TABLE 21.2.   Advancing maturity of forage effect on digestible dry matter for horses

Forage stage	DDM[a] (%)	DM basis (%)				
		CP	Cellulose	Calcium	Phosphorus	Magnesium
			Alfalfa[b]			
Prebud	83[c]	30.9	15.2[d]	. . .[d]	. . .	. . .
Bud (1%)	76	26.9	20.9	2.2	.33	. . .
Bud (62%)	72	25.2	21.7	1.6	.24	.23
Bloom (11%)	66	21.3	27.3	1.3	.24	. . .
Bloom (46%)	62	19.1	28.5	1.2	.22	.29
Bloom (96%)	59	16.9	31.4	1.1	.20	. . .
			Orchardgrass[e]			
First growth						
Boot	71	22.8	25.9	0.9	.43	.14
Head emergence	63	17.9	28.8	0.8	.42	.14
Preanthesis	59	15.3	30.3	0.8	.31	.14
Postanthesis	55	13.4	31.3	1.1	.24	.14
First aftermath (42 d)	67	20.4	26.5	0.7	.24	.18
First aftermath (56 d)	61	17.1	28.1	0.6	.19	.26
Second aftermath (42 d)	67	21.2	26.2	0.7	.28	.28
Second aftermath (56 d)	63	18.9	27.1	0.6	.26	.31
			Corn silage[f]			
Blister	54	11.7	30.8	0.38	.28	.28
Early milk	54	12.0	28.8	0.31	.27	.26
Milk–early dough	54	11.6	25.1	0.26	.25	.22
Dough-dent	53	10.9	22.9	. . .	.23	.22
Glaze	51	10.2	20.0	0.28	.31	.22
Flint	51	10.0	18.8	0.23	.31	.24
Postfrost	51	9.6	19.3	0.14	.30	.19
Mature	52	10.9	16.9	0.09	.33	.16

[a]DDM estimated from ruminant data using the following equation: Equine % DDM = 19.92 + 1.48% crude protein + 0.24% ruminant DDM. Vander Noot and Trout (1971).

[b]California data from Weir et al. (1960).

[c]Ruminant DDM estimated from TDN using the equation DDM = (TDN − 16.0) ÷ 0.69. Unpublished Rutgers data on alfalfa from Evans, Rutgers.

[d]For alfalfa, fiber reported as crude fiber. Mineral values from Morrison (1959).

[e]New Jersey data from Colburn et al. (1968). Unpublished Rutgers mineral data from Evans, Rutgers.

[f]Ohio data from Johnson and McClure (1968).

trient requirements to meet their physiological conditions.

Horses can sleep while standing, and this phenomenon has resulted in much discussion about the comfort and metabolism or energy requirement in the standing and lying positions. Metabolism in horses while standing and lying has been reported to be the same (Brody 1945).

PROTEIN

Crude protein (CP) and digestible protein (DP) requirements of horses decline from birth to maturity both as a percentage of diet intake and in total grams per day (Table 21.1). Rate of growth, pregnancy, and lactation increase the DP requirement, but work does not. In the horse, digestibility of protein changes in proportion to diet protein (Fig. 21.2), and the protein requirement for the horse should be expressed as DP. The slope of the regression line in Fig. 21.2 indicates that the coefficient of true digestibility is 85% for

forage protein, and the constant indicates that obligatory fecal loss or metabolic fecal protein is 2.8% of the ingested diet dry matter. With 6% and 16% diet protein, the respective total fecal losses are 3.7% and 5.2%. Coefficients of true digestibility for forage and grain proteins were 85% and 95%, respectively. The respective values for the percent DP and the coefficients of (apparent) digestibility, when protein in all-forage and all-grain diets is increased from 6% to 20%, are given in Table 21.3.

Forages and other feedstuffs are analyzed for protein, and the resulting value can be equated with the CP requirement (Tables 21.1, 21.3). The CP requirement for the 265-kg growing horse with a mature weight of 400 kg is 600 g. Division of 600 g by 5000 g of ingested diet results in a figure of 12% CP. When regression is used (Fig. 21.2, Table 21.3), the diet protein should be 12% with forage and about 2% lower with an all-grain diet.

If a lower-energy diet is fed at a higher lev-

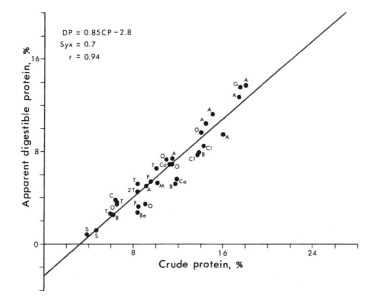

Fig. 21.2. Regression of the percentage apparent DP on the percentage crude protein content in all-forage diets fed to horses. The number of data points (in parentheses) follows the author-date citation: Darlington and Hershberger 1968 (9); Fonnesbeck et al. 1967 (12); Vander Noot and Gilbreath 1970 (4); and Slade et al. 1970 (8). Means were plotted for alfalfa (A), bermudagrass (Be), smooth bromegrass (B), cereal hay (C), reed canarygrass (Ca), tall fescue (F), green chop (G), meadow hay (M), orchardgrass (O), straw (S), and timothy (T).

el of intake to the mature horse working at a medium level, the CP requirements given in Table 21.1 may be too low to meet both the obligatory and the endogenous losses in forage diets. It appears that the process of digestion in higher-fiber forage diets results in an increased requirement, not physical work of the horse per se. Blood hematocrit may be used to indicate the adequacy of diet protein (Biddle et al. 1975).

Protein quality or balance of amino acids in the diet may be deficient in the presence of adequate DP in nonruminant animals such as the horse. The more critical times in the life cycle of the horse are during rapid growth of the foal when it has a decreasing supply of milk and must depend on other available feedstuffs, during pregnancy when improper diets are fed to prevent fattening, and during lactation when milk production is high. Based upon recent amino acid data for different feedstuffs and the amino acid requirements for growing swine published by the National Research Council (NRC), low-protein forages and several cereal grains may be borderline in some amino acids for the rapidly growing foal (NRC 1977, 1979). Barley, corn, oat, and sorghum (*Sorghum bicolor* [L.] Moench) grains, and more mature alfalfa with less than 15% protein may be borderline in both lysine and methionine. Threonine, isoleucine, and tryptophan are suspect also for several of these feedstuffs. Inclusion of soybean meal or high-protein forage to provide 10%-30% of the total diet protein should correct most amino acid deficiencies.

Forages contain varying amounts of nonprotein nitrogen (NPN); in an eight-wk trial 2.5%-4% urea in the diet did not appear to be detrimental to the horse (Slade et al. 1970; Nelson and Tyznik 1971). Many questions remain concerning the utilization of NPN and the contribution of the protein digested in the cecum and large intestine to the nutrition of the horse (Reitnour et al. 1970; Hintz et al. 1971).

MINERALS

Large variations exist in mineral composition within and among forage species. These variations in most of the minerals considered essential for animals have been recorded for alfalfa (Bear and Wallace 1950), and other mineral composition data for many feedstuffs have been reported by the NRC (1977). Re-

TABLE 21.3. Digestibility of dietary protein, horses

Dietary crude protein (%)	Forage[a]		Grain[b]	
	DP (%)	Coefficient of digestibility (% × 100)	DP (%)	Coefficient of digestibility (% × 100)
6	2.3	38	3.5	58
8	4.0	50	5.4	67
10	5.7	57	7.3	73
12	7.4	62	9.2	77
14	9.1	65	11.1	79
16	10.8	67	13.0	81
18	12.5	69	14.9	83
20	14.2	71	16.8	84

[a]DP calculated from equation presented in Fig. 21.2.
[b]Equation, DP = 0.95% crude protein − 2.2, based upon data from Olson (1969).

cent mineral composition data (Ralston and LeRoy 1991) have reconfirmed the large variations in mineral contents in forages. Limited experimental data on the mineral requirements for the horse are available (Squibb 1958; Morrison 1959; Olson 1969; Cunha 1971; NRC 1973, 1978; Maynard et al. 1979). The complications of mineral interrelationships have been discussed for other animals, and some of these may apply to the horse (Tillman 1966). A chemical analysis for minerals in a diet gives the amount present, but it does not give the amount biologically available to the horse.

Salt. Most forages supply a deficient amount of sodium (Na) to the horse. The mature horse at medium work may lose 60 g NaCl daily in the sweat and 35 g in the urine (NRC 1973, 1978). Providing self-feeding salt for horses on pasture or on a variable work schedule provides an adequate supply.

Phosphorus and Potassium. Forage contents of phosphorus (P) and potassium (K) decline with advancing maturity of the forage (Table 21.2) and with lower soil concentrations. Grains usually have higher amounts of P than do forages, but they are deficient in K for the horse. A large amount of P in forage and grain is present as phytate P, and the bioavailability of P in this form is expected to be lower than that of inorganic orthophosphate in the horse.

It is still necessary to meet the NRC diet level requirements for P and Ca. But new data do suggest that a range in mineral ingredients and forages can influence cation-anion balance in a given diet (Petito and Evans 1981). Cation-anion balance is more than the balance calculated as Na plus K less chlorine (Cl) (Jasaitis et al. 1987), but this limited-type research in the horse with Na-K-Cl diet variation has produced metabolic acidosis (Stutz et al. 1991) and a negative Ca balance (Wall et al. 1991). Diet sources and/or levels of protein, P, and Ca influence bone integrity perhaps via acid-base metabolism as influenced by cations and anions in the diet in the horse (Thompson et al. 1988) and in the rat (Petito and Evans 1984).

Calcium and Magnesium. Legumes contain more calcium (Ca) than do grasses, and grasses are higher in Ca than are most grains. Mixtures of legumes and grains tend to provide a good balance of P, Ca, and magnesium (Mg) in the diet. Cattle have adapted or changed their

rate of absorption of Ca with time when diet Ca exceeded the requirement (Garces and Evans 1971). It may be that the excessive Ca in legumes undergoes the same treatment in the horse when P and Mg are adequate in the diet. Most reports suggest that a proper ratio be maintained between Ca and P and that diet Ca should exceed diet P. The Ca:P ratio in horse milk is 1.7:1.0 and appears to be adequate for the horse or foal (Maynard et al. 1979). The Mg content of forage is closely related to soil temperature, and for this reason it increases with advancing maturity in spring and summer forages but does not increase in forages harvested in the fall (Table 21.2) (Rook and Storry 1962).

Sulfur. The NRC has concluded that inorganic sulfur (S) is not essential in a diet for the horse. However, even if adequate S amino acids are fed, it has increased both weight gain and femur ash in the rat (Evans and Davis 1976).

Trace Elements. Large variations in trace mineral concentrations in feedstuffs have been related to plant species, soil type, area, mineral content, and growing season (Bear and Wallace 1950) (Table 21.4). The NRC has recommended 0.1 mg iodine (I), less than 0.05 ppm cobalt (Co), 5-8 ppm copper (Cu), and less than 40 ppm iron (Fe) for the horse (NRC 1973, 1978). It has been suggested that most forages contain adequate zinc (Zn) and manganese (Mn), but this is not always true (Table 21.4). In addition, Cu and Fe should be maintained at NRC levels to ensure growth and skeletal development in nonmature horses (Ott and Asquith 1989). Molybdenum (Mo) and selenium (Se) should also be considered (Olson 1969; Cunha 1971; Maynard et al. 1979).

VITAMINS

Green leafy forage harvested at an early stage of maturity usually will provide adequate amounts of carotene and B vitamins for the horse. Forages vary in vitamin composition (NRC 1973, 1978). Adequate levels of vitamin A activity or carotene in forages ingested by the mare and her foal are critical, because the vitamin A level may enhance the foal's ability to develop expected favorable immunologic activity (Stowe 1982). In both first-growth and regrowth orchardgrass at all levels of N fertilization studied, carotene has varied with percent protein: carotene, mg/454 g = 0.79 (% protein) − 3.12 (Evans et al. 1965).

TABLE 21.4. Mineral composition and variation in alfalfa

Mineral	Number of observations	Range	Mean
Sodium (%)	43	0.01–0.33	0.16
Chlorine (%)	110	0.06–0.54	0.28
Phosphorus (%)	882	0.01–0.97	0.26
Potassium (%)	442	0.22–3.37	1.77
Calcium (%)	879	0.15–2.99	1.64
Magnesium (%)	389	0.03–0.84	0.32
Sulfur (%)	76	0.20–0.73	0.36
Iodine (ppm)	56	0.06–0.29	0.12
Cobalt (ppm)	113	0.02–0.31	0.13
Copper (ppm)	58	4–38	14
Iron (ppm)	265	40–1640	240
Zinc (ppm)	67	10–29	17
Manganese (ppm)	231	8–100	52

Source: NRC (1958).

It may be that carotene in lower-protein versus higher-protein forage has a lower availability in the horse (Fonnesbeck and Symons 1967).

Limited data on ascorbic acid for horses are available, and the choline content of forages for horses should be evaluated (NRC 1958; Stillions et al. 1971a). It has been indicated that mature horses do not have a diet vitamin B_{12} requirement; however, this does not mean that the growing horse does not require diet B_{12} (Stillions et al. 1971b). It may be that when a forage supplies the minimum adequate DP to the horse, more consideration should be given to fat-soluble vitamins than to water-soluble ones.

MAXIMUM FORAGE DIETS

In addition to the feeding value of forage, quality horse pastures should form a high-quality turf and withstand close grazing (Fig. 21.3). Kentucky bluegrass (*Poa pratensis* L.) sod was not as wear resistant as tall fescue (*Festuca arundinacea* Schreb.) sod (Seaney and Reid 1981). Orchardgrass stores its energy reserves above the soil surface and does not withstand as close grazing as timothy (*Phleum pratense* L.), which stores its reserves below the soil surface. Intermediate grasses are Kentucky bluegrass, smooth bromegrass (*Bromus inermus* Leyss.), and reed canarygrass. Horse pastures should include both legumes and grasses that are adapted to a given location. In most cases horse pastures should be clipped, and at times the droppings should be spread to aid in maintaining uniform fertility and in reducing parasite populations.

Pastures vary greatly in productivity. Hectares required to graze one horse during a growing season range from 0.2 to more than 1.6 (Seaney and Reid 1981). Horse prefer-

ences in US studies ranked Kentucky bluegrass first among the grasses and alfalfa first among the legumes (Merritt et al. 1969); however, in British studies creeping red fescue (*F. rubra* L.) and tall fescue were rated over Kentucky bluegrass and timothy (Archer 1980). Legume pastures usually require rotational grazing, and N fertilizer should be considered when legumes make up less than 25% of the pasture (Lacefield et al. 1981).

A hectare of improved pasture may meet nutritional needs for one horse (Evans et al. 1990) when good crop management is applied. But a pasture that provides a quality environment for feed and exercise (Raub et al. 1989) and minimizes denuding of the vegetation may require 3 ha or more per mature horse.

Maintenance at Mature Weight. In most cases, a diet of low-quality mature forage containing 35%–40% cellulose and 8%–10% protein on a dry basis should provide the maintenance nutrient requirements for the mature horse. Usually, salt and water are provided free-choice. In areas known to have problems related to nutrition, diet supplements may be required. Should the distended intestinal tract associated with a low-quality forage diet be undesirable, one of two actions could be taken. The horse could be fed the low-quality forage several times per day, or

Fig. 21.3. Mare and foal on extensive pasture.

the total diet fiber could be reduced by feeding a limited amount of grain mixture to reduce forage intake and reach the desired body condition.

Work. Of all the required nutrients, only that for energy varies greatly with the intensity of work. However, a supply of nutrients lost in sweat should be provided (e.g., water and Na). The working horse has a reduced eating time and a higher energy requirement. After work the horse should not be allowed to eat too fast or to overeat. Voluntary intake decreases as forage maturity or fiber content increases (Table 21.2). Some work can be supported by a low-quality forage in the mature horse; the condition of the horse should indicate the relationship between diet energy and work intensity. If the horse is losing body condition, a lower-fiber forage and/or a 8%-10% protein-grain mixture should be fed in an amount that maintains the desired condition.

Pregnancy. Nutrient requirements for the last quarter of pregnancy differ from work requirements (i.e., energy is similar to the maintenance requirement, but requirements for all other nutrients are greater than maintenance). A large area of low-quality pasture is desirable for the mature pregnant mare. This provides exercise, and a high-fiber forage reduces energy intake. However, depending on the size of the mare, in addition to this high-fiber forage up to 1 kg/d of protein supplement fortified with minerals and vitamins should be provided, as discussed in the protein section.

Lactation. As feed, the average quality of hay presently fed in the US does not provide adequate nutrition for the lactating mare without some grain mixture added to the diet. By harvesting forage at an earlier stage of maturity so that fiber will not restrict energy intake, all-forage diets can be consumed in amounts that will meet the nutrient requirements of most mares (Table 21.2). The small mare (pony) and the mare in poor condition are exceptions. In most cases a 14% protein forage would provide adequate protein, but energy intake as limited by forage fiber would not be adequate. DDM levels provided by alfalfa and orchardgrass at various stages of maturity are given in Table 21.2.

Growth. Like the lactating mare, the growing foal needs a high-quality, low-fiber diet that does not restrict voluntary intake. The diet should supply adequate amounts of protein containing a proper balance of amino acids and minerals and vitamins that are biologically available from plant sources. Initially the foal consumes mare's milk, which is 25% protein, 15% fat, 55% lactose, 5% mineral (0.9% Ca, 0.54% P, 0.9% K), and provides energy of 4.5 kcal/g (Maynard et al. 1979).

At a few weeks of age the foal starts to nibble available feedstuffs (i.e., forage as hay or pasture and/or grain that is provided for the mare). When teething occurs, the diet should be provided in a form that enables adequate voluntary intake. As the foal grows, the high-energy, 25% protein milk diet is replaced by other feedstuffs. Digestible protein for the growing foal (from before weaning to past 1 yr of age) should be decreased from 17% to about 10%. Creep pastures for foals should be developed. The level of protein required depends on the anticipated mature weight of the horse and its stage of growth (Table 21.1). Even past 1 yr, and at the expense of some overfeeding of protein, the growing horse should have good quality forage to provide adequate energy.

When forages are borderline in quality, the protein supplement with minerals and vitamins for the pregnant mare or the grain mixture for the lactating mare may be used to provide a balanced diet. In areas where the recurrence of anemia and poor skeletal development are frequent, the supply and availability of diet Co, Fe, and Cu should be evaluated (Ott and Asquith 1989; Tillman 1966; Evans and Abraham 1973; NRC 1973, 1978). The fast-gaining horse of larger, mature weight requires a large voluntary intake of forage, indicating a requirement for a low-fiber, immature forage (Table 21.1).

QUESTIONS

1. Using the same fibrous diet, will the mature bovine or the mature horse digest more fiber? Why?
2. What is digestible protein? Metabolic fecal protein? Nonprotein N?
3. Distinguish between apparent and true digestibility of protein.
4. What is the percentage of Ca in calcium carbonate and dicalcium phosphate? Also what is the pH and acidity potential for each ingredient?
5. A 385-kg horse consumed 6.9 kg of forage. Express body weight in kilograms to the 0.75 power of body weight and calculate the diet intake of $g/W^{0.75}$.
6. How do lignin, cellulose, and hemicellulose content influence the digestibility of energy in the horse?
7. Using the information in this chapter and your

local forage data, balance diets for a mare at stages of maintenance, work, pregnancy, lactation, and growth.
8. Describe and discuss the use of a creep pasture for foals.

REFERENCES

Archer, M. 1980. Grassland management for horses. Vet. Rec. 107:171-74.

Bear, FE, and A Wallace. 1950. Alfalfa, its mineral requirements and chemical composition. N.J. Agric. Exp. Stn. Bull. 748.

Biddle, GN, JL Evans, and JR Trout. 1975. Labile nitrogen reserves and plasma nitrogen fractions in growing cattle. J. Nutr. 105:1584-91.

Blaxter, KL. 1961. Efficiency of feed conversion by different classes of livestock in relation to food production. Fed. Proc. 20 (part 3, suppl. 7):268-74.

Brody, S. 1945. Bioenergetics and Growth. New York: Hafner.

Colburn, MW, and JL Evans. 1967. Chemical composition of the cell wall constituent and acid detergent fiber fractions of forages. J. Dairy Sci. 50:1130-35.

Colburn, MW, JL Evans, and CH Ramage. 1968. Apparent and true digestibility of forage nutrients by ruminant animals. J. Dairy Sci. 51:1450-57.

Crowell, P. 1951. Cavalcade of American Horses. New York: Bonanza Books.

Cunha, TJ. 1971. The mineral needs of the horse. Feedstuffs 43(46):34,36,38, and 51.

Darlington, JM, and TV Hershberger. 1968. Effect of forage maturity on digestibility, intake and nutritive value of alfalfa, timothy and orchardgrass by equines. J. Anim. Sci. 27:1572-76.

Evans, JL, and PA Abraham. 1973. Anemia, iron storage and ceruloplasmin in copper nutrition in the growing rat. J. Nutr. 103:196-201.

Evans, JL, and GK Davis. 1976. In Molybdenum in the Environment. New York: Marcel Dekker.

Evans, JL, J Arroyo-Aguilu, MW Taylor, and CH Ramage. 1965. Date of harvest of New Jersey forages as related to the nutrition of ruminant animals. N.J. Agr. Exp. Stn. Bull. 814.

Evans, JW, A Borton, HF Hintz, and LD Van Vleck. 1990. The Horse. 2d ed. San Francisco, Calif.: WH Freeman.

Fonnesbeck, PV. 1968a. Digestion of soluble and fibrous carbohydrate of forage by horses. J. Anim. Sci. 27:1336-44.

———. 1968b. Consumption and excretion of water by horses receiving all hay and hay-grain diets. J. Anim. Sci. 27:1350-56.

———. 1969. Partitioning the nutrients of forage for horses. J. Anim. Sci. 28:624-33.

Fonnesbeck, PV, and LD Symons. 1967. Utilization of the carotene of hay by horses. J. Anim. Sci. 26:1030-38.

Fonnesbeck, PV, RK Lydman, GW Vander Noot, and LD Symons. 1967. Digestibility of the proximate nutrients of forages by horses. J. Anim. Sci. 26:1039-45.

Garces, MA, and JL Evans. 1971. Calcium and magnesium absorption in growing cattle as influenced by age of animal and source of dietary nitrogen. J. Anim. Sci. 32:789-93.

Haenlein, GFW, RD Holdren, and YM Yoon. 1966. Comparative response of horses and sheep to different physical forms of alfalfa hay. J. Anim. Sci. 25:740-43.

Hintz, HF, DE Hogue, EF Walker, Jr., JE Lowe, and HF Schryver. 1971. Apparent digestion in various segments of the digestive tract of ponies fed diets with varying roughage-grain ratios. J. Anim. Sci. 32:245-48.

Hughes, HD, and ME Heath. 1962. Hays and pasture for horses. In Forages: The Science of Grassland Agriculture, 2nd ed. Ames: Iowa State Univ. Press, 671-83.

Jasaitis, DK, JE Wohlt, and JL Evans. 1987. Influence of feed ion content on buffering capacity of ruminant feedstuffs in vitro. J. Dairy Sci. 70:1391-1403.

Johnson, RR, and KE McClure. 1968. Corn plant maturity. IV. Effects on digestibility of corn silage in sheep. J. Anim. Sci. 27:535-40.

Lacefield, GD, JK Evans, and JP Baker. 1981. Horse pastures. Univ. of Ky Agric. Coop. Ext. Serv. Agron-81.

Maynard, LA, JK Loosli, HF Hintz, and RG Warner. 1979. Animal Nutrition. 7th ed. New York: McGraw-Hill.

Merritt, TL, JB Washko, and RH Swain. 1969. The selection and management of forage species for horses. Pa. State Univ. Anim. Sci. Mimeo H-69-1.

Morrison, FB. 1959. Feeds and Feeding. 22d ed. Ithaca, N.Y.: Morrison.

National Research Council (NRC). 1958. Composition of Cereal Grains and Forages. Publ. 585. Washington, D.C.: National Academy of Science.

———.1973, 1978. Nutrient Requirements of Horses. Washington, D.C.: National Academy of Science.

———. 1977. Nutrient Requirements of Poultry. Washington, D.C.: National Academy of Science.

———. 1979. Nutrient Requirements of Swine. Washington, D.C.: National Academy of Science.

Nelson, DD, and WJ Tyznik. 1971. Protein and nonprotein nitrogen utilization in the horse. J. Anim. Sci. 32:68-73.

Olson, NO. 1969. The nutrition of the horse. Nutrition of Animals of Agricultural Importance 17(2):921-60. New York: Pergamon Press.

Ott, EA, and RL Asquith. 1989. The influence of mineral supplementation on growth and skeletal development of yearling horses J. Anim. Sci. 67:2831-40.

Petito, SL, and JL Evans. 1981. In DD Hemphell (ed.), Trace Substances in Environmental Health, vol. 15. Columbia: Univ. of Missouri, 217-25.

———. 1984. Calcium status of the growing rat as affected by diet acidity from ammonium chloride, phosphate, and protein. J. Nutr. 114:1049-59.

Ralston, SL, and R LeRoy. 1991. Nutrient analysis of hays fed to horses in New Jersey. In Proc. 12th Equine Nutr. Physiol. Symp. Univ. of Calgary. 12:251.

Raub, RH, SG Jackson, and JP Baker. 1989. The effect of exercise on bone growth and development in weanling horses. J. Anim. Sci. 67:2508-14.

Reitnour, CM, JP Baker, GE Mitchell, Jr., CO Little, and DD Kratzer. 1970. Amino acids in equine cecal contents, cecal bacteria and serum. J. Nutr. 100:349-54.

Rook, JAF, and JE Storry. 1962. Magnesium in the nutrition of farm animals. Nutr. Abstr. Rev. 32:1055-77.

Seaney, RR, and WS Reid. 1981. Pasture improvement and management for horses. Cornell Coll. Agric. Life Sci. Info. Bull. 171, 1-12.

Simpson, GG. 1961. Horses: The Story of the Horse Family in the Modern World and through Sixty Million Years of History. Doubleday, N.Y.: American Museum of Natural History and Science.

Slade, LM, DW Robinson, and KE Casey. 1970. Nitrogen metabolism in nonruminant herbivores. I. The influence of nonprotein nitrogen and protein quality on the nitrogen retention of adult mares. J. Anim. Sci. 30:753-60.

Squibb, RL. 1958. Fifty years of research in America on the nutrition of the horse. J. Anim. Sci. 17:1007-14.

Stillions, MC, SM Teeter, and WE Nelson. 1971a. Ascorbic acid requirement of mature horses. J. Anim. Sci. 32:249-51.

———. 1971b. Utilization of dietary vitamin B_{12} and cobalt by mature horses. J. Anim. Sci. 32:252-55.

Stowe, HD. 1982. Vitamin A profiles of equine serum and milk. J. Anim. Sci. 54:76-81.

Stutz, WA, DR Topliff, DW Freeman, WB Tucker, and JW Breazile. 1991. Effects of dietary cation-anion balance on blood parameters in exercising horses. Proc. 12th Equine Nutr. Physiol. Symp. Univ. of Calgary. 2:135-43.

Thompson, KN, SG Jackson, and JP Baker. 1988. The influence of high planes of nutrition on skeletal growth and development of weanling horses. J. Anim. Sci. 66:2459-67.

Tillman, AD. 1966. Recent developments in beef cattle feeding. In Proc. Pfizer Res. Conf. 14:7-18.

Vander Noot, GW, and EB Gilbreath. 1970. Comparative digestibility of components of forages by geldings and steers. J. Anim. Sci. 31:351-55.

Vander Noot, GW, and JR Trout. 1971. Prediction of digestible components of forages by equines. J. Anim. Sci. 33:38-41.

Wall, DL, DR Topliff, DW Freeman, DG Wagner, and JW Breazile. 1991. Effects of dietary cation-anion balance on urinary mineral excretion in exercised horses. Proc. 12th Equine Nutr. Physiol. Symp. Univ. of Calgary. 12:121-26.

Weir, WC, LG Jones, and JH Meyer. 1960. Effect of cutting interval and stage of maturity on the digestibility and yield of alfalfa. J. Anim. Sci. 19:5-19.

Yoshida, S, H Sudo, K Noro, and C Nokaya. 1967. Measurement of drinking water volume in horses. Exp. Rep. Equine Health Lab. 4:29-36.

22 Forages For Sheep, Goats, and Rabbits

DONALD G. ELY
University of Kentucky

Sheep

THE INDUSTRY

DEVELOPMENT of the world sheep (*Ovis aries*) industry has paralleled the colonization of the western hemisphere, Australia, New Zealand, and other countries because sheep have the ability to convert noncompetitive plant materials into essential food, fiber, and by-products for human use. The same principle applies today in that sheep can be raised entirely from forages. Therefore, production costs are low when compared with other livestock enterprises.

There are approximately 890 million sheep in the world (USDA 1991). Their location extends from the Arctic circle to the tips of South America and New Zealand, with the highest concentrations in the temperate climates and sparsely settled areas of the southern hemisphere and in the tropical areas of South America, Africa, and India. The 10 leading countries have over 78% of the world's sheep (Table 22.1). Australia and the former USSR produce 33% of the total. Sheep pro-

duce 4.1% of the world's meat, 1.6% of the milk, and 1.63 million mt (3.6 billion lb) of wool annually (FAO 1991).

Historically, the economic viability of the sheep industry has been influenced by wars, politics, supply and demand, meat and fiber substitutes, short-term fads, competition for land, labor, capital, climate, and traditions.

Sheep in the US number just over 9 million head (USDA 1991). Texas has approximately 19% of this total (Table 22.2). About half of the US sheep population is maintained under western range conditions. The remaining half is found in the fenced areas of Texas and in small flocks (20 to 200 hd) of the Midwest and East. Very few sheep are found in the Southeast.

Sheep lead all farm animals in their ability

DONALD G. ELY is Professor of Animal Sciences, University of Kentucky. He holds the MS degree from Oklahoma State University and the PhD from the University of Kentucky. His research encompasses nutrient requirements of growing-finishing lambs, supplementation of grazing ruminant diets, carcass composition, and milk production studies.

TABLE 22.1. Leading sheep-producing countries of the world, by rank

Country	No. of head (*thousand*)
Australia	162,744
Former USSR	134,000
China	112,820
New Zealand	57,000
India	55,700
Iran	45,000
Turkey	40,553
South Africa	32,580
United Kingdom	29,954
Argentina	27,552
World total	889,242

Source: FAO (1991).

TABLE 22.2.   Ten leading states in stock sheep, by rank

State	No. of head (*thousand*)
Texas	1710
California	840
Wyoming	810
Colorado	645
South Dakota	543
Montana	502
Utah	440
Oregon	390
New Mexico	337
Iowa	261
US total	9079

Source: USDA (1994).

to produce marketable products (lamb and wool) from forage alone (Terrill and Price 1985). The advantage is even greater on low-quality (Ragland et al. 1990) or sparse forage. Sheep are particularly adept in the use of a variety of forages—from high-quality grass and legume forage in improved pastures to forbs and browse of varying quality in extensive range conditions to harvested hays and silages with concentrates fed in confinement (Glimp 1991). Supplementation of forage diets with concentrates may be needed by ewes only during a 6-wk late gestation and 8-wk lactation period or when rapid growth of lambs on pasture is desired (Ely et al. 1979). Therefore, the expected level of production of the ewe is closely associated with that of the lamb, rather than a level where supplementation may be required throughout the year (Glimp 1991).

Grassland sheep production is practiced on extensive rangelands, semiarid rangelands, and improved (cultivated) pastures. Shrubs, trees, annual grasses, perennial grasses, and annual herbs and forbs are the main vegetative inhabitants of extensive rangelands. Although supplementation with protein and/or energy might increase total production, most sheep operations in these areas are so extensively managed it is not economically feasible to do so. During periods of vegetative growth, sheep in these operations selectively graze prostrate-growing grasses and annual forbs. As herbage quantity and quality decline, extensively managed ewes become less selective and revert to browsing on shrubs and trees, especially during winter maintenance periods (Glimp 1991).

Semiarid rangelands receive between 400 and 700 mm rainfall annually. Ecosystems include some land capable of small-grain production along with steep-sloped and/or shallow and rocky areas incapable of cultivation.

Sheep and cattle graze these areas, particularly winter wheat (*Triticum aestivum* L. emend. Thell.). Although rainfall is adequate to support relatively good plant growth once or twice during the year, the lack of consistently high forage quality throughout the year usually limits maximum lamb and wool production. Supplementation schemes are built around small-grain grazing or the use of sorghum-sudangrass (*Sorghum bicolor* [L.] Moench) adapted to specific areas. Hay and/or grain supplementation may be required in winter.

Sheep produced on farms that have improved or cultivated pastures are usually rotated with corn (*Zea mays* L.), soybeans (*Glycine max* [L.] Merr.), alfalfa (*Medicago sativa* L.), and other crops. In these situations, sheep compete economically for the land use because yields of forage dry matter (DM) are high. Seasonal periods of very high-quality forage production should coincide with the sheep's peak nutrient needs, and pasture growth should be distributed throughout the year. For example, ewe stocking rates are set by predetermining how many ewes can be supported during the period of lowest pasture DM production. The periods when forage growth is greatest and forage quality is highest should coincide with the lactation and lamb growth phase of the annual production cycle. Surplus pasture is either grazed by cattle or harvested as hay or silage for feeding during winter or drought periods.

SHEEP CHARACTERISTICS

Compared with other ruminant species, sheep are relatively small and easily handled. They have a small, narrow mouth and tongue. The upper lips are divided by a vertical cleft that allows mobility and selective grazing. They prefer to consume broadleaf plants (legumes and weeds) and the finer plant parts, although they will eat large leaves, flowers, and seeds. They can graze, in moderate amounts and during certain seasons, some plants that may be toxic to other ruminants.

The characteristic short gestation (147 d), high fecundity, and short growing period of sheep add both efficiency and flexibility to their production. Other advantages of sheep compared with other ruminants are production of two marketable products (lamb and wool), multiple births, low investment per dollar of income, slaughter lamb production from forages, and production of high-quality wool from forages alone. Sheep consume a wide array of forages (including weeds), contribute to reduced mowing requirements, and recycle soil nutrients necessary for plant growth through their fecal and urine excretions. Furthermore, the adaptability and flexibility of sheep to highly extensive or intensive methods, or combinations thereof, permit the use of a great variety of forages.

The life cycle of the sheep is considered to be the productive life of the ewe. This normally ranges from five to seven lamb crops per lifetime. More specifically, nutrition is based on the ewe's annual production phases: flushing (FL), breeding (BR), early gestation (EG), late gestation (LG), lactation (LA), and maintenance (dry, nonpregnant; MA). The months when each phase occurs vary widely and are based on whether lambing occurs in fall (October), winter (January-February), or spring (April). Likewise, the extent of forage use in an individual sheep production scheme depends on the climate, lambing time, and primary objective of the operation (wool, feeder lamb, or slaughter lamb production). Regardless of the annual production scheme, ewe weight changes will follow the basic pattern of Figure 22.1 (SID 1987). Weight loss during LA will be approximately 80% of that shown in Figure 22.1 if lambs are early weaned at 60 d of age. Weight changes of ewes rearing a single lamb would be approximately two-thirds of those rearing twins, whether early weaned or not.

Energy is the most commonly deficient nutrient in the ewe's diet (Ball et al. 1991). The energy requirements (TDN = total digestible nutrients) of a 70-kg ewe (NRC 1985) change during the year because of different requirements for each phase of production (Morrical 1992) (Fig. 22.2). Even when these require-

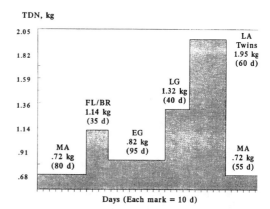

Fig. 22.2. Daily total digestible nutrient requirements of a 70-kg ewe in different stages of annual production.

ments are met, weights of ewes fluctuate throughout the year according to Figure 22.1. If the sheep operation is built around forage feeding, energy is likely to be the first limiting nutrient, especially during LG and LA, because forages are not as energy dense as concentrates. To avoid any nutritional deficiencies, the producer must match each production phase during the year to the quantity and quality of forage available.

The annual production program usually begins with the FL phase (Kruesi 1985). To obtain maximum lambing rates, ewes have to be gaining weight at conception. To gain this weight, ewes cannot be in more than moderate body condition when FL begins. Providing each ewe with 0.23-0.45 kg/d of grain, in addition to pasture, or moving the flock to a fresh pasture of high quality (no grain) will stimulate the desired weight gain (Figs. 22.1 and 22.2). If grain is used to flush, it should be continued through the first 3 wk of BR. If fresh pasture is used, the amount and quality must be adequate to provide the necessary energy for the duration of the FL and BR phases. Grass, grass-legume mixed, and annual forage crop pastures, in the vegetative stage of growth, can adequately flush ewes.

Energy requirements during EG are low (Fig. 22.2) because ewe gains are only slightly above MA (Fig. 22.1). Ewes can serve as scavengers to clean fencerows, corn and soybean fields, and other areas during this period. Quality of forages is of less importance during EG as long as the quantity is adequate.

Fetal growth increases dramatically during the last 4-6 wk of gestation. Consequently, nu-

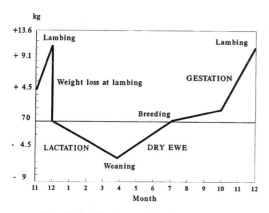

Fig. 22.1. Yearly weight changes of a 70-kg ewe rearing twins.

trient requirements, especially energy, increase accordingly (Fig. 22.2). Feeding ewes in LG is not simply a matter of feeding more but also is a matter of feeding more "energy-dense" diets. Even if the highest-quality pasture, hay, or silage is fed, some concentrated energy source (grain) should be provided to meet the requirements for the ewe and her fetus(es). Example diets for ewes in LG are presented in Table 22.3.

Nutrient requirements are greatest during LA (Fig. 22.2). Still, weight losses can be expected (Fig. 22.1). Therefore, feeding in LA is for milk production, not body condition. Example diets for LA are presented in Table 22.4. The highest-quality pasture or harvested forage should be saved for this production phase. Even then, extra energy will be required to maximize milk production. If a non-legume hay is fed, supplementation with soybean meal will be required to meet the ewe's protein requirements.

The nutrient requirements for the ewe after weaning (MA) are low (Fig. 22.2). This is the phase when low quality pasture can be provided to keep the ewe from becoming too obese for the upcoming FL period. Ewes can again serve as scavengers: cleaning fence rows, grazing after weaned lambs, or grazing behind cattle.

Preweaning growth of lambs on pasture depends primarily on milk production of the ewe and supplemental feed supply in the form of forage or concentrates. If lambs are to be marketed as feeders, they are usually weaned at 4-7 mo of age. In this extensive rangeland production system, their preweaning diet consists of milk and a sparse forage supply. If lambs are to be marketed for slaughter directly off the ewe at 4-6 mo, their preweaning diet typically contains milk, high-quality forage, and a supplemental, energy-dense creep mixture. The same diet can be fed if lambs are early weaned at 2 mo and finished for slaughter off pasture. Although available forage quality and quantity are exceptionally high in

these situations, lambs still have to be supplemented with some concentrate to produce top quality slaughter lambs.

Ely et al. (1979, 1980) evaluated postweaning performance of lambs grazing pure stands of cool-season grasses and grass-legume mixtures. Lambs were either unsupplemented, supplemented with shelled corn at 1% of their body weight daily, or provided shelled corn ad libitum. Grass-legume mixtures produced faster lamb growth than pure stands of grass, but in all cases energy supplementation was required to efficiently produce slaughter lambs directly from high-quality pasture. The level of supplementation depends on how soon after weaning the producers want to get the lambs to slaughter weight. This work shows early-weaned lambs can make efficient use of high-quality forages, but forage intake limits maximum performance from pasture alone.

Dry matter intake of forages depends on sheep size, sex, age, and stage of production; on palatability and availability of the forage; and on environmental factors (Terrill and Price 1985). These authors reviewed many sources of data and conclude that variation in intake was similar for low- and high-grade dried forages: 55.7 g/kg body weight $(BW)^{0.75}$ for shadscale (*Altriplex confertifolia* [Torr. & Frem.] S. Wats.) to 82.2 g/kg $BW^{0.75}$ for immature orchardgrass (*Dactylis glomerata* L.). However, the voluntary intake of forage, as consumed in pasture, varied from 69.6 g/kg $BW^{0.75}$ for shadscale to 481.0 g/kg $BW^{0.75}$ for lush ladino clover (*Trifolium repens* L.). Therefore, daily forage consumption ranged from 0.8 kg for a 25-kg lamb grazing shadscale to 16.9 kg for a 115-kg ram grazing ladino clover.

High-quality forages are usually highly palatable, which stimulates intake. However, highest-quality forages are not always the most palatable (Ball et al. 1991). Intake is usually higher on leafy forages with fine stems, good green color, and no smell of mold or mustiness. These same forages have a high

TABLE 22.3.   Example diets for a 70-kg ewe in late gestation

Ingredient	Diet			
	1	2	3	4
		(kg/d)		
Grass hay	1.5[a]	. . .	. . .	. . .
Alfalfa hay	. . .	1.8[b]	. . .	0.5
Pasture	. . .	. . .	Ad libitum[c]	. . .
Corn silage	. . .	. . .	. . .	Ad libitum
Shelled corn	0.6	0.6	0.5	0.2

[a]Early bloom.
[b]Midbloom.
[c]Free-choice access to small-grain pasture, spring or fall growth of cool-season grasses, or turnips.

TABLE 22.4.   Example diets for a 70-kg ewe nursing twins

Ingredient	Diet			
	1	2	3	4
	(kg/d)			
Grass hay	2.3[a]	. . .	. . .	. . .
Alfalfa hay	. . .	2.3[b]	0.9	0.9
Pasture	. . .	. . .	Free-choice[c]	. . .
Corn silage	. . .	. . .	. . .	Free-choice
Shelled corn	0.7	0.7	0.7	0.5
Soybean meal	0.2	. . .	. . .	

[a]Early bloom.
[b]Midbloom.
[c]Free-choice access to small-grain pastures, spring or fall growth of cool-season grasses, or turnips.

protein content but low concentrations of fiber and lignin. Animal performance may be satisfactory when restricted to a less palatable forage if the nutritive value is sufficiently high. When no choice is available, sheep will consume low-quality forage. If given a choice, however, sheep may choose a more palatable forage with a lower nutritive value that will ultimately result in lower animal performance.

In general, sheep tend to select leaf in preference to stem and young green material in preference to older dry material (Arnold 1981). Consequently, the diet consumed usually contains proportionately more nitrogen (N) and metabolizable energy and less fiber than would be predicted from whole-plant analysis. However, early in the growing season, forages may have such a high water content that sheep (especially lambs) are unable to consume enough energy to support maximum performance, even though the energy digestibility may be 70%-85% (Church 1991). Therefore, neither intake nor digestibility, alone, may be a true indicator of forage quality.

Plant characteristics, such as taste, odor, and feel, will affect diet selection even though they may be unrelated to digestibility (Arnold 1981; Hodgson 1982). The rate of intake depends on (1) the potential rate at which it can be eaten (ease of fracture, particle size, water content, and degree of satiation) (Arnold 1981), (2) accessibility of forage (height, density, and position in the sward in relation to other components) (Arnold 1981; Hodgson 1982), and (3) acceptability by the animal (taste, odor, and surface characteristics) (Arnold 1981; Provenza 1991). Taste is the most important sense to sheep, and bitter is the most unpleasant taste. Smell is supplementary to the other senses. Sight is related to relocating plant species preferred by the other senses (Terrill and Price 1985) and is important in tall pastures where sheep have to select from different horizontal planes

(Kenney and Black 1984). Touch to the lips is also of supplementary value.

Forages that can be eaten faster are generally preferred (Kenny and Black 1984). Sheep discriminate less among forages that have high intake rates than among those with low rates. Obviously, quality of the forage, and ultimately intake, influence the performance of sheep assigned to consume a given forage. For example, growing lambs require high-quality forage for maximum production, whereas dry, open ewes can perform adequately on lower-quality forages because their productive function is simply for MA.

Role of Digestibility. Any sheep older than 60 d normally has developed a digestive tract capable of efficiently digesting cellulose (fiber). The enlarged four-compartment stomach performs sequentially to allow this type of digestion. Herbage taken in through the mouth is swallowed through the esophagus directly into the reticulo-rumen (compartments 1 and 2) with very little chewing. After 4-12 h of soaking, the consumed forage is regurgitated, rechewed to allow bacterial entry, and reswallowed before it is digested by the microbial population inhabiting the reticulo-rumen.

The microorganisms secrete the enzyme cellulase, which digests the cellulose to volatile fatty acids. These are absorbed through the rumen wall into the bloodstream, where they serve as an energy source to body cells, either directly or after transformation into glucose.

The microbes and any undigested dietary material pass out of the reticulo-rumen, through the omasum (compartment 3), and into the abomasum (compartment 4) and upper small intestine. Degradation of microbial cells begins and ends in the small intestine, from which free amino acids are liberated and absorbed into the blood. Any starches or sugars that escape degradation in the reticulo-rumen are also digested in and absorbed from the small intestine. Even with this unique digestive capability, digestion depends on the type of feed consumed (forage versus concentrate; immature herbage versus mature). For example, digestion of immature leafy grass DM may be 80%-90%, whereas mature stemmy material of the same species may be digested less than 50%.

Fiber (cellulose) is the primary energy source for sheep. Its digestibility ultimately governs the animal's nutritional status and leads to the conclusion that digestible energy is generally the first nutrient that limits performance of sheep consuming forage (Ball et al. 1991). Animal production decreases rapidly as vegetative herbage proceeds to maturation because as cell walls become older lignin constitutes a larger proportion of the wall. This physiological change renders plant cell walls less digestible. In turn, retention time of ingesta in the digestive tract gradually increases as forages mature, which reduces subsequent forage intake.

Digestibility can also be affected by the forage species consumed. Cool-season perennial grasses including orchardgrass, tall fescue (*Festuca arundinacea* Schreb.), Kentucky bluegrass (*Poa pratensis* L.), smooth brome-grass (*Bromus inermis* Leyss.), and timothy (*Phleum pratense* L.) are generally more digestible than warm-season grasses like bermudagrass (*Cynodon dactylon* [L.] Pers.) and bahiagrass (*Paspalum notatum* Flugge). Cool-season annuals, such as rye (*Secale cereale* L.) and cereal grain forage, are even more digestible than cool-season perennials at the same stage of maturity. Legumes are generally of higher quality and more digestible (especially protein) than grasses because of the high leaf:stem ratio and digestibility of cell walls. Leaves are more digestible than stems because they contain less lignin.

Herbage of cool- and warm-season perennial species is most digestible in the spring, lowest in the mid- to late summer, and intermediate in the fall. Even DM digestibility of rotationally stocked alfalfa can decline from a high of 69% in April to a low of 57% in July (Hoveland et al. 1988).

Drought stress has little effect on forage digestibility, as long as the plants retain the leaves. In fact, a moderate drought stress may even increase animal performance above the normal, if adequate forage is available, because digestibility may be increased. Conversely, excessive rainfall may decrease the forage DM content, have little effect on digestibility, but reduce DM intake (DMI). To maintain desired performance, sheep may need to be supplemented with protein during excessively dry periods but with energy, in the form of a concentrate, during excessive rainfall periods.

Assuming other chemical elements are present in amounts to produce adequate forage growth, N fertilization will increase the crude protein content and DM production of grasses. However, N fertilization generally has little effect on the DM digestibility of the leafy portions of grasses.

Utilization Efficiency. The annual diet of the world's sheep contains 85%-90% forage—more than any other class of livestock. Feed represents 50%-60% of the total annual variable costs, making it essential that nutritional needs of ewes be met at the lowest possible cost without interfering with production. Forages contribute most of the energy in the sheep's diet.

The major factors affecting the feeding value of herbage are the rate and extent of energy digestion. These combine to govern voluntary intake and, thus, the amount of energy that can be consumed. Available digestible energy in herbage decreases as maturity stage advances (Fig. 22.3).

With typical management practices, forages can supply the energy requirements for ewes at MA (dry, open), during FL/BR, and during EG; for yearling replacements; and for mature rams. Energy requirements of lactating ewes, growing lambs, and finishing lambs cannot be met simply by feeding even the highest-quality forage. High-quality forages alone approach, but do not meet, the energy requirements of ewes in LG (Table 22.3).

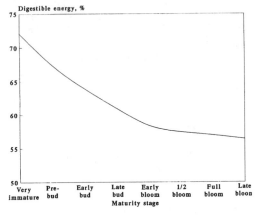

Fig. 22.3. Changes in alfalfa digestible energy with maturity (Kruesi 1985).

The main factor affecting the efficiency of utilization of forage by lambs being finished for slaughter is the amount of digestible energy consumed. Even with the highest-quality forage, supplementation with a concentrate may be necessary to produce most efficient gains. Lambs should weigh at least 27 kg before allowing forage to be the main source of dietary energy. To finish lambs on pasture and supplemental energy, they must be taught, by ewes, to graze. Forage should always be in the vegetative stage. Highest gains are obtained when the stocking rate is low to medium (10-12 lambs/ha). Supplementation with energy (concentrate) at 1%-2% of body weight daily will maximize both forage use and lamb gain.

FORAGE CONSTITUENTS AND CONTRIBUTIONS

The work of many scientists around the world shows the cellulose, hemicellulose, and pectin concentrations in grass leaves to be 15%-30%, 10%-25%, and 1%-2%, respectively. In contrast, legume leaves contain 6%-12% cellulose, 4%-10% hemicellulose, and 4%-8% pectin. The stems of grasses and legumes have a similar fiber content (Church 1991).

Forage mineral concentrations are influenced by soil fertility, fertilization, plant species, and climate. Compared with grasses, legumes have more calcium (Ca), magnesium (Mg), and sulfur (S) and frequently more copper (Cu). In general, the mineral content is highest in leaves and decreases with maturity.

Concentration and digestibility of both protein and nonstructural carbohydrates (cell contents) decline as the plant matures, but concentration of structural carbohydrates (cellulose, hemicellulose, and lignin) increases. Energy digestibility may decline 0.5-1.0 percentage units per day as plants progress to physiological maturity. The decreased digestibility, especially of grasses, is often attributed to the increase in stem and leaf sheath tissues, which decline more rapidly than leaf blade digestibility. There is also a decrease in the leaf:stem ratio with advancing maturity. Similar changes occur in legumes but at a slower rate. Legumes lose their lower leaves with advancing maturity, which contributes to the decrease in the leaf:stem ratio. But there is less lignification of legume stems, so the change in digestibility is not as dramatic as with grasses during the heading stage.

Hay. Even though pasture is the most economical nutrient source for sheep, hay feeding is essential in many areas of the world for the successful completion of the annual nutrition program. This is especially so if the productive function of the sheep occurs when pasture is unavailable (see Tables 22.3 and 22.4 for examples). Then the quality of hay should be appropriate to match the requirements of the sheep.

The Hay Marketing Task Force of the American Forage and Grassland Council (AFGC) has developed the hay-grading system illustrated in Table 22.5. A relative feeding value (RFV) index has been developed to predict intake and energy value of a forage based on its contents of neutral detergent (NDF) and acid detergent fiber (ADF). Full-bloom alfalfa is assigned an RFV of 100. The RFV decreases as forage quality decreases (higher NDF and ADF reduce daily DMI and digestibility). Furthermore, high DMI is indicative of high forage quality. High protein levels are desirable, but protein is not used to calculate RFVs.

High-quality hay is fine stemmed, is leafy, and contains high levels of crude protein and digestible energy, but it has low fiber concentrations. This hay is green, has a minimum of large stems, and is devoid of weeds. Its sweet smell is similar to that of lawn clippings. The best time to harvest legumes for relatively high DM yields and high RFVs is early bloom. Grasses should be harvested at the boot stage (just before seed-head emergence). Exceptions include sericea lespedeza (*Lespedeza cuneata*

TABLE 22.5. Quality standards for legume, grass, or grass-legume hay

Standard	CP	ADF	NDF	DDM	DMI	RFV
	(%)	(%)	(%)	(%)	(kg/d)	
Prime	>19.0	<31.0	<40.0	>65.0	>3.0	151
1	17.2	31.4	40.5	62.6	2.6-3.0	125-151
2	14.2	36.4	47.5	58.6	2.3-2.5	103-124
3	11.1	41.4	54.6	56.6	2.0-2.2	87-102
4	8.1	43.4	61.6	53.6	1.8-1.9	75-86
5	<8.0	>45.0	>65.0	<53.0	<1.8	<75

Note: CP = crude protein; ADF = acid detergent fiber; NDF = neutral detergent fiber; DDM = digestible dry matter; DMI = dry matter intake; RFV = relative feeding value (dimensionless).

[Dum.-Cours.] G. Don.) (30-40 cm), summer annuals (75-90 cm), and hybrid bermudagrass at 4- to 5-wk intervals or 6-8 cm.

In reality, hay may not be harvested at the highest quality. Factors influencing quality include plant species and cultivar, weed infestation, insect damage, disease, weather at harvest, harvesting techniques, fertilization, and stage of maturity at harvest. The most important producer-controlled factor is the maturity stage. An example of the yield and chemical composition changes of alfalfa harvested at different maturity stages is shown in Table 22.6 (Wier et al. 1960). Yield increases as maturity at harvest is later. Total digestible nutrient and crude protein yield/ha may also be greater, but digestible protein/ha will be lower with later maturity harvests because the protein is not highly digestible. Although quantity of hay produced per unit of land is an important consideration, even more important is the quantity of available and digestible nutrients that can be used to support the desired level of animal performance.

Given a choice, sheep will select high-quality hay. However, all do not need the highest-quality hay to meet a specific productive function. Lower-quality grass and legume hays can be economically fed to ewes at MA, ewes in EG, replacement yearlings, and mature rams. Higher-quality hays should be reserved for those animals with higher requirements for DM, crude protein, digestible energy, minerals, and vitamins, i.e., ewes in LG, ewes in LA, growing and finishing lambs, and replacement ewe lambs scheduled to lamb first at 12-14 mo of age.

The economic benefits for grinding or pelleting high-quality hays for sheep are minimal. Low-quality hay may be processed to encourage consumption, when mixed with other dietary ingredients, and to reduce wastage. Digestibility is normally not altered by processing.

Silage. The most common silages fed to sheep are corn, alfalfa, sorghum, and grass. Silage quality can be no better than the quality of the forage as it goes into the silo. To maximize yield of digestible nutrients, legumes and grasses should be harvested in the early bloom stage. Nonstructural carbohydrate levels are high and fiber is relatively low at this stage. To obtain maximum digestible energy per unit of land, corn should be at the hard dent stage (33%-37% DM) and sorghum at the late milk-early dough stage (28%-30% DM) at harvest time.

Ewes in EG, LG, and LA make the most efficient use of silage. Still, they may require supplementation with a concentrate (grain) and/or protein (legume hay). Some example diets for each production stage are presented in Table 22.7. Replacement ewes and rams, and mature rams, can also use large amounts of silage because of their relatively low nutrient requirements.

Pasture. Grazing sheep consume herbage from permanent pastures, cropland, and temporary pastures. Permanent pastures are found on marginal land in mountainous and semiarid areas that is unsuitable for cropping. Forage on these lands include native grasses, wild legumes, and browse plants. Native pastures in the plains include the gramas (*Bouteloua* spp.), buffalograss (*Buchloe dactyloides* [Nutt.] Engelm.), and wheatgrasses (*Agropyron* spp.). Permanent pastures in the central and eastern US are predominantly Kentucky bluegrass, tall fescue, orchardgrass, smooth bromegrass, reed canarygrass (*Phalaris arundinacea* L.), or timothy. Bermudagrass, carpetgrass (*Axonopus compressus* [Sw.] Beauv.), and dallisgrass (*Paspalum dilatatum* Poir.) are important in the South.

Pastures in rotation with cropland, i.e., cropland pastures, are more productive than permanent pastures. The cropping sequence in the rotation varies from 1 yr to several. Clovers (*Trifolium* spp.) are valuable for cropland pastures. Red (*T. pratense* L.), alsike (*T. hybridum* L.), and sweetclover (*Melilotus* spp.) are used in short-term rotations. Ladino clover is probably the most-used legume in

TABLE 22.6.   Calculated yield of TDN and digestible protein from alfalfa cut at four stages of maturity

	Maturity stage			
Item	Prebud	Bud	0.1 bloom	0.5 bloom
Season yield, kg/ha	13260	16430	19130	19260
TDN, %[a]	66.1	60.4	57.2	54.7
TDN yield, kg/ha	8765	9926	10940	10533
CP yield, kg/ha[a]	3542	3908	3947	3652
CP digestibility, %	78.8	74.6	72.8	70.3
DP, kg/ha[a]	2791	2916	2873	2567

[a]TDN = total digestible nutrients; CP = crude protein; DP = digestible protein.

TABLE 22.7.   Example corn silage diets for ewes

Feedstuff	Production stage[a]		
	EG[b]	LG[b]	LA[b]
	(kg/d)		
Corn silage	2.8	2.8	3.7
Shelled corn	. . .	0.5	0.9
Alfalfa hay	0.9[c]	0.9[d]	0.9[d]

[a]70-kg ewe.
[b]EG = first 15 wk gestation; LG = last 6 wk gestation; LA = first 8 wk lactation.
[c]Mid- to late bloom maturity stage.
[d]0.1-bloom maturity stage.

pasture mixes, especially in combination with Kentucky bluegrass. Other valuable legumes for grass-legume pastures are alfalfa, lespedeza, birdsfoot trefoil (*Lotus corniculatus* L.), and subterranean clover (*T. subterraneum* L.). Orchardgrass, smooth bromegrass, and tall fescue are widely used with red clover and alfalfa in grass-legume mixtures.

Wheat, rye, and winter barley (*Hordeum vulgare* L.) make valuable late fall, winter, and/or early spring temporary pastures. They should be grazed so they are not allowed to joint. Sudangrass or sorghum-sudan hybrids grow rapidly to provide a high carrying capacity during the dry periods of summer. Turnips (*Brassica rapa* L.) can be grazed economically in the late fall and early winter.

Outhouse and Rhykerd (1982) conclude the technology is available so sheep can graze pasture throughout the year. They indicate wheat, oat (*Avena sativa* L.), rape (*B. napus* L.), Kentucky bluegrass, tall fescue, turnips, bluegrass, orchardgrass, smooth bromegrass, timothy, birdsfoot trefoil, lespedeza, and sudangrass are candidates for use at strategic times in a year-round grazing program. When properly managed, these pastures could support 20-25 ewes and their lambs or 50-60 weaned lambs per hectare.

Crop residues, such as those in cornfields and grainfields, and beet (*Beta vulgaris* L.), carrot (*Daucus carota* L.), or onion (*Allium cepa* L.) tops can be grazed by sheep. Alfalfa regrowth is a high-quality forage. Sheep are used to graze weeds and grasses along ditches, fencerows, and on horse farms. Although these forages alone may not meet the requirements for the high productivity desired of ewes in LG or LA, or for finishing lambs, they are adequate for ewes at MA or those in EG.

In contrast to crops grown for seed, pasture plants are not allowed to go to seed because of continual or frequent removal. This management forces plants to depend on any ungrazed leaves and stored food reserves in roots, rhizomes, stolons, or stem bases for regrowth.

While close grazing reduces leaf area to the extent that erect-growing grasses and legumes are unable to produce maximum DM yields, similar grazing management is much less harmful to the more prostrate clover species. Conversely, undergrazing results in the accumulation of large amounts of dead leaves, which sheep avoid if possible. Heavy shading, an outcome of undergrazing, reduces tiller growth and DM production. Unless leaf growth is used fairly early, quality will become suboptimal and DM production reduced. Therefore, management of sheep pastures requires a compromise as to plant height and closeness of grazing so production of high-quality forage extends over a long grazing season.

Grazing management maximizes animal production while maintaining or improving forage quality. Two basic grazing management systems exist in the US: continuous and rotational stocking. Since sheep are selective grazers, the producer must control where, when, and how long they graze an area to prevent them from overgrazing the preferred plants and leaving the less desirable ones.

Continuous stocking is traditional and simple and requires the least amount of labor and fence. However, the wide fluctuation in quantity of forage available during the season permits overgrazing areas containing preferred plants and undergrazing those areas where less palatable, but perhaps as nutritious, plants grow (Fig. 22.4). Avoiding these plants allows them to become mature, more unpalatable, and less digestible.

Although rotational stocking requires more labor and fence than continuous, DM production per unit of land area is greater, carrying

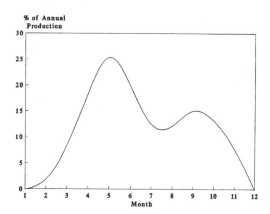

Fig. 22.4.   Seasonal growth of cool-season grasses.

capacity is increased, and forage quality over the complete grazing season is uniformly higher quality. Rotation through a series of pastures at 1- to 2-d intervals (minimum 21-d rest) provides high-quality feed through the season, extends the grazing season, increases DM production, shifts the pasture to low-growing grasses and legumes, and distributes feces and urine more uniformly.

Alternate stocking is a form of rotational stocking in which the use of two paddocks is alternated every 2-3 wk throughout the growing season. Although this regrowth time is adequate in spring, it is too short in summer. When sheep are moved into the rested pasture, they consume the shortest grass first, leaving the taller, less palatable herbage to become lower quality. Then growth becomes uneven, and the nutrient status of the animals fluctuates up and down during the 2- to 3-wk period.

Pasture is the most economical source of nutrients for sheep. A comparison with hay is shown in Table 22.8. The main economic advantage of pasture is lower fertilizer and machinery costs. Pasture and hay yields are compared in Table 22.9. Although DM yield is higher with hay, the quality of forage is usually lower than that of forage consumed by direct grazing (indicated by TDN values). This, plus the lower annual cost for pasture, produces a significant economic advantage for pasture when expressed either as cost/kg of DM or monthly cost/ewe.

FORAGE-LIVESTOCK SYSTEMS

Animal Health. Sheep prefer to consume broadleaf plants, like legumes, weeds, and browse, grown on higher elevations. Their grazing habits are often maligned as the primary causes of denudation and erosion of vegetated land, especially when the land has

TABLE 22.8. Annual production costs of pasture and hay

Item	Pasture	Hay
	($/ha)	
Seed	3.60	10.80
Lime	37.20	37.20
Fertilizer[a]	74.40	168.00
Machinery[b]	10.80	171.00
Fencing[c]	52.80	0.00
Taxes	40.80	52.80
Interest (at 12%)	26.40	45.60
	$246.00	$485.40

Source: Kruesi (1985).
[a]Pasture = 0-100-200 kg N-P-K; Hay = 330-100-200 kg N-P-K.
[b]Includes fuel, repairs, and operator.
[c]Materials for electric fencing and labor at $1.27/m plus temporary electric subdivision fencing at $0.27/m.

TABLE 22.9. Comparison of pasture and hay yield

Item	Pasture	Hay
Yield, kg DM/ha[a]	10,890	13,200
Average quality (% TDN)	60-65	45-55
Annual cost	$246.00	$485.40
Cost/kg of DM	$ 0.045	$ 0.066
Monthly cost/ewe	$ 2.40	$ 3.60

Source: Kruesi (1985).
[a]Assuming some inefficiencies of animal harvest, yield of pasture is set at 82.5% of hay yield.

been overstocked. Sheep prefer low-growing, broadleaf plants that can be grazed close to the ground at repeated intervals (e.g., legumes). Because weeds are readily consumed, sheep increase the uniformity of the herbage produced on the land. Grazing sheep with other animal species reduces animal losses from poisonous plants because lambs learn to eat what their mothers eat (Provenza 1991). They sample novel "harmful" plants avoided by their mothers (Mirza and Provenza 1990), a tactic that enables them to avoid a lethal dose of toxin and to adjust intake in accordance with toxicity. They do not need to maximize intake of any particular nutrient on a daily basis because of their ability to tolerate short-term deficiencies of energy, N, and various minerals and vitamins (Booth 1985).

Ewes in moderate condition can survive weight loss up to 30% for 6 mo without any effect on future production (Glimp 1991). Level of feed intake over extended periods for survival is only about 60% of National Research Council (NRC 1985) recommendations for dry mature ewes (Glimp 1991). Moderate levels of lamb and wool production can generally be maintained at 75%-80% of NRC (1985) requirements.

If the forage remaining in pastures becomes dry, protein may be the first nutrient that needs to be supplemented. Then, as the supply of DM becomes limited, the primary supplement should be energy, with protein supplied once the energy needs are met. The minerals most likely to be deficient in forage for sheep are phosphorus (P) and selenium (Se).

Although ewes can use forage all year, the LG production phase may require supplementation with energy to prevent ketosis (pregnancy toxemia), especially when excessively fat or thin ewes are carrying twins. Energy derived from low-quality pasture or hay alone may be inadequate to meet high demands due to rapid fetal growth. To compensate, energy from body stores of the ewe is transferred to the fetuses, during which toxicity may develop. Liberal amounts of high-quality forage should be supplemented with energy to prevent ketosis. Ewes may also develop listeriosis if they consume large amounts of moldy silage or hay during LG and LA.

If ewes are nutritionally flushed and bred on pasture, care should be taken in selection of the plant species they are to consume. Flushing can be accomplished on alfalfa pasture, although it may contain levels of coumestans, which suppress ovulation rate. Red and subterranean clovers contain a fertility depressant, formononetin. These legumes should not be grazed immediately before and during breeding. It is best to keep ewes off grass-legume pasture for at least 30 d prior to FL/BR if the mixture is 30% or more red clover. Grazing pure stands of ladino clover may delay conception, as can alfalfa, birdsfoot trefoil, and red clover. For maximum reproductive performance, ewes should be flushed and bred on grass pastures, grass-ladino clover pastures, or annual forage crops.

Ball et al. (1991) summarizes other potential nutritional disorders of sheep. Nitrate poisoning is most likely to occur when sheep graze sudangrass, sorghum-sudangrass hybrids, millet (*Setaria* spp.), corn, wheat, or oat during periods of low soil moisture or low humidity after a heavy application of N fertilizer. Prussic acid poisoning can occur in sheep grazing johnsongrass (*S. halepense* [L.] Pers.), sorghum, or sorghum-sudangrass hybrids after frost or when grazing regrowth after a drought. Plants should regrow for at least a week after either of these occurrences before sheep are allowed to graze these forages.

Tall fescue toxicity can occur in sheep grazing endophyte-infected forage during summer. To reduce the severity, the growth of toxic fescue forage should be kept short or diluted by interseeding with legumes.

Internal parasites pose a significant problem for grazing sheep, especially lambs. The most effective method of prevention is a strict rotational stocking program in which pastures are rested a minimum of 21 d. Parasites can be minimized or prevented (1) by breeding ewes so their lambs graze forage at times other than during the peak parasite season and (2) by allocating early-weaned lambs to parasite-free pastures that are rotationally stocked.

Economic Issues. Sheep producers have little control over the gross return they receive for their products, but they have a lot of control over their production costs. Since feed costs constitute the greatest percentage of variable costs, the most logical way to reduce feed costs is to limit use of expensive harvested feeds and to exploit pasture use. Chappell et al. (1992) devised a forage-based sheep production plan whereby the ewe flock is divided equally so lambing occurs in April and October. By sorting lambs on weight, and feeding accordingly, a consistent supply of lambs can be marketed for slaughter every month of the year. Ely et al. (1992) support this plan with a monthly nutrition program built around rotationally stocking pastures consisting of cool-season grasses with and without legumes, summer annuals, turnips, and winter wheat. Rotational stocking of these forages extracts as much of the nutrients from the forages as possible (Fig. 22.5).

Gerrish (1992) analyzed 10 yr of data from Missouri lamb sales and calculated an average annual price of $1.30/kg. After determining average costs of production for winter (January-February), fall (October), and spring (April) lambing, he calculated the selling price when the percentage of lamb crop marketed was 130%, 150%, and 170% (Fig. 22.6). Selling prices needed to offset inputs were significantly lower with spring lambing because of the greater use of forage diets. Although fall required a higher selling price than spring lambing, it was considerably lower than winter lambing. The higher selling prices needed for winter and fall lambing are a direct result of feeding more expensive harvested feeds. These data emphasize how use of forage and pasture can lower the selling price for market lambs and encourage the economic viability of sheep production.

Environmental Issues. The primary concerns of the human population regarding sheep production and the environment revolve around

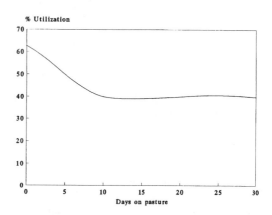

Fig. 22.5. Impact of length of grazing period on utilization rate (Gerrish 1992).

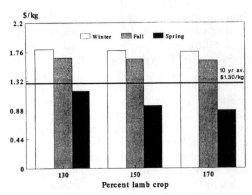

Fig. 22.6. Breakeven price of lambs produced during different seasons (Gerrish 1992).

soil erosion, groundwater quality, and the wildlife habitat. If pastures are maintained in healthy, vigorous condition, soil erosion is virtually eliminated. The vegetative cover breaks the impact of rain drops, and healthy root growth facilitates rapid water infiltration, both of which discourage runoff. A water source for sheep in each pasture prevents the establishment of travel lanes where erosion might develop. Any water running off healthy pastures typically has low levels of nitrates,

phosphates, and enterobacterial species.

Few pesticides are required if pastures are properly grazed. Weeds are controlled biologically by the sheep because they consume a variety of plant species. Most problem insects are also controlled through timely grazing and appropriate stocking rates. The natural reduction of these biological pests from the system, by the sheep, greatly reduces the need for pesticides and the potential for groundwater contamination.

Sheep are opportunistic creatures relative to the harvest of solar energy through plant biomass produced from the soil. Yet they are synergistic in their abilities to assimilate quality products from forage for human use (Parker 1990). Efficient use of solar energy, recycling nutrients to the soil, use of noncompetitive renewable resources (ligno-cellulosic), contribution to soil and water conservation, low capital investment requirements, and adding flexibility are favorable characteristics of forage farming with sheep. These attributes have been available to mankind since the sheep was discovered as a source of food and fiber almost 11,000 yr ago. They are still available and will undoubtedly continue to be so in the future.

Goats

Goats (*Capra hircus*) produce meat, milk, fiber, and skins throughout the world. Although over 500 million of these highly adaptable ruminants populate the world, they are most widely distributed in dry climates where the terrain is rough and vegetation is sparse. Their socioeconomic contributions in lesser-developed countries are significant.

The approximately 3 million head in the US include Angora (mostly extensively managed), dairy (intensively managed), and Spanish or "meat" goats (mostly extensively managed). The Angora, and to some extent the meat goat, are valuable for the sale of mohair and slaughter animals. Their stabilizing influence on the vegetation, by retarding encroachment of brush species, is unparalleled. The dairy goat industry is expanding but is still minor compared with the dairy cattle industry. Most large goat dairies are located near metropolitan areas, but smaller herds (5-20 head) produce milk for family, neighbors, and friends.

Goats consume a wide variety of low-quality, woody plants not readily eaten by sheep or cattle. They can also be highly discriminatory among plant fractions or feed particles that appear identical to humans. The goat is active and inquisitive in its foraging behavior (Huston 1991) and is noted for its preference of woody or brushy plants. A common stance for browsing is standing only on the hind legs with the front feet and head hidden in the lower tree branches. Resultant trees and shrubs have a flattened-underline appearance 120-180 cm above the ground. Even though they thrive on woody plants, goats still require adequate nutrition for reproduction and growth.

The diet selected by the goat is usually more nutrient dense than the average of the available vegetation. Grasses and herbaceous legumes have a higher concentration of nutrients than shrubs and tree leaves and often comprise the major portion of the goat's diet

(Huston 1991). At other times, the diet will contain the most tender and palatable leaves from trees, leaf tips from shrubs, fallen mast (oak acorns), flowers and/or buds from xeric plants, and plant seed heads. All of these minor plant parts contain concentrations of one or more required nutrients. If the vegetation contains a variety of plants, the diet will be complex. The diet composition will change daily as plants emerge, mature, and drop leaves. Compared with less adaptable livestock in this situation, the goat's diet will be more consistent in nutrient composition. However, on pastures with less plant diversity, diets selected by goats, sheep, and cattle will be similar.

Goats eat the same amount of forage as sheep of equal size. The browse that goats prefer may contain higher levels of crude protein and P than grasses. Goats are proficient in selecting plants and plant parts that are of high nutritional value. They are also proficient in avoiding plants containing inhibitors of digestion, such as lignin, oil, tannins, or sil-

ica (Terrill and Price 1985).

Vegetation in arid areas is highly seasonal: high quality and palatable in spring and early summer but low quality and unpalatable during dormancy. Nondairy goats can usually satisfy their nutrient requirements from vegetation alone (perennial warm-season grasses, forbs, legumes, and browse; long season or evergreen shrubs; and annual plants). Under drought or overgrazing conditions, options for diet selection become limited, and diet quality follows the pattern of limited options. If only the production component is present, diet quality fluctuates with the nutrient content of this component. Then supplemental concentrates are required to bring the diet quality up to a level that achieves the desired productivity.

Forage produced on monoculture pastures in temperate climates is either grazed directly at high stocking rates or harvested for hay. These forages are generally higher quality than range forages, including legumes, and typify forage fed to dairy goats.

Rabbits

The rabbit (*Oryctolagus cuniculus*) was domesticated about 2000 yr ago. Approximately 30 million kg of domestic rabbit meat are consumed annually by Americans. This meat is produced mainly by backyard producers that have only a few does and by a few large commercial producers.

Rabbits cannot compete with poultry or swine when feed grains are low priced. However, as the world's human population increases and grains become more expensive, rabbits may become a more competitive food source because of their ability to convert forages to meat.

By the time the rabbit is 3-5 wk of age, it has a developed cecum and colon inhabited by anaerobic microorganisms (Varela-Alvarez 1991). In contrast to swine and poultry, this microbial population can digest roughage (forage). In fact, the rabbit can graze natural, unharvested forages for much of its nutrient needs.

In commercial operations, forages are used as a source of fiber or bulk (Varela-Alvarez 1991). They can be incorporated directly into a completely pelleted feed or fed separately

from the pellet. When given apart from the pellet, it is better to offer forages in the form of hay.

Alfalfa is the most widely used hay. However, early cuttings, with more than 20% crude protein, should be avoided because of the problem with diarrhea. Three types of pellets are manufactured: all-grain, all-hay, and complete pellets (forage plus grain). Because the complete diet for the rabbit should contain 30%-50% roughage, the all-grain pellet should be supplemented with hay and the all-hay pellet with grain for maximum production. To maximize consumption, rabbits should be fed late in the day.

QUESTIONS

1. Explain why sheep lead all farm animals in their ability to produce marketable products from forage alone.
2. List five characteristics of sheep that give them an economic advantage over other ruminants.
3. What is meant by "matching the ewes' production phase to the quantity and quality of forage available"? Give specific examples.
4. Highest-quality forage should be reserved for

which two production phases of the ewe?

5. Describe a system of producing slaughter lambs on pasture.
6. List the plant characteristics that determine intake levels by sheep. Describe their effects.
7. What are three main factors affecting forage digestibility in sheep?
8. Describe the composition of two hays, one with an RFV of 90 while the other has an RFV of 75.
9. Describe a system in which ewes remain on pasture throughout the year; indicate the forage(s) to be grazed during each performance phase of the ewes' annual management system.
10. How do sheep contribute to environmental cleanliness?
11. Explain why the diet selected by goats is not equal to the average of the available vegetation.
12. Can rabbits compete economically as a meat producer with poultry and swine? Why or why not?

REFERENCES

Arnold, GG. 1981. Grazing behavior. In FHW Morley (ed.), Grazing Animals. Amsterdam: Elsevier, 79-104.

Ball, DM, CS Hoveland, and GD Lacefield. 1991. Southern Forages. Norcross, Ga.: Potash and Phosphate Institute, Chaps. 16 and 21.

Booth, DA. 1985. Food-conditioning, eating preferences, and aversions with interoceptive elements: Conditional appetites and satieties. In NS Bravemann and P Bronstein (eds.), Experimental Assessments and Clinical Applications of Conditioned Food Aversions. New York: New York Academy of Science, 22-41.

Chappell, GLM, DG Ely, and DK Aaron. 1992. Integrated sheep production: Sheep as a business. (The K-100 system). Univ. Ky. Sheeprofit Day Prog. Rep. 342, 25-29.

Church, DC. 1991. Livestock Feeds and Feeding. 3d ed. Englewood Cliffs, N.J.: Prentice-Hall, 55-100.

Ely, DG, BP Glenn, M. Mahyuddin, JD Kemp, FA Thrift, and WP Deweese. 1979. Drylot vs. pasture: Early-weaned lamb performance to two slaughter weights. J. Anim. Sci. 48:32-37.

Ely, DG, WP Deweese, RJ Thomas, BP Glenn, TW Robb, and RC Buckner. 1980. Supplemental energy for early-weaned lambs grazing grass-legume mixtures in spring. Univ. Ky. Sheeprofit Day Prog. Rep. 248, 11-12.

Ely, DG, GLM Chappell, and DK Aaron. 1992. Nutrition programs in an integrated sheep production system. Univ. Ky. Sheeprofit Day Prog. Rep. 342, 39-46.

Food and Agriculture Organization (FAO) of the United Nations. 1991. FAO Production Yearbook. Rome, 45:194.

Gerrish, JR. 1992. Intensive grazing management of sheep. 14th Annu. Iowa Sheep Symp., Des Moines, Iowa, 95-116.

Glimp, HA. 1991. Nutrition of the ewe. In DC Church (ed.), Livestock Feeds and Feeding, 3d ed. Englewood Cliffs, N.J.: Prentice-Hall, 306-22.

Hodgson, J. 1982. Influence of sward characteristics on diet selection and herbage intake by the grazing animal. In JB Hacker (ed.), Nutritional Limits to Animal Production from Pastures. Farnham Royal, UK: Commonwealth Agriculture Bureaux, 153-66.

Hoveland, CS, NS Hill, RS Lowrey, Jr., SL Fales, ME McCormick, and AE Smith. 1988. Steer performance on birdsfoot trefoil and alfalfa pasture in central Georgia. J. Prod. Agric. 1:343-46.

Huston, JE. 1991. Goats and goat nutrition. In DC Church (ed.), Livestock Feeds and Feeding, 3d ed. Englewood Cliffs, N.J.: Prentice-Hall, 350-67.

Kenny, PA, and JL Black. 1984. Factors affecting diet selection by sheep. I. Potential intake rate and acceptability of food. Aust. J. Agric. Res. 35:551-63.

Kruesi, WK. 1985. The Sheep Raiser's Manual. Charlotte, Vt.: Williamson Publishing, Chap. 3.

Mirza, SN, and FD Provenza. 1990. Preference of the mother affects selection and avoidance of foods by lambs differing in age. Appl. Anim. Behav. Sci. 28:255-63.

Morrical, D. 1992. Basic ewe nutrition. Shepherd 37(10):10-13.

National Research Council (NRC). 1985. Nutrient Requirements of Sheep. Washington, D.C.: National Academy Press.

Outhouse, JB, and CL Rhykerd. 1982. Pasture and forages for sheep. Sheep Breeder and Sheepman 102(9):92-97.

Parker, CF. 1990. Role of animals in sustainable agriculture. In CA Edwards, R Lal, P Madden, RH Miller, and G House (eds.), Sustainable Agricultural Systems. Ankeney, Iowa: Soil and Water Conservation Society, 238-45.

Provenza, FD. 1991. Behavior and nutrition are complementary behaviors. In Proc. Grazing Livest. Nutr. Conf. Okla. State Univ. MP-133, 157-69.

Ragland, KK, DG Ely, DK Aaron, VL Owens, and WP Deweese. 1990. Digestibility and nitrogen metabolism in Hampshire and Polypay sheep. Univ. Ky. Sheeprofit Day Prog. Rep. 325, 3-10.

Sheep Industry Development (SID). 1987. Sheep Production Handbook. Denver, Colo.: Double Quick Printing.

Terrill, CE, and DA Price. 1985. Sheep, goats, and rabbits: Efficient users of forages. In ME Heath, RF Barnes, and DS Metcalfe (eds.), Forages: The Science of Grassland Agriculture, 4th ed. Ames: Iowa State Univ. Press, 578-89.

USDA. 1991. Agricultural Statistics. Washington, D.C.: US Gov. Print. Off., 271, Table 422.

Varela-Alvarez, H. 1991. Feeding rabbits. In DC Church (ed.), Livestock Feeds and Feeding, 3d ed. Englewood Cliffs, N.J.: Prentice-Hall, 452-65.

Wier, WC, LC Jones, and JH Meyer. 1960. Effect of cutting interval and stage of maturity on the digestibility and yield of alfalfa. J. Anim. Sci. 19:5-19.

23

Forages for Swine, Poultry, and Other Species

JAMES R. FOSTER
Purdue University

Swine Industry

NATURE AND CHANGES

S WINE production units continue to increase in size. As volume of production has increased, swine producers have substituted capital for labor. This has resulted in a continued trend toward rearing of hogs in environmentally regulated facilities with increased labor efficiency and less use of pasture.

Many swine producers report that the move from pasture to enclosed facilities has failed to improve performance of their hogs, but this shift has permitted an increase in volume of production with the same or less labor.

Prior to about 1950, pasture was considered essential in meeting the nutritional requirements of swine, particularly in regard to reproductive performance. Research prior to 1950 clearly demonstrated that rations containing 5% or less alfalfa (*Medicago sativa* L.) meal were nutritionally inadequate for gestation and lactation under drylot conditions (Ross et al. 1942; Krider et al. 1946). However, the control rations usually consisted of corn (*Zea mays* L.), expeller-processed soy-

bean (*Glycine max* [L.] Merr.) meal, salt, limestone, and 5% alfalfa meal.

After 1950 the use of synthetic vitamins became nearly universal in swine ration formulation. In addition, research identified the need for certain trace minerals and other nutrients and resulted in a better understanding of the complete nutritional needs of the sow.

Most research since 1960 has shown little if any effect on reproductive performance with dehydrated alfalfa meal or sun-cured alfalfa in gestation rations (see Foster 1985).

Today, pasture or other forages are not considered essential to meet the nutritional needs of swine. Instead, the question is how to meet these nutritional requirements most economically. In many swine operations forages can help meet the nutrient requirements of swine economically, particularly during gestation.

FORAGE CONSTITUENTS AND CONTRIBUTIONS

The most common forages used in swine production are legume pasture and dehydrated alfalfa meal. Additional possibilities include other forms of pasture (rape [*Brassica napus* L.], winter rye [*Secale cereale* L.], and certain grasses, such as smooth bromegrass [*Bromus inermis* Leyss.]) and silages (corn or grass-legume).

James R. Foster, Professor of Animal Sciences, is currently serving as Extension Swine Specialist, Purdue University. Indiana born and reared, he obtained training in animal husbandry at Purdue and earned the PhD degree in animal nutrition from Iowa State University.

Pastures. The most common pastures for swine in the US are alfalfa, ladino clover (*Trifolium repens* L.), or a combination of the two (see Fig. 23.1). Analyses of these pastures indicate alfalfa has slightly higher dry matter (DM), protein, and energy; however, feeding trials have not shown a consistent advantage of one over the other (Foster 1985). Both plants are good sources of calcium (Ca) and most of the required vitamins except D and B_{12}. Generally, vitamin D is not a problem when swine are exposed to sunlight.

Danielson et al. (1969) used a double indicator method (chromium oxide-chlorophyll technique) to estimate alfalfa pasture intake by pigs on a growing-finishing ration. Estimated intake values ranged from a low of 5% to a high of 12% of the total DM intake.

Legume pasture used to a lesser extent than alfalfa and ladino clover include red clover (*Trifolium pratense* L.), lespedeza (*Kummerowia* spp.), birdsfoot trefoil (*Lotus corniculatus* L.), and sweetclover (*Melilotus alba* Medik.).

RAPE (*Brassica napus* L.). Rape is an annual, usually seeded in the spring in the Corn Belt and in the fall in the South (Chap. 36, Vol. 1). Frequently, it is seeded alone; however, a mixture of oat (*Avena sativa* L.) and rape allows grazing a few days earlier than does rape alone. As the oat pasture decreases in quantity, rape continues to grow and furnish high-quality pasture—even into the cool fall months. Rape can be seeded in an emergency and within about 8-10 wk will provide a pasture for swine that compares favorably with alfalfa. Its nutritional value is nearly equal to that of legumes. Although not considered a serious problem, white hogs on rape pasture may experience some sunburning, particularly when the plant is wet from rain or a heavy dew.

WINTER RYE (*Secale cereale* L.). Winter rye has very rapid growth in fall and early spring and is adapted to many regions of the nation. In Kentucky, bred sows on winter rye pasture during gestation were fed levels of 0.91, 1.36, or 1.81 kg of supplemental feed daily and compared with those in drylot fed 2.7 kg feed/sow/d (Barnhart and Cathey 1956, 1957; Barnhart et al. 1958; Selke et al. 1960). The reproductive performance of sows fed 0.91 kg of feed on winter rye pasture was essentially equal to that of those fed in drylot, but sows on winter rye pasture required less supplemental feed. The winter rye pasture had a carrying capacity of approximately 12 sows per hectare. In another year, more than 1.33

Fig. 23.1.   High-quality legume pasture can be used to meet more than half the nutrient requirements for gestating sows. Good pasture can accommodate up to 24 sows/ha. *Purdue Univ. photo.*

kg of feed were needed for sows on winter rye pasture. Feed costs per live pig farrowed were reduced about 30% when sows were on winter rye pasture. These costs did not include a charge for the pasture.

Ambient temperatures, the quality of the winter rye pasture, and size and condition of the sow will determine the amount of concentrate needed for gestating sows to ensure good reproductive performance. This amount may vary from slightly less than 1 kg to nearly 2 kg per sow per day of a 15% protein ration. Gilts require slightly more feed than older sows because they are still growing. Cold weather increases the total feed requirements by 10%-20% at 0°C compared with 25°C.

OTHER GRASSES. Krider and Terrill (1950) compared smooth bromegrass with alfalfa and a mixture of the two. Growing-finishing pigs self-fed to 90 kg required 5.5% more concentrate feed per unit of gain on smooth bromegrass than on alfalfa. Pigs on smooth bromegrass consumed 33% more protein supplement and 13% more minerals per unit of gain compared with those on alfalfa. Pigs gained slightly faster on alfalfa. A pasture mixture of smooth bromegrass-alfalfa was fully equal to the alfalfa pasture for self-fed pigs.

Stored Forages

ALFALFA MEAL. This widely used forage product is a good source of most vitamins. However, because of the relatively high fiber and low energy content of alfalfa meal, rate of gain and feed efficiency generally decline as the level of alfalfa meal in growing-finishing swine rations increases. The most common form used in swine rations is dehydrated alfalfa meal. Also, in some instances alfalfa hay has been ground and fed as alfalfa meal. For example, Bohman et al. (1953, 1955) fed alfalfa meal at 0%, 10%, 20%, 30%, 50%, and 60% of the ration. Growing-finishing swine utilized up to 50% alfalfa in their ration, although gain and feed efficiency were depressed somewhat as the percentage of alfalfa in the ration increased. Pelleting the 50% alfalfa ration improved daily gain by 13% and feed efficiency by 25% compared with the ground mixture.

Becker et al. (1956) conclude that dehydrated alfalfa meal at levels of 10% and 20% in protein-deficient rations provided limited supplementary protein to improve performance. Grabouski and Danielson (1978) attribute the growth-depressing property of alfalfa meal, fed to growing-finishing swine at 25% of the diet, to a lowered energy content of the diet and a lower intake rate during the 30- to 55-kg growth period. However, addition of lard at levels from 1.25% to 3.75% of the diet to supplement energy content failed to overcome this growth depression.

Heitman and Meyer (1959) fed alfalfa meal as a source of energy for swine at 5%, 20%, and 40% of the diet; the alfalfa was harvested at three stages of maturity (16% bud, 3% bloom, and 34% bloom) and prepared through three methods (sun-cured, dehydrated, and dehydrated reground). The rate of gain was lower on rations containing sun-cured meal compared with that on others, even though feed consumption was not affected by method of preparation. The stage of maturity had no effect. Poor utilization of alfalfa as a source of energy is indicated by a metabolizable energy content of the hays averaging only about 55% of that of barley (*Hordeum* spp.) grain. The average replacement value of 1 kg alfalfa meal was calculated to be 0.28 kg of concentrate.

Cromwell et al. (1984) report that the Ca in alfalfa meal is poorly available, whereas the phosphorus (P) in alfalfa meal is highly available for the growing pig. Therefore, although alfalfa meal contains 1.3% Ca, only about 0.3% should be considered as available Ca. On the other hand, all of the P in alfalfa meal (0.3%) can be considered available. Stahly and Cromwell (1986) conclude that the nutritional value of dehydrated alfalfa meal for growing pigs was greater in a moderately cold (10.0°C) versus a warm (22.5°C) or hot (35.0°C) thermal environment.

Silage. Good corn or grass-legume silage, properly supplemented, may constitute the major part of the ration for pregnant sows and gilts (Conrad and Beeson 1954; Catron et al. 1955; Johnson et al. 1957; Becker et al. 1959; Bowden and Clarke 1963). Advantages of feeding silage to sows include

1. The cost per pig at farrowing may be reduced.

2. Sows are prevented from getting too fat.

3. Farms that have beef and dairy cattle often have silage available.

4. Nutritionally, silage is the next best forage feed to summer pasture.

Corn, legume (alfalfa), or grass-legume (alfalfa-smooth bromegrass) silages have been used successfully in swine operations.

CORN SILAGE. Corn silage is an excellent source of many vitamins needed by swine, but it is deficient in protein, energy, and certain minerals. Generally, corn silage is full-fed once or twice daily and supplemented with 1.0-1.5 kg/sow/d concentrate. A wide variation in silage intake has been reported. In many trials, sows have consumed 4.5-6.5 kg daily of a good quality silage. However, Becker et al. (1959) report that sows consumed only 1.5 kg/d silage during gestation. In this case, additional concentrate was needed. Feed costs during gestation on a per pig-weaned basis were reduced 28% by feeding corn silage (Conrad and Beeson 1954). For best results, corn silage should be made when the ears are formed but when the plant is still green and not frosted. Silage made from corn nearing maturity is less palatable. Fine-chopped silage is best for sow feed because it minimizes waste.

GRASS-LEGUME SILAGE. The most common grass-legume silage is alfalfa or a combination of alfalfa-smooth bromegrass. Nutritionally, grass-legume silage has more protein and Ca but less energy than corn silage. Shelled corn sometimes is added at the time of ensiling to increase the energy value. Consumption of grass-legume silage by feeder pigs was very low when a concentrate was self-fed (Bowden and Clarke 1963). Reduction of the concentrate intake by 25% resulted in a 200% increase in silage consumption, but it was insufficient to overcome the reduction in weight gain from concentrate restriction. Selke et al. (1960) found that pregnant sows self-fed alfalfa silage consumed 6.33 kg/d silage with 1.33 kg concentrate. Reproductive performance was about equal to that of sows fed on winter rye pasture and somewhat better than for sows fed 2.67 kg concentrate daily in drylot.

LOW-MOISTURE SILAGE. Low-moisture silage is defined as a legume, grass, or cereal cut near its peak nutritional stage of maturity, wilted to 40%-60% moisture, finely chopped, and ensiled. It is sometimes referred to as *haylage*. Nutritionally, energy is the limiting factor in the use of low-moisture silage for swine.

Good reproductive performance has been reported by Hoagland et al. (1963) using four parts of haylage mixed with one part ground corn and stored in an airtight silo. The mixture at 45.5% moisture, 19.5% crude protein

(CP), and 20.0%-25.0% fiber was full-fed to both gilts and sows starting about 2 wk prior to the start of breeding and continuing through gestation. Average daily consumption of the haylage-corn mixture was 3.8 kg for gilts and 5.3 kg for sows. Sows made satisfactory gains during the middle third of gestation when fed only the haylage-corn mixture. Gilts required some supplemental feed throughout gestation to produce satisfactory gains. Digestibility of the CP in the haylage-corn mixture was 55%-60% for sows.

SWINE CHARACTERISTICS

The nutrient requirements for breeding swine have been summarized by the National Research Council (NRC 1988). Fairly reliable information is available regarding the nutrient content of ingredients, and most of the needed vitamins are available in synthetic form. Thus, a nutritionally adequate ration can be formulated without depending on forages as a source of nutrients. The nutritional content of forages commonly used for swine is shown in Table 23.1. However, values vary considerably among samples, and those listed represent averages from a large number of analyses.

Life Cycle Nutrition

GROWING-FINISHING. The digestive system of the growing-finishing pig in the US is not well adapted for utilization of forages. However, breeds native to China have a much greater ability to utilize forage during growing-finishing than do our hogs. Research to evaluate the use of pasture for growing-finishing hogs is summarized in Table 23.2. Although feed concentrates required per unit of gain vary widely among drylots, and pasture varies widely among experiments, these results suggest a 2%-3% feed saving by the use of legume pasture compared with drylot. Assuming an average difference of 0.1 kg feed per kg of gain, in favor of pasture, the feed-saving value of a hectare of good legume pasture can be calculated as follows:

50 pigs/ha gaining 80 kg = 4000 kg pork

4000 kg pork × 0.10 = 400 kg feed saved/ha

There appears to be no advantage in rate of gain for pigs fed on pasture.

Brumm and Peo (1985) note no improvement in daily gain of purchased feeder pigs

TABLE 23.1.   Nutrient composition of forages

	Pasture				Silage			Alfalfa meal (dehy)
	Alfalfa	Ladino clover	Rape	Winter rye	Corn	Grass-legume	Low moisture	
Dry matter (%)	24	20	16	17	26	28	55	93
Digestible energy (kcal/kg)	660	525	525	575	880	750	975	1400
Crude protein (%)	6.0	5.1	3.0	5.0	2.1	5.0	6.3	17.0
Fat (%)	0.9	1.0	0.6	0.8	0.8	1.1	1.4	3.0
Calcium (%)	0.5	0.3	0.2	0.1	0.05	0.04	0.5	1.3
Phosphorus (%)	0.1	0.07	0.06	0.1	0.05	0.09	0.12	0.23
Sodium (%)	0.04	0.02	0.01	...	0.01	0.04	0.05	0.09
Chlorine (%)	0.11	...	...	...	0.05	0.11	0.15	0.46
Copper (mg/kg)	2.5	...	0.6	1.1	0.6	2.9	3.4	9.9
Iron (mg/kg)	0.01	0.01	0.003	...	...	0.003	0.01	0.05
Zinc (mg/kg)	3.6	...	...	...	2.5	...	...	16
Vitamin A equivalent (IU/g)	104	118	...	155	7	45	31	268
Vitamin D (IU/g)	0	0	0	0	0.1	0	0.2	...
Riboflavin (mg/kg)	5.0	4.0	...	...	...	...	...	12.3
Niacin (mg/kg)	8.2	...	...	...	5.7	5.7	6.0	46
Pantothenic acid (mg/kg)	4.0	...	...	...	...	...	...	30

Sources: Based on averages of several references, mainly Morrison (1956) and Crampton and Harris (1969).

when alfalfa meal was included in the receiving diet for 2 wk. However, the addition of 10% midbloom, third-cutting alfalfa to the receiving diet for 2 wk did improve daily gain from purchase to market.

GESTATION. The digestive system of swine 8 mo of age or older is sufficiently developed to permit utilization of relatively large amounts of forage. In addition, the swine producer is not striving for maximum gains during gestation, and hence the energy requirements are somewhat less for the pregnant sow than for the growing-finishing hog. Barnhart and Cathey (1957) and Eyles (1959) demonstrate that concentrate feed can be saved during the gestation period by the use of good pasture, the amount depending primarily on quality of the pasture. For example, sows on good qual-

ity legume pasture can meet more than half of their nutrient requirements from pasture alone (Table 23.3).

Good pasture for swine is deficient in energy, P, salt, and vitamin B_{12}, and protein may be borderline. These requirements can be met by feeding about 1 kg corn and 150 g protein supplement/sow/d. Most commercial protein supplements fed at this level will also contain adequate amounts of minerals and vitamin B_{12} to meet these needs.

Figuring 24 sows for each hectare and a daily rental charge for the pasture of $1.00/ha, the daily feed costs for each sow would be about $0.04 for pasture and $0.15 for supplemental feed. This total daily feed cost of $0.19/sow compares with a daily feed cost of $0.25-$0.30 for a sow in drylot.

Danielson and Noonan (1975) fed up to 97%

TABLE 23.2.   Effect of pasture on performance of growing-finishing swine

Location	Comparison	Number of pigs	Daily gain (kg)	Daily feed (kg)	Feed/gain
Nebraska	Drylot	80	0.73	2.38	3.28
	Alfalfa	80	0.65	1.97	3.02
Illinois	Drylot	224	0.64	1.79	3.04
	Alfalfa-ladino	224	0.63	1.66	2.99
Florida	Concrete	17	0.76	2.71	3.56
	Legume (full-fed)	16	0.79	2.74	3.45
	Legume (limit-fed)	17	0.63	2.13	3.38
Florida	Concrete	12	0.73	2.47	3.39
	Millet (full-fed)	15	0.70	2.34	3.36
	Millet (limit-fed)	15	0.54	1.83	3.42
Florida	Concrete	40	0.64	2.13	3.33
	Oat-wheat (winter)	40	0.68	2.27	3.34

Sources: Nebraska – Hudman and Peo (1957); Illinois – Terrill et al. (1958); Florida – Wallace and Combs (1958), Christmas et al. (1959), Wallace et al. (1960).

TABLE 23.3.   Percentage of sow nutritional requirement for gestation obtained from high-quality alfalfa–ladino clover pasture

Nutrient	Daily requirement[a]	Daily intake[b]	% of requirement
Protein (g)	280	385	138
Digestible energy (kcal)	6600	4200	64
Calcium (g)	17	28	165
Phosphorus (g)	13	5.6	43
NaCl (g)	10	4.2	42
Vitamin A (IU)	9000	777,000	8633
Vitamin D (IU)	900	. . .[c]	. . .[c]
Riboflavin (mg)	11	31.5	286
Niacin (mg)	80	57.4	72
Pantothenic acid (mg)	50	28	56
Vitamin B_{12} (μg)	55	0	0

[a]Adapted from Luce et al. (1990), Pork Industry Handbook Fact Sheet PIH-23.
[b]Based on daily pasture consumption of 7 kg.
[c]Sows exposed to sunlight will have enough vitamin D produced by irradiation of the sun on the skin to meet this requirement.

alfalfa hay in the diet for gestating gilts. In three studies the gestation diet containing the highest level of alfalfa hay produced the greatest percentage of gilts farrowing. The authors conclude that high levels of alfalfa hay could be fed during gestation economically.

A study by Nuzback et al. (1984) indicates that sun-cured alfalfa is equally digested as meal or pellets and increased utilization by gestating sows could be expected with decreased particle size.

LACTATION. The trend toward enclosed facilities for swine in central farrowing units has lessened the use of pasture for the lactation period. Different pastures were compared with a confinement system on gestation-lactation performance of swine from breeding to weaning (Handlin et al. 1972). Pastures were either oat, red clover-orchardgrass (*Dactylis glomerata* L.), ladino-orchardgrass, rape, or millet (*Pennisetum* spp.).

The litter and individual pig weights at birth were similar for all treatments. However, pigs produced in the confinement system consumed significantly less creep feed and were lighter in weight at weaning time. Red clover-orchardgrass produced more grazeable forage than the other forages in this experiment.

Pollmann et al. (1981) studied dietary addi-tions of alfalfa and tallow on sow reproductive performance through three successive reproductive cycles. Sun-cured alfalfa was fed at a level of 50% of the gestation diet followed by a lactation diet containing 8% tallow. Thirty-four percent more sows fed the alfalfa treatment completed the three gestation-lactation cycles than did those fed the control diet with no alfalfa. Birth weights were about 5% lower, but survival rate increased by about 8% at 14 d postpartum for the alfalfa-fed group. The addition of tallow did not alter survival rate.

SUPPLEMENTATION OF PASTURE. General recommendations for supplementing swine on high-quality legume pasture are outlined in Table 23.4. The amount of supplemental feed will be higher on lower-quality pasture and during stages of the life cycle when energy needs are high, such as lactation and growing-finishing.

SUSTAINABILITY. Swine operations that utilize pasture help promote sustainable agriculture concepts. The gestation stage of the life cycle likely fits this system best. However, there are successful farrow to finish operations, large and small, that utilize pasture. Such operations, if well managed, may reduce manure management problems often associated with intensive, centralized operations, and improve overall herd health, thus reducing medication costs.

TABLE 23.4.   Suggested feeding program for swine on high-quality pasture

Life cycle stage	No./ha	Feed/day (kg)	Protein level in concentrate (%)
Gestation	24	1	13
Lactation (litters)	12	4–5	13
Growing	75	Full feed	15
Finishing	50	Full feed	12

Forages for Poultry

CHICKENS (*Gallus domesticus*)

Several years ago most broilers and layers were maintained on green pastures. With that production system, chickens consumed a large supply of xanthophylls, the yellow and red pigments found in grasses and legumes. The xanthophylls imparted a bright yellow color to the skin, shanks, and beaks of broilers and to the yolks of eggs.

Today, broiler and egg production is a highly intensified industry, and it is no longer practical for chickens to roam the range. They are largely confined in houses under controlled environmental conditions and fed an all-mash or crumbles diet. Many consumers, however, still associate the yellow skin color of broilers with healthy chickens. Also, there is a demand for orange-pigmented yolks for use in egg noodles, yellow cake mixes, and other products produced by food processors.

Alfalfa, in the form of dehydrated alfalfa meal, is the main forage fed to poultry, largely to provide a source of pigment for producing yellow in the skin and shanks of broilers and in egg yolks (Scott et al. 1982). Because alfalfa meal, the most plentiful source of xanthophylls among the common poultry feedstuffs, is relatively low in energy, poultry producers limit the amount of alfalfa meal to about 2.5%-5.0% of the diet, just enough to increase the pigmentation of the shanks and eggs. Levels above 5.0% significantly increase the feed required per unit of gain in broilers or per dozen eggs produced by laying hens. Most commonly 17% dehydrated alfalfa meal is used, which contains at least 17% CP and about 775 kcal/kg of metabolizable energy.

GEESE (*Anser anser/cynoides*)

Geese are the best of all domestic poultry foragers. Although geese can go on pasture as early as the first week, nearly 100% of their diet can be forage after they are 5 to 6 wk old.

Geese are very selective and tend to pick out the most palatable forages. They will reject alfalfa and tough grasses and select the more succulent clovers and grasses. A 1-ha pasture will support 50-100 birds, depending on the size of the geese and the quality of the pasture (see Fig. 23.2).

Geese can be used to control grassy weeds in crops such as strawberry or cotton. The geese will selectively eat grasses and other weeds with little effect on the more mature leaves of the crop plants.

QUESTIONS

1. Compare alfalfa and ladino clover pasture for use in swine production.
2. What factors should be considered in supplementing pasture with grain for gestating sows?
3. What are the advantages of feeding silage to sows?
4. Compare the nutritional value of corn silage with grass-legume silage.
5. Outline a year-round pasture program for brood sows.
6. Why do sows apparently benefit from pasture more than do growing-finishing swine?
7. Explain how the feed-saving value of a hectare of pasture can be calculated for growing-finishing swine.
8. Why is alfalfa meal used in poultry rations?
9. Explain how the use of forages for geese differs from forage use for chickens.

REFERENCES

Barnhart, CE, and TW Cathey. 1956. Winter Pasture versus a Drylot for Bred Sows. Ky. Livest. Field Day Rep.

———. 1957. Legume Pasture for Bred Sows. Ky. Livest. Field Day Rep.

Barnhart, CE, MD Whiteker, and CW Nichols. 1958. Winter Pasture versus a Drylot for Bred Gilts. Ky. Livest. Field Day Rep.

Becker, DE, LJ Hanson, AH Jensen, SW Terrill, and HW Norton. 1956. Dehydrated alfalfa meal as a dietary ingredient for swine. J. Anim. Sci. 15:820-29.

Fig. 23.2. Weed grasses may be controlled in a strawberry patch by properly managed geese. Shelter should be provided outside the field, and food and water should be placed at the edges of the patch to avoid the excessive damage to the strawberry plants shown at left. *Univ. of Illinois photo.*

Becker, DE, SW Terrill, and AH Jensen. 1959. Silage for Sows during Gestation. Univ. Ill. Mimeo AS-507.

Bohman, VR, JR Kidwell, and JA McCormick. 1953. High levels of alfalfa in the rations of growing-fattening swine. J. Anim. Sci. 12:876-80.

Bohman, VR, JE Hunter, and JA McCormick. 1955. The effect of graded levels of alfalfa and aureomycin upon growing-fattening swine. J. Anim. Sci. 14:499-506.

Bowden, DM, and MF Clarke. 1963. Grass-legume forage fed fresh and as silage for market hogs. J. Anim. Sci. 22:934-39.

Brumm, MC and ER Peo, Jr. 1985. Effect of receiving diets containing alfalfa and certain feed additives on performance of feeder pigs transported long distances. J. Anim. Sci. 61:9-17.

Catron, D, G Ashton, V Speer, CC Culbertson, and EL Quaife. 1955. Silage for Sows. Iowa State Coll. Mimeo AH-680.

Christmas, RB, GE McCabe, HD Wallace, GE Combs, and AZ Palmer. 1959. Millet Pasture vs. Concrete for Growing-Finishing Swine. Fla. Agric. Exp. Stn. Anim. Husb. Nutr. Mimeo Ser. 59-6.

Conrad, JH, and WM Beeson. 1954. Grass Silage and Corn Silage as a Feed for Brood Sows. Purdue Univ. Mimeo AH-133.

Crampton, EW, and LE Harris. 1969. Applied Nutrition, 2d ed. San Francisco, Calif.: WH Freeman.

Cromwell, GL, TS Stahly, and HJ Monegue. 1984. Biological Availability of the Calcium and Phosphorus in Dehydrated Alfalfa Meal for Growing Pigs. Ky. Swine Res. Rep.

Danielson, DM, and JJ Noonan. 1975. Roughages in swine gestation diets. J. Anim. Sci. 41:94-99.

Danielson, DM, JE Butcher, and JC Street. 1969. Estimation of alfalfa pasture intake and nutrient utilization by growing-finishing swine. J. Anim. Sci. 28:6-12.

Eyles, DE. 1959. Feeding and management of pregnant sows on pasture. Anim. Prod. 1:41-50.

Foster, JR. 1985. Forages for swine and poultry. In ME Heath, RF Barnes, and DS Metcalfe (eds.), Forages: The Science of Grassland Agriculture, 4th ed. Ames: Iowa State Univ. Press, 590-96.

Grabouski, PH, and DM Danielson. 1978. Alfalfa in swine finishing diets: Feedlot performance. J. Anim. Sci. (suppl. 1): 218.

Handlin, DL, JW Nickles, GR Craddock, and WE Johnson. 1972. Forages for gestation and lactation periods. J. Anim. Sci. 34:348.

Heitman, H, Jr., and JH Meyer. 1959. Alfalfa meal as a source of energy by swine. J. Anim. Sci. 18:796-804.

Hoagland, JM, HW Jones, and RA Pickett. 1963. Haylage as a Gestation Ration for Sows and Gilts. Purdue Univ. Res. Prog. Rep. 75.

Hudman, DB, and ER Peo, Jr. 1957. Protein Levels for Growing-Finishing Swine on Pasture and in Drylot. Nebr. Swine Prog. Rep. 348.

Johnson, CW, VC Speer, GC Ashton, CC Culbertson, and DV Catron. 1957. Supplementary plane of nutrition for sows fed corn silage. J. Anim. Sci. 16:600-606.

Krider, JL, and SW Terrill. 1950. Comparisons of pastures and supplements for growing-fattening pigs. J. Anim. Sci. 9:289-99.

Krider, JL, BW Fairbanks, RF VanPoucke, DE Becker, and WE Carroll. 1946. Sardine condensed fish solubles and rye pasture for sows during gestation and lactation. J. Anim. Sci. 5:256-63.

Luce, WG, GR Hollis, DC Mahan, and ER Miller. 1990. Swine Diets. Pork Ind. Handb. Fact Sheet PIH-23.

Morrison, FB. 1956. Feeds and Feeding. 22d ed. Ithaca, N.Y.: Morrison, 1000-1114.

National Research Council (NRC), Subcommittee on Swine Nutrition. 1988. Nutrient Requirements of Swine. Washington, D.C.: National Academy of Science.

Nuzback, LJ, DS Pollmann, and KC Behnke. 1984. Effect of particle size and physical form of sun-cured alfalfa on digestibility for gravid swine. J. Anim. Sci. 58:378-85.

Pollmann, DS, M Danielson, MA Crenshaw, and ER Peo, Jr. 1981. Long-term effects of dietary additions of alfalfa and tallow on sow reproductive performance. J. Anim. Sci. 51:294-99.

Ross, OB, PH Phillips, and G Bohstedt. 1942. The effect of diet on brood sow performance. J. Anim. Sci. 1:353.

Scott, ML, MC Nesheim, and RJ Young. 1982. Nutrition of the Chicken. 3d ed. Ithaca, N.Y.: ML Scott, 406-61.

Selke, M, CE Barnhard, R Ayer, and CH Nichols. 1960. Balbo Rye Pasture and Legume Silage versus a Drylot for Brood Sows. Ky. Livest. Field Day Rep.

Stahly, TS, and GL Cromwell. 1986. Responses to dietary additions of fiber (alfalfa meal) in growing pigs housed in a cold, warm or hot thermal environment. J. Anim. Sci. 63:1870-76.

Terrill, SW, AH Jensen, and DE Becker. 1958. Pasture versus Drylot for Growing-Finishing Swine. Univ. Ill. Ext. Serv. Agric. Econ. AS-477.

Wallace, HD, and GE Combs. 1958. Pasture vs. Concrete for Growing-Finishing Swine. Fla. Agric. Exp. Stn. Anim. Husb. Nutr. Mimeo Ser. 59-3.

Wallace, HD, GE MaCabe, AZ Palmer, M Koger, JW Carpenter, and GE Combs. 1960. Pasture vs. Concrete for Growing-Finishing Swine with Special Emphasis on the Carcasses Produced. Fla. Agric. Exp. Stn. Anim. Husb. Nutr. Mimeo Ser. 60-11.

APPENDIX 1 # The Metric System

The International System of Units[1] (SI) is based on seven base units as follows:

Measure	Base Unit	Symbol
length	meter	m
mass	kilogram	kg
time	second	s
electric current	ampere	A
thermodynamic temperature	kelvin	K
amount of substance	mole	mol
luminous intensity	candela	cd

Following are prefixes and their meanings, suggested for use with units of the metric system:

Multiples and Submultiples	Prefix	Meaning	Symbol
$1,000,000,000,000 = 10^{12}$	tera	one trillion times	T
$1,000,000,000 = 10^{9}$	giga	one billion times	G
$1,000,000 = 10^{6}$	mega	one million times	M
$1,000 = 10^{3}$	kilo	one thousand times	k
$100 = 10^{2}$	hecto	one hundred times	h
$10 = 10^{1}$	deca	ten times	da

The Unit = 1

$0.1 = 10^{-1}$	deci	one tenth of	d
$0.01 = 10^{-2}$	centi	one hundredth of	c
$0.001 = 10^{-3}$	milli	one thousandth of	m
$0.000,001 = 10^{-6}$	micro	one millionth of	μ
$0.000,000,001 = 10^{-9}$	nano	one billionth of	n
$0.000,000,000,001 = 10^{-12}$	pico	one trillionth of	p

CONVERSION FROM A METRIC UNIT TO THE ENGLISH EQUIVALENT

Metric Unit	English Unit Equivalent
Length	
kilometer (km)	0.621 mile (mi)
meter (m)	1.094 yard (yd)
meter (m)	3.281 foot (ft)
centimeter (cm)	0.394 inch (in.)
millimeter (mm)	0.039 inch (in.)
Area	
kilometer^2 (km^2)	0.386 mile^2 (mi^2)
kilometer^2 (km^2)	247.1 acre (A)
hectare (ha)	2.471 acre (A)

[1]Additional information maybe found in the 1979 Standard for Metric Practice. *In* Annual Book of ASTM Standards, pt. 41, Designation 380-79, pp. 504-45. Philadelphia: Amer. Soc. for Testing and Materials (ASTM).

	Metric Unit	English Unit Equivalent

Volume

Metric Unit	English Unit Equivalent
meter^3 (m^3)	1.308 yard^3 (yd^3)
meter^3 (m^3)	35.316 foot^3 (ft^3)
hectoliter (hL)	3.532 foot^3 (ft^3)
hectoliter (hL)	2.838 bushel (US) (bu)
liter (L)	1.057 quart (US liq.) (qt)

Mass

Metric Unit	English Unit Equivalent
ton (mt) = 1000 kg	1.102 ton (t)
quintal (q)	220.5 pound (lb)
kilogram (kg)	2.205 pound (lb)
gram (g)	0.00221 pound (lb)

Yield or Rate

Metric Unit	English Unit Equivalent
ton/hectare (mt/ha)	0.446 ton/acre
hectoliter/hectare (hL/ha)	7.013 bushel/acre
kilogram/hectare (kg/ha)	0.892 pound/acre
quintal/hectare (q/ha)	89.24 pound/acre
quintal/hectare (q/ha)	0.892 hundredweight/acre

Pressure

Metric Unit	English Unit Equivalent
pascal (Pa)	1.45×10^{-4} pound/inch^2 (psi)
megapascal (MPa)	9.87 atmosphere (atm)
megapascal (MPa)	10 bar (bar)
bar (10^6 dynes/cm^2)	14.5 pound/inch^2 (psi)

Note: Bar is a metric term, but megapascal is the SI unit.

Temperature

Metric Unit	English Unit Equivalent
Celsius (C)	1.80C + 32 = Fahrenheit (F)
Kelvin (K) = C + 273.15	

Radiation (Light)

Metric Unit	English Unit Equivalent
lux (lx) = 1 lumen/m^2	0.0929 foot-candle (ft-c)
watt (W) = J/s	none
photon (photon)	none

Note: Lux is a metric term, but it is a measure of illumination as seen by the human eye and not energy. Light energy, more correctly called *radiant energy*, is specific to the radiation source because the discrete particles (photons) at each wavelength in the output spectrum have a specific energy level. For example, about 5% of the solar energy at the earth's surface is ultraviolet (<400 nm) and about 45% is far-red and infra-red radiation (>700 nm) that is not detectable by the human eye. *Total radiation* or *radiant flux density* usually indicates the complete solar spectrum and is expressed as W/m^2. The portion of the spectrum between 400 and 700 nm wavelengths is active in photosynthesis and is called *photosynthetically active radiation* (PAR). The quantity of PAR available to a plant or leaf is the *photosynthetic photon flux density* (PPFD). It is expressed in photons because photosynthesis is a photon-driven process. At high noon on a clear summer day PPFD is about 2000 µmol photons/m^{2}s.

Time

Metric Unit	English Unit Equivalent
second (s)	second (s)

Botanical and Common Names of Grasses, Legumes and Other Plants

LISTED BELOW ARE THE SCIENTIFIC AND COMMON NAMES for several grasses, legumes, and other plants mentioned in Volume 1 and Volume 2 of this book. To complement the alphabetical listing according to common names in Volume 1, we have organized this alphabetical listing according to scientific names. Other common names are in parentheses. Abbreviations and symbols used are: syn. = synonym, indicating more than one scientific name is in use; X before the genus name indicates a hybrid between two genera; x before the species epithet or subspecies name indicates a hybrid between two species or two subspecies.

Scientific names continue to change as new technologies and approaches to classification are developed. The USDA has developed the Germplasm Resources Information Network (GRIN) database and the Missouri Botanical Garden is leading a major project on Flora of North America (FNA). Both have developed on-line databases of scientific names that are accessible through internet. In addition, the Crop Science Society of America, the Weed Science Society of America, and the American Phytopathological Society have published listings of scientific names for the most common species.

The editors of *Forages* have worked closely with the editors of the *Cool-Season Grass Monograph* being published by the American Society of Agronomy to be consistent in names and naming. In both books we have used the USDA GRIN database as the primary resource. Our choice for a scientific name is based on usage and other data bases, but should not be a judgement of absolute correctness among several taxonomic philosophies. We do, however, encourage continued research and international dialogue to systematically develop more consistency and universal acceptance of scientific names.

SCIENTIFIC NAME	COMMON NAMES
Abutilon theophrasti Medicus	velvetleaf
Acacia greggii A. Graycatclaw	acacia
Acacia koa A. Gray	koa
Aeschynomene L.	jointvetch (aeschynomene)
A. americana L.	American
A. falcata (Poir.) DC.	Australian
Agropyron L.	wheatgrass
A. cristatum (L.) Gaertn.	fairway crested
A. desertorum (Fisch. ex Link) Schult.	standard crested
A. fragile (Roth) Candargy	Siberian crested
Agrostis L.	bentgrass
A. canina L.	velvet
A. hyemalis (Walt.) Britton et al.	winter (ticklegrass)
A. nebulosa Boiss. & Reut.	cloud
A. stolonifera L. var. *palustris* (Huds.) Farw. (syn. *A. palustris* Huds.)	creeping

A. tenuis Sibth. .... colonial
Agrostis gigantea Roth (syn. *A. alba* L.) .... redtop
Alopecurus L. .... foxtail
    A. arundinaceus Poir. .... creeping
    A. pratensis L. .... meadow
Alysicarpus vaginalis (L.) DC. .... alyceclover
Amaranthus retroflexus L. .... redroot pigweed
Ambrosia L. .... ragweed
    A. artemisiifolia L. .... common
    A. bidentata Michx. .... lanceleaf
    A. psilostachya DC. .... western
Ammophila Host .... beachgrass
    A. arenaria (L.) Link .... European
    A. breviligulata Fern. .... American
Andropogon gayanus Kunth .... gambagrass
Andropogon L. .... bluestem
    A. gerardii var. *paucipilus* (Nash) Fern. .... sand
    (syn. *A. hallii* Hack)
    A. gerardii Vitman .... big (turkey foot)
Andropogon virginicus L. .... broomsedge
Anthoxanthum odoratum L. .... vernalgrass, sweet
Anthyllis vulneraria L. .... kidneyvetch
Arachis L. .... peanut
    A. glabrata Benth. .... perennial (rhizoma)
    A. hypogaea L. .... peanut
    A. pintoi Krap. & Greg., nom. nud. .... pinto
Arctagrostis latifolia (R. Br.) Griseb. .... polargrass
Aristida L. .... wiregrass (three-awn)
    A. longiseta Steud. .... red
    A. purpurea Nutt. .... purple
    A. stricta Michx. .... pineland
Arrhenatherum Beuv. or *Danthonia* Lam. & DC. .... oatgrass
    A. elatius (L.) J.S. & C. Presl .... tall
    A. elatius var. *bulbosum* (Willd.) Spenner .... bulbous
Artemisia L. .... sagebrush
    A. filifolia Torr. .... sand
    A. nova A. Nels. .... black
    A. tridentata Nutt. .... big
    A. tripartita Rydb. .... threetip
Astragalus L. or *Oxytropis* DC. .... milkvetch
    A. cicer L. .... cicer
    A. falcatus Lam. .... sicklepod (sickle)
Astrebla pectinata (Lindl.) F. Muell. .... mitchellgrass
Atriplex confertifolia (Torr. & Frem.) S. Wats. .... shadscale (shadscale saltbush)
Atriplex canescens [Pursh] Nutt. .... four-wing saltbush
Atriplex spp. .... saltbush
Avena L. .... oat
    A. barbata Pott ex Link .... slender
    A. fatua L. .... wild
    A. sativa L. (syn. *A. byzantina* K. Koch) .... cultivated, white or red
    A. sterilis L. .... animated
Axonopus Beauv. .... carpetgrass
    A. affinis Chase .... common
    A. compressus (Sw.) Beauv. .... tropical
Balsamorhiza sagittata (Pursh) Nutt. .... balsamroot, arrowleaf
Bambusa bambos Druce .... bamboo
Beckmannia syzigachne (Steud.) Fern. .... sloughgrass, American
Beta vulgaris L. .... sugarbeet
Blepharoneuron tricholepis (Torr.) Nash .... pine dropseed

Bothriochloa O. Kuntze .... Old World bluestem
  B. barbinodis (Lag.) Herter .... cane
  B. caucasia (Trin.) C.E. Hubb. .... caucasian
  B. ischaemum (L.) Keng .... Turkestan or yellow (yellow beardgrass)
  B. saccharoides (Sw.) Rydb. .... silver (silver beardgrass)
Bouteloua Lag. .... grama
  B. curtipendula (Michx.) Torr. .... sideoats
  B. eriopoda (Torr.) Torr. .... black
  B. filiformis (Fourn.) Griffiths .... slender
  B. gracilis (H.B.K.) Lag. ex Steud. .... blue
  B. hirsuta Lag. .... hairy
  B. rothrockii Vasey .... Rothrock
Brachiaria brizantha (A. Rich.) Stapf .... palisadegrass
Brachiaria decumbens Stapf .... signalgrass
Brachiaria humidicola (Rendle) Schweick. .... koroniviagrass (creeping signalgrass)
Brachiaria mutica (Forssk.) Stapf .... paragrass
Brachiaria ruziziensis Germ. & Evrard .... ruzigrass
Brassica napus L. .... rape (winter rape)
Brassica oleracea L. .... kale
Brassica oleracea L. var. *capitata* L. .... cabbage
Brassica oleracea var. *gongyloides* L. .... kohlrabi
Brassica pekinensis L. .... Chinese cabbage
Brassica rapa L. .... turnip
Briza L. .... quakinggrass
  B. maxima L. .... big
  B. minor L. .... little
Bromus L. .... bromegrass
  B. anomalus Rupr. .... nodding
  B. arvensis L. .... field
  B. biebersteinii Roem. & Schult. .... Bieberstein
  B. carinatus Hook. & Arn. .... California
  B. ciliatus L. .... fringed
  B. inermis Leyss. .... smooth
  B. japonicus Thunb. ex Murr. .... Japanese (Japanese chess)
  B. marginatus Nees ex Steud. .... mountain
  B. pumpellianus Scribn. .... pumpelly
  B. riparius Rehm. (formerly *B. erectus* Huds.) .... meadow
  B. rubens L. .... red
  B. tectorum L. .... downy (cheatgrass)
Bromus catharticus Vahl .... rescuegrass
    (syn. *B. unioloides* Kunth or *B. willdenowii* Kunth)
Bromus commutatus Schrad. .... hairy chess
Bromus mollis L. .... soft chess
Bromus rigidus Roth .... ripgutgrass
Bromus secalinus L. .... chess (cheat)
Buchloe dactyloides (Nutt.) Engelm. .... buffalograss
Cajanus cajan (L.) Millsp. .... pigeonpea
Calamagrostis canadensis (Michx.) Beauv. .... bluejoint reedgrass (bluetop)
Calamagrostis rubescens Buckl. .... pinegrass
Calamovilfa Hack. .... sandreed
  C. gigantea (Nutt.) Scribn. & Merr. .... big
  C. longifolia (Hook.) Scribn. .... prairie
Calopogonium mucunoides Desv. .... calopa
Carex filifolia Nutt. .... threadleaf sedge
Cassia tora L. .... sickle senna
Cenchrus ciliaris L. .... buffelgrass
Cenchrus setigerus Vahl .... birdwoodgrass

Centrosema pascuorum Mart. ex Benth. . . . . centurion
Centrosema pubescens Benth. . . . . centro
Ceratoides lanata (Pursh) J. Howell . . . . winterfat
Chamaecrista fasciculata (Michx.) E. Greene . . . . partridge pea
Chamaecrista rotundifolia (Pers.) Greene . . . . roundleaf cassia
Chasmanthium latifolium (Michx.) Yates . . . . uniola, broadleaf
Chenopodium album L. . . . . common lambsquarters
Chloris gayana Kunth . . . . rhodesgrass
Chloris distichophylla Lag. . . . . weeping chloris
Cicer arietinum L. . . . . chickpea (garbanzo)
Cirsium L. . . . . thistle
   C. arvense (L.) Scop. . . . . Canada
   C. nutans L. . . . . musk
Clitoria ternatea L. . . . . butterfly pea
Cnidoscows texanus (Muella-Arg.) Small . . . . bullnettle
Coffee arabica L. . . . . coffee
Condalia lycioides (A. Gray) Weberb. . . . . graythorn
Cornus spp. . . . . dogwood
Coronilla varia L. . . . . crownvetch
Cortaderia selloana (Schult.) Asch. & Graebn. . . . . pampasgrass
Critesion Rafin. . . . . barley
   C. brachyantherum (Nevski) Barkw. & D.R. Dewey . . . . . . meadow
   C. jubatum (L.) Nevski . . . . foxtail
Crotalaria L. . . . . crotalaria (rattle-box)
   C. brevidens var. *intermedia* (Kotschy) Polhill . . . . slenderleaf
   C. incana L. . . . . shak
   C. juncea L. . . . . sunn hemp
   C. lanceolata E. Mey. . . . . lance
   C. pallida Aiton . . . . smooth
   C. spectabilis Roth . . . . showy
Curcurbita spp. . . . . pumpkin
Cuscuta spp. . . . . dodder
Cyamopsis tetragonoloba (L.) Taub. . . . . guar
Cynodon dactylon (L.) Pers. . . . . bermudagrass
Cynodon nlemfuensis Vanderyst . . . . stargrass
   C. aethiopicus Clayton & Harlan . . . . giant
   C. plectostachyus (K. Schum.) Pilger . . . . Naivasha
Cynosurus cristatus L. . . . . crested dogtail
Cyperus esculentus L. . . . . yellow nutsedge
Dactylis glomerata L. . . . . orchardgrass (cocksfoot)
Dalea alopecuroides Willd. . . . . dalea
Danthonia californica Boland . . . . California oatgrass
Danthonia spicata (L.) Beauv. . . . . povertygrass
Datura stramonium L. . . . . jimsonweed
Deschampsia Beauv. . . . . hairgrass
   D. beringensis Hult. . . . . Bering
   D. caespitosa (L.) Beauv. . . . . tufted
Descurainia pinnata (Walt.) Britt. . . . . tansymustard
Desmanthus illinoensis [Michx.] MacM. . . . . Illinois bundleflower
D. virgatus (L.) Willd. . . . . desmanthus
Desmodium Desv. . . . . desmodium
   D. heterocarpon (L.) DC. . . . . carpon
   D. heterophyllum (Willd.) DC. hetero
   D. intortum (Mill) Urb. . . . . greenleaf
   D. uncinatum (Jacq.) DC. . . . . silverleaf
Desmodium Desv. . . . . beggarweed (tickclover)
   D. incanum DC. . . . . creeping (kaimi clover)
   D. tortuosum (Sw.) DC. . . . . Florida (tall tickclover)
Dichanthium annulatum (Forssk.) Stapf . . . . shedagrass (Kleberg bluestem,
   hindigrass, marvelgrass)

Dichanthium aristatum (Poir.) C.E. Hubb. . . . . angletongrass
Dichanthium cariscosum (L.) A. Camus . . . . Nadi bluegrass
Digitaria californica (Benth.) Henr. . . . . cottontop (Arizona cottontop)
Digitaria eriantha Steud. . . . . pangolagrass
    (formerly *D. decumbens* Stent)
Digitaria L. . . . . crabgrass
    D. ciliaris (Retz.) Koel . . . .   southern
    D. sanguinalis (L.) Scop. . . . .   hairy (large)
Digitaria spp. . . . . woollyfinger
Digiteria eriantha Steud. . . . . digitgrass (fingergrass)
Diospyros virginiana L. . . . . persimmon
Distichlis spicata subsp. *stricta* (Torr.) Thorne . . . . inland saltgrass
Echinochloa crus-galli var. *frumentacea* (Link) W. Wight . . . . Japanese millet
    (syn. *E. frumentacea* Link)
Echinochloa polystachya (H.B.K.) Hitchc. . . . . alemangrass
Echinochloa pyramidalis (Lam.) Hitchc. & Chase . . . . antelopegrass
Ehrharta calycina Smith . . . . veldtgrass
Eleocharis spp. . . . . rushes
Elymus elymoides (Rafin.) Swezey . . . . squirreltailgrass
    (formerly *Sitanion hystrix* [Nutt.] J.G. Smith)
Elymus L. . . . . wildrye, wheatgrass
    E. canadensis L. . . . .   Canada wildrye
    E. glaucus Buckl. . . . .   blue wildrye
    E. sibiricus L. . . . .   Siberian wildrye
    E. lanceolatus (Scribn. & Smith) Gould . . . .   thickspike wheatgrass
    (formerly *Elytrigia dasystachya* [Hook.] A. & D. Löve)
    E. macrourus (Turcz.) Tzvelev . . . .   arctic wheatgrass
    E. trachycaulus (Link) Gould ex Shinners
     subsp. *trachycaulus* . . . .   slender wheatgrass
    E. trachycaulus subsp. *subsecundus* (Link) Gould . . . .   bearded wheatgrass
Elytrigia repens (L.) Nevski . . . . quackgrass (couchgrass)
    (formerly *Agropyron repens* [L.] Beauv.)
Ephedra spp. . . . . Mormon tea
Eragrostis Wolf . . . . lovegrass
    E. curvula (Schrad.) Nees var. *conferta* Nees . . . .   Boer
    (formerly *E. chloromelas* Steud.)
    E. curvula (Schrad.) Nees var. *curvula* Nees . . . .   weeping
    E. lehmanniana Nees . . . .   Lehmann
    E. superba Peyr. . . . .   Wilman
    E. trichodes (Nutt.) Wood . . . .   sand
Eremochloa ophiuroides (Munro) Hack. . . . . centipedegrass
Eriochloa polystachya Kunth . . . . caribgrass
Eupatonium capillifolium (Lam.) Small . . . . dogfennel
Festuca L. . . . . fescue
    F. arizonica Vasey . . . .   Arizona
    F. arundinacea Schreb. . . . .   tall
    F. idahoensis Elmer . . . .   Idaho
    F. ovina L. . . . .   sheep
    F. ovina var. *duriuscula* (L.) Koch . . . .   hard
    F. pratensis Huds. . . . .   meadow
    F. rubra L. . . . .   red
    F. rubra subsp. *commutata* Gaud. . . . .   chewings
    F. viridula Vasey . . . .   greenleaf
Gliricidia sepium (Jacq.) Kunth ex Wap. . . . . gliricida
Glottidium vesicarium (Jacq.) Harper . . . . bagpod
Glyceria grandis S. Watson ex A. Gray . . . . American mannagrass
Glycine max (L.) Merr. . . . . soybean
Gossypium spp. . . . . cotton
Gramineae or *Poaceae* . . . . grasses
Grayia spp. . . . . hopsage

Gutierrezia dracunculoides (DC.) Blake .... common broomweed
Halogeton glomeratus (Bieb.) C.A. Mey. .... halogeton
Haplopappus tenuisectus (Greene) Blake ex Benson .... burroweed
Hedysarum coronarium L. .... sulla (sulla sweetvetch)
Helenium omarum (Raf.) H. Rock .... bitter sneezeweed
Helianthus mollis Lam. .... ashy sunflower
Hemarthria altissima (Poir.) Stapf & C.E. Hubb. .... limpograss
Heteropogon contortus (L.) Beauv. ex Roem. & Schult. ...... tanglehead
Hilaria belangeri (Steud.) Nash .... curly mesquite
Hilaria jamesii (Torr.) Benth. .... galletagrass
Hilaria rigida (Turb.) Benth. .... big galletagrass
Hilaria mutica (Buckl.) Benth. .... tobosagrass
Holcus L. .... velvetgrass
   H. lanatus L. .... common (Yorkshire fog)
   H. mollis L. .... German
Hordeum L. .... barley
   H. bulbosum L. .... bulbous
   H. vulgare L. .... common
Humulus lupulus L. .... hop
Hymenachne amplexicaulis (Rudge) Nees .... hymenachne
Hyparrhenia rufa (Nees) Stapf .... jaragua grass
Imperata cylindrica (L.) Raeusch. .... cogongrass (bladygrass)
Indigofera hirsuta L. .... hairy indigo
Iva xanthifolia Nutt. .... marshelder
Juncus spp.; *Eleocharis* spp. .... rushes
Juniperus virginiana L. .... cedar, eastern red
Koeleria macrantha (Ledeb.) Schult. .... junegrass (prairie junegrass)
   (formerly *K. cristata* Pers.)
Kummerowia (formerly *Lespedeza*) .... annual lespedeza
   K. stipulacea (Maxim.) Makino .... korean
   K. striata (Thunb.) Schindler .... striate (common)
Lablab purpureus (L.) Sweet .... lablab bean (hyacinth bean)
Larrea tridentata (Sesse & Mocino ex DC.) Coville .... creosotebush
Lathyrus L. .... pea (peavine)
   L. cicer L. .... falcon-pea (flatpod peavine)
   L. hirsutus L. .... roughpea (caleypea,
                                                      singletarypea)
   L. maritimus (L.) Bigel. .... beach
   L. sativus L. .... grasspea
   L. sylvestris L. .... flatpea (Wagner pea)
   L. tingitanus L. .... Tangier (Tangier pea)
Leguminosae or *Fabaceae* .... legumes
Lens culinaris Medikus (syn. *L. esculenta* Moench) .... common lentil
Leptochloa dubia (Kunth) Nees .... green sprangletop
Lespedeza (see *Kummerowia*) .... lespedeza
   L. cuneata (Dum.-Cours.) G. Don .... sericea
Leucaena leucocephala (Lam.) de Wit .... leucaena (koa haole, ipil-ipil)
   L. retusa Gray................................. littleleaf lead-tree
Leymus Hochst. .... wildrye
   L. angustus (Trin.) Pilger .... Altai
   L. cinereus (Scribn. & Merr.) A. Löve .... Great Basin
   L. condensatus (Presl) A. Löve .... giant
   L. mollis (Trin.) Pilger subsp. *mollis* .... dune
   L. triticoides (Buckl.) Pilger .... beardless
Lolium L. .... ryegrass
   L. multiflorum Lam. .... annual (Italian)
   L. perenne L. .... perennial
   L. remotum Schrank .... flax

L. rigidum Gaud.	Wimmera
L. subulatum Vis.	dalmatian
Lolium temulentum L.	darnel
Lolium persicum Boiss. & Hohen. ex Boiss.	Persian darnel
Lotononis bainesii Baker	lotononis
Lotus L.	trefoil
L. corniculatus L.	birdsfoot
L. tenuis Waldst. & Kit. ex Willd.	narrowleaf birdsfoot
L. uliginosus Schkuhr. (syn. *L. pedunculatus* Cav.)	big
Lotus crassifolius (Benth.) Greene	big deervetch
Lotus purshianus (Benth.) Clem. & Clem.	Spanish clover
Lupinus L.	lupine
L. albus L.	white
L. angustifolius L.	blue
L. luteus L.	yellow
L. nootkatensis Donn ex Sims	Nootka
L. perennis L.	sundial
L. subcarnosus Hook.	Texas (bluebonnet)
Lycopersicon esculentum L.	tomato
Maclura pomifera [Raf.] Schneid.	osage orange
Macroptilium atropurpureum (Mocino & Sesse ex DC.) Urb.	siratro
Macroptilium lathyroides (L.) Urb.	phasey bean
Macroptilium longepedunculatum (Benth.) Urb.	llanos macro
Macrotyloma axillare (E. Meyer) Verdc.	perennial horse gram (axillaris)
Medicago L.	alfalfa
M. sativa nothosubsubsp. *varia* (Martyn) Arcang. (syn. *M. media* Pers.)	variegated
M. sativa L.	alfalfa (lucerne)
M. sativa subsp. *falcata* (L.) Arcang. (syn. *M. falcata* L.)	yellow
Medicago L.	burclover
M. arabica (L.) Huds.	spotted
M. minima (L.) Bartal.	little
M. polymorpha L. (syn. *M. hispida* Gaertn.)	California
Medicago L.	medic
M. lupulina L.	black
M. scutellata (L.) Mill.	snail
Medicago orbicularis (L.) Bartal.	buttonclover
Melilotus indica (L.) All.	sourclover
Melilotus Mill.	sweetclover
M. alba Medik.	white
M. alba var. *annua* Coe	Israel (hubam)
M. dentata (Waldst. & Kit.) Pers.	Banat
M. indica (L.) All.	annual yellow
M. officinalis Lam.	yellow
M. suaveolens Ledeb.	Daghestan
Melinis minutiflora Beauv.	molassesgrass
Metrosideros collina A. Gray	Ohia (Ohia lehua)
Mucuna pruriens var. *utilis* (Wallich ex Wight) Baker ex Burck (formerly *Stizolobium decringianum* Bart.)	Florida velvetbean
Muhlenbergia Schreb.	muhly
M. montana (Nutt.) Hitch.	mountain
M. porteri Scribn.	bush
M. pungens Thurb.	sandhill
M. wrightii Vasey	spike
Muhlenbergia schreberi J.F. Gmel.	nimblewill
Neonotonia wightii (Wight & Arn.) Lackey	perennial soybean (glycine)
Nephrolepsis exaltata Schott	staghorn fern

Onobrychis Mill. . . . . sainfoin
   O. arenaria (Kit.) DC. . . . . Siberian
   O. transcaucasica Grossh. . . . . Russian
   O. viciifolia Scop. . . . . common
Ornithopus sativus Brot. . . . . serradella
Oryza sativa L. . . . . rice
Oryzopsis Michx. . . . . ricegrass
   O. bloomeri (Bolander) Ricker . . . . bloomer
   O. hymenoides (Roem. & Schult.) Ricker . . . . indian
Oryzopsis miliacea (L.) Asch. & Schweinf. . . . . smilograss
Oxytropis riparia Litv. . . . . ruby milkvetch
Panicum antidotale Retz. . . . . blue panic
Panicum clandestinum L. . . . . deertonguegrass
Panicum coloratum L. . . . . kleingrass
Panicum coloratum var. *makarikariensis* Goosens . . . . makarikarigrass
Panicum maximum Jacq. . . . . guineagrass
Panicum maximum Jacq. var. *trichoglume* . . . . green panic
Panicum miliaceum L. . . . . proso millet (hog millet, broom
                                              millet, broomcorn)
Panicum obtusum Kunth . . . . vine mesquitegrass
Panicum ramosum L. . . . . browntop millet
   (formerly *Brachiaria ramosa* [L.] Stapf)
Panicum repens L. . . . . torpedograss
Panicum virgatum L. . . . . switchgrass
Pascopyrum smithii (Rydb.) A. Löve . . . . western wheatgrass
   (formerly *Elymus smithii* [Rydb.] Gould)
Paspalum conjugatum Bergius . . . . hilograss
Paspalum dilatatum Poir. . . . . dallisgrass
Paspalum distichum L. . . . . knotgrass
Paspalum L. . . . . paspalum
   P. laeve Michx. . . . . field
   P. malacophyllum Trin. . . . . ribbed
   P. plicatulum Michx. . . . . brownseed
   P. stramineum Nash . . . . sand
Paspalum nicorae Parodi . . . . brunswickgrass
Paspalum notatum Flügge . . . . bahiagrass
Paspalum scrobiculatum L. . . . . scrobicgrass
Paspalum urvillei Steud. . . . . vaseygrass
Pennisetum americanum (L.) Leeke . . . . pearlmillet
   (formerly *P. glaucum* [L.] R. Br.)
Pennisetum clandestinum Hochst. ex Chiov. . . . . kikuyugrass
Pennisetum flaccidum Griseb. . . . . flaccidgrass
Pennisetum orientale L. Rich . . . . Oriental pennisetum
Pennisetum purpureum Schumach. elephantgrass (napiergrass)
Phalaris aquatica L. x *P. arundinacea* L. . . . . ronphagrass
Phalaris arundinacea var. *picta* (L.) Asch. & Graebn. . . . . . . ribbongrass (gardener's garters)
Phalaris caroliniana Walt. . . . . maygrass
Phalaris coerulescens Desf. . . . . sunolgrass
Phalaris L. . . . . canarygrass
   P. aquatica L. . . . . bulbous
   P. arundinacea L. . . . . reed
   P. canariensis L. . . . . annual
   P. minor Retz. . . . . littleseed
Phalaris stenoptera Hack. (syn. *P. aquatica* L.) . . . . hardinggrass
Phalaris tuberosa var. *hirtiglumis* Batt. & Trab. . . . . koleagrass
Phaseolus L. or *Vigna* Savi . . . . bean
   P. acutifolius A. Gray . . . . Texas
   P. acutifolius var. *latifolius* Freem. . . . . tepary

Phleum pratense L. . . . . timothy (herdgrass)
　　P. alpinum L. . . . . alpine
　　P. bertolonii DC. . . . . turf
Phragmites australis (Cav.) Steudel . . . . common reed
Picea spp. . . . . spruce
Pisum sativum L. subsp. *sativum* var. *arvense* (L.) Poir. . . . . . field pea
Poa L. . . . . bluegrass
　　P. ampla Merr. . . . . big
　　P. annua L. . . . . annual
　　P. arachnifera Torr. . . . . Texas
　　P. bulbosa L. . . . . bulbous (winter)
　　P. canbyi (Scribn.) Howell . . . . canby
　　P. compressa L. . . . . Canada
　　P. fendleriana (Steud.) Vasey . . . . mutton
　　P. glaucantha Gaudin . . . . upland
　　P. interior Rydb. . . . . inland
　　P. nevadensis Vasey ex Scribn. . . . . Nevada
　　P. pratensis L. . . . . Kentucky
　　P. secunda J.S. Presl . . . . Sandberg
　　P. trivialis L. . . . . rough (rough stalked)
Prosopis juliflora (Sw.) DC. . . . . mesquite
　　P. glandulosa Torr. var. *glandulosa* . . . . honey
　　P. pallida (Willd.) Kunth . . . . algaroba
　　P. velutina Woot. . . . . velvet
Prunus spp. . . . . plum
Psathyrostachys juncea (Fisch.) Nevski . . . . Russian wildrye
Pseudoroegneria spicata (Pursh) A. Löve . . . . bluebunch wheatgrass
　　(formerly *Elytrigia spicata* [Pursh] D.R. Dewey)
Pseudotsuga spp. . . . . fir
Psidium guajava L. . . . . guava
Puccinellia nuttalliana (Schult.) Hitchc. . . . . nuttall alkaligrass (salt meadow
　　　　　　　　　　　　　　　　　　　　　　　　　　　　　grass)
Pueraria lobata (Willd.) Ohwi . . . . kudzu
　　P. phaseoloides (Roxb.) Benth. . . . . tropical (puero)
Redfieldia flexuosa (Thurb.) Vasey . . . . blowoutgrass
Rhus spp. . . . . sumac
Rhynchelytrum repens (Willd.) C. E. Hubb. . . . . natalgrass
Rosa multiflora Thunb. ex Murr. . . . . multiflora rose
Rubus allegheniensis Porter . . . . Allegheny blackberry
Rumex crispus L. . . . . curly dock
Saccharum officinarum L. . . . . sugarcane
Salix spp. . . . . willow
Sarcobatus vermiculatus (Hook.) Torr. . . . . greasewood
Schizachyrium Nees . . . . bluestem
　　S. scoparium (Michx.) Nash . . . . little
　　S. stoloniferum Nash . . . . creeping
Scolochloa festucacea (Willd.) Link . . . . rivergrass
Secale cereale L. . . . . rye
Sesbania sesban (L.) Merr. . . . . sesbania
Sesbania exaltata (Raf.) Rybd. ex A.W. Hill . . . . hemp sesbania
Setaria faberi Herrm. . . . . giant foxtail
Setaria italica (L.) Beauv. . . . . foxtail millet (Italian)
Setaria leucopila (Scribn. & Merv.) K. Schumi . . . . plains bristlegrass
　　(syn. *S. macrostachya* H.B.K.)
Setaria sphacelata (Schum.) Stapf & C.E. Hubb. ex
　　M.B. Moss. . . . . setariagrass (golden timothy)
Simsia calva Engeln. & Gray . . . . awnless bush sunflower
Solanum carolinense L. . . . . horsenettle

Solanum tuberosum L. . . . . potato
Sorghastrum nutans (L.) Nash . . . . indiangrass
Sorghum arundinaceum (Desv.) Stapf . . . . tunisgrass
    (formerly *S. virgatum* [Hack.] Stapf)
Sorghum bicolor (L.) Moench . . . . sorghum (broomcorn, milo)
Sorghum x *drummondii* [Steudel] Millsp. & Chase . . . . sudangrass
    (formerly *S. sudanense* [Piper] Stapf)
    (syn. *S. bicolor* (L.) Moensch)
Sorghum halepense (L.) Pers. . . . . johnsongrass
Sorghum x *almum* Parodi . . . . columbusgrass
Spartina Schreb. . . . . cordgrass
   S. foliosa Trin. . . . . spike (California)
   S. pectinata Link . . . . prairie
Sporobolus africanus (Poir.) Robyns & Tourn. . . . . rattailgrass (smutgrass)
Sporobolus asper (Michx.) Kunth . . . . longleaf rushgrass (tall
                                             dropseed)
Sporobolus Munro ex Scribn. . . . . sacaton
   S. airoides (Torr.) Torr. . . . . alkali
   S. wrightii Munro ex. Scribn. . . . . big
Sporobolus R. Br. or Blepharoneuron Nash . . . . dropseed
   S. asper (Michx.) Kunth . . . . tall
   S. asper var. *hookeri* (Trin.) Vasey . . . . meadow
   S. contractus Hitchc. . . . . spike
   S. cryptandrus (Torr.) A. Gray . . . . sand
   S. flexuosus (Thurb.) Rydb. . . . . mesa
   S. giganteus Nash . . . . giant
   S. heterolepis (A. Gray) A. Gray . . . . prairie
   S. junceus (Michx.) Kunth . . . . blue (pineywoods)
Stellaria media (L.) Vill. . . . . common chickweed
Stenotaphrum secundatum (Walt.) Kuntze . . . . St. Augustine grass
Stipa californica Merr. & Davy . . . . California needlegrass
Stipa comata Trin. & Rupr. . . . . needlegrass (needle-and-thread)
Stipa falcata Hughes . . . . speargrass
Stipa leucotricha Trin. & Rupr. . . . . Texas needlegrass
Stipa robusta (Vasey) Scribn. . . . . sleepygrass
Stipa spartea Trin. . . . . porcupinegrass
Stipa viridula Trin. . . . . green needlegrass
Stiporyzopsis caduca (Beal) B.L. Johnson & Rogler . . . . Mandan ricegrass
Strophostyles helvola (L.) Ell. . . . . trailing wildbean
Stylosanthes guianensis (Aubl.) Sw. stylo
   S. hamata (L.) Taub. . . . . Caribbean stylo
   S. humilis Kunth . . . . townsville stylo
   S. scabra Vog. . . . . shrubby stylo
Symphoricarpos orbiculatus Moench . . . . coralberry (buckbrush)
Taraxacum officinala Weber . . . . dandelion
Thinopyrum intermedium subsp. *barbulatum* . . . . pubescent wheatgrass
    (Schur) Barkworth & D.R. Dewey
    (formerly *Elytrigia trichophora* [Link] Nevski)
Thinopyrum intermedium (Host) Barkworth and D.R. Dewey intermediate wheatgrass
    (formerly *Elytrigia intermedia* [Host] Nevski)
Thinopyrum ponticum (Podp.) Barkworth & D.R. Dewey . . . . tall wheatgrass
    (formerly *Elytrigia elongata* [Host] Nevski)
Tridens flavus (L.) Hitchc. . . . . purpletop
Trifolium L. . . . . clover
   T. agrarium L. . . . . hop
   T. alexandrinum L. . . . . berseem (Egyptian)
   T. ambiguum Bieb. . . . . kura
   T. arvense L. . . . . rabbitfoot

T. bifidum A. Gray	pinhole
T. campestre Schreb.	large hop
T. carolinianum Michx.	carolina
T. cilolatum Benth.	tree
T. cyathiferum Lindl.	cup
T. dasyphyllum Torr. & A. Gray	whiproot
T. dubium Sibth.	small hop
T. fragiferum L.	strawberry
T. glomeratum L.	cluster
T. gracilentum Torr. & A. Gray	pin-point
T. gymnocarpon Nutt.	hollyleaf
T. hirtum All.	rose
T. hybridum L.	alsike
T. incarnatum L.	crimson
T. lappaceum L.	lappa
T. longipes Nutt. ex Torr. & A. Gray	longstalk
T. medium L.	zigzag
T. michelianum Savi	bigflower (Mikes)
T. microcephalum Pursh.	maiden
T. microdon Hook. & Arnott	squarehead (tomcal)
T. nigrescens Viv.	ball
T. pannonicum Jacq.	Hungarian
T. parryi A. Gray	Parry
T. pratense L.	red
T. purshianus (Benth.) Clem. & Clem.	Spanish
T. reflexum L.	buffalo
T. repens L.	white (white dutch, ladino)
T. resupinatum L.	Persian
T. striatum L.	striate (knotted)
T. subterraneum L.	subterranean
T. variegatum Nutt. ex Torr. & A. Gray	whitetip
T. vesiculosum Savi	arrowleaf
T. wormskioldii Lehm.	seaside
Trigonella foenum-graecum L.	fenugreek
Tripsacum dactyloides (L.) L.	eastern gamagrass
Trisetum flavescens (L.) Beauv.	yellow trisetum
Triticum aestivum L. emend. Thell.	wheat (club, spelt)
Triticum monococcum L.	einkorn
Triticum tauschii (Coss.) Schmal.	goatgrass
(syn. *Aegilops tauschii* subsp. *tauschii*)	
Triticum durum Desf.	durum (emmer)
X *Tritiosecale* Wittmack	triticale
Uniola paniculata L.	sea-oats
Urochloa mosambicensis (Hack.) Dandy	urochloagrass
Vernonia baldwinii Torr.	western ironweed
Vicia faba L.	broadbean (horsebean)
Vicia L.	vetch
V. angustifolia L.	narrowleaf
V. articulata Hornem.	single-flowered
V. benghalensis L.	purple
V. cracca L.	bird
V. ervilia (L.) Willd.	bitter
V. grandiflora Scop.	bigflower (grandiflora, showy)
V. monantha Retz.	bard
V. pannonica Crantz	Hungarian
V. sativa L.	common
V. sativa subsp. *cordata* (Wulfen ex Hoppe) Asch. & Graebn.	cordateleaf

V. villosa Roth .... hairy (winter, Russian)
Vigna Savi .... bean
   V. aconitifolia (Jacq.) Marechal .... mat (moth)
   V. angularis (Willd.) Ohwi & H. Ohashi .... adsuki
   V. radiata (L.) Wilcz. .... mung
   V. umbellata (Thunb.) Ohwi & H. Ohashi .... rice
Vigna parkeri .... creeping vigna
Vigna unguiculata (L.) Walp. .... cowpea
Vitis spp. .... grape
Vulpia myuros (L.) C. Gmelin .... rattail fescue
Vulpia octoflora (Walter) Rydb. .... sixweeks fescue
Zea mays L. .... corn (maize)
Zea mays subsp. *mexicana* (Schrad.) Iltis .... teosinte
Zizania palustris L. .... cultivated wildrice
Zornia latifolia .... zornia
Zoysia japonica Steud. .... zoysia (Japanese lawngrass)
Zoysia matrella (L.) Merr. .... manilagrass
Zoysia tenuifolia Willd. ex Thiele .... mascarenegrass

REFERENCES

Barkworth, ME and DR Dewey. 1985. Genomically based genera of the perennial Triticeae of North America: Identification and membership. Amer J Bot 72:767-76.

Brako, L, AY Rossman and DF Farr. 1995. Scientific and common names of 7,000 vascular plants in the United States. St. Paul, MN: Amer Phytopath Soc.

Cronquist, A. 1981. An intergrated system of classification of flowering plants. New York: Columbia Univ Press.

Encke, F, G Buchheim, and S Seybold. 1984. Zander-Handworterbuch der Pflanzennamen. 13th ed. Stuttgart, Germany: Eugen Ulmer.

Hitchcock, AS (rev. by A Chase). 1951. Manual of the Grasses of the United States. USDA Misc Publ 200. Reprinted by Dover Press.

LH Bailey Hortorium Staff. 1976. Hortus third. New York: Macmillan.

Patterson, DT, Chairman. 1984. Composite list of weeds. Weed Science 32, Suppl 2:1-137.

Terrell, EE, SR Hill, JH Wiersema, and WE Rice. 1986. A checklist of names of 3000 vascular plants of economic importance. USDA Agr Hdbk 505 (rev. ed.).

Voss, EG, et al., editors. 1983. International code of botanical nomenclature. Regnum Vegetabile 111. Utrecht, Netherlands: Bohn, Scheltema, and Holkema.

INDEX

The index lists common names of grasses, legumes, and other plants. The corresponding botanical names are listed in the appendix.